The Theory of Groups

THE THEORY OF GROUPS

Marshall Hall, Jr.

PROFESSOR OF MATHEMATICS
CALIFORNIA INSTITUTE OF TECHNOLOGY

Chelsea Publishing Company
New York, N.Y.

SECOND EDITION

First published 1959.

The Second Edition is textually virtually the same as the 1966, 1967, and 1968 (tenth) printings of the First Edition.

CIP

Library of Congress Cataloging in Publication Data

Hall, Marshall, 1910-

The theory of groups.

Reprint of the 1968 (10th printing) ed. published by Macmillan, New York

Bibliography: p.

Includes indexes.

1: Groups, Theory of. I. Title.

QA171.H27 1976 512'.22 75-42306

ISBN 0-8284-0288-4

Printed in the United States of America

DEDICATED TO

PHILIP HALL

PREFACE

The present volume is intended to serve a dual purpose. The first ten chapters are meant to be the basis for a course in Group Theory, and exercises have been included at the end of each of these chapters. The last ten chapters are meant to be useful as optional material in a course or as reference material. When used as a text, the book is intended for students who have had an introductory course in Modern Algebra comparable to a course taught from Birkhoff and MacLane's *A Survey of Modern Algebra.* I have tried to make this book as self-contained as possible, but where background material is needed references have been given, chiefly to Birkhoff and MacLane.

Current research in Group Theory, as witnessed by the publications covered in *Mathematical Reviews* is vigorous and extensive. It is no longer possible to cover the whole subject matter or even to give a complete bibliography. I have therefore been guided to a considerable extent by my own interests in selecting the subjects treated, and the bibliography covers only references made in the book itself. I have made a deliberate effort to curtail the treatment of some subjects of great interest whose detailed study is readily available in recent publications. For detailed investigations of infinite Abelian groups, the reader is referred to the appropriate sections of the second edition of Kurosch's *Theory of Groups* and Kaplansky's monograph *Infinite Abelian Groups.* The monographs *Structure of a Group and the Structure of its Lattice of Subgroups* by Suzuki and *Generators and Relations for Discrete Groups* by Coxeter and Moser, both in the Ergebnisse series, are recommended to the reader who wishes to go further with these subjects.

This book developed from lecture notes on the course in Group Theory which I have given at The Ohio State University over a period of years. The major part of this volume in its present form was written at Trinity College, Cambridge, during 1956 while I held a Fellowship from the John Simon Guggenheim Foundation. I give my thanks to the Foundation for the grant enabling me to carry out this work and to the Fellows of Trinity College for giving me the privileges of the College.

I must chiefly give my thanks to Professor Philip Hall of King's College, Cambridge, who gave me many valuable suggestions on the preparation of my manuscript and some unpublished material of his own. In recognition of his many kindnesses, this book is dedicated to him.

I wish also to acknowledge the helpfulness of Professors Herbert J. Ryser and Jan Korringa and also Dr. Ernest T. Parker in giving me their assistance on a number of matters relating to the manuscript.

Marshall Hall, Jr.

Columbus, Ohio

CONTENTS

1. INTRODUCTION

1.1. Algebraic Laws.

A large part of algebra is concerned with systems of elements which, like numbers, may be combined by addition or multiplication or both. We are given a system whose elements are designated by letters a, $b, c, \cdots$. We write $S = S(a, b, c, \cdots)$ for this system. The properties of these systems depend upon which of the following basic laws hold:

Closure Laws.	$A0$. *Addition is well defined.*	$M0$. *Multiplication is well defined.*

These mean that, for every ordered pair of elements, a, b of S, $a + b = c$ exists and is a unique element of S, and that also $ab = d$ exists and is a unique element of S.

Associative Laws.	$A1$. $(a + b) + c = a + (b + c)$	$M1$. $(ab)c = a(bc)$
Commutative Laws.	$A2$. $b + a = a + b$	$M2$. $ba = ab$
Zero and Unit.	$A3$. 0 *exists such that* $0 + a = a + 0 = a$ *for all* a.	$M3$. 1 *exists such that* $1a = a1 = a$ *for all* a.
Negatives and Inverses.	$A4$. *For every* a, $-a$ *exists such that* $(-a) + a = a + (-a) = 0$.	$M4$.* *For every* $a \neq 0$, a^{-1} *exists such that* $(a^{-1})a = a(a^{-1}) = 1$.
Distributive Laws.	$D1$. $a(b + c) = ab + ac$.	
	$D2$. $(b + c)a = ba + ca$.	

Definition: *A system satisfying all these laws is called a field. A system satisfying* $A0$, -1, -2, -3, -4, $M0$, -1, *and* $D1$, -2 *is called a ring.*

It should be noted that $A0$–$A4$ are exactly parallel to $M0$–$M4$ except for the nonexistence of the inverse of 0 in $M4$. In the distributive laws, however, addition and multiplication behave quite differently.

---

* The statement of $M4$ given here applies to a system in which both addition and multiplication are defined. If addition is not defined, there is no 0 in S, and the law can be rephrased: "For every a, a^{-1} exists such that $(a^{-1})a = a(a^{-1}) = 1$."

This parallelism between addition and multiplication is exploited in the use of logarithms, where the basic correspondence between them is given by the law:

$$\log(xy) = \log x + \log y.$$

In general an n-ary operation in a set S is a function $f = f(a_1, \cdots, a_n)$ of n arguments $(a_1, \cdots, a_n)$ which are elements of S and whose value $f(a_1, \cdots, a_n) = b$ is a unique element of S when f is defined for these arguments. If, for every choice of $a_1, \cdots, a_n$ in S, $f(a_1, \cdots, a_n)$ is defined, we say that the operation f is *well defined* or that the set S is *closed* with respect to the operation f.

In a field F, addition and multiplication are well-defined binary operations, while the inverse operation $f(a) = a^{-1}$ is a unary operation defined for every element except zero.

1.2. Mappings.

A very fundamental concept of modern mathematics is that of a *mapping* of a set S into a set T.

DEFINITION: *A mapping α of a set S into a set T is a rule which assigns to each x of the set S a unique y of the set T.* Symbolically we write this in either of the notations:

$$\alpha: x \rightarrow y \quad \text{or} \quad y = (x)\alpha.$$

The element y is called the *image* of x under α. If every y of the set T is the image of at least one x in S, we say that α is a mapping of S *onto* T.

The mappings of a set into (or onto) itself are of particular importance. For example a rotation in a plane may be regarded as a mapping of the set of points in the plane onto itself. Two mappings α and β of a set S into itself may be combined to yield a third mapping of S into itself, according to the following definition.

DEFINITION: *Given two mappings α, β, of a set S into itself, we define a third mapping γ of S into itself by the rule: If $y = (x)\alpha$ and $z = (y)\beta$, then $z = (x)\gamma$. The mapping γ is called the product of α and β, and we write $\gamma = \alpha\beta$.*

Here, since $y = (x)\alpha$ is unique and $z = (y)\beta$ is unique, $z = [(x)\alpha]\beta = (x)\gamma$ is defined for every x of S and is a unique element of S.

THEOREM 1.2.1. *The mappings of a set S into itself satisfy $M0$, $M1$, and $M3$ if multiplication is interpreted to be the product of mappings.*

Proof: It has already been noted that $M0$ is satisfied. Let us consider $M1$. Let α, β, γ be three given mappings. Take any element x of S and let $y = (x)\alpha$, $z = (y)\beta$, and $w = (z)\gamma$. Then $(x)[(\alpha\beta)\gamma] = z(\gamma) = w$, and $(x)[\alpha(\beta\gamma)] = y(\beta\gamma) = w$. Since both mappings, $(\alpha\beta)\gamma$ and $\alpha(\beta\gamma)$, give the same image for every x in S, it follows that $(\alpha\beta)\gamma = \alpha(\beta\gamma)$.

As for $M3$, let 1 be the mapping such that $(x)1 = x$ for every x in S. Then 1 is a unit in the sense that for every mapping α, $\alpha 1 = 1\alpha = \alpha$.

In general, neither $M2$ nor $M4$ holds for mappings. But $M4$ holds for an important class of mappings, namely, the one-to-one mappings of S onto itself.

DEFINITION: *A mapping α of a set S onto T is said to be one-to-one (which we will frequently write* 1–1) *if every element of T is the image of exactly one element of S.* We indicate such a mapping by the notation: $\alpha : x \leftrightarrows y$, where x is an element of S, and y is an element of T. We say that S and T have the same cardinal number* of elements.

THEOREM 1.2.2. *The one-to-one mappings of a set S onto itself satisfy $M0$, $M1$, $M3$, and $M4$.*

Proof: Since Theorem 1.2.1 covers $M0$, $M1$, and $M3$, we need only verify $M4$. If $\alpha : x \leftrightarrows y$ is a one-to-one mapping of S onto itself, then by definition, for every y of S there is exactly one x of S such that $y = (x)\alpha$. This assignment of a unique x to each y determines a one-to-one mapping $\tau : y \leftrightarrows x$ of S onto itself. From the definition of τ we see that $(x)(\alpha\tau) = x$ for every x in S and $y(\tau\alpha) = y$ for every y in S. Hence, $\alpha\tau = \tau\alpha = 1$, and τ is a mapping satisfying the requirements for α^{-1} in $M4$.

We call a one-to-one mapping of a set onto itself a *permutation.* When the given set is finite, a permutation may be written by putting the elements of the set in a row and their images below them. Thus $\alpha = \begin{pmatrix} 1, 2, 3 \\ 2, 3, 1 \end{pmatrix}$ and $\beta = \begin{pmatrix} 1, 2, 3 \\ 1, 3, 2 \end{pmatrix}$ are two permutations of the set $S(1, 2, 3)$. Their product is defined to be the permutation

* For a discussion of cardinal numbers, see Birkhoff and MacLane [1], p. 356. This number and others like it throughout the book refer to the Bibliography.

$\alpha\beta = \begin{pmatrix} 1, 2, 3 \\ 3, 2, 1 \end{pmatrix}$. Note that the product rule for permutations given here is obtained by multiplying from left to right. Some authors define the product so that multiplication is from right to left.

1.3. Definitions for Groups and Some Related Systems.

We see that, as single operations, the laws governing addition and multiplication are the same. Of these, all but the commutative law are satisfied by the product rule for the one-to-one mappings of a set onto itself. The laws obeyed by these one-to-one mappings are those which we shall use to define a group.

DEFINITION (FIRST DEFINITION OF A GROUP): *A group G is a set of elements $G(a, b, c, \cdots)$ and a binary operation called "product" such that:*

$G0$. Closure Law. *For every ordered pair a, b of elements of G, the product $ab = c$ exists and is a unique element of G.*

$G1$. Associative Law. $(ab)c = a(bc)$.

$G2$. Existence of Unit. *An element 1 exists such that $1a = a1 = a$ for every a of G.*

$G3$. Existence of Inverse. *For every a of G there exists an element a^{-1} of G such that $a^{-1}a = aa^{-1} = 1$.*

These laws are redundant. We may omit one-half of $G2$ and $G3$, and replace them by:

$G2.^*$ *An element 1 exists such that $1a = a$ for every a of G.*

$G3.^*$ *For every a of G there exists an element x of G such that $xa = 1$.*

We can show that these in turn imply $G2$ and $G3$. For a given a let

$$xa = 1 \quad \text{and} \quad yx = 1,$$

by $G3.^*$

Then we have

$$ax = 1(ax) = (yx)(ax) = y[x(ax)] = y[1x] = yx = 1,$$

so that $G3$ is satisfied. Also,

$$a = 1a = (ax)a = a(xa) = a1,$$

so that $G2$ is satisfied.

The uniqueness of the unit 1 and of an inverse a^{-1} are readily established (see Ex. 13). We could, of course, also replace $G2$ and $G3$

by the assumption of the existence of 1 and x such that: $a1 = a$ and $ax = 1$. But if we assume that they satisfy $a1 = a$ and $xa = 1$, the situation is slightly different.*

There are a number of ways of bracketing an ordered sequence $a_1a_2 \cdots a_n$ to give it a value by calculating a succession of binary products. For $n = 3$ there are just two ways of bracketing, namely, $(a_1a_2)a_3$ and $a_1(a_2a_3)$, and the associative law asserts the equality of these two products. An important consequence of the associative law is the *generalized associative law.*

All ways of bracketing an ordered sequence $a_1a_2, \cdots a_n$ *to give it a value by calculating a succession of binary products yield the same value.*

It is a simple matter, using induction on n, to prove that the generalized associative law is a consequence of the associative law (see Ex. 1).

Another definition may be given which does not explicitly postulate the existence of the unit.

Definition (Second Definition of a Group): *A group G is a set of elements $G(a, b, \cdots)$ such that*

1) *For every ordered pair a, b of elements of G, a binary product ab is defined such that $ab = c$ is a unique element of G.*

2) *For every element a of G a unary operation "inverse," a^{-1}, is defined such that a^{-1} is a unique element of G.*

3) *Associative Law.* $(ab)c = a(bc)$.

4) *Inverse Laws.* $a^{-1}(ab) = b = (ba)a^{-1}$.

It is an easy task to show that any set which satisfies the axioms of the first definition also satisfies those of the second. To show the converse, assume the axioms of the second definition and consider the relation:

$$a^{-1}a = [(a^{-1}a)b]b^{-1} = (a^{-1}a)(bb^{-1}) = a^{-1}[a(bb^{-1})] = bb^{-1}.$$

When $a = b$, we see that $a^{-1}a = aa^{-1}$, and consequently the element $a^{-1}a = aa^{-1}$ is the same for every a in G. Let us call this element "1," so that $G3$ is satisfied. Also,

$$1b = (a^{-1}a)b = a^{-1}(ab) = b,$$

and

$$b1 = b(aa^{-1}) = (ba)a^{-1} = b,$$

* H. B. Mann [1].

and $G2$ is satisfied. Therefore the two definitions of a group are equivalent.

There is a third definition of a group as follows:

DEFINITION (THIRD DEFINITION OF A GROUP): *A group G is a set of elements $G(a, b, \cdots)$ and a binary operation a/b such that:*

$L0$. For every ordered pair a, b of elements of G, a/b is defined such that $a/b = c$ is a unique element of G.

$L1$. $a/a = b/b$.

$L2$. $a/(b/b) = a$.

$L3$. $(a/a)/(b/c) = c/b$.

$L4$. $(a/c)/(b/c) = a/b$.

In terms of this operation, let us define a unary operation of inverse b^{-1} by the rule

$$b^{-1} = (b/b)/b.$$

Here

$$(b^{-1})^{-1} = (b^{-1}/b^{-1})/b^{-1} = (b^{-1}/b^{-1})/[(b/b)/b] = b/(b/b) = b,$$

using in turn $L3$ and $L2$. We now define a binary operation of product by the rule

$$ab = a/b^{-1}.$$

Then $a/b = a/(b^{-1})^{-1} = ab^{-1}$. Let us write 1 for the common value of $a/a = b/b$ as given by $L1$. Then $L1$ becomes $aa^{-1} = 1$, whence also for any a, $1 = a^{-1}(a^{-1})^{-1} = a^{-1}a$. Thus $G3$ of the first definition holds. In $b^{-1} = (b/b)/b$, put $b = 1$, whence $1^{-1} = 11^{-1}$, and so $1 = 1/1 = 11^{-1} = 1^{-1}$. $L2$ now becomes $a1^{-1} = a1 =$ a. By definition $b^{-1} = 1/b = 1b^{-1}$, and with $b = a^{-1}$, this gives $(a^{-1})^{-1} = 1(a^{-1})^{-1}$, or $a = 1a$. Thus $G2$ of the first definition holds. $L3$ now becomes $1(bc^{-1})^{-1} = cb^{-1}$, whence $(bc^{-1})^{-1} = cb^{-1}$. In $L4$, put $a = x$, $b = 1$, $c = y^{-1}$; whence $(xy)(1y)^{-1} = x1^{-1} = x$ or $(xy)y^{-1} = x$. Now, for any x, y, z, put $a = xy$, $b = z^{-1}$, $c = y$. Then $ac^{-1} = (xy)y^{-1} = x$, and $L4$ becomes $(ac^{-1})(bc^{-1})^{-1} = ab^{-1}$, whence $(ac^{-1})(cb^{-1}) = ab^{-1}$. But in terms of x, y, z this becomes $x(yz) = (xy)z$, the associative law $G1$. Thus this definition of group implies the first definition. But in terms of the first definition if we put $ab^{-1} = a/b$, we easily find that the laws $L0$, -1, -2, -3, -4 are satisfied, and therefore the definitions are equivalent.

There are systems which satisfy some but not all the axioms for a group. The following are the main types:

DEFINITION: *A quasi-group Q is a system of elements $Q(a, b, c, \cdots)$ in which a binary operation of product ab is defined such that, in $ab = c$, any two of a, b, c determine the third uniquely as an element of Q.*

DEFINITION: *A loop is a quasi-group with a unit 1 such that $1a = a1 = a$ for every element a.*

DEFINITION: *A semi-group is a system $S(a, b, c, \cdots)$ of elements with a binary operation of product ab such that $(ab)c = a(bc)$.*

A group clearly satisfies all these definitions. We may, with Kurosch, further define a group as a set which is both a semi-group and a quasi-group. As a semi-group $G0$ and $G1$ are satisfied. Let t be the unique element such that $tb = b$ for some particular b, and let y be determined by b and a so that $by = a$. Then $(tb)y = by$ and $t(by) = by$, or $ta = a$ for any a, and $G2^*$ is satisfied. In a quasi-group $G3^*$ is also satisfied. But we have already shown that these properties define a group.

We call a system with a binary product and unary inverse satisfying

$$a^{-1}(ab) = b = (ba)a^{-1}$$

a quasi-group with the inverse property, this law being the inverse property. We must show that the product defines a quasi-group. If $ab = c$, we find $b = a^{-1}(ab) = a^{-1}c$, and $a = (ab)b^{-1} = cb^{-1}$. Thus a and b determine c uniquely; and also given c and a, there is at most one b, and given c and b, there is at most one a. Write $a(a^{-1}c) = w$. Then $a^{-1}[a(a^{-1}c)] = a^{-1}w$, whence $a^{-1}c = a^{-1}w$. Then $(a^{-1})^{-1}(a^{-1}c) = (a^{-1})^{-1}(a^{-1}w)$ whence $c = w$. Hence $a(a^{-1}c) = c$, and similarly, $(cb^{-1})b = c$, and the system is a quasi-group. We note that an inverse quasi-group need not be a loop. With three elements a, b, c and relations $a^2 = a$, $ab = ba = c$, $b^2 = b$, $bc = cb = a$, $c^2 = c$, $ca = ac = b$, we find that each element is its own inverse, and we have a quasi-group with inverse property but no unit.

1.4. Subgroups, Isomorphisms, Homomorphisms.

A subset of the elements of a group G may itself form a group with respect to the product as defined in G. Such a set of elements H is called a *subgroup*.

In any group G the unit 1 satisfies $1^2 = 1$. Conversely, if x is an element of G such that $x^2 = x$, then $x = x^{-1}(x^2) = x^{-1}x = 1$. Thus the unit of a subgroup H, since it satisfies $x^2 = x$, must be the same as the unit of the whole group G.

THEOREM 1.4.1. *A non-empty subset H of a group G is a subgroup if the two following conditions hold:*

$S1$. *If $h_1 \in H$, $h_2 \in H$, then $h_1h_2 \in H$.*
$S2$. *If $h_1 \in H$, then $h_1^{-1} \in H$.*

Proof: The two properties given guarantee the validity of $G0$, $G2$, $G3$ in H. And since products in H agree with those in G, $G1$ is also satisfied in H.

There are various relationships between pairs of groups which are worth considering. The first such relationship is that of *isomorphism.*

DEFINITION: *A one-to-one mapping $G \leftrightarrows H$ of the elements of a group G onto those of a group H is called an isomorphism if whenever $g_1 \leftrightarrows h_1$ and $g_2 \leftrightarrows h_2$, then $g_1g_2 \leftrightarrows h_1h_2$.*

EXAMPLE 1. Since all the permutations of a set form a group (Theorem 1.2.2), any set of permutations satisfying $S1$ and $S2$ will form a group which is a subgroup of the full group of permutations. For example, let us consider the following two such subgroups:

$$\begin{array}{cc} G_1 & G_2 \\ x_1 = \begin{pmatrix} 1, 2, 3 \\ 1, 2, 3 \end{pmatrix} & y_1 = \begin{pmatrix} 1, 2, 3, 4, 5, 6 \\ 1, 2, 3, 4, 5, 6 \end{pmatrix} \\ x_2 = \begin{pmatrix} 1, 2, 3 \\ 2, 3, 1 \end{pmatrix} & y_2 = \begin{pmatrix} 1, 2, 3, 4, 5, 6 \\ 2, 3, 1, 6, 4, 5 \end{pmatrix} \\ x_3 = \begin{pmatrix} 1, 2, 3 \\ 3, 1, 2 \end{pmatrix} & y_3 = \begin{pmatrix} 1, 2, 3, 4, 5, 6 \\ 3, 1, 2, 5, 6, 4 \end{pmatrix} \\ x_4 = \begin{pmatrix} 1, 2, 3 \\ 1, 3, 2 \end{pmatrix} & y_4 = \begin{pmatrix} 1, 2, 3, 4, 5, 6 \\ 4, 5, 6, 1, 2, 3 \end{pmatrix} \\ x_5 = \begin{pmatrix} 1, 2, 3 \\ 3, 2, 1 \end{pmatrix} & y_5 = \begin{pmatrix} 1, 2, 3, 4, 5, 6 \\ 5, 6, 4, 3, 1, 2 \end{pmatrix} \\ x_6 = \begin{pmatrix} 1, 2, 3 \\ 2, 1, 3 \end{pmatrix} & y_6 = \begin{pmatrix} 1, 2, 3, 4, 5, 6 \\ 6, 4, 5, 2, 3, 1 \end{pmatrix} \end{array}$$

If we map x_i of G_1 onto y_i of G_2, we find that products correspond in every instance. Hence G_1 and G_2 are isomorphic.

More generally we may have a mapping (usually many to one) of the elements of one group G onto those of another group H, which we call a *homomorphism* if the mapping preserves products.

DEFINITION: *A mapping $G \to H$ of the elements of a group G onto those of a group H is called a homomorphism if whenever $g_1 \to h_1$ and $g_2 \to h_2$, then $g_1g_2 \to h_1h_2$.*

In the homomorphism $G \to H$ let 1 be the identity of G and let $1 \to e$, where e is in H. Then $1^2 \to e^2$. Since $1^2 = 1$, then $e^2 = e$. We see that e is therefore the identity of H. Also if $g \to h$ and $g^{-1} \to k$, then $gg^{-1} \to hk$, and so $1 \to hk = e$. Therefore $k = h^{-1}$ and the mapping takes inverses into inverses. We may observe that a one-to-one homomorphism is an isomorphism.

EXAMPLE 2. If G_1 is the permutation group above and H is the multiplicative group of the two real numbers 1, -1, then we have a homomorphism:

$$\begin{array}{ll} x_1 \to 1 & x_4 \to -1 \\ x_2 \to 1 & x_5 \to -1 \\ x_3 \to 1 & x_6 \to -1 \end{array}$$

Not only are permutation groups of interest in themselves, but also every such group is isomorphic to a permutation group.

THEOREM 1.4.2 (CAYLEY). *Every group G is isomorphic to a permutation group of its own elements.*

Proof: For each $g \in G$, define the mapping $R(g) : x \to xg$ for all $x \in G$. For a fixed g this is a mapping of the elements of G onto themselves, since for a given y, $yg^{-1} \to (yg^{-1})g = y$. It is also one-to-one, since from $x_1g = x_2g$ it follows that $x_1 = x_2$. Thus $R(g)$ is a permutation for each g. The mapping $R(g_1)R(g_2)$ is the mapping $x \leftrightarrows x(g_1)g_2 = x(g_1g_2)$, and so, $R(g_1)R(g_2) = R(g_1g_2)$. Moreover, in $R(g_1)$, $1 \leftrightarrows g_1$, and in $R(g_2)$, $1 \leftrightarrows g_2$. Hence if $g_1 \neq g_2$, then $R(g_1) \neq R(g_2)$. Thus the mapping $g \leftrightarrows R(g)$ is an isomorphism. We observe in addition that $R(1) = I$, the identical mapping, and that $R(g^{-1})R(g) = I$, so that $R(g^{-1}) = [R(g)]^{-1}$.

The permutations $R(g) : x \leftrightarrows xg$ are called the *right regular representation* of G. We may also consider the permutations $L(g) : x \leftrightarrows gx$, the *left regular representation* of G. We find that $L(g)$ is anti-isomorphic to G. This means that the mapping $L(g)$ is one-to-one and that it *reverses* multiplication, i.e., $L(g_1g_2) = L(g_2)L(g_1)$.

If we have a set of subgroups H_j of G where j ranges over a system of indices J, then the set of elements of G, each of which belongs to every H_j, will satisfy $S1$ and $S2$ and so be a subgroup H called the *intersection* of the H_j. We write this: $H = \bigcap_j H_j$. Moreover, the set of all finite products, $g_1g_2 \cdots g_s$, where each g_i belongs to some H_j also satisfies $S1$ and $S2$. This set forms a subgroup T called the *union* of the H_j, written $T = \bigcup_j H_j$. For the intersection and union of two subgroups H and K we write $H \cap K$ and $H \cup K$, respectively. This notation is in agreement with that of lattice theory and will be considered more fully in Chap. 8.

An arbitrary set of elements in a group is called a *complex*. If A and B are two complexes in a group G, we write AB for the complex consisting of all elements ab, $a \in A$, $b \in B$, and call AB the product of A and B. We easily verify the associative law $(AB)C = A(BC)$ for the multiplication of complexes.

If K is any complex in a group G, we designate by $\{K\}$ the subgroup consisting of all finite products $x_1 \cdots x_n$, where each x_i is an element of K or the inverse of an element of K. We say that $\{K\}$ is *generated* by K. It is easy to see that $\{K\}$ is contained in any subgroup of G which contains K.

1.5. Cosets. Theorem of Lagrange. Cyclic groups. Indices.

Given a group G and a subgroup H. The set of elements hx, all $h \in H$, $x \in G$, x fixed, is called a *left coset* of H and we write Hx to designate this set. Similarly, the set of elements xh, all $h \in H$, is called a *right coset* xH of H.

THEOREM 1.5.1. *Two left (right) cosets of H in G are either disjoint or identical sets of elements. A left (right) coset of H contains the same cardinal number of elements as H.*

Proof: If cosets Hx and Hy have no element in common, there is nothing to prove. Hence, suppose $z \in Hx$, $z \in Hy$. Then $z = h_1x = h_2y$. Here $x = h_1^{-1}h_2y$ and $hx = hh_1^{-1}h_2y = h'y$, whence $Hx \subseteq Hy$. Similarly, $hy = hh_2^{-1}h_1x = h''x$, whence $Hy \subseteq Hx$. Here $Hx = Hy$; that is, the sets are identical. A similar proof holds for right cosets.

The correspondences $h \leftrightarrows hx$, $h \leftrightarrows xh$, $h \in H$, show that H, Hx, and xH contain the same cardinal number of elements.

The element $x = x1 = 1x$ belongs to the cosets xH and Hx and is called the *representative* of the coset. From Theorem 1.5.1, any element $u \in Hx$ may be taken as the representative, since $Hu = Hx$. Thus $H = H1 = 1H$ is one of its own cosets, and it is usually convenient (and under certain conventions compulsory) to take the identity as the representative of a subgroup regarded as one of its own cosets. We write

$$G = H + Hx_2 + \cdots + Hx_r, \tag{1.5.1}$$

to indicate that the cosets $H, Hx_2, \cdots, Hx_r$ are disjoint and exhaust G. Here the indicated addition is only a convenient notation and not to be regarded as an operation.

Since $(Hx)^{-1}$ (the set of inverses of the elements of the form hx) is equal to $x^{-1}H$ and $(yH)^{-1} = Hy^{-1}$, there is a one-to-one correspondence between left and right cosets of H. Thus, from (1.5.1),

$$G = H + x_2^{-1}H + \cdots + x_r^{-1}H. \tag{1.5.2}$$

The cardinal number r of right or left cosets of a subgroup H in a group G is called the *index* of H in G and is written $[G:H]$. The *order* of a group G is the cardinal number of elements in G. The identity alone is a subgroup, and its cosets consist of single elements. Thus the order of a group is the index of the identity subgroup.

Theorem 1.5.2 (Theorem of Lagrange). *The order of a group G is the product of the order of a subgroup H and the index of H in G.*

Proof: Each of the $r = [G:H]$ disjoint cosets of H in G contains the same number of elements as H, which is the order of H.

If H is a subgroup of G, and K is a subgroup of H, let

$$G = H + Hx_2 + \cdots + Hx_s,$$
$$H = K + Ky_2 + \cdots + Ky_r.$$

Then, for $g \in G$, $g = hx_j$, $h \in H$ in a unique way, and $h = ky_i$, $k \in K$ uniquely. Thus the cosets of K in G are given by Ky_ix_j $i = 1, \cdots r$, $j = 1, \cdots, s$. For two such cosets to be equal, they would have to belong to the same coset of H and so have the same x_j. Multiplying by x_j^{-1} on the right, we see that they would also have to have the same y_i. Thus the cosets of K in H are given by Ky_ix_j, and these are all different. We have thus proved the theorem:

THEOREM 1.5.3. *If* $G \supseteq H \supseteq K$, *then* $[G:K] = [G:H][H:K]$.

A group G is *cyclic* if every element in it is a power b^i of some fixed element b. If we write $(b^{-1})^r = b^{-r}$, then by the associative law and induction we can show $b^m b^t = b^{m+t}$ for any integral exponents m, t. If all powers of b are distinct, then the cyclic group is of infinite order and is isomorphic with the additive group of all integers, these being the exponents of the generator b. If not all powers are distinct, let $b^m = b^t$ with $m > t$. Then $b^{m-t} = 1$, with $m - t$ positive. Let $n > 0$ be the least positive integer, with $b^n = 1$. Then we readily see that the elements of the group are $1, b, \cdots, b^{n-1}$ and that with $0 \leq r$, $s < n$, $b^r b^s = b^{r+s}$ if $r + s < n$, while $b^r b^s = b^{r+s-n}$ if $r + s \geq n$. From this we may verify directly that for each positive n there is, to within isomorphism, a unique cyclic group of order n. This is also the additive group of integers modulo n. Thus, for a cyclic group generated by an element b, its order will either be infinite or some positive integer n, in which case n is the smallest positive integer such that $b^n = 1$. We define the *order of an element* b as the order of the cyclic group $\{b\}$ which it generates.

The nature and number of subgroups of a group G are surely of great value in describing G itself. But if G contains no subgroup except itself and the identity, then there are no proper subgroups which describe its structure. In this case we can give a very simple direct description of G.

THEOREM 1.5.4. *Let G be a group, not the identity alone. Then G has no subgroup except itself and the identity if, and only if, G is a finite cyclic group of prime order.*

Proof: Under the hypothesis if $b \neq 1$ is an element of G, then the cyclic group generated by b is not the identity and must be the entire group G. If b is of infinite order, then b^2 generates a proper subgroup, the elements b^{2j}. Hence b is of finite order, n, and $b^n = 1$. If n is not a prime, then $n = uv$ with $u > 1$, $v > 1$. Here the powers of b^u generate a proper subgroup of order v. Hence n is a prime and G is a cyclic group of prime order. But from the Theorem of Lagrange a group of prime order cannot contain a subgroup different from the identity and the whole group.

There is a basic relation on indices of subgroups.

Theorem 1.5.5. *Inequality on indices.* $[A \cup B:B] \geq [A:A \cap B]$.

Proof: Call $A \cap B = D$ and let $A = D1 + Dx_2 + \cdots + Dx_r$. Then we assert that the cosets $B1, Bx_2, \cdots, Bx_r$ are all distinct in $A \cup B$. For if $Bx_i = Bx_j$, $j \neq i$, then $x_j = bx_i$ with $b \in B$. But here x_i and x_j both belong to A, and so for this b also $b \in A$, whence $b \in A \cap B = D$; so the cosets Dx_j and Dx_i have in common the element $x_j = bx_i$ contrary to assumption. Hence there are at least as many distinct cosets of B in $A \cup B$ as there are of $A \cap B$ in A, proving the inequality.

THEOREM 1.5.6. EQUALITY OF INDICES. *If* $[A \cup B:B]$ *and* $[A \cup B:A]$ *are finite and relatively prime, then* $[A \cup B:B] = [A:A \cap B]$ *and* $[A \cup B:A] = [B:A \cap B]$.

Proof: By Theorem 1.5.3,

$$[A \cup B:A \cap B] = [A \cup B:B][B:A \cap B] = [A \cup B:A][A:A \cap B].$$

By Theorem 1.5.5, $[A \cup B:B] \geq [A:A \cap B]$, but also from the above relation $[A \cup B:B]$ divides $[A:A \cap B]$, since it is relatively prime to $[A \cup B:A]$. Hence $[A \cup B:B] = [A:A \cap B]$ and similarly $[A \cup B:A] = [B:A \cap B]$.

1.6. Conjugates and Classes.

Let G be a group and S any set of elements in G. Then the set S' of elements of the form $x^{-1}sx$, $s \in S$, x fixed, is called the *transform* of S by x and is written in either of the forms $S' = x^{-1}Sx$ or $S' = S^x$.

LEMMA 1.6.1. *S and S^x contain the same number of elements.*

Proof: $s \leftrightarrows x^{-1}sx$ is a 1–1 correspondence, since $s \rightarrow x^{-1}sx = s'$ is a mapping and so is $s' \rightarrow xs'x^{-1} = x(x^{-1}sx)x^{-1} = s$.

If S and S' are two sets in G, H is some subgroup of G, and some $x \in H$ exists such that $S' = S^x$, we say that S and S' are *conjugate under* H. If $S' = x^{-1}Sx$, then $S = (x^{-1})^{-1}S'x^{-1}$. Moreover, if $S'' = y^{-1}S'y$, then $S'' = y^{-1}x^{-1}Sxy = (xy)^{-1}S(xy)$. Since trivially $S = 1^{-1}S1$, we see that the relation of being conjugate under H is an equivalence relation, being reflexive, symmetric, and transitive. We call the set of all S' conjugate to a given S a *class* of conjugates. From $(x^{-1}sx)^{-1} = x^{-1}s^{-1}x$ and $x^{-1}s_1x \cdot x^{-1}s_2x = x^{-1}(s_1s_2)x$, we deduce:

LEMMA 1.6.2. *Any set conjugate to a subgroup is also a subgroup.*

If $x^{-1}Sx = S$, then $S = xSx^{-1}$. If also $y^{-1}Sy = S$, then $S = (xy)^{-1}S(xy)$. Hence the set of $x \in H$, such that $S^x = S$, is a subgroup of H which we shall call the *normalizer of* S *in* H, and we designate this as $N_H(S)$. Again the set of $x \in H$, such that $x^{-1}sx = s$ for all $s \in S$, may similarly be shown to be a subgroup of H which we call the *centralizer of* S *in* H and designate $C_H(S)$ [or $Z_H(S)$ if we follow the German spelling]. Note that if S consists of a single element, the centralizer and normalizer are identical; moreover, always $C_H(S) \subseteq N_H(S)$. When $H = G$ it is customary to speak merely of the normalizer or centralizer of S. The centralizer Z of G in G is called the *center of* G.

THEOREM 1.6.1. *The number of conjugates of* S *under* H *is the index in* H *of the normalizer of* S *in* H, $[H:N_H(S)]$.

Proof: Write $N_H(S) = D$ for brevity and let

$$H = D + Dx_2 + \cdots + Dx_r. \quad r = [H:N_H(S)].$$

Then $x^{-1}Sx = y^{-1}Sy$, $x, y \in H$ if, and only if, $S = (yx^{-1})^{-1}S(yx^{-1})$; that is, $yx^{-1} \in D$ or $y \in Dx$. Hence two conjugates of S under H are the same if, and only if, the transforming elements belong to the same left coset of D. Hence the number of distinct conjugates is the index of D in H, as was to be shown.

If S consists of a single element s, the conjugates under G form a *class*. Thus the classes of elements in G are a partitioning of the elements of G, and we write

$$G = C_1 + C_2 + \cdots + C_s, \tag{1.6.1}$$

the C_i being disjoint classes and every element being in exactly one class. The identity 1 is always a class. From Theorem 1.6.1 the number of elements in a class C_i is the index of a subgroup and hence a divisor of the order of the group.

1.7. Double Cosets.

Given a group G and two subgroups H and K, not necessarily distinct, the set of elements HxK, where x is some fixed element of G, is called a *double coset*. As with ordinary cosets, we may prove:

LEMMA 1.7.1. *Two double cosets* HxK *and* HyK *are either disjoint or identical.*

Proof: Here, if $z = h_1xk_1 = h_2yk_2$, $hxk = hh_1^{-1}h_2yk_2k_1^{-1}k$, whence $HxK \subseteq HyK$, and similarly, $HyK \subseteq HxK$.

A double coset HxK contains all left cosets of H of the form Hxk and all right cosets of K of the form hxK. Moreover, it is clear that HxK consists of complete left cosets of H and of complete right cosets of K.

THEOREM 1.7.1. *The number of left cosets of H in HxK is $[K:K \cap x^{-1}Hx]$, and the number of right cosets of K in HxK is $[x^{-1}Hx:K \cap x^{-1}Hx]$.*

Proof: We put the elements of HxK into a 1–1 correspondence with the elements $x^{-1}HxK$ by the rule $hxk \leftrightarrows x^{-1}hxk$. This correspondence gives a 1–1 correspondence between the left cosets Hxk of H in HxK and the left cosets $x^{-1}Hx \cdot k$ of $x^{-1}Hx$ in $x^{-1}HxK$, and also between the right cosets hxK of K in HxK and the right cosets $x^{-1}hxK$ of K in $x^{-1}HxK$. Let us write $x^{-1}Hx = A$, and $A \cap K = D$. Now if $A = 1 \cdot D + u_2D + \cdots + u_rD$, $r = [A:D]$, then $u_i \in A$, whence $K, u_2K, \cdots, u_rK$ are right cosets of K in AK. They are distinct since if $u_iK = u_jK$, then $u_i^{-1}u_j \in K$, but since $u_i, u_j \in A$, this would mean that $u_i^{-1}u_j \in D$, and thus $u_iD = u_jD$ contrary to assumption. Every right coset of K in AK is of the form aK, where $a \in A$ is of the form u_id with $d \in D$. But $u_idK = u_iK$. Thus the number of right cosets of K in AK is $[A:D] = [x^{-1}Hx:x^{-1}Hx \cap K]$, and by the 1–1 correspondence, this is the number of right cosets of K in HxK. In the same way it may be shown that the number of left cosets of A in AK is $[K:D] = [K:x^{-1}Hx \cap K]$ and this, by the 1–1 correspondence, is the number of left cosets of H in HxK.

1.8. Remarks on Infinite Groups.

Many of the theorems on groups do not involve the issue as to whether or not the groups are finite. But in some instances the facts are essentially different for finite and infinite groups, and occasionally when the facts are similar, the methods of proof differ.

An infinite group G may have certain finite properties. Some important properties of this kind are:

1) G is finitely generated.

2) G is *periodic*, that is, the elements of G are of finite order.

3) G satisfies the *maximal condition:* Every ascending chain of distinct subgroups $A_1 \subset A_2 \subset A_3 \subset \cdots$ is necessarily finite.

4) G satisfies the *minimal condition:* Every descending chain of distinct subgroups $A_1 \supset A_2 \supset A_3 \supset \cdots$ is necessarily finite.

An infinite group G is said to have a property *locally* if this property holds for every finitely generated subgroup. A family H_i of homomorphic images of a group G is said to be a *residual family* for G, if for every $g \neq 1$ of G there is at least one H_i in which the image of this g is not the identity. We say that G has a property *residually* if there is a residual family for G of homomorphic images all having the property.

THEOREM 1.8.1. *A group G satisfies the maximal condition if, and only if, G and every subgroup of G are finitely generated.*

Proof: Let H be a subgroup of G which is not finitely generated. We may construct recursively an infinite ascending chain of distinct subgroups of H, $\{h_1\} \subset \{h_1, h_2\} \subset \cdots \subset \{h_1, \cdots, h_i\} \subset \cdots$, by choosing h_1 arbitrarily, and recursively h_i, an element of H not in $\{h_1, \cdots, h_{i-1}\}$. Such an h_i always exists, since H cannot be the finitely generated group $\{h_1, \cdots, h_{i-1}\}$. Conversely, suppose that G and all its subgroups are finitely generated. Then let $B_1 \subseteq B_2 \subseteq B_3 \subseteq \cdots$ be an ascending chain of subgroups in G. We shall show that after a certain point in this chain all subgroups are equal, and so there is not an infinite ascending chain of distinct subgroups. The set of all elements b, such that $b \in B_i$ for some B_i in the chain, forms a subgroup B of G, since if $b \in B_i$ and $b' \in B_j$, then both b and b' belong to any B_k with $k \geq i$, $k \geq j$, and so also their product and their inverses are in B_k.

By hypothesis B is finitely generated, say, by elements $x_1, \cdots, x_n$. Let B_{j_1} be the first B_i containing x_1 and generally B_{j_k} be the first B_i containing x_k, for $k = 1, \cdots, n$. Then if m is the largest of $j_1, \cdots j_n$, B_m will contain all of $x_1, \cdots, x_n$, and so $B = B_m = B_{m+1} = \cdots$, and all further groups in the chain are equal to B. We shall see later that there are groups which are finitely generated but which have subgroups that are not finitely generated.

THEOREM 1.8.2. *A group G which satisfies the minimal condition is periodic.*

Proof: If G contains an element b of infinite order, then $\{b\} \supset \{b^2\} \supset \{b^4\} \supset \cdots \supset \{b^{2^i}\} \supset \cdots$ is an infinite descending chain of distinct subgroups.

In an infinite group we cannot use finite induction on its order, and so some substitute is needed to replace this method of proof which is so valuable for finite groups. One way to make this replacement is to appeal to certain very general axioms on sets and ordering. Suppose that we have an ordering relation $a \leq b$ on the elements of a set S of objects $\{a, b, c, \cdots\}$. The ordering may satisfy some of the following axioms:

$O1$) If *$a \leq b$, and $b \leq a$, then $a = b$.*

$O2$) If *$a \leq b$, and $b \leq c$, then $a \leq c$.*

$O3$) *Either $a \leq b$ or $b \leq a$ for any two a,b.*

$O4$) *Any nonempty subset T of S has a first element x_1, i.e., an element x_1 such that $x_1 \leq t$ for every $t \in T$.*

If the first two axioms hold, we say that the ordering is a *partial ordering.* If the first three axioms hold, we say that the ordering is a *simple ordering.* If all four axioms hold, we say that the ordering is a *well-ordering.* We may appeal to the axiom of well-ordering: *Every set S may be well-ordered.* Let us write $a < b$ to mean $a \leq b$ but $a \neq b$.

In a well-ordered set we may prove propositions by the method of *transfinite induction.* This proceeds as follows: Designate the first element of S as 1. Then, if $P(a)$ is a proposition about the elements of S and if $P(1)$ is true, and if the truth of $P(x)$ for all $x < a$ implies the truth of $P(a)$, we conclude that $P(b)$ is true for all $b \in S$. Let T be the subset of S, such that $P(t)$ is false for $t \in T$. If T is nonempty, it contains a first element c. But then either $c = 1$ or $P(x)$ is true for all $x < c$. In either event this would lead to the truth of $P(c)$ contrary to the choice of c in T. Hence T must be empty and $P(b)$ true for all $b \in S$. We note in passing that in a well-ordered set any descending sequence $a_1 > a_2 > a_3 > \cdots$ is necessarily finite since it must contain a first element.

Another axiom, logically equivalent to the axiom of well-ordering is Zorn's lemma. This again deals with ordering in sets.

LEMMA 1.8.1. (ZORN'S LEMMA). *Given a partially ordered set S. Suppose that every simply ordered subset of S has an upper bound (lower bound) in S. Then S has a maximal (minimal) element.* Here if U is a subset of S, then an *upper bound* b of U is an element such that $b \geq u$ for all $u \in U$. A *maximal element* w has no upper bound different from itself. Reversing the inclusion, we similarly define *lower bound* and *minimal element.*

Suppose we consider subgroups of a group G partially ordered by inclusion $A \subseteq B$ if A is a subgroup of B. Then the set of all elements in a simply ordered subset of subgroups will itself form a subgroup, since if g_1 is in one of the groups and g_2 is in another, then both g_1 and g_2 are in the greater of the two subgroups and so also are their product and their inverses. For this reason Zorn's lemma is well suited to proofs in group theory or to abstract algebra in general.

Both the Axiom of Well-Ordering and Zorn's lemma are logically equivalent to:

Axiom of Choice. *For any family F of subsets $\{S_i\}$ of a set S, there is a choice function $f(S_i)$ defined for the subsets of F whose values are elements of S, such that $f(S_i) = a_i \in S_i$, the subsets S_i not being void.*

In certain arguments the Axiom of Choice appears to lead to paradoxes, and for this reason it is suspect. All three principles are surely valid if the set S is countable, i.e., its objects may be put into a one-to-one correspondence with the natural numbers $1,2,3, \cdots$. Presumably, they are valid for other sets S and possibly for all well-defined sets, though it might be remarked that as yet no one has actually constructed a well-ordering of the set of all real numbers. When using any one of these principles in this book, it is to be understood that by "Every set S" we mean "Every set S for which the axiom is valid."

A useful application of these methods is the following:

Theorem 1.8.3. *Let g be an element of a group G and H a subgroup of G which does not contain g. Then there is a subgroup M containing H which is maximal with respect to the property of not containing g.*

Proof: We use Zorn's lemma. Subgroups containing H but not containing g form a partially ordered set under inclusion. The elements of a simply ordered set of these groups themselves form a group which contains H but not g. Hence a maximal group M exists containing H but not g.

Using this theorem, we easily derive the following:

Theorem 1.8.4. *Let G be a finitely generated group and H a proper subgroup of G. Then there exists a maximal subgroup M of G containing H.*

Proof: Let G be generated by $x_1, \cdots, x_m$ and let y_1 be the first of

$x_1, \cdots, x_m$ not contained in H. Let $M_1 \supseteq H$ where M_1 is maximal with respect to the property of not containing y_1. Then any subgroup of G containing M_1 properly contains y_1, and so also $\{M_1, y_1\} = H_1$. If $H_1 = G$, then M_1 is the maximal subgroup sought. If not, choose $M_2 \supseteq H_1$ where M_2 is maximal with respect to the property of not containing y_2, the first of $x_1, \cdots, x_m$ not contained in H_1. Since $G = \{x_1, \cdots, x_m\}$, by continuing this process we must reach an $M_i \supseteq H_{i-1} \supseteq \cdots \supseteq H$, where $\{M_i, y_i\} = G$ and $M_i = M$ is the maximal subgroup sought.

1.9. Examples of Groups.

The one-to-one mappings of a set onto itself which preserve some property usually form a group. Many of the most interesting groups arise naturally in this way. The symmetries of a geometric figure are of this kind. These are the congruent (i.e., distance-preserving) mappings of the figure onto itself. The first two examples below are groups of symmetries.

Example 1. Dihedral groups. The symmetries of a regular polygon of $n \geqq 3$ sides form a group of order $2n$. These are determined entirely by the way in which the vertices are mapped onto themselves. Let the vertices be numbered $1, 2, \cdots, n$ in a clockwise manner. The vertex 1 may be mapped onto any vertex $1, 2, \cdots$, n, and the remaining vertices placed in either a clockwise or a counterclockwise direction. All symmetries are generated by the rotation

$$a = \begin{pmatrix} 1, 2, 3 & \cdots & n-1 & n \\ 2, 3, 4 & \cdots & n & 1 \end{pmatrix},$$

and the reflection

$$b = \begin{pmatrix} 1, 2, & 3 & \cdots & n-1, n \\ 1, n, & n-1 & \cdots & 3\ , 2 \end{pmatrix}.$$

Here $a^n = 1$, $b^2 = 1$, $ba = a^{-1}b$. Moreover, these relations determine the group completely, since every element generated by a and b is of the form $a^{i_1}b^{j_1} \cdots a^{i_r}b^{j_r}$ and since $ba^i = a^{-i}b$, as we may show from the last relation, every element can be put into the form a^i or a^ib with $i = 0, 1, \cdots, n-1$; these are the $2n$ different elements of the group. These relations also define a group for $n = 2$ which is of order 4. This is called the *four group*.

Example 2. Symmetries of the Cube. The symmetries of the

cube are determined by the mappings of the eight vertices onto themselves. Let these be numbered as in the figure. The symmetries include the rotation

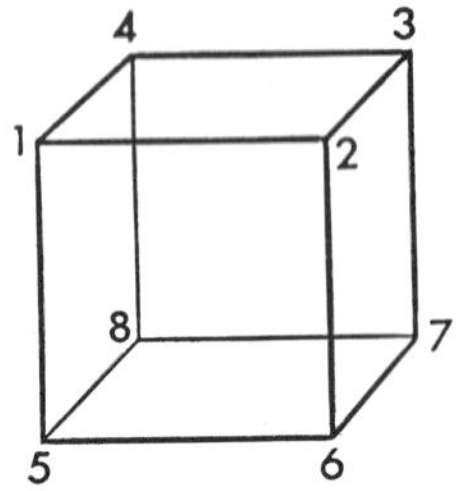

Fig. 1. Symmetries of the cube.

$$a = \begin{pmatrix} 1, 2, 3, 4, 5, 6, 7, 8 \\ 2, 3, 4, 1, 6, 7, 8, 5 \end{pmatrix},$$

$$b = \begin{pmatrix} 1, 2, 3, 4, 5, 6, 7, 8 \\ 1, 4, 8, 5, 2, 3, 7, 6 \end{pmatrix},$$

and the reflection

$$c = \begin{pmatrix} 1, 2, 3, 4, 5, 6, 7, 8 \\ 5, 6, 7, 8, 1, 2, 3, 4 \end{pmatrix}.$$

The elements a and b generate a group G_1 which takes every vertex into every other vertex. We may see this from the diagram

$$1 \xrightarrow{a} 2 \xrightarrow{a} 3 \xrightarrow{a} 4 \xrightarrow{b} 5 \xrightarrow{a} 6 \xrightarrow{a} 7 \xrightarrow{a} 8 \xrightarrow{a} 5 \xrightarrow{b} 2 \xrightarrow{a^{-1}} 1.$$

In this diagram $i \xrightarrow{x} j$ means that the element x takes i into j. From this we may read that ba^2 takes 4 into 7. The elements that fix 1 form a subgroup H_1, and we may write

$$G_1 = H_1 + H_1x_2 + H_1x_3 + H_1x_4 + H_1x_5 + H_1x_6 + H_1x_7 + H_1x_8,$$

where x_i is an element taking 1 into i. We may take the x's as follows: $x_2 = a$, $x_3 = a^2$, $x_4 = a^3$, $x_5 = a^3b$, $x_6 = a^3ba$, $x_7 = a^3ba^2$, $x_8 = a^3ba^3$. Since there are only eight letters in all, and all elements in a coset H_1x_i take 1 into the same i, this includes all conceivable cosets of H_1, and the index of H_1 in G_1 is 8.

A rotation of the cube fixing the vertex 1 must permute the three adjacent vertices cyclically. Thus H_1 is of order 3 containing only $1, b, b^2$, and so G_1 is of order 24. The reflection c is not in G_1 but as $c^2 = 1$, $ca = ac$ and $cb = a^2ba^2c$, we see that G, the group generated by a, b, c satisfies $G = G_1 + G_1c$ and is of order 48. G is the full group of symmetries of the cube.

Example 3. What is the order of a group G generated by elements a and b subject only to the relations

$$a^7 = 1, \quad b^3 = 1, \quad ba = a^rb?$$

Every element of G may be expressed by a finite sequence of a's and

b's. From the relation $ba = a^r b$ we may ultimately express an element in a form in which no b is followed by an a. Thus every element may be put in the form

$$g = a^i b^j, \quad i = 0, \quad 1, \cdots, 6; \quad j = 0, 1, 2.$$

From this we see that the order of G is, at most, 21. But the actual order depends on the value of r in the relation $ba = a^r b$. We see that $ba^2 = a^r ba = a^r(a^r b) = a^{2r}b$, and similarly, $ba^i = a^{ir}b$. Thus $b^2 a = ba^r b = a^{r^2}b^2$. From this we get $b^2 a^i = a^{ir^2}b^2$. Hence $b^3 a = ba^{r^2}b^2 = a^{r^3}b^3$. But as $b^3 = 1$, this gives $a = a^{r^3}$; but also $a^7 = 1$. Of the values $r = 1, 2, 3, 4, 5, 6$, we find that $r = 3, 5, 6$ lead to $a = 1$, and the group is merely the cyclic group of order 3 given by $b^3 = 1$. But $r = 1, 2, 4$ do not lead to this result. If $r = 1$, then $ba = ab = c$ is an element of order 21. Conversely, in a cyclic group of order 21, with $c^{21} = 1$, if we put $b = c^7$, $a = c^{-6}$, we have $a^7 = 1$, $b^3 = 1$, $ba = ab = c$. With $r = 2$ the following permutations work:

$$a = \begin{pmatrix} 1, 2, 3, 4, 5, 6, 7 \\ 2, 3, 4, 5, 6, 7, 1 \end{pmatrix},$$

$$b = \begin{pmatrix} 1, 2, 3, 4, 5, 6, 7 \\ 1, 5, 2, 6, 3, 7, 4 \end{pmatrix}.$$

For $r = 4$ the same permutation for a and the inverse of the second permutation will do for b. This example is meant to show that an apparently slight change in defining relations can make a major difference in the group defined.

Example 4. Let us find the group of permutations on the seven letters A, B, C, D, E, F, G which permute among themselves the columns of the following diagram, the order of the letters in a column being immaterial:

$$\begin{matrix} A, B, C, D, E, F, G \\ B, C, D, E, F, G, A \\ D, E, F, G, A, B, C. \end{matrix}$$

We see at once that the permutation

$$a = \begin{pmatrix} A, B, C, D, E, F, G \\ B, C, D, E, F, G, A \end{pmatrix}$$

permutes the columns cyclically. Thus, if G is the entire group of permutations and H is the subgroup taking the first column into itself, the coset Ha^i, $i = 1, \cdots 6$ consists of all elements mapping the first column onto the $i + 1$st. Hence

$$G = H + \cdots + Ha^6, \quad [G:H] = 7.$$

Within H there may be elements permuting A, B, D cyclically. If we try this, we find that this does not appear to determine the mapping of the remaining letters, but if we also assume that C is mapped into itself, a permutation is completely determined which works

$$b = \begin{pmatrix} A, B, C, D, E, F, G \\ B, D, C, A, F, G, E \end{pmatrix}.$$

Thus, if K is the subgroup which fixes the first column and also the letter A,

$$H = K + Kb + Kb^2, \quad b^3 = 1, \quad [H:K] = 3.$$

Within K let us seek an element which interchanges B and D. Take

$$c = \begin{pmatrix} A, B, C, D, E, F, G \\ A, D, C, B, F, E, G \end{pmatrix}.$$

Hence, if T is the subgroup fixing A, B, and D,

$$K = T + Tc, \quad c^2 = 1, \quad [K:T] = 2.$$

Within T, the three letters A, B, D are fixed and the letter C can go into any one of the four possibilities C, E, F, G. Each one of these choices leads to exactly one permutation:

$$\begin{pmatrix} A, B, C, D, E, F, G \\ A, B, C, D, E, F, G \end{pmatrix},$$

$$\begin{pmatrix} A, B, C, D, E, F, G \\ A, B, E, D, C, G, F \end{pmatrix},$$

$$\begin{pmatrix} A, B, C, D, E, F, G \\ A, B, F, D, G, C, E \end{pmatrix},$$

$$\begin{pmatrix} A, B, C, D, E, F, G \\ A, B, G, D, F, E, C \end{pmatrix}.$$

Thus T is a group of order 4, K is of order 8, H is of order 24, and G is of order 168. If the seven letters $A, \cdots, G$ are regarded as points and the columns as lines, the diagram represents the finite projective plane with seven points, and the group G is its collineation group.

EXAMPLE 5. THE QUATERNION GROUP. The following group of order 8 is in many ways an exceptional group. Its unusual properties will be discussed later in § 12.5. Here we are interested in presenting it in terms of its multiplication table, or Cayley table, as it is called after the English mathematician Cayley. In the row with x_i on the left and in the column with x_j at the top we enter the product $x_k = x_i x_j$.

	x_1	x_2	x_3	x_4	x_5	x_6	x_7	x_8
x_1	x_1	x_2	x_3	x_4	x_5	x_6	x_7	x_8
x_2	x_2	x_5	x_4	x_7	x_6	x_1	x_8	x_3
x_3	x_3	x_8	x_5	x_2	x_7	x_4	x_1	x_6
x_4	x_4	x_3	x_6	x_5	x_8	x_7	x_2	x_1
x_5	x_5	x_6	x_7	x_8	x_1	x_2	x_3	x_4
x_6	x_6	x_1	x_8	x_3	x_2	x_5	x_4	x_7
x_7	x_7	x_4	x_1	x_6	x_3	x_8	x_5	x_2
x_8	x_8	x_7	x_2	x_1	x_4	x_3	x_6	x_5.

In this table, from the fact that every x_i occurs exactly once in every row and in every column, we see that in a product $ab = c$ any two of a, b, c determine the third uniquely. Thus the preceding table is the multiplication table of a quasi-group. By inspection we see also that $x_1 x_i = x_i x_1 = x_i$ in every instance whence $x_1 = 1$ is a two-sided unit and the table determines a loop. But both these properties are preserved if we replace the last two rows by

x_7	x_7	x_4	x_2	x_1	x_3	x_8	x_6	x_5
x_8	x_8	x_7	x_1	x_6	x_4	x_3	x_5	x_2.

But the table as given is alleged to be the multiplication table of a group, and for this it is necessary to verify the associative law $(ab)c = a(bc)$ for products.

A full verification of the associative law in this case would involve potentially $8^3 = 512$ verifications. Even though it is easy to see that $(ab)c = a(bc)$ whenever any one of a, b, or c is the identity, this still leaves 343 verifications. Here we appeal to the converse of the Cayley Theorem 1.4.1.

THEOREM 1.9.1 (CONVERSE OF CAYLEY'S THEOREM). *A loop is a group if the right regular mappings $x \to xg$ form a group.*

Proof: Here in $R(g)R(h)$ we have $1 \to g \to gh$. But also in $R(gh)$ we have $1 \to gh$, and this is the only mapping taking 1 into gh. Hence $R(g)R(h) = R(gh)$, whence, for every x, $(xg)h = x(gh)$ and the associative law holds.

In this case we write $a = x_2$, $b = x_3$ and calculate

$$A = R(a) = \begin{pmatrix} x_1, & x_2, & x_3, & x_4, & x_5, & x_6, & x_7, & x_8 \\ x_2, & x_5, & x_8, & x_3, & x_6, & x_1, & x_4, & x_7 \end{pmatrix},$$

$$B = R(b) = \begin{pmatrix} x_1, & x_2, & x_3, & x_4, & x_5, & x_6, & x_7, & x_8 \\ x_3, & x_4, & x_5, & x_6, & x_7, & x_8, & x_1, & x_2 \end{pmatrix}.$$

Here $A^4 = B^4 = 1$, $A^2 = B^2$, $BA = A^3B$, and we easily see that these generate a group of order 8 that is indeed the right regular representation of the given loop and which is therefore a group. The second rows of the permutations are the columns of the multiplication table. In terms of the generators a and b we have $x_1 = 1$, $x_2 = a$, $x_3 = b$, $x_4 = ab$, $x_5 = a^2 = b^2$, $x_6 = a^3$, $x_7 = b^3 = a^2b$, $x_8 = a^3b$; also, $a^4 = 1$, $b^4 = 1$, $b^2 = a^2$, $ba = a^3b$.

EXERCISES

1. Show that from the associative law $(ab)c = a(bc)$ it follows that all methods of bracketing $a_1a_2 \cdots a_n$, without altering the order of the factors, yield the same product.
2. Show that $(ab)^{-1} = b^{-1}a^{-1}$ in any group and that more generally $(a_1a_2 \cdots a_{n-1}a_n)^{-1} = a_n^{-1}a_{n-1}^{-1} \cdots a_2^{-1}a_1^{-1}$.
3. Show that a and a^{-1} are of the same order.
4. Show that ab and ba are of the same order. (Hint: ab and ba are conjugate elements.)
5. If $a^m = 1$, and $b^n = 1$, where m and n are positive integers, and if $ba = ab$, show that $(ab)^k = 1$, where k is the least common multiple of m and n. Find an example with $ba \neq ab$ where this is untrue.
6. If a group G has only one element a of order 2, show that for every x in G, $xa = ax$.
7. Show that the only finite group with exactly two classes of elements is the group of order 2.
8. If $p < q$ are primes, show that a group of order pq cannot have two distinct subgroups of order q.
9. If H is a proper subgroup of the finite group G, show that the conjugates of H do not include all elements of G.

10. Show that the loops of orders 1, 2, 3, 4 are groups, but find a loop of order 5 which is not a group.
11. Show that the double coset HxK contains precisely those right cosets of K which have at least one element in common with Hx.
12. (William Scott.) Show that a system with a binary product and a unit 1 such that $1a = a1 = a$ for all a will be associative if we take as a law the equality of two distinct bracketings of $a_1a_2 \cdots a_n$.
13. From the axioms of the first definition of a group prove the uniqueness of the unit 1 and the uniqueness of the inverse a^{-1}.
14. If A and B are two finite subgroups of a group G, show that the complex AB contains exactly $[A:1][B:1]/[A \cap B:1]$ distinct elements.

2. NORMAL SUBGROUPS AND HOMOMORPHISMS

2.1. Normal Subgroups.

A subgroup H of a group G is said to be a *normal subgroup* if $x^{-1}Hx = H$ for all $x \in G$. In the terminology of §1.6 a subgroup H of G is a normal subgroup if $N_G(H) = G$.

LEMMA 2.1.1. *A subgroup H of G is a normal subgroup of G if and only if every left coset Hx is also a right coset xH.*

Proof: If $x^{-1}Hx = H$ for all x, then $Hx = xH$ and conversely, if $Hx = yH$ then $x \in yH$ so that $yH = xH$. Hence $Hx = xH$ for all $x \in G$ and so $x^{-1}Hx = H$.

COROLLARY 2.2.1. *A subgroup of index 2 is necessarily a normal subgroup.*

For, if $G = H + Hx$, then $G = H + xH$.

For finite groups, $x^{-1}Hx \subseteq H$ implies $x^{-1}Hx = H$, since $x^{-1}Hx$ and H have the same number of elements, but the inclusion need not imply equality for infinite groups. However, if $x^{-1}Hx \subseteq H$ and $xHx^{-1} \subseteq H$, then $H = x(x^{-1}Hx)x^{-1} \subseteq xHx^{-1} \subseteq H$, whence $H = xHx^{-1}$ and similarly $H = x^{-1}Hx$. Thus $x^{-1}Hx \subseteq H$ for all x is sufficient for H to be normal.

A group G that contains no proper normal subgroup is said to be a *simple* group. The term "simple" must be understood in a purely technical sense. The groups without any proper subgroups are, by Theorem 1.5.4, the finite cyclic groups of prime order, and these are simple groups both in the technical sense and in the more ordinary sense of being uncomplicated. But there are many other simple groups, one of these being the group of order 168 given in the fourth example in §1.8. The determination of the finite simple groups is an unsolved problem. It has been conjectured that a simple finite group, except for those of prime order, is necessarily of even order, but even this seems to be an unusually difficult problem.

2.2. The Kernel of a Homomorphism.

Suppose the group H is a homomorphic image of the group G. Consider the set T of elements $t \in G$ consisting of all elements of G mapped onto the identity of H.

$$G \to H, \quad T \to 1. \tag{2.2.1}$$

As noted in §1.4, $1 \to 1$, whence $1 \in T$. If $t \to 1$, $t^{-1} \to u$, then $1 = tt^{-1} \to u$. But $1 \to 1$, whence $u = 1$, and so $t^{-1} \to 1$ and $t^{-1} \in T$. Also if $t_1 \to 1$, $t_2 \to 1$, then $t_1t_2 \to 1$, whence $t_1t_2 \in T$. Hence T is a subgroup of G. Moreover if $x \in G$, $t \in T$, then $x \to y$, $t \to 1$, $x^{-1} \to y^{-1}$, and $x^{-1}tx \to y^{-1}1y = 1$, whence $x^{-1}tx \in T$, and so T is a normal subgroup of G. The set T is called the *kernel* of the homomorphism $G \to H$.

THEOREM 2.2.1 (FIRST THEOREM ON HOMOMORPHISMS). *In the homomorphism $G \to H$ the set T of elements of G mapped onto the identity of H is a normal subgroup of G. Two elements of G have the same image in H if, and only if, they belong to the same coset of T.*

Proof: We have already shown that T is a normal subgroup of G. Suppose $x \to u$, $y \to u$, $x, y \in G$, $u \in H$. Then $xy^{-1} \to 1$ and $xy^{-1} \in T$, whence $x \in Ty$ and x and y are in the same coset of T. Conversely, if $x \in Ty$, then $x = ty$, and if $y \to u$, then (since $t \to 1$) we have $x \to u$ and x and y have the same image in H.

2.3. Factor Groups.

In the preceding section it was shown that the kernel of a homomorphism of a group G is a normal subgroup T. Conversely, it is true that every normal subgroup T is the kernel of a homomorphism and, in fact, of a unique homomorphism. Suppose

$$G \underset{*}{=} T + Tx_2 + \cdots + Tx_r, \tag{2.3.1}$$

where T is a normal subgroup of G. We shall take the cosets Tx_i as the elements of a system H. We define a product in H as

$$(Tx_i)(Tx_j) = Tx_k, \tag{2.3.2}$$

if $x_ix_j \in Tx_k$ in G.

It is necessary to show that the product is uniquely defined. Let

t_1x_i and t_2x_j be arbitrary elements of Tx_i and Tx_j, respectively. Here $t_1x_it_2x_j = t_1x_it_2x_i^{-1}\cdot x_ix_j = t_3x_ix_j$, since T is a normal subgroup. But if $x_ix_j \in Tx_k$, then also $t_3x_ix_j \in Tx_k$. Thus all products of an element in Tx_i and an element in Tx_j yield elements of the same coset Tx_k. Thus the product in (2.3.2) depends solely on the cosets and not on the choice of the representatives; hence the product in H is well defined.

Since T is a normal subgroup $T^2 = T$, $Tx_i = x_iT$. Hence, in H, T is a unit as $T\cdot Tx_i = Tx_i$, $Tx_i\cdot T = Tx_iT = TTx_i = Tx_i$. Moreover, the product is associative, since $(Tx_iTx_j)Tx_k = Tx_ix_jx_k = Tx_i(Tx_jTx_k)$. If $x_i^{-1} \in Tx_j$, then Tx_iTx_j contains $x_ix_i^{-1} = 1$, and so $Tx_iTx_j = T$, whence in H, Tx_j is the inverse of Tx_i, as we may also readily verify that $Tx_jTx_i = T$. This completes the verification that H is a group which we call the *factor group of G with respect to T*. We write $H = G/T$.

THEOREM 2.3.1 (SECOND THEOREM ON HOMOMORPHISMS). *Given a group G and a normal subgroup T. Then if $H = G/T$, there is a homomorphism $G \to H$ whose kernel is T. This homomorphism is given by $g \to Tx_i$ if $g \in Tx_i$ in G.*

Proof: Consider the mapping $g \to Tx_i$ of G onto H when $g \in Tx_i$ in G. If $g_1 \in Tx_i$, $g_2 \in Tx_j$, then (as we showed) $g_1g_2 \in Tx_k$, where $x_ix_j \in Tx_k$. Hence $g_1g_2 \to Tx_k = Tx_iTx_j$. Thus the mapping of G onto H preserves products and so is a homomorphism. Since T is the identity of G/T, then $g \to 1$ ($= T$ in H) if, and only if, $g \in T$ in G, whence T is the kernel of the homomorphism. This completes the proof.

THEOREM 2.3.2 (THIRD THEOREM ON HOMOMORPHISMS). *If $G \to K$ is a homomorphism of G onto K, and T is the kernel of the homomorphism, then K is isomorphic to $G/T = H$. If $x \to x^*$ in the homomorphism $G \to K$, then $x^* \leftrightarrows Tx$ is an isomorphism between K and H.*

Proof: Since elements of G in the same coset of T have the same image in K, the correspondence $x^* \leftrightarrows Tx$ is one to one. But if $x \to x^*$, $y \to y^*$, then $xy \to x^*y^*$. But $xy \in Txy$, whence $x^*y^* \leftrightarrows Txy = TxTy$. Thus the correspondence $x^* \leftrightarrows Tx$ preserves products and is an isomorphism of K and $H = G/T$.

Let us summarize the content of these three main theorems on homomorphisms. We have shown that the kernel of any homomorphism is a normal subgroup, that any normal subgroup is the kernel of a homomorphism whose image is unique (to within isomorphism), and that this image is the factor group of the given group with respect to the normal subgroup.

THEOREM 2.3.3. *If A and B are subgroups of a group G and either one of them is a normal subgroup, then $A \cup B = AB$.*

Proof: We must show that every finite product $x_1x_2 \cdots x_s$ with $x_i \in A$ or B can be put in the form ab. Now if B is normal a product $ba = aa^{-1}ba = ab'$, while if A is normal $ba = bab^{-1}b = a'b$ then we can rewrite the product so that no b precedes an a. The product now takes the form $a_1a_2 \cdots a_jb_{j+1} \cdots b_s = ab$, where a_i, $a \in A$ and b_i, $b \in B$.

THEOREM 2.3.4. *Let T be a normal subgroup of G. There is a 1–1 correspondence between subgroups K^* of $H = G/T$ and subgroups K of G such that $G \supseteq K \supseteq T$, where K consists of all elements of G mapped onto elements of K^*. If K^* is normal in H, then K is normal in G and conversely. Also $[G:K] = [H:K^*]$.*

Proof: Trivially, the image in H of a subgroup of G is a subgroup. Now if K^* is a subgroup of H, the inverse image K of K^* in G will contain T, the inverse image of 1. Also the inverse image satisfies the requirements for a subgroup.

Hence the inverse image of a subgroup K^* of H is a unique subgroup K such that $G \supseteq K \supseteq T$, and the same K^* is the unique image of K in the homomorphism $G \to H$. Hence $K \leftrightarrows K^*$ is a 1–1 correspondence between $G \supseteq K \supseteq T$ and $H \supseteq K^* \supseteq 1$. If K^* is normal in H then $x^{-1}Kx \to x^{*-1}\, K^*x^* = K^*$, whence $x^{-1}Kx \subseteq K$ for any x, and so K is normal in G. Again, if K is normal, the normality of its image K^* is trivial. Finally, the inverse image of a coset K^*g^* is seen to be a coset Kg, whence $[G:K] = [H:K^*]$.

If an arbitrary subgroup A has the image A^*, then the inverse image of A^* is readily seen to be $A \cup T = AT$.

2.4. Operators.

A mapping $\alpha: g \to g^\alpha$ of a group G into itself is called an *endomorphism* of G or an *operator* on G, if $(xy)^\alpha = x^\alpha y^\alpha$. Thus an endomorphism is a homomorphism of G into itself. An *automorphism* is a 1–1 endomorphism mapping G onto itself. If $g^\alpha = h^\alpha$ implies $g = h$, the endomorphism is an isomorphism, which in a finite group is necessarily an automorphism. But an infinite group may be isomorphic to a proper subgroup. Thus $x \to 2x$ is an endomorphism which is an isomorphism of the additive group of integers but not an automorphism.

A subgroup H of G is said to be *admissible* with respect to endomorphisms α_i if $H^{\alpha_i} \subseteq H$ for all α_i. It follows immediately from the definitions that *unions and intersections of admissible subgroups are admissible subgroups.* Again it is clear that an operator α may also be regarded as an operator in an admissible subgroup. But it can happen that two operators which are different for an entire group may agree in their effect on an admissible subgroup. Moreover if $G \to K$ is a homomorphism of G onto K whose kernel T is admissible with respect to an endomorphism α, then we may define a corresponding operator in K. We put

$$(Tx)^\alpha = Tx^\alpha. \tag{2.4.1}$$

This is a natural definition since applying the endomorphism to the coset Tx gives only elements belonging to Tx^α. We readily verify that this defines an operator in K and also that if $x \to x^*$ in the homomorphism $G \to K$, then $x^\alpha \to x^{*\alpha}$.

Two groups A and B are *operator isomorphic* if there is a 1–1 correspondence $A \leftrightarrows B$ and also $\alpha_i \leftrightarrows \beta_i$ between the groups and the operators on them such that $a \leftrightarrows b$ is an isomorphism and $a^{\alpha_i} \leftrightarrows b^{\beta_i}$ in this isomorphism. Thus an operator isomorphism is stronger than an isomorphism.

Theorem 2.4.1. *Given a group G and a set Ω of operators on G. Suppose A is an admissible subgroup of G and T an admissible normal subgroup. Then $A \cap T$ is an admissible normal subgroup of A and the factor groups $A \cup T/T$ and $A/A \cap T$ are operator isomorphic.*

Proof: $A \cap T$ as an intersection of Ω subgroups (i.e., admissible under Ω) will be an Ω subgroup of A. If $u \in A \cap T$, $a \in A$, then $a^{-1}ua \in A$. Also, since T is normal in G and $u \in T$, $a^{-1}ua \in A \cap T$, and so $A \cap T$ is normal in A.

Let us write $D = A \cap T$.

$$A = D + Da_2 + \cdots + Da_r. \tag{2.4.2}$$

Then we assert

$$(2.4.3) \qquad A \cup T = T + Ta_2 + \cdots + Ta_r,$$

using the same coset representatives in (2.4.3) as we did in (2.4.2). Here, if $Ta_i = Ta_j$, then $a_i a_j^{-1} \epsilon T$. But $a_i a_j^{-1} \epsilon$ A, whence $a_i a_j^{-1} \epsilon A \cap T = D$, contrary to (2.4.2). Hence the cosets Ta_i in (2.4.3) are all distinct. Moreover, since T is a normal subgroup, $A \cup T = TA$, and so any coset of T in $A \cup T$ is of the form $Ta = Tda_i$, with $a = da_i$ from (2.4.2). But as $d \epsilon T$, $Tda_i = Ta_i$, and so the cosets in (2.4.3) will exhaust $A \cup T$. The correspondence

$$(2.4.4) \qquad Da_i \leftrightarrows Ta_i$$

is a 1–1 correspondence between the cosets in (2.4.2) and (2.4.3), and thus a 1–1 correspondence between the elements of A/D and those of $A \cup T/T$. Also, if $a_i a_j = da_k$ with $d \epsilon D$, since $D \subseteq T$, we shall have both $Da_iDa_j = Da_k$ and $Ta_iTa_j = Ta_k$. Thus the rule (2.4.4) is an isomorphism between the factor groups A/D and $A \cup T/T$. An operator $\alpha \epsilon \Omega$ determines an operator in A/D and also one in $A \cup T/T$ by the rules $(Da_i)^\alpha = Da_i^\alpha$ and $(Ta_i)^\alpha = Ta_i^\alpha$. For the operators given in this way, it is immediate that (2.4.4) determines an operator isomorphism. This completes the proof.

We may easily verify that a subgroup K of a group G is a normal subgroup if, and only if, it is admissible under the family of inner automorphisms of G. In terms of operators we define two successively stronger forms of normality for subgroups. A subgroup admissible under all automorphisms of a group is called a *characteristic subgroup,* and a subgroup admissible under all endomorphisms is called a *fully invariant* subgroup. Thus the center Z of a group G is a characteristic subgroup, since if $zg = gz$ for all $g \epsilon G$, then for an automorphism α, we have $z^\alpha g^\alpha = g^\alpha z^\alpha$, and as g runs over all elements of G, g^α will also run over all elements of G, and we conclude that $z^\alpha \epsilon Z$. But the center is not necessarily a fully invariant subgroup. As an example, consider the group G of order 16 defined by the relations $a^4 = 1$, $b^2 = c^2 = 1$, $ba = a^{-1}b$, $ca = ac$, $cb = bc$. Here the center Z is of order 4 and generated by a^2 and c. But the mapping $a \to b$, $b \to b$, $c \to b$ defines an endomorphism of G mapping the element of the center c onto the element b, which is not in the center. But an endomorphism preserves the form of an element, whence the subgroups generated by all x^3, $x \in G$ or by all $x^{-1}y^{-1}xy$, x, $y \in G$ will be fully invariant.

A particularly useful property of these stronger forms of normality is the fact that, although a normal subgroup H of a normal subgroup K of a group G is not in general a normal subgroup, it follows from the definitions that a characteristic subgroup of a characteristic subgroup is characteristic, and a fully invariant subgroup of a fully invariant subgroup is fully invariant. Also a characteristic subgroup of a normal subgroup is a normal subgroup.

2.5. Direct Products and Cartesian Products.

Given two groups A and B, we may form from these the set of ordered pairs (a, b), $a \in A$, $b \in B$. These ordered pairs will be the elements of a new group, the *direct product* $A \times B$, if we define our product by the rule

$$(a_1, b_1)(a_2, b_2) = (a_1a_2, b_1b_2). \tag{2.5.1}$$

The verification that the product rule (2.5.1) satisfies the group axioms with $(1, 1)$ as the identity element is straightforward, depending only on the validity of these axioms for A and B. Moreover, the correspondence $(a, b) \leftrightarrows (b, a)$ shows that $A \times B$ and $B \times A$ are isomorphic, so that we may speak of the direct product of two groups without specifying their order. The correspondence $a \leftrightarrows (a, 1)$ is an isomorphism between A and the set of elements in $A \times B$, with the second component the identity. Similarly, $b \leftrightarrows (1, b)$ is an isomorphism between B and the subgroup of elements $(1, b)$. Let us identify A and B with these subgroups. With this identification we say that $G = A \times B$ is the direct product of its subgroups A and B. Since $(a, 1)(1, b) = (a, b) = (1, b)(a, 1)$, it follows that in $A \times B$ every element of A *permutes* (or *commutes*) with every element of B; that is, $ab = ba$ for $a \in A$, $b \in B$.

In the direct product, $(a, b)^{-1} = (a^{-1}, b^{-1})$. Hence $(a_1, b_1)^{-1}(a_2, 1)(a_1, b_1) = (a_1^{-1}a_2a_1, 1)$, and so A is a normal subgroup of $A \times B$. Similarly, B is a normal subgroup of $A \times B$. The only element simultaneously of the forms $(a, 1)$ and $(1, b)$ is $(1, 1)$, whence $A \cap B = 1$. Moreover, $A \cup B$ includes all products $(a, 1)(1, b) = (a, b)$, whence $A \cup B = A \times B$. These relations between A and B characterize $A \times B$.

THEOREM 2.5.1. *A group G is isomorphic to the direct product of two subgroups A and B if A and B are normal subgroups such that $A \cap B = 1$, $A \cup B = G$.*

Proof: We have already noted that in the direct product $A \times B$, the subgroups A and B have these properties. Suppose conversely that A and B are normal subgroups of G, with $A \cap B = 1$, $A \cup B = G$. Consider an element $a^{-1}b^{-1}ab = a^{-1}(b^{-1}ab) = (a^{-1}b^{-1}a)b$, where $a \in A$, $b \in B$. Since A and B are normal subgroups, from the first bracketing it is an element of A, and from the second, an element of B. This is an element of $A \cap B = 1$, whence $a^{-1}b^{-1}ab = 1$, and so $ab = ba$. From Theorem 2.3.3, $G = A \cup B = AB$, whence every element g can be put in the form $g = ab$. Moreover, this form is unique since $a_1b_1 = a_2b_2$ implies $a_2^{-1}a_1 = b_2b_1^{-1} \in A \cap B = 1$, whence $a_1 = a_2$, $b_1 = b_2$. If $g = ab$, let us put $g \leftrightarrows (a, b)$. If $g_1 = a_1b_1$, $g_2 = a_2b_2$, then $g_1g_2 = a_1b_1a_2b_2 = (a_1a_2)(b_1b_2)$. Thus the correspondence between G and $A \times B$ is not only one-to-one but also preserves products, and we have established an isomorphism between G and $A \times B$.

We may generalize the preceding ideas to define a product of any number of groups, finite or infinite. Suppose we are given an indexed system of groups A_i where i runs over some index system I (we shall assume that I is well ordered for some of our theorems). We construct *formal products* $\prod_{i \in I} a_i$. A formal product is simply a choice of one element a_i from each of the groups A_i. All formal products form a group called the *Cartesian product* of the A_i, where the product rule is

$$\prod_{i \in I} a_i \cdot \prod_{i \in I} b_i = \prod_{i \in I} c_i, \qquad c_i = a_ib_i, \tag{2.5.2}$$

for every $i \in I$.

The subgroup of the Cartesian product in which $a_i = 1$ for all but a finite number of indices is called the *direct product* of the A_i. Clearly, the direct and Cartesian product coincide when the number of factors is finite. In both cases the elements $\prod_i a_i$, where $a_i = 1$ for $i \neq j$, form a normal subgroup isomorphic to A_j, and identifying A_j with this subgroup in every case, we observe that $A_j \cap \left(\bigcup_{i \neq j} A_i\right) = 1$. Here $\bigcup_{i \in I} A_i$ is the direct product.

THEOREM 2.5.2. *A group G is isomorphic to the direct product of subgroups A_i, $i \in I$, if*

1) *Every A_i is a normal subgroup.*
2) $A_j \cap \left(\bigcup_{i \neq j} A_i\right) = 1$ *for every* $j \in I$.
3) $G = \bigcup_{i \in I} A_i$.

Proof: The proof follows very closely that of Theorem 2.5.1. From 1) and 2) every a_j permutes with every finite product of a_i's with $i \neq j$. Also from 1), 2), and 3) every $g \in G$ is expressible as a finite product of elements from the A_i, and apart from order, has a unique form as a product using at most one factor from each A_i. This gives us an isomorphism between G and the direct product of the A_i. An element of G can be put in the form $g = 1$ or $g = b_1 \cdots b_m$, $b_k \neq 1$, $k = 1, \cdots, m$, where the b's are from different A_i's. Here g corresponds to the element $\prod_{i \in I} a_i$, where $a_i = b_k$ if there is a $b_k \in A_i$ in the product form for g, and $a_i = 1$ otherwise. This correspondence yields the isomorphism between G and the direct product of the A_i.

EXERCISES

1. Show that every dihedral group is homomorphic to the group of order 2.
2. Show that if $p < q$ are primes, then in a group of order pq a subgroup of order q is normal. (See Ex. 8 in Chap. 1.)
3. Show that the subgroups of the quaternion group are all normal.
4. In a cube let x, y, z be the three lines joining mid-points of opposite faces. Show that the symmetries G of the cube permute these lines in a permutation group H of order 6. Show that H is a homomorphic image of G.
5. Consider the 1–1 mappings $x \leftrightarrows ax + b$, a,b, real, $a \neq 0$ of the real numbers onto themselves. Show that these form a group G in which the translations $T: x \leftrightarrows x + t$ form a normal subgroup. What is the factor group G/T?
6. For each element b of a group G define an operator of conjugation by $b: g \to g^b = b^{-1}gb$. Which subgroups are admissible with respect to all such operators? If T is a normal subgroup of G, show that the operator induced in $H = G/T$ is also a conjugation.

3. ELEMENTARY THEORY OF ABELIAN GROUPS

3.1. Definition of Abelian Group. Cyclic Groups.

A group G which satisfies the commutative law

$$G4, \; ba = ab, \quad \text{for all } a, b \in G,$$

is called an *Abelian group* after the mathematician Abel. We also say that elements a and b *permute* if $ba = ab$.

In §1.5 we defined a cyclic group as a group generated by a single element (say, b), with all its elements being powers of b. Since $b^i b^j = b^j b^i = b^{i+j}$ for any integers i, j, we see that every cyclic group is Abelian. We also noted in §1.5 that, within isomorphism, there is a unique cyclic group of infinite order and a unique cyclic group of each finite order n. It is also true that every subgroup of a cyclic group is cyclic. We prove this in a precise form.

THEOREM 3.1.1. *Every subgroup of an infinite cyclic group different from the identity is an infinite cyclic group of finite index, and there is a unique subgroup for each finite index. Every subgroup of a finite cyclic group of order n is a cyclic group of order dividing n, and there is a unique subgroup of each order dividing n.*

Proof: Given a cyclic group G generated by an element b and a subgroup H of G. If H is not the identity and if $b^i \in H$, then $b^{-i} \in H$, and one or the other of these exponents is positive. Suppose m is the least positive exponent of any element occurring in H, and let b^t be any element of H. Then, choosing r appropriately, we have $t = mr + s$ with $0 \leq s < m$. Here $b^t = (b^m)^r b^s$. Since both b^t and b^m belong to H, it follows that b^s also belongs to H. But if s is anything except 0 in the range $0 \leq s < m$, this would conflict with our definition of m as the least positive exponent of b occurring for an element of H. Hence $s = 0$, and $b^t = (b^m)^r$, and all elements of H are powers of b^m, whence H is cyclic. Since for any x which is an integer we have $x = km + i$, where i is one of $0, 1, \cdots, m - 1$, we readily verify that

(3.1.1) $$G = H + Hb + \cdots + Hb^{m-1}.$$

The equation (3.1.1) contains all possible cosets of H and these are different, since $b^i = hb^j$ with $i \neq j$ in the range from 0 to $m - 1$ would give a smaller positive power of b in H, this being either b^{i-j} or b^{j-i}. Hence $[G:H] = m$. Here m is the smallest positive power of b contained in H and also is the index of H in G. Thus, if G is infinite, since for any positive m the elements $(b^m)^r$ form a subgroup, there is a unique subgroup of index m. If G is finite, of order n, then $b^n = 1$, and so $n = mr$, and m is a divisor of n. Here, for any m dividing n, if $n = mr$ we have the elements $1, b^m, b^{2m}, \cdots, b^{(r-1)m}$ forming a subgroup of order r and index m. Since $n = mr$ can be any factorization of n into two factors, we see that there is one, and only one, subgroup of each order r dividing n.

3.2. Some Structure Theorems for Abelian Groups.

An infinite Abelian group may have a very complicated structure. As a relatively simple example, the multiplicative group of all complex numbers except zero contains elements of infinite order and also of every finite order.

If $a^n = 1$, $b^m = 1$ in an Abelian group, then $(a^{-1})^n = 1$ and $(ab)^{mn} = 1$, whence the elements of finite order in any Abelian group A form a subgroup F. Every endomorphism α of A maps an element of finite order onto an element of finite order. Thus, in the sense of §2.4, F is a fully invariant subgroup of A. In §1.8 we introduced the term *periodic group* (the term *torsion group* is used in certain applications) for a group all of whose elements are of finite order. In contrast a group in which no element except the identity is of finite order is called an *aperiodic group* (or *torsion-free group*).

THEOREM 3.2.1. *Given an Abelian group A. Let F be the subgroup of elements of finite order. Then A/F is aperiodic.*

Proof: Suppose to the contrary that $x \neq 1$ in A/F is of finite order m. Then in the homomorphism $A \to A/F$ let $u \to x$. Then $u^m \to x^m = 1$, whence $u^m \in F$ and u^m is of some finite order, say, n. Here $(u^m)^n = 1$ and u itself is of finite order. Thus $u \in F$ and $u \to 1$ although we assumed $x \neq 1$.

This theorem reduces the problem of constructing all Abelian groups to three more explicit problems:

1) The determination of all periodic Abelian groups.

2) The determination of all aperiodic Abelian groups.

3) The construction of an Abelian group A with a given periodic group F as a subgroup, such that the factor group A/F shall be isomorphic to a given aperiodic group H. No one of these is completely settled, but it appears that we know most about the first and least about the last.

We shall say that a set of elements a_i in an Abelian group A is *independent* if a finite product $\prod_i a_i^{e_i} = 1$ only when $a_i^{e_i} = 1$ for every i. If the a_i are independent and also generate A, we say that the a_i form a *basis* for A. Thus elements a_i form a basis for A if, and only if, A is the direct product of the cyclic groups generated by the a_i.

Suppose an Abelian group A is generated by elements $a_1, \cdots, a_r$. Then every element of A is of the form $a_1^{u_1} \cdots a_r^{u_r}$, where the u_i are integers. If

$$a_1^{x_1} \cdots a_r^{x_r} = 1 \tag{3.2.1}$$

is a relation on these generators, we say that

$$a_1^{-x_1} \cdots a_r^{-x_r} = 1 \tag{3.2.2}$$

is its inverse relation. From a set S of relations holding in A we may derive others by taking the product of relations of S and inverses of relations of S. Two sets of relations S_1 and S_2 are said to be *equivalent* if the relations of each set may be derived in this way from the relations of the other set. This is easily seen to be a true equivalence. We say that a set S is a set of *defining relations* for A if every relation holding in A may be derived from those of S. It may be shown that an arbitrary set S of relations on generators $a_1, \cdots, a_r$ is a set of defining relations for that Abelian group A generated by $a_1, \cdots, a_r$ in which the relations derived from S hold, but no others hold. The group A may, of course, reduce to the identity element alone.

THEOREM 3.2.2. *An Abelian group generated by a finite number r of elements has a basis of, at most, r elements.*

Proof: The theorem is trivially true for $r = 1$, since then the group is cyclic. Suppose that A is generated by $a_1, \cdots, a_r$. Our proof will be based on induction on r, and for fixed r on the smallest positive integer m such that $x_i = m$ in a relation

$$a_1^{x_1} \cdots a_r^{x_r} = 1. \tag{3.2.3}$$

If there is only the relation with all $x_i = 0$, then A is the direct product of the infinite cyclic groups $\{a_i\}$ and our theorem is true. Otherwise, some relation or its inverse will contain some positive exponents. Let us renumber the a's, if necessary, so that the smallest positive exponent in a relation is $x_1 = m$. If $m = 1$, then we have

$$a_1 = a_2^{-x_2} \cdots a_r^{-x_r}, \tag{3.2.4}$$

and A is generated by the $r - 1$ elements $a_2, \cdots, a_r$, and by induction our theorem is true. Now suppose $x_1 = m > 1$ in the relation

$$a_1^m a_2^{x_2} \cdots a_r^{x_r} = 1. \tag{3.2.5}$$

Let $y_1, \cdots, y_r$ be the exponents in a further relation. Then, for any integer k, from this relation and (3.2.5) we may derive a relation with exponents $y_1 - km$, $y_2 - kx_2, \cdots, y_r - kx_r$. We may choose k so that $0 \leq y_1 - km < m$. But since m was the smallest positive exponent in any relation, we must have $y_1 - km = 0$, and so the relation with exponents $y_1, \cdots, y_r$ can be derived from (3.2.5) and the relation with exponents $0, y_2 - kx_2, \cdots, y_r - kx_r$. Thus the set of all relations for A is equivalent to the set S consisting of (3.2.5) and relations involving only $a_2, \cdots, a_r$.

In (3.2.5) let $x_2 = k_2m + s_2, \cdots, x_r = k_rm + s_r$, where we choose k_i, $i = 2, \cdots, r$ so that $0 \leq s_i < m$. If we take a new element

$$a_1^* = a_1 a_2^{k_2} \cdots a_r^{k_r}, \tag{3.2.6}$$

then $a_1^*, a_2, \cdots, a_r$ also generate A, and in terms of these generators, (3.2.5) becomes

$$a_1^{*m} a_2^{s_2} \cdots\cdot a_r^{s_r} = 1. \tag{3.2.7}$$

Here if any s is different from zero, it is a positive number less than m and we may apply our induction. But if $s_2 = \cdots = s_r = 0$, then (3.2.7) becomes

$$a_1^{*m} = 1, \tag{3.2.8}$$

and since (3.2.5) and relations involving only $a_2, \cdots, a_r$ were a defining set of relations for A in terms of generators $a_1, a_2, \cdots, a_r$, it follows that (3.2.8) and relations involving only $a_2, \cdots, a_r$ are a defining set of relations in terms of generators $a_1^*, a_2, \cdots, a_r$. Hence A is the direct product of the cyclic group of order m generated by a_1^* and the group generated by the $r - 1$ elements $a_2, \cdots, a_r$, which by our induction is the direct product of, at most, $r - 1$ cyclic groups. Thus we have proved our theorem in all cases.

To study periodic Abelian groups we need a lemma which holds in any group.

LEMMA 3.2.1. *Let x be an element of order mn in any group where m and n are relatively prime integers. Then x has a unique representation $x = yz = zy$, where y is of order m and z of order n. Both y and z are powers of x.*

Proof: We write (a,b) for the greatest common divisor of two integers. The statement that m and n are relatively prime is that $(m,n) = 1$. From the Euclidean algorithm, integers u and v exist such that $um + vn = 1$, and hence $x = x^{vn}x^{um} = x^{um}x^{vn}$. Put $y = x^{vn}$, $z = x^{um}$. Then $x = yz = zy$ and $y^m = x^{vnm} = 1$, and $z^n = x^{umn} = 1$. Thus the exact order of y is some divisor m_1 of m, and of z some divisor n_1 of n. But from $x = yz = zy$ it will follow that the order of x is a divisor of m_1n_1. Since this order was mn, it follows that $m_1 = m$ is the order of y and $n_1 = n$ is the order of z. If x had a second representation $x = y_1z_1 = z_1y_1$ with y_1 of order m and z_1 of order n, let us note first that y_1 and z_1 permute with x, since $xy_1 = y_1z_1y_1 = y_1x$ and $xz_1 = z_1y_1z_1 = z_1x$. But then y_1 and z_1 permute with y and z, which are powers of x. Now $yz = x = y_1z_1$ leads to $w = y_1^{-1}y = z_1z^{-1}$. But y and y_1 are permuting elements of order m, and z and z_1 are permuting elements of order n. Hence the element w satisfies $w^m = 1$ and also $w^n = 1$, and since $(m, n) = 1$, this yields $w = 1$; so, $y_1 = y$, $z_1 = z$, proves the uniqueness of the representation.

By repeated application of this lemma we find:

LEMMA 3.2.2. *Let x be an element of order $n = n_1n_2 \cdots n_r$ where $(n_i,n_j) = 1$ for $i \neq j$. Then x has a unique representation $x = x_1x_2 \cdots x_r$ where $x_jx_i = x_ix_j$ and x_i is of order n_i. Every x_i is a power of x.*

In particular, if $n = p_1^{e_1} \cdots p_r^{e_r}$, where $p_1, \cdots, p_r$ are distinct primes, we may apply this lemma with $n_i = p_i^{e_i}$.

In a periodic Abelian group A consider the set of elements P whose orders are powers of a fixed prime p, where we include the identity as being of order $p^0 = 1$. If $x^{p^a} = 1$, $y^{p^b} = 1$, then $(xy)^{p^c} = 1$ with $c = \max(a,b)$ and $(x^{-1})^{p^a} = 1$. Hence P is a subgroup which we call the *Sylow p-subgroup*, $S(p)$. We call P an Abelian p-group.

THEOREM 3.2.3. *A periodic Abelian group is the direct product of its Sylow subgroups, $S(p)$.*

Proof: Clearly, $\prod_p S(p)$, the direct product of the Sylow subgroups of A, is a subgroup of A. But, from Lemma 3.2.2, if $x \in A$ is of order $n = p_1^{e_1} \cdots p_r^{e_r}$, then, $x = x_1 x_2 \cdots x_r$ with $x_i \in S(p_i)$; so, every element of x of A belongs to the direct product of the Sylow subgroups, whence this direct product must be the entire group A.

3.3. Finite Abelian Groups. Invariants.

A finite Abelian group is, of course, periodic and finitely generated. Applying the results of the preceding section, we may say the following:

THEOREM 3.3.1. *A finite Abelian group of order $n = p_1^{e_1} \cdots p_r^{e_r}$ is the direct product of Sylow subgroups $S(p_1), \cdots, S(p_r)$. Here $S(p_i)$ is of order $p_i^{e_i}$ and is the direct product of cyclic groups of orders $p_i^{e_{i1}}, \cdots, p_i^{e_{is}}$ where $e_{i1} + \cdots + e_{is} = e_i$.*

Proof: In the Abelian group of order n, we know that the orders of the elements are divisors of n, whence a Sylow subgroup belonging to a prime not dividing n can consist only of the identity. Thus, if $p_1, \cdots, p_r$ are the distinct primes dividing n, the group is the direct product $S(p_1) \times \cdots \times S(p_r)$. But this much does not tell us the orders of the $S(p_i)$, some of which might trivially be the identity. Since $S(p_i)$ is the identity or the direct product of cyclic groups of orders $p_i^{e_{i1}}, \cdots, p_i^{e_{is}}$, the order of $S(p_i)$ will be the product of these orders (say, $p_i^{t_i}$ with $t_i = e_{i1} + \cdots + e_{is}$), and the order of the entire group will be the product of the orders of the $S(p_i)$. But because of the unique factorization of the integer n, it must follow that $p_i^{t_i} = p_i^{e_i}$ in each case. As a consequence of this and Theorem 3.1.1 we have the following corollary.

COROLLARY 3.3.1. *An Abelian group of order n contains an element of order p if p is a prime dividing n.*

A finite Abelian p-group, $A(p)$, can usually be written as a direct product of cyclic groups in several ways. For example, if $a^8 = 1$,

$b^4 = 1$, the group $A(2) = \{a\} \times \{b\}$ is of order 32. If we put $c = ab$ and $d = a^4b$, then $c^8 = 1$, $d^4 = 1$, $a = c^5d^{-1}$, $b = c^4d$. We readily verify that $A(2) = \{c\} \times \{d\}$. In this case $A(2)$ is a direct product of cyclic groups in two different ways, but the number of factors and their orders are the same. This is true in general for finite Abelian p-groups, but since the cyclic group of order 6 is the direct product of the cyclic groups of orders 2 and 3, it is not true for finite Abelian groups which are not p-groups. If A is an Abelian p-group which is the direct product of cyclic groups of orders $p^{e_1}, \cdots, p^{e_r}$, then these numbers are called the *invariants* of the group. In the special, but important, case in which all the invariants are $p, \cdots, p$, we say that A is an *elementary Abelian group.* Clearly, the invariants of an Abelian group A determine A to within isomorphism, but they are invariants in a stronger sense, as given precisely by the following theorem:

THEOREM 3.3.2. *If a finite Abelian p-group A is the direct product of cyclic groups in two ways, $A = A_1 \times \cdots \times A_r = B_1 \times \cdots \times B_s$, then the number of factors is the same in both cases, $r = s$, and the orders of $A_1, \cdots, A_r$ are the same as those of $B_1, \cdots, B_s$ in some arrangement.*

Proof: We use induction on the order of A, the theorem being trivial when A is of order p.

If A is any Abelian p-group, let us write A_p for the subgroup of elements x of A satisfying $x^p = 1$ and A^p for the subgroup of elements of the form y^p, $y \in A$. Let A have a basis $a_1, \cdots, a_r$, where a_i is of order p^{e_i}, $i = 1, \cdots, r$, and let us number the a's so that $e_1 \geqq e_2 \geqq \cdots \geqq e_r$. Then we may easily verify that A_p has a basis $a_1^{p^{e_1-1}}, \cdots, a_r^{p^{e_r-1}}$ and is of order p^r. If A is elementary Abelian, then $A^p = 1$. Otherwise, let e_m be the last exponent greater than 1, i.e., $e_1 \geqq \cdots \geqq e_m > e_{m+1} = \cdots = e_r = 1$. Then A^p has a basis $a_1^p, \cdots, a_m^p$, as may easily be shown.

Let A have a second basis, $b_1, \cdots, b_s$, where b_i is of order p^{f_i}, $i = 1, \cdots, s$ and $f_1 \geqq f_2 \geqq \cdots \geqq f_s$. Then A_p is of order p^r from the basis $a_1, \cdots, a_r$ and of order p^s from the basis $b_1, \cdots, b_s$, whence $r = s$. If A is elementary Abelian, this completes the proof. If not, let $f_1 \geqq f_2 \geqq \cdots \geqq f_n > f_{n+1} = \cdots = f_s = 1$. Then A^p has invariants $p^{e_1-1}, \cdots, p^{e_m-1}$ and also invariants $p^{f_1-1} \cdots, p^{f_n-1}$. By induction $m = n$ and $e_1 - 1 = f_1 - 1, \cdots, e_m - 1 = f_m - 1$. From this and the fact that $s = r$, it follows that $e_1 = f_1, \cdots, e_r = f_r$, proving our theorem.

COROLLARY 3.3.2. *If two finite Abelian p-groups do not have the same invariants, they are not isomorphic.*

THEOREM 3.3.3. *An Abelian group A with invariants* $p^{e_1}, \cdots, p^{e_r}$ $e_1 \geqq \cdots \geqq e_r$ *has a subgroup K with invariants* $p^{k_1}, \cdots, p^{k_t}$ $k_1 \geqq \cdots \geqq k_t$ *if, and only if,* $t \leq r$ *and* $k_1 \leq e_1, \cdots, k_t \leq e_t$.

Proof: We prove first that the exponents of the invariants of a subgroup K of A satisfy the inequalities of the theorem, proceeding by induction on the order of A, the theorem being trivial if A is of order p.

Since K_p is a subgroup of A_p, it follows that $t \leq r$, proving the theorem if A is elementary Abelian. Otherwise, let $e_1 \geqq \cdots \geq e_m > e_{m+1} = \cdots = e_r = 1$ and $k_1 \geqq \cdots \geqq k_u > k_{u+1} = \cdots = k_t = 1$. Then K^p is a subgroup of A^p and the invariants of K^p are $p^{k_1-1}, \cdots, p^{k_u-1}$, and those of A^p are $p^{e_1-1}, \cdots, p^{e_m-1}$. By induction $u \leq m$ and $k_i - 1 \leq e_i - 1$, $i = 1, \cdots, u$. Hence $k_i \leq e_i$, $i = 1, \cdots, u$ and as $k_{u+1} = \cdots = k_t = 1$, also $k_i \leq e_i$, $i = u + 1, \cdots, t$, whence $k_i \leq e_i$, $i = 1, \cdots, t$.

If the inequalities of the theorem hold, then there is one subgroup of A with the given invariants which we can take as a basis for appropriate powers of the first t basis elements of A. But it is not in general true that, given A and a subgroup K, we can choose a basis for A and a basis for K so that the basis for K consists of powers of elements in the basis for A. (See Ex. 5.)

EXERCISES

1. An Abelian group A is generated by elements a, b, c with defining relations $a^3b^9c^9 = 1$ and $a^9b^{-3}c^9 = 1$. Find a basis for A and the orders of the basis elements.
2. Show that a finite Abelian p-group is generated by its elements of highest order.
3. An Abelian group has invariants p^3, p^2. How many subgroups of order p^2 does it contain?
4. Give two examples of Abelian p-groups which contain exactly $p^2 + p + 1$ subgroups of order p.
5. Let A be the Abelian group generated by a and b with defining relations $a^{p^3} = 1$, $b^p = 1$. Let K be the subgroup generated by the element $x = a^pb$. Show that it is not possible to choose a basis for A and a basis for K so that the basis element for K is a power of a basis element for A.

4. SYLOW THEOREMS

4.1. Falsity of the Converse of the Theorem of Lagrange.

According to the Theorem of Lagrange, the order of a subgroup of a finite group is a divisor of the order of the group. But, conversely, a group of order n need not have a subgroup of order m if m is a divisor of n. In particular the following permutation group of order 12 will be found to have no subgroup of order 6:

$$\begin{pmatrix}1, 2, 3, 4\\1, 2, 3, 4\end{pmatrix} \qquad \begin{pmatrix}1, 2, 3, 4\\1, 3, 4, 2\end{pmatrix}$$

$$\begin{pmatrix}1, 2, 3, 4\\2, 1, 4, 3\end{pmatrix} \qquad \begin{pmatrix}1, 2, 3, 4\\1, 4, 2, 3\end{pmatrix}$$

$$\begin{pmatrix}1, 2, 3, 4\\3, 4, 1, 2\end{pmatrix} \qquad \begin{pmatrix}1, 2, 3, 4\\3, 2, 4, 1\end{pmatrix}$$

$$\begin{pmatrix}1, 2, 3, 4\\4, 3, 2, 1\end{pmatrix} \qquad \begin{pmatrix}1, 2, 3, 4\\4, 2, 1, 3\end{pmatrix}$$

$$\begin{pmatrix}1, 2, 3, 4\\2, 3, 1, 4\end{pmatrix} \qquad \begin{pmatrix}1, 2, 3, 4\\2, 4, 3, 1\end{pmatrix}$$

$$\begin{pmatrix}1, 2, 3, 4\\3, 1, 2, 4\end{pmatrix} \qquad \begin{pmatrix}1, 2, 3, 4\\4, 1, 3, 2\end{pmatrix}$$

It does, however, have subgroups of orders 2, 3, and 4.

Thus, in general, if m divides n, we cannot be sure that a group of order n contains a subgroup of order m. But it is true that if m is a prime or prime power, then such subgroups exist. The existence and number of such subgroups is the subject of the *Sylow theorems* which follow. We begin with a theorem which will serve as a starting point for the *Sylow theorems.*

THEOREM 4.1.1. *If the order of a group G is divisible by a prime p, then G contains an element of order p.*

Proof: Let $n = mp$ be the order of G. Here, if $m = 1$, G is the cyclic group of order p and the theorem is true. We proceed by induction on m. If G contains a proper subgroup H whose index $[G:H]$ is not divisible by p, then the order of H is divisible by p, and so by induction H contains an element of order p. Now suppose that every proper subgroup of G has an index divisible by p. Then, from §1.6, $n = n_1 + n_2 + \cdots + n_s$, where each n_i is the number of conjugates in a class of elements of G. Each $n_i \neq 1$ is the index of a proper subgroup in G, and hence by hypothesis, divisible by p. Here $n_1 = 1$, the identity being a class. Hence the number of $n_i = 1$ is a multiple of p. An element a_i is a class in G if, and only if, it belongs to the center Z of G. Thus the center Z is of order divisible by p. Then for $z \in Z$ and any $g \in G$, we have $zg = gz$. Hence, *a fortiori*, the elements of Z permute with each other and Z is an Abelian group. But now from the corollary to Theorem 3.3.1, Z contains an element of order p.

4.2. The Three Sylow Theorems.

From Theorem 4.1.1 we are guaranteed the existence of at least one subgroup of order p whenever p divides the order of G. We shall show that if G is of order $n = p^m s$, then there will also be subgroups of orders $p^2, p^3, \cdots, p^m$.

THEOREM 4.2.1 (FIRST SYLOW THEOREM). *If G is of order $n = p^m s$ where $p \nmid s$, p a prime, then G contains subgroups of orders p^i, $i = 1, \cdots, m$, and each subgroup of order p^i, $i = 1, \cdots, m - 1$, is a normal subgroup of at least one subgroup of order p^{i+1}.*

Proof: The proof is by induction on i. As previously stated, G contains a subgroup of order p. Let P be a subgroup of order p^i, $i \geq 1$. Write G in terms of double cosets of P, $G = P + Px_2P + \cdots + Px_rP$, and let there be a_j right cosets of P in Px_jP. Then $[G:P] = a_1 + a_2 + \cdots + a_r$, where $a_j = [x_j^{-1}Px_j : x_j^{-1}Px_j \cap P]$, and $a_1 = 1$ for the double coset $P \cdot 1 \cdot P = P$. Now $a_j = 1$ or a power of p. Since $p \mid [G:P]$, the number of a_j's equal to 1 must be a multiple of p. If $a_j = 1$, then $x_j^{-1}Px_j = P$ and x_j, and the coset $Px_j = x_jP$ must belong to the normalizer K of P. Conversely, if $x_j \in K$, then $x_j^{-1}Px_j = P$ and $a_j = 1$. Thus $[K:P]$ is the number of a_j's $= 1$ and so $p \mid [K:P]$. Hence the factor group K/P has order $[K:P]$ divisible by p. Thus K/P contains a subgroup J^* of order p. By Theorem 2.3.4 $J^* = J/P$, where $J \subseteq K$, and $[J:P] = [J^*:1] = p$, and so J is a subgroup of order p^{i+1}, containing P as a normal subgroup.

DEFINITION: *A group P is a p-group if every element of P except the identity has order a power of a prime p.*

DEFINITION: *A subgroup S of a group G is a Sylow subgroup of G if it is a p-group and is not contained in any larger p-group which is a subgroup of G.*

In terms of these definitions we may express some of the consequences of the first Sylow theorem.

COROLLARY 4.2.1. *Every finite group G of order $n = p^m s$, $p \nmid s$, p a prime, contains a Sylow subgroup of order p^m, and every p-group which is a subgroup of G is contained in a Sylow subgroup of G.*

Every group of order p^m is a p-group. From Theorem 4.1.1, if the order of a group is divisible by two different primes, it cannot be a p-group. Hence every finite p-group is of order a power of p, say, p^m.

COROLLARY 4.2.2. *Every subgroup of a p-group P of order p^m is contained in a maximal subgroup of order p^{m-1}, and all the maximal subgroups of P are normal subgroups.*

THEOREM 4.2.2 (SECOND SYLOW THEOREM). *In a finite group G, the Sylow p-subgroups are conjugate.*

Proof: Let P_1 and P_2 be two Sylow p-subgroups. Then $G = P_1P_2 + P_1x_2P_2 + \cdots + P_1x_sP_2$. Let there be b_i right cosets of P_2 in $P_1x_iP_2$. Here $b_i = [x_i^{-1}P_1x_i : x_i^{-1}P_1x_i \cap P_2]$ and is 1 or a power of p. But $b_1 + \cdots + b_s = [G:P_2]$ is not a multiple of p. Hence, for some i, $b_i = 1$ and $x_i^{-1}P_1x_i = P_2$.

THEOREM 4.2.3 (THIRD SYLOW THEOREM). *The number of Sylow p-subgroups of a finite group G is of the form $1 + kp$ and is a divisor of the order of G.*

Proof: This is trivial if there is only one Sylow p-subgroup. Let S_0 be one Sylow p-subgroup and $S_1, \cdots, S_r$ the remaining ones. These fall into a number of disjoint conjugate sets with respect to transformation by elements of S_0. By the second Sylow theorem, S_i is the only Sylow p-subgroup in its normalizer K_i. Hence the normalizer of S_i in S_0 ($i \neq 0$) is a proper subgroup of S_0, and so the number of conjugates of S_i under S_0 is a power of p, p^e, $e \geq 1$. Hence $r = p^{e_1} + \cdots + p^{e_s} = kp$, and there are $1 + r = 1 + kp$ Sylow p-subgroups of G. The number of Sylow p-subgroups is, by the second Sylow theorem, the index of the normalizer of S_0, and so a divisor of the order of G.

THEOREM 4.2.4. *Let K be the normalizer of the Sylow p-subgroup P in the finite group G. Then if H is any subgroup $G \supseteq H \supseteq K \supseteq P$, it follows that H is its own normalizer in G.*

Proof: Suppose $x^{-1}Hx = H$. Then $H \supseteq x^{-1}Px = P'$, which must be a Sylow p-subgroup of H. Hence, for some $u \in H$, $u^{-1}P'u = P$, whence $u^{-1}x^{-1}Pxu = P$ and $xu \in K$. Hence $x \in H$, and H is its own normalizer.

The following theorem, apart from its own interest, has some important applications which will be made in later chapters.

THEOREM 4.2.5 (BURNSIDE). *If in the finite group G a p-group h is normal in one Sylow p-subgroup but not in another which contains it, then there exist $r > 1$, $r \not\equiv 0 \pmod p$ conjugate groups $h = h_1, \cdots, h_r$ which are all normal in $H = h_1 \cup h_2 \cup \cdots \cup h_r$ but not all normal in any Sylow p-subgroup of G. Then $h_1, \cdots, h_r$ are a complete set of conjugates of each other in N_H, the normalizer of H.*

Proof: Let N_h be the normalizer of h. Let Q be a Sylow p-subgroup of G such that h is a non-normal subgroup of Q and so that $D = N_h \cap Q$ is maximal. Let q be the normalizer of D in Q, and N_D the normalizer of D in G. We assert $Q \supseteq q \supset D \supset h$, for h is normal and of index p is some subgroup of Q, but h is non-normal in Q. Hence $Q \supset D \supset h$. Also, D, a proper subgroup of Q, is properly contained in its normalizer q in Q. Hence $Q \supseteq q \supset D \supset h$. Now, since $D = N_h \cap Q$, h is not normal in q and *a fortiori* not normal in N_D. Let $h = h_1, \cdots, h_s$, $s > 1$ be the conjugates of h in N_D. Since h is normal in D, and N_D induces automorphisms in D, every h_i is

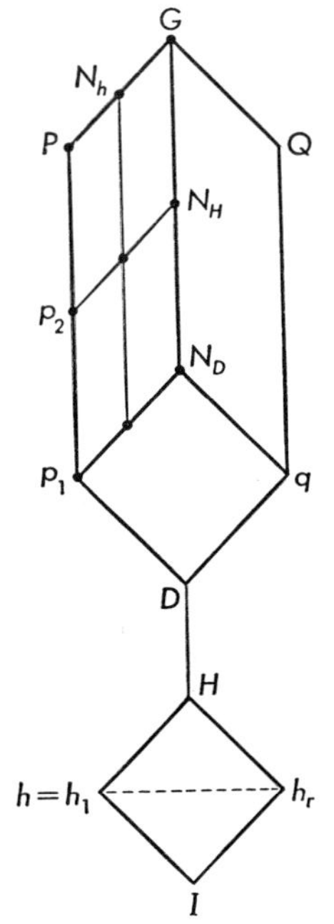

Fig. 2. A theorem of Burnside.

also normal in D, and so *a fortiori* in $H = h_1 \cup h_2 \cup \cdots \cup h_s \subseteq D$. The normalizer N_H of H contains N_D, since the elements of N_D transform H into itself.

Let p_1 be a Sylow subgroup of $N_h \cap N_D$ and $P_1 \supseteq p_1$ be a Sylow subgroup of N_h. By hypothesis P_1 is a Sylow subgroup of G. Then $D \subset p_1$, since D is not its own normalizer in P_1. Now $N_h \cap N_D \subseteq N_D \subseteq N_H$, and let $p_2 \supseteq p_1$ be a Sylow subgroup of N_H, and finally let $P \supseteq p_2$ be a Sylow subgroup of G. If $P \not\subseteq N_h$, then $P \cap N_h \supseteq p_1 \supset D$, contrary to the maximal property of D. Hence $P \subseteq N_h$ and so $N_h \cap N_H \supseteq P \cap N_H = p_2$, since p_2 was a Sylow subgroup of N_H.

Let $h = h_1, \cdots, h_s, \cdots, h_r$ be the conjugates of h in N_H (and hence all normal subgroups of H). The normalizer of h in N_H is $N_H \cap N_h$ and so the number of conjugates of h in N_H is $r = [N_H : N_H \cap N_h]$. But $N_H \cap N_h \supseteq p_2$, a Sylow subgroup of N_H. Hence $r \not\equiv 0 \pmod{p}$.

If all $h_1, \cdots, h_r$ were normal subgroups of some Sylow subgroup S_p, then $S_p \subseteq N_H$, and every Sylow subgroup of N_H contains and normalizes all the h's. But $q \subseteq N_D \subseteq N_H$ is a p-group of N_H, which does not normalize h_1.

4.3. Finite p-Groups.

From the Sylow theorems a group G of order $n = p_1^{e_1} \cdots p_r^{e_r}$ contains for each i a subgroup of order $p_i^{e_i}$, and all subgroups of this order are isomorphic, as they are conjugate. Thus the problem of constructing finite groups may be regarded as having two parts: 1) constructing groups of prime power order, and 2) combining groups of prime power orders dividing a number n to form a group of order n. When all the Sylow subgroups are cyclic (and this will certainly be the case when all $e_i = 1$), we can solve the second problem; the solution is given in Chap. 9 (Theorem 9.4.3). Thus, although neither of these problems is in any sense solved in general, we must solve the first problem to have the subgroups to use in the second problem. It seems to be true that the difficulties of combining Sylow subgroups to form a group rest very heavily on the complexities of the prime power groups, the p-groups, as we shall call them.

A first fact about p-groups of great value is the following:

THEOREM 4.3.1. *The center of a finite p-group is greater than the identity alone.*

Proof: If P is a finite p-group, let us write P as a sum of classes:

$$P = C_1 + C_2 + \cdots + C_r. \tag{4.3.1}$$

Here C_1 consists of the identity alone. Let h_i be the number of elements in C_i, which by Theorem 1.6.1 is the index of a subgroup of P and so is either 1 for an element of the center or is otherwise a power of p. But if P is of order p^m we must have

$$p^m = h_1 + h_2 + \cdots + h_r. \tag{4.3.2}$$

Here $h_1 = 1$, and consequently in (4.3.2) the remaining h's cannot all be proper powers of p and so there must be further h's equal to 1, so that the center of P is greater than the identity alone.

We restate Corollary 4.2.2 as a theorem.

THEOREM 4.3.2. *Every proper subgroup of a p-group P of order p^m is contained in a maximal subgroup of order p^{m-1}, and all the maximal subgroups of P are normal subgroups.*

A further consequence of the (first) Sylow theorem 4.2.1 is that no proper subgroup of a p-group is its own normalizer. This fact even has a converse, which we now prove.

THEOREM 4.3.3. *In a finite group G the property that no proper subgroup is its own normalizer holds if, and only if, G is the direct product of its Sylow subgroups.*

Proof: Suppose that no proper subgroup of G is its own normalizer. By Theorem 4.2.4, K, the normalizer of a Sylow subgroup P, is its own normalizer, whence by the assumption, K must be the entire group G. Thus P is a normal subgroup of G. From this and Theorem 2.5.2, the union of the Sylow subgroups is the direct product of the Sylow subgroups, and so G is the direct product of its Sylow subgroups. Now suppose $G = P_1 \times \cdots \times P_r$, where P_i is a group of order $p_i^{e_i}$ and $p_i \neq p_j$ for $i \neq j$. Now if $g = g_1 g_2 \cdots g_r$, with $g_i \in P_i$, the conditions of Lemma 3.2.2 hold, and each g_i is a power of g. Thus, when an element g occurs in a subgroup H of G, each of its components g_i is also an element of H. Thus H must itself be a direct product $H = H_1 \times \cdots \times H_r$ where $H_i = H \cap P_i$ is a subgroup of P_i. If H is a proper subgroup of G, then some H_j is a proper subgroup of P_j, and by replacing this H_j by a larger subgroup of P_j in which it is normal, we get a subgroup larger than H in which H is normal.

THEOREM 4.3.4. *If A is a normal subgroup of order p contained in the p-group P, then A is in the center of P.*

Proof: A, being of order p, is cyclic and is generated by an element a, the elements of A being $1, a, \cdots, a^{p-1}$. Since A is normal, the conjugates of the element a are contained in the set $a, a^2, \cdots, a^{p-1}$. But the number of conjugates of a is the index of its centralizer and so is 1 or a power of p. But as the number of conjugates is at most $p - 1$, the only possibility is 1, whence a and so A is in the center of P.

4.4. Groups of Orders p, p^2, pq, p^3.

A group of prime order p cannot have a proper subgroup and so must be a cyclic group, generated by any element different from the identity. We have already shown in Theorem 1.5.4 that a group G without any proper subgroups is cyclic of prime order.

A group G of order p^2, if it is not cyclic, will contain two distinct subgroups of order p, say $\{a\}$ and $\{b\}$, where $a^p = 1$, $b^p = 1$, and $\{a\} \cap \{b\} = 1$. Since these are both maximal subgroups, by Corollary 4.2.2, they will both be normal, whence, by Theorem 3.2.1, $G = \{a\} \times \{b\}$; and so, G is an Abelian group with a, b as a basis.

Suppose G is of order pq, where $p < q$ are primes. By the third Sylow theorem, the number of subgroups of order q is of the form $1 + kq$ and divides p, whence it must be 1, and the unique subgroup of order q will be normal, say $\{b\}$, with $b^q = 1$. The number of subgroups of order p is of the form $1 + kp$ and divides q, whence it is 1 or q. If the number is 1, we have for some a a normal subgroup $\{a\}$ with $a^p = 1$, and G as the direct product of $\{a\}$ and $\{b\}$. But here $c = ab$ is of order pq and G is cyclic. There remains the case with $1 + kp = q$ subgroups of order p, where a subgroup $\{a\}$ of order p is not normal. Then we have

$$a^p = 1, \quad b^q = 1,$$

and since $\{b\}$ is normal, $a^{-1}ba = b^r$ for some r. Here if $r = 1$, G is Abelian and is the cyclic group above. Hence $r \neq 1$. Then $a^{-1}b^ia = b^{ir}$ for any i, and in particular $a^{-1}b^ra = b^{r^2}$, whence $a^{-2}ba^2 = a^{-1}b^ra = b^{r^2}$. More generally we find $a^{-j}ba^j = b^{r^j}$, proceeding by induction. Thus for $j = p$ we have $b = a^{-p}ba^p = b^{r^p}$, whence $r^p \equiv 1 \pmod{q}$. That this necessary condition on r is also sufficient may be verified by establishing the general rule

$$(a^u b^v)(a^x b^y) = a^{u+x} b^{vr^x + y},$$

for multiplying any two elements and proving that this rule defines a group of order pq. This is a special case of a more general rule which will be established in Theorem 6.5.1.

For groups of order p^3, there are three Abelian types, with invariants respectively (p^3), (p^2, p), and (p, p, p). In finding non-Abelian groups, we handle the cases $p = 2$ and p-odd separately. First let $p = 2$ and consider non-Abelian groups of order 8. There can be no element of order 8, since then the group would be cyclic. If all elements are of order 2, then $(ab)^2 = 1$, or $abab = 1$, $ba = a^2bab^2 = ab$, and the group is Abelian. Hence there must be an element of order 4, say, $a^4 = 1$. If $b \notin \{a\} = A$, then $G = A + Ab$ and $b^2 \in A$. If $b^2 = a$ or a^3, then b is of order 8 and G is cyclic. Hence $b^2 = 1$ or a^2. Also $b^{-1}ab \in A$, since A is normal, and $b^{-1}ab = a$ or a^3, since it is an element of order 4. But with $b^{-1}ab = a$, G will be Abelian. Hence $b^{-1}ab = a^3$. Thus we have found two non-Abelian groups, the dihedral group with defining relations

$$a^4 = 1, \quad b^2 = 1, \quad b^{-1}ab = a^3,$$

and the quaternion group with defining relations

$$a^4 = 1, \quad b^2 = a^2, \quad b^{-1}ab = a^3.$$

It is easily verified that these relations do define two groups of order 8 and that they are not isomorphic to each other.

Finally, consider non-Abelian groups of order p^3, p an odd prime. Since G is not cyclic, it contains no element of order p^3. Let us first suppose that G contains an element of order p^2, $a^{p^2} = 1$. Then $\{a\} = A$, as a maximal subgroup is normal. Let $b \notin A$. Then $G = A + Ab + \cdots + Ab^{p-1}$, and $b^p \in A$, $b^{-1}ab = a^r$. Here $r \neq 1$, since G is non-Abelian. Since we find by induction on j that $b^{-j}ab^j = a^{r^j}$, and since b^p as an element of A permutes with a, we have $a = b^{-p}ab^p = a^{r^p}$, whence $r^p \equiv 1 \pmod{p^2}$. From the Fermat theorem, $r^p \equiv r \pmod p$, and so $r \equiv 1 \pmod p$. Write $r = 1 + sp$. Then, with j chosen so that $js \equiv 1 \pmod p$, we have

$$b^{-j}ab^j = a^{(1+sp)^j} = a^{1+sjp} = a^{1+p}.$$

Since $(j, p) = 1$, $b^j \notin A$, we may replace b by b^j to get

$$G = A + Ab + \cdots + Ab^{p-1},$$

where $b^{-1}ab = a^{1+p}$.

Now $b^p \in A$, whence $b^p = a^t$. Here t must be a multiple of p since b is not of order p^3. Write $b^p = a^{up}$. Then, using the rule $a^ib = ba^{i(1+p)}$, we calculate and find

$$\begin{aligned}(ba^{-u})^p &= b^pa^{-u[1+(1+p)+(1+p)^2+\cdots+(1+p)^{p-1}]}\\ &= b^pa^{-up-up(1+2+\cdots+p-1)}\\ &= b^pa^{-up} = 1.\end{aligned}$$

Here we use the fact that $1 + 2 + \cdots + p - 1 = p(p - 1)/2$ is a multiple of p since p is odd. Now with $b_1 = ba^{-u}$, we have the relations $a^{p^2} = 1$, $b_1{}^p = 1$, $b_1{}^{-1}ab_1 = a^{1+p}$. This last follows since $b_1{}^{-1}ab_1 = a^u(b^{-1}ab)a^{-u}$.

As a last case suppose that G contains no element of order p^2. The center Z must be of order p, since if it were of order as much as p^2, G would be Abelian. G/Z will be of the type $x^p = 1$, $y^p = 1$, $yx = xy$. If in the homomorphism $G \to G/Z$, $a \to x$, $b \to y$, then $a^p = 1$, $b^p = 1$, $a^{-1}b^{-1}ab = c \in Z$. If $a^{-1}b^{-1}ab = 1$, since a, b, and Z generate G, G would be Abelian. Hence $c \neq 1$ is a generator for Z and our relations become

$$a^p = 1, \quad b^p = 1, \quad c^p = 1, \quad ab = bac, \quad ac = ca, \quad bc = cb.$$

Table of Defining Relations.

I. G order p.
   1) Cyclic. $a^p = 1$.

II. G order p^2.
   1) Cyclic. $a^{p^2} = 1$.
   2) Elementary Abelian. $a^p = 1$, $b^p = 1$, $ba = ab$.

III. G order pq, $p < q$.
   1) Cyclic. $a^{pq} = 1$.
   2) Non-Abelian. $a^p = 1$, $b^q = 1$, $a^{-1}ba = b^r$, $r^p \equiv 1 \pmod q$, $r \not\equiv 1 \pmod q$, p divides $q - 1$.

The solutions of $z^p \equiv 1 \pmod q$, $z \not\equiv 1 \pmod q$ are $r, r^2, \cdots, r^{p-1}$, and all yield the same group, since replacing a by a^j as a generator of $\{a\}$ replaces r by r^j.

IV. G order p^3.
   Abelian.
   1) $a^{p^3} = 1$

2) $a^{p^2} = 1,\ b^p = 1,\ ba = ab.$
3) $a^p = b^p = c^p = 1,\ ba = ab,\ ca = ac,\ cb = bc.$
Non-Abelian order $2^3 = 8$.
4) Dihedral. $a^4 = 1,\ b^2 = 1,\ ba = a^{-1}b.$
5) Quaternion. $a^4 = 1,\ b^2 = a^2,\ ba = a^{-1}b.$
Non-Abelian order p^3, p odd.
4) $a^{p^2} = 1,\ b^p = 1,\ b^{-1}ab = a^{1+p}.$
5) $a^p = 1,\ b^p = 1,\ c^p = 1,\ ab = bac,\ ca = ac,\ cb = bc.$

EXERCISES

1. Show that if H is a normal subgroup of the finite group G, and if $[G:H]$ is prime to p, then H contains every Sylow p-subgroup of G.
2. Show that in a group G, a normal subgroup K of order p^a is contained in every Sylow p-subgroup of G.
3. Show that a group of order p^2q, where p and q are distinct primes, must contain a normal Sylow subgroup.
4. Show that a group of order 200 must contain a normal Sylow subgroup.
5. How many elements of order 7 are there in a group of order 168 which contains no normal subgroup?
6. The following table lists the number of distinct groups of each order from 1 through 20. Verify this for all orders except 16.

Order	1	2	3	4	5	6	7	8	9	10	11	12	13	14	15	16	17	18	19	20
Number	1	1	1	2	1	2	1	5	2	2	1	5	1	2	1	14	1	5	1	5

5. PERMUTATION GROUPS

5.1. Cycles.

In Chap. 1 it was noted in the Theorem of Cayley that every group may be written as a permutation group. As noted there, the same group may be written in terms of permutations in various ways. For a permutation π we write $(x_i)\pi = x_j$ to mean that π carries x_i into x_j.

A *finite cycle* is a permutation π on a finite set of letters $x_1, x_2, \cdots, x_n$ such that $(x_1)\pi = x_2, \cdots, (x_{n-1})\pi = x_n, (x_n)\pi = x_1$.

An *infinite cycle* is a permutation π on an infinite set of letters x_i, $i = -\infty, \cdots, +\infty$ such that $(x_i)\pi = x_{i+1}$, $i = -\infty, \cdots, +\infty$.

We write $(x_1, x_2, \cdots, x_n)$ for a finite cycle and $(\cdots x_{-1}, x_0, x_1, \cdots)$ for an infinite cycle. It is clear that the cycle $(x_2, \cdots, x_n, x_1)$ is the same permutation as $(x_1, x_2, \cdots, x_n)$.

THEOREM 5.1.1. *Given any permutation π on a set of letters S. The set S may be divided into disjoint subsets such that π is a cycle on each subset.*

Proof: Let x_1 be any letter of the set S. If $(x_1)\pi = x_1$, then (x_1) is a cycle by itself. If $(x_1)\pi \neq x_1$, write $(x_1)\pi = x_2$. Now write $(x_2)\pi = x_3, \cdots, (x_i)\pi = x_{i+1}$, continuing indefinitely unless a letter is repeated. If $(x_1)\pi = x_2, \cdots, (x_{n-1})\pi = x_n$ are all different, but $(x_n)\pi$ is a letter already used; then $(x_n)\pi = x_i$, for some $i = 1, \cdots, n$. If $i = 2, \cdots, n$, then also $(x_{i-1})\pi = x_i$, contrary to the assumption $x_n \neq x_{i-1}$. Hence $(x_n)\pi = x_1$ and we have a finite cycle $(x_1, \cdots, x_n)$ as the effect of π on the letters $x_1, \cdots, x_n$. If $(x_1)\pi^i = x_{i+1}$ are all different $i = 1, \cdots$, let x_0 be the letter such that $(x_0)\pi = x_1$. Continuing, define in succession $x_{-1}, x_{-2}, \cdots$, by $(x_{i-1})\pi = x_i$, $i = 0, -1, -2, \cdots$. These will all be different since π cannot take two different letters into the same letter. Thus each x of S is part of a set of letters permuted by π in a cycle. But clearly any letter determines the entire cycle since in $(x)\pi = y$ either letter x or y determines the other uniquely. Hence the different cycles are disjoint.

We may thus write a permutation π as a succession of cycles, and since the cycles are on disjoint sets of letters, clearly the order of writing the cycles is immaterial. It is often customary to omit the cycles of length one, it being understood that all letters omitted are fixed.

Thus $\pi = (1)(2)(3, 4, 5) = (3, 4, 5)$. With this convention a permutation may be regarded as the group product of its cycles, whenever the number of cycles is finite.

THEOREM 5.1.2. *The order of a permutation π is the least common multiple of the lengths of its cycles.*

Proof: In the cycle $(x_1, \cdots, x_n)$, $(x_i)\pi^j = x_{i+j}$, where $i + j$ is reduced modulo n. Hence $(x_i)\pi^t = x_i$ if, and only if, t is a multiple of n. Hence $(x_i)\pi^m = x_i$ for all $x_i \in S$ if, and only if, m is a multiple of the lengths of all the cycles of π. Here $\pi^m = 1$. If π contains a cycle of infinite length, or arbitrarily long cycles, then π is of infinite order.

A useful computational form is the following:

LEMMA 5.1.1. *If*

$$T = (a_{11} \cdots a_{1r})(a_{21}, \cdots, a_{2s}) \cdots (a_{m1}, \cdots, a_{mt})$$

and

$$S = \begin{pmatrix} a_{11} \cdots a_{1r} a_{21} \cdots a_{2s} \cdots a_{m1} \cdots a_{mt} \\ b_{11} \cdots b_{1r} b_{21} \cdots b_{2s} \cdots b_{m1} \cdots b_{mt} \end{pmatrix}$$

then $$S^{-1}TS = (b_{11} \cdots b_{1r})(b_{21} \cdots b_{2s}) \cdots (b_{m1} \cdots b_{mt}).$$

For a typical element b_{jk} we have

$$b_{j,k} \xrightarrow{S^{-1}} a_{j,k} \xrightarrow{T} a_{j,k+1} \xrightarrow{S} b_{j,k+1}$$

and so $b_{jk} \rightarrow b_{j,k+1}$ under $S^{-1}TS$.

The group of all permutations on a set of letters is called the *symmetric group.* The symmetric group on n letters is often designated as S_n.

THEOREM 5.1.3. *Two permutations are conjugate in a symmetric group if, and only if, they have the same number of cycles of each length.*

Necessity of the condition follows from the rule above. For sufficiency suppose

$$T = (a_{11} \cdots a_{1r})(a_{21} \cdots a_{2s}) \cdots (a_{m1} \cdots a_{mt}),$$

and $$R = (b_{11} \cdots b_{1r})(b_{21} \cdots b_{2s}) \cdots (b_{m1} \cdots b_{mt}),$$

including even cycles of length one. Since by hypothesis T and R have the same number of cycles of each length, we may assume the cycles to be lettered as given here. Then

$$Q = \begin{pmatrix} a_{11} & \cdots & a_{1r} & \cdots & a_{m1} & \cdots & a_{mt} \\ b_{11} & \cdots & b_{1r} & \cdots & b_{m1} & \cdots & b_{mt} \end{pmatrix}$$

is such that $Q^{-1}TQ = R$. Note that this theorem does not impose any conditions of finiteness, and "same number" refers to the cardinal number involved. We must include cycles of length one since if the number of letters involved is infinite, T and R could have the same number of cycles of lengths greater than one and yet fix a different number of letters. Thus $T = (0, 1)(2, 3)(4, 5) \cdots$ and $R = (0)(1, 2)(3, 4)(5, 6) \cdots$ are *not* conjugate in the symmetric group on the letters $0, 1, 2, 3, \cdots$.

5.2. Transitivity.

THEOREM 5.2.1. *Let G be a permutation group on letters $x_1 \cdots x_n$. Let S be any subset of these letters. Then the permutations of G, fixing all the letters of S, form a subgroup K. The permutations permuting the letters of S among themselves form a subgroup H which contains K as a normal subgroup.*

Proof: If two elements a and b permute the letters of S among themselves, or fix the letters of S, so does the product ab and the inverse a^{-1}. Hence there is a subgroup H permuting the letters of S and a subgroup K fixing the letters of S. If $h \in H$, $k \in K$; then $h^{-1}kh$ fixes the letters of S whence K is a normal subgroup of H.

DEFINITION: A permutation group G on letters $x_1, \cdots, x_n$ is *transitive* on a subset S of $x_1, \cdots, x_n$ if for every $\sigma \in G$ and $x_i \in S$, $(x_i)\sigma \in S$, and if for $x_i, x_j \in S$ there is a $\sigma \in G$ with $(x_i)\sigma = x_j$. The letters of S constitute a *set of transitivity.*

THEOREM 5.2.2. *If for a fixed letter x_1 the set S consists of all $x_i = (x_1)\sigma$, $\sigma \in G$, then S is a set of transitivity.*

Proof: If $(x_1)\sigma = x_i$, $(x_1)\tau = x_j$, then $(x_i)\sigma^{-1}\tau = x_j$. Moreover, if $(x_1)\sigma = x_i$, $(x_i)\rho = x_k$, then $(x_1)\sigma\rho = x_k$.

THEOREM 5.2.3. *If S is a set of transitivity for a permutation group G and x_1 is a letter of S, for each $x_i \in S$ choose $\sigma_i \in G$ with $(x_1)\sigma_i = x_i$.*

Let H be the subgroup of G fixing x_1. Then $G = H\sigma_1 + \cdots + H\sigma_i + \cdots$.

Proof: If $g = h\sigma_i$ with $h \in H$, then $(x_1)g = x_i$, whence the cosets $H\sigma_i$ are distinct. Moreover, let g be any element of G. Then $(x_1)g = x_i$ for some $x_i \in S$. Then $(x_1)g\sigma_i^{-1} = x_1$, whence $g\sigma_i^{-1} \in H$, $g = h\sigma_i \in H\sigma_i$, and so the cosets $H\sigma_i$ exhaust G.

COROLLARY 5.2.1. *If S is a set of transitivity for G which contains exactly r letters, then H, the subgroup fixing one letter of S, is of index r in G.*

DEFINITION: *A group G is k-ply transitive on letters of a set S if it is transitive on S and if any ordered set of k different letters of S is taken into an arbitrary ordered set of k different letters of S by some element of G.*

The analogue of Theorem 5.2.2 holds for k-ply transitive groups. If G takes a fixed set of k letters $x_1, x_2, \cdots, x_k$ into an arbitrary ordered set $y_1, y_2, \cdots, y_k$ of letters of S, then G is k-ply transitive on the letters of S. Also the subgroup of G, which fixes $r < k$ letters of S, will be $(k - r)$-ply transitive on the remaining letters of S. Also, if G is r-ply transitive and if a subgroup H fixing r letters is itself s-ply transitive, then G is $(r + s)$-ply transitive.

5.3. Representations of a Group by Permutations.

It has been noted that an abstract group may be represented in more than one way as a permutation group. We shall call a group of permutations P a representation of G if there is a mapping of G onto P, $g \rightarrow \pi(g)$, $g \in G$, $\pi(g) \in P$ such that $\pi(g_1)\pi(g_2) = \pi(g_1g_2)$. Note that P is necessarily a homomorphic image of G. If P is in fact isomorphic to G, we shall say that P is a *faithful* representation of G. Just as all homomorphic images of G are given by factor groups modulo a normal subgroup of G, all transitive permutation representations of G may be found in terms of left cosets of subgroups.

Since the non-Abelian group of order 6 may be faithfully represented as a transitive permutation group on three letters and also on six letters, we must distinguish as permutation groups certain groups which are isomorphic as abstract groups.

DEFINITION: *A permutation group P_1 on a set S_1 is isomorphic as a permutation group to a permutation group P_2 on a set S_2 if there is an* *isomorphism* $\pi_{P_1} \leftrightarrows \pi_{P_2}$ *between* P_1 *and* P_2 *and a one-to-one correspondence* $x_i \leftrightarrows y_i$ *between* S_1 *and* S_2 *such that* $(x_i)\pi_{P_1} = x_j$ *if, and only if,* $(y_i)\pi_{P_2} = y_j$.

THEOREM 5.3.1. *Given a group* G *and a subgroup* H.
a) *For each* $g \in G$ *there is a permutation of the set of left cosets of* H:

$$\pi(g) = \begin{pmatrix} Hx \\ Hxg \end{pmatrix}, \ x \in G.$$

b) $g \rightarrow \pi(g)$ *is a representation of* G *as a transitive permutation group on the set of distinct left cosets of* H, *and* $\pi(g)$ *fixes* H *if, and only if,* $g \in H$.

Conversely suppose $g \rightarrow \pi(g)$ *is a representation of* G *as a transitive permutation group* P *on a set of elements* S.

c) *If* s_1 *is a particular element of* S, *the* g*'s such that* $\pi(g)$ *fixes* s_1 *are a subgroup* H *of* G.

d) *The elements of* S *may be put into a one-to-one correspondence with the left cosets of* H *so that* P *is isomorphic as a permutation group to the group of permutations* $\pi(g)$ *given in* a) *and* b).

Proof: a) $Hx \rightarrow (Hx)g = Hxg$ maps each left coset Hx onto a unique left coset Hxg. Since $(Hxg^{-1})g = Hx$, $\pi(g) = \begin{pmatrix} Hx \\ Hxg \end{pmatrix}$ is a permutation of the set of distinct left cosets of H.

b) Since $(Hxg_1)g_2 = Hx(g_1g_2)$, it follows that $\pi(g_1)\pi(g_2) = \pi(g_1g_2)$, and so $g \rightarrow \pi(g)$ is a representation of G. $H \rightarrow Hg = H$ if, and only if, $g \in H$. Otherwise expressed, $\pi(g)$ fixes H if, and only if, $g \in H$. Since $H \rightarrow Hx$ by $\pi(x)$, the representation is transitive.

c) We verify directly that those g's such that $(s_1)\pi(g) = s_1$ are a subgroup H, since if g_1 and g_2 have this property, so do g_1g_2 and g_1^{-1}.

d) The set of g's such that $(s_1)\pi(g) = s_i$ is not vacuous, since P is transitive. If one of these g's is designated as x_i, it follows immediately that the entire set is the left coset Hx_i, H being the subgroup found in c) which fixes s_1. Conversely, all the elements of a left coset Hx have the property that their corresponding permutations all map s_1 onto the same image. This establishes a one-to-one correspondence, $s_i \leftrightarrows Hx_i$, between elements of S and left cosets of H. Let P_1 be the permutation group of left cosets of H as given by a) and b), with $\pi_1(g) = \begin{pmatrix} Hx \\ Hxg \end{pmatrix}$, $g \in G$ the permutations of P_1. In P if

$(s_i)\pi(g) = s_j$, then $(s_1)[\pi(x_i)\pi(g)] = s_j$, whence $x_ig \in Hx_j$, and hence $(Hx_i)g = Hx_j$; conversely, this relation implies $(s_i)\pi(g) = s_j$. Thus $s_i\pi(g) = s_j$ if, and only if, $Hx_i\pi_1(g) = Hx_j$. In particular $\pi(g)$ is the identity if, and only if, $\pi_1(g)$ is the identity. Thus P and P_1 are homomorphic images of G, both with the same kernel, and $\pi(g) \leftrightarrows \pi_1(g)$ is an isomorphism between P and P_1. And with $s_i \leftrightarrows Hx_i$, a one-to-one correspondence between S and the set of left cosets of H, we have established that P is isomorphic as a permutation group to P_1, since $(s_i)\pi(g) = s_j$ if, and only if, $Hx_i\pi_1(g) = Hx_j$.

In the light of this theorem we may speak of any transitive permutation representation of a group G as the representation on a subgroup H. If H is the identity, then the representation is the right regular representation given in §1.4.

THEOREM 5.3.2. *In the representation $g \to \pi(g)$ of Theorem 5.3.1, the elements mapped onto the identity form the largest normal subgroup of G contained in H, and so the representation is faithful if, and only if, H contains no normal subgroup of G greater than the identity.*

Proof: For what g is $\pi(g)$ the identity? Here $Hxg = Hx$ for all $x \in G$. Hence $x^{-1}Hxg = x^{-1}Hx$ or $g \in x^{-1}Hx$. Then $g \in \bigcap_x x^{-1}Hx = N$. Here N is clearly a normal subgroup of G contained in H. Moreover, any normal subgroup of G contained in H is contained in every $x^{-1}Hx$ and so in N. Thus N is the largest normal subgroup of G contained in H. Conversely, if $g \in N$, then $Hxg = Hx$ for every x, and so $\pi(g) = 1$. $N = 1$ is the necessary and sufficient condition that $g \to \pi(g)$ be a faithful representation of G.

COROLLARY 5.3.1. *The only faithful transitive representation of an Abelian group is the regular representation.*

THEOREM 5.3.3. *Two faithful representations of G on subgroups H_1 and H_2 are isomorphic as permutation groups if, and only if, there is an automorphism α of G such that $H_1{}^\alpha = H_2$.*

Proof: If α is an automorphism of G such that $H_1{}^\alpha = H_2$, then

$$H_1x \rightleftarrows H_1{}^\alpha x^\alpha = H_2x^\alpha$$

is a one-to-one correspondence between the cosets of H_1 and H_2 such that, if $g \to \pi_1(g)$ is the representation on H_1 and $g \to \pi_2(g)$ on H_2, then

$$\pi_1(g) \rightleftarrows \pi_2(g^\alpha).$$

On the other hand suppose there is a permutation isomorphism

$$\pi_1(g) \rightleftarrows \pi_2(g^*).$$

Since the representations are faithful, this defines a one-to-one correspondence $g \rightleftarrows g^*$ which will be an automorphism β of G. In the permutation isomorphism $\pi_1(g) \rightleftarrows \pi_2(g^*)$, we shall have $H_1 \leftrightarrows H_2u$. Hence, if $H_1g = H_1$,

$$H_2ug^\beta = H_2u$$

or

$$u^{-1}H_2ug^\beta = u^{-1}H_2u,$$

and conversely. Hence if $g \in H_1$, then $g^\beta \in u^{-1}H_2u$, and conversely. Or $H_1{}^\beta = u^{-1}H_2u$ or $H_2 = uH_1{}^\beta u^{-1} = H_1{}^\alpha$, where α is an automorphism of G.

5.4. The Alternating Group A_n.

Consider the polynomial in n variables $\Delta = \prod_{i<j} (x_i - x_j); i, j \leq n;$ $n \geqslant 2$. If $x_1, x_2, \cdots, x_n$ are replaced by a permutation of themselves, then Δ is replaced either by Δ or $-\Delta$. Writing Δ out,

$$\begin{aligned}\Delta = (x_1 - x_2)(x_1 - x_3) \cdots (x_1 - x_n) \cdot \\ (x_2 - x_3) \cdots (x_2 - x_n) \\ \cdots\cdots\cdots\cdots\cdots \\ \cdot \quad (x_{n-1} - x_n),\end{aligned}$$

we see that the interchange (x_1, x_2) replaces $x_1 - x_2$ by $x_2 - x_1 = -(x_1 - x_2)$, interchanges the remaining terms of the first row with the terms of the second row, and leaves the remaining terms unchanged. Thus the permutation (x_1, x_2) replaces Δ by $-\Delta$. We shall call a permutation *even* if it leaves Δ unchanged, and *odd* if it replaces Δ by $-\Delta$.

THEOREM 5.4.1. *The even permutations on $x_1, x_2, \cdots, x_n$ form a normal subgroup of index two in the symmetric group S_n. This group is called the alternating group A_n.*

Proof: We may verify directly that the product of two even permutations is even, the product of two odd permutations is even, and that the product of an even and odd permutation in either order is odd. We note that the identity is an even permutation.

Hence the even permutations of S_n form a subgroup A_n. Since (x_1, x_2) is an odd permutation, the coset $A_n\ (x_1, x_2)$ consists entirely of odd permutations. But if π is any permutation, then one of π, $\pi\cdot(x_1, x_2)$ is even and the other is odd. Since $\pi = [\pi\cdot(x_1, x_2)]\cdot(x_1, x_2)$, we see that A_n and $A_n(x_1, x_2)$ exhaust the elements of S_n, and $S_n = A_n + A_n(x_1, x_2) = A_n + (x_1, x_2)A_n$. Thus A_n is of index 2 in S_n and so is a normal subgroup.

A cycle of length two (x_i, x_j) is called a *tranposition.* Hence all transpositions in S_n are conjugate (Theorem 5.1.3) to (x_1, x_2). But whatever π is, π and π^{-1} have the same parity, and so $\pi^{-1}(x_1, x_2)\pi = (x_i, x_j)$ is odd. We may also compute directly that every transposition (x_i, x_j) is an odd permutation.

Any cycle of length n is the product of $n - 1$ transpositions, since $(x_1, x_2, \cdots x_n) = (x_1, x_2)(x_1, x_3) \cdots (x_1, x_n)$. Thus (Theorem 5.1.1) any finite permutation may be written as a product of transpositions. The product of an even number of transpositions is an even permutation, of an odd number, odd. Hence, though a permutation may be written in many ways as a product of transpositions, the number of transpositions involved will always have the same parity.

THEOREM 5.4.2. *A_n, $n \geq 3$, is $(n - 2)$-ply transitive.*

Proof: Let $y_1, \cdots, y_{n-2}, y_{n-1}, y_n$ be an arbitrary ordering of $x_1, \cdots, x_{n-2}, x_{n-1}, x_n$. Then if

$$u = \begin{pmatrix} x_1, & \cdots, & x_{n-2}, & x_{n-1}, & x_n \\ y_1, & \cdots, & y_{n-2}, & y_{n-1}, & y_n \end{pmatrix},$$

and

$$v = \begin{pmatrix} x_1, & \cdots, & x_{n-2}, & x_{n-1}, & x_n \\ y_1, & \cdots, & y_{n-2}, & y_n, & y_{n-1} \end{pmatrix},$$

we have $v = u(y_{n-1}, y_n)$ and one of u, v is even, the other odd. Hence A_n is $(n - 2)$-ply but not n-ply transitive. Clearly, it could not be $(n - 1)$-ply transitive without also being n-ply transitive.

In the group of permutations on an infinite set of ω letters we may define the alternating group A_ω as consisting of those permutations which may be written as the product of an even number of transpositions. A_ω will be a subgroup of index two in the group H_ω of those permutations each of which displaces only a finite number of letters. From Theorem 5.1.3, H_ω will be a normal subgroup of S_ω, and A_ω will be a normal subgroup of S_ω of index 2 in H_ω.

THEOREM 5.4.3. *The alternating group A_n is a simple group for any value of n, finite or infinite, except $n = 4$.*

A_2 is the identity. A_3 is the cyclic group of order 3 and, so, simple. The group A_4 must be treated separately. We may suppose that there are at least 5 letters.

LEMMA 5.4.1. *A_n, $n \geq 3$ is generated by all cycles (a, b, c) of length three.*

Proof: A_n is surely generated by all elements which are the product of two transpositions. If the two transpositions are identical, their product is 1. If they have one letter in common, say (a, b) and (a, c), we have $(a, b)(a, c) = (a, b, c)$. If they have no letter in common, $(a, b)(c, d) = (a, b)(a, c)(c, a)(c, d) = (a, b, c)(c, a, d)$, proving the lemma.

We shall prove that a normal subgroup G, greater than the identity and contained in A_n, $n \geq 5$ must contain all cycles of length three and hence be equal to A_n. This will be established by treating a number of cases. Note that since $G \subseteq A_n$, every element of G can be written as a product of a finite number of finite cycles.

CASE 1. *G contains a cycle of length three (a, b, c).*

Here any other cycle of length three (x, y, z) belongs with (a, b, c) in an alternating group A_r on a finite number r of letters, where we may take $r \geq 5$. Since A_r is $r - 2 \geq 3$-ply transitive, (a, b, c) and (x, y, z) are conjugate in A_r and *a fortiori* in A_n. But G, being normal, must contain all conjugates of (a, b, c) in A_n, and so all cycles of length three, whence by the Lemma 5.4.1. $G = A_n$.

CASE 2. *G contains an element g with a cycle of length $s \geq 4$.*

Write

$$g = (a_1, a_2, \cdots, a_r) \cdots (c_1, c_2, \cdots, c_{s-3}, c_{s-2}, c_{s-1}, c_s).$$

Here $t = (c_{s-2}, c_{s-1}, c_s) \in A_n$, and

$$t^{-1}gt = (a_1, a_2, \cdots, a_r) \cdots (c_1, c_2, \cdots, c_{s-3}, c_{s-1}, c_s, c_{s-2}).$$

But $gt^{-1}g^{-1}t = (c_{s-3}, c_s, c_{s-2})$ will belong to G since G is normal.

We have thus reduced Case 2 to Case 1. We now consider cases in which the lengths of the cycles are not greater than 3.

CASE 3. *Some $g \in G$ has two or more cycles of length 3.*

$$g = (a_1, a_2, a_3)(b_1, b_2, b_3) \cdots (c_1, \cdots c_r)$$

Take $t = (a_3, b_1, b_2) \in A_n$. Here

$$h = t^{-1}gt = (a_1, a_2, b_1)(b_2, a_3, b_3) \cdots (c_1, \cdots, c_r) \in G$$

and

$$gh^{-1} = (a_2, b_2, a_3, b_1, b_3) \in G$$

which reduces to Case 2.

CASE 4. *Some $g \in G$ has one or more cycles of length three and its remaining cycles of length two.*

$$g = (x_1, x_2)(y_1, y_2) \cdots (z_1, z_2)(a, b, c) \cdots (d, e, f).$$

Here

$$g^2 = (a, c, b) \cdots (d, f, e) \in G.$$

This reduces either to Case 1 or Case 3.

CASE 5. *Some $g \in G$ contains only cycles of length two and has at least four of these.*

$$g = (x, y)(z, u) \cdots (a, b)(c, d) \in G.$$

Take $t = (y, a)(b, c) \in A_n$.

$$\begin{aligned} h = t^{-1}gt &= (x, a)(z, u) \cdots (y, c)(b, d) \in G. \\ gh &= (x, c, b)(y, a, d) \in G. \end{aligned}$$

This reduces to Case 3.

CASE 6. *$g \in G$ contains only two cycles of length two.*

$$g = (a, b)(c, d) \in G.$$

Here, since we assume $n \geq 5$, there will be some letter e of the permutation set, $e \neq a, b, c, d$.

Here

$$\begin{aligned} t &= (a, b, e) \in A_n \\ h = t^{-1}gt &= (b, e)(c, d) \in G, \\ gh &= (a, e, b) \in G \end{aligned}$$

and this reduces to Case 1.

The alternating group A_4 on 1, 2, 3, 4 contains a normal subgroup of order 4 whose elements are (1), (12)(34), (13)(24), and (14)(23).

5.5. Intransitive Groups. Subdirect Products.

If a permutation group G is intransitive, let $S_i(x_{i1}, \cdots)$, $i \in I$, an index system, be the various sets of letters on which it is transitive. If we suppress all letters except those of the set S_i, then these permutations of the set S_i themselves form a group G_i. For each $i \in I$ an element g of G will determine a $g_i \in G_i$, namely, the permutation of the letters of S_i which g induces. We can moreover write

$$g = \prod_i g_i, \tag{5.5.1}$$

regarding g as an element of the Cartesian product of the G_i, since within G the group operations agree with those in the Cartesian product $\prod_i G_i$. Thus an intransitive group may be regarded as a subgroup of the Cartesian product of transitive groups. Here we say that G is the *subdirect product* of groups G_i. More precisely, a group is said to be a subdirect product of groups G_i if (1) G is a subgroup of the Cartesian product of the G_i; and (2) for each $g_j \in G_j$ there is at least one $g \in G$ which has g_j as its jth component. Here the second condition requires that all elements of the groups G_i actually occur in this representation of G.

If in the subdirect product $G \subseteq \prod_i G_i$ all components g_i may occur independently, then G is the entire Cartesian product. This will not be true in general, and the following theorem describes the kind of dependence which arises between the components of a subdirect product. Let G_i and G_j be two components or possibly the groups determined by disjoint sets of components G_i, $i \in I_1$, G_j, $j \in I_2$, $I_1 \cap I_2 = 0$. Suppressing all components except G_i and G_j, the elements of G determine a group G^* which is the subdirect product of G_i and G_j. We may describe the interdependence of the components G_i and G_j in G by describing exactly the induced subdirect product G^* of G_i and G_j.

THEOREM 5.5.1. *Let G^* be the subdirect product of the groups G_i and G_j and let H_{ij} and H_{ji} be the subgroups of G_i and G_j, respectively, of elements of one factor occurring in G^* with the identity of the other factor. Then H_{ij} is normal in G_i and H_{ji} is normal in G_j, and there is an isomor-*

phism between the factor groups $G_i/H_{ij} \cong K \cong G_j/H_{ji}$ *such that* (g_1, g_2), $g_1 \in G_i$, $g_2 \in G_j$ *is an element of* G^* *if, and only if,* g_1 *and* g_2 *have the same image* k *in the homomorphisms* $G_i \to K$, $G_j \to K$.

Proof: If $(h, 1)$ are the elements H_{ij} of G_i occurring with the identity of G_j in G^*, then we easily verify that H_{ij} is a normal subgroup of G_i and similarly that elements H_{ji} of the type $(1, h)$ in G^* are a normal subgroup of G_j. Moreover, for $g_1 \in G_i$, the set of elements $g_2 \in G_j$ occurring with a fixed g_1 is seen to be a coset of H_{ji}. In the same way the set of g_1's occurring with a fixed g_2 is seen to be a coset of H_{ij}. Still further, if (g_1, g_2) belongs to G^*, then all elements of the form $(H_{ij}g_1, H_{ji}g_2)$ belong to G^* and no other pair (g'_1, g'_2) of G^* involves any one of these elements as a component. Hence for each (g_1, g_2) of G^* there is determined a one-to-one correspondence $H_{ij}g_1 \rightleftarrows H_{ji}g_2$ between a coset of H_{ij} in G_i and a coset of H_{ji} in G_j.

If (g_1, g_2) and (g_3, g_4) belong to G^*, then (g_1g_3, g_2g_4) also belongs to G^*, and so this correspondence preserves products and must therefore be an isomorphism between the factor groups G_i/H_{ij} and G_j/H_{ji}. Here if we write $G_i/H_{ij} = K = G_j/H_{ji}$, then if (g_1g_2) belongs to G^*, we see that g_1 and g_2 belong to corresponding cosets and so have the same image k in the homomorphic image K of both G_i and G_j.

Conversely, if two groups G_i and G_j have normal subgroups H_{ij} and H_{ji}, respectively, such that $G_i/H_{ij} = K = G_j/H_{ji}$, then all pairs (g_1, g_2) with $g_1 \in G_i$, $g_2 \in G_j$ such that $g_1 \to k$, $g_2 \to k$ in the homomorphisms $G_i \to K$, $G_j \to K$ will form a subdirect product G^* as above.

5.6. Primitive Groups.

Suppose G is a permutation group $G \neq 1$ on letters which can be divided into disjoint sets $S_1, \cdots, S_m$ such that every permutation of G either maps all letters of a set S_i onto themselves or onto the letters of another set S_j. Except for the trivial cases in which there is only one set or in which every set consists of a single letter, we say that G is *imprimitive,* and we call $S_1, \cdots, S_m$ the sets of imprimitivity. Thus an intransitive group is *a fortiori* imprimitive. If G is not imprimitive, we say that G is *primitive.* Thus a primitive group is a transitive group whose letters cannot be divided into proper sets permuted among themselves.

THEOREM 5.6.1. *Let* G *be a transitive but imprimitive group. Let*

S_1 be one of the sets of imprimitivity and y_1 one of the letters of S_1, and H the subgroup of elements fixing y_1. Then the elements of G taking S_1 into itself form a subgroup K properly contained between G and H. The number of sets of imprimitivity is the index $[G:K]$, and each set of imprimitivity has the same number of letters $[K:H]$. Conversely, if G is a transitive group and H is the subgroup fixing a letter y_1, and if there is a subgroup K such that $G \supset K \supset H$, then G is imprimitive and one of its sets of imprimitivity consists of the $[K:H]$ letters into which elements of K take y_1. There are $[G:K]$ sets of imprimitivity corresponding to left cosets of K. Thus a permutation group G is primitive if, and only if, the subgroup H fixing a letter is a maximal subgroup.

Proof: Suppose that G is transitive and imprimitive. Let $S_1, \cdots, S_m$ be the sets of imprimitivity for G, and let H be the subgroup fixing a letter y_1 of S_1. Then, if

$$G = H + Hx_2 + \cdots + Hx_n, \tag{5.6.1}$$

we may, by Theorem 5.3.1, regard the letters permuted by G $y_1, y_2, \cdots, y_n$ as being the left cosets Hx_i of (5.6.1) permuted by the rule $\pi(g): Hx_i \to Hx_ig$ for each $g \in G$. If $y_1, y_2, \cdots, y_t$ are the letters of S_1, then the elements of G taking these letters into themselves form a subgroup K. An element fixing y_1 must take all of S_1 into itself, whence $H \subset K$, the inclusion being proper since an element taking y_1 into y_2 will belong to K but not to H. Now K is transitive on the letters of S_1. Thus

$$K = H + Hx_2 + \cdots + Hx_t, \tag{5.6.2}$$

and we see that the number t of letters in S_1 is $[K:H]$. Since S_1 does not contain all letters permuted by G, K will be a proper subgroup of G. Now if S_i is any one of the sets of imprimitivity, there is a permutation of G taking y_1 into a letter of S_i, whence all of S_1 is mapped onto all of S_i, and so S_i has the same number of letters as S_1. Moreover, in the permutation $Hx_i \to Hx_ig$, we also have $Kx_i \to Kx_ig$, whence the sets of imprimitivity are seen to be the left cosets of K in (5.6.1), and so their number is $[G:K]$.

Conversely, suppose that G is a transitive group given by the permutations $Hx_i \to Hx_ig$ of the cosets of the subgroup H fixing a letter y_1, and suppose there is a subgroup K such that $G \supset K \supset H$. Then the cosets of K consist of sets of cosets of H, and these will form a system of imprimitivity for G. Hence G is primitive if, and only if, the subgroup H is maximal.

We may make a few elementary remarks which follow from the definition of primitivity and this theorem. A doubly transitive group is surely primitive, since if S_1 is any set of letters which are part of the letters permuted by a doubly transitive group G, then there is a permutation which takes one letter of S_1 into itself and a second letter of S_1 into a letter outside of S_1. Thus S_1 cannot be a set of imprimitivity. Secondly, a group of degree n (the degree of a permutation group is the number of letters it permutes) can have a set of imprimitivity of t letters only if t is a divisor of n, since in Theorem 5.6.1 $n = [G:H]$ and $t = [K:H]$. Thus a group of prime degree is certainly primitive. Now in a p-group every subgroup is contained in a maximal subgroup of index p, which is normal (Corollary 4.2.2). Thus a permutation group which is a p-group is imprimitive unless it is on p letters, in which case it is the cyclic group of order p.

THEOREM 5.6.2. *Let G be a permutation group on n letters which is primitive, and let H be a transitive subgroup of G on m letters, fixing the remaining $n - m$ letters. Then* (1) *if H is primitive, G is $n - m + 1$ fold transitive;* (2) *in any event G is doubly transitive.*

Proof: H is transitive on a set of m of the n letters of G. Each of the conjugates of H is transitive on some set of m letters, and since G is transitive, every letter occurs in at least one of these sets. If these sets are either disjoint or identical, then they would be sets of imprimitivity for G. Hence H has conjugates which displace some of but not all the same letters as H. Let H' be one of those which has the largest number of letters in common with H. Let us write

$$\begin{aligned} H&:(a_1, \cdots, a_r, \quad c_1, \cdots, c_s); \\ H'&:(b_1, \cdots, b_r, \quad c_1, \cdots, c_s), \quad r + s = m. \end{aligned} \tag{5.6.3}$$

By this we understand that the c's are the letters which both H and H' displace, the group H also displacing r letters a_i, and H' also displacing r letters b_i. We assert that when H is primitive, then $r = 1$, and if H is imprimitive and $r > 1$, then $a_1, \cdots, a_r$ are a set of imprimitivity for H. Consider an element h' of H'.

$$h' = \begin{pmatrix} b_1, & \cdots, & b_u, & b_{u+1}, & \cdots, & b_r, & c_1, & \cdots, & c_{r-u}, & c_{r-u+1}, & \cdots, & c_s \\ b\,, & \cdots, & b\,, & c\,\,\,, & \cdots, & c\,, & b\,, & \cdots, & b\,\,\,, & c\,\,\,, & \cdots, & c \end{pmatrix}, \tag{5.6.4}$$

where this indicates primarily the number u of b's mapped onto b's, b's mapped onto c's, c's onto b's, and c's mapped onto c's. We note that the number $r - u$ of b's mapped onto c's must be the same as the number of c's mapped onto b's, since there must be (r) b's in the second row of h' in (5.6.4). Hence $h'^{-1}Hh'$ displaces (r) a's, $(r - u)b$'s, and $(s - r + u)c$'s, and so will have $s + u$ letters in common with H. Thus, if $r > 1$ and H' is primitive, we can choose an element h', taking some of but not all the b's into themselves, whence $1 \leq u < r$; thus $h'^{-1}Hh'$ has $s + u$ letters in common with H, which is more than s but not all $r + s = m$. In any event we must have $r = 1$ when H is primitive, and if $r = 1$ whether H is primitive or not, then $H \cup H'$ is doubly transitive on $m + 1$ letters and, so, primitive. We can continue with this group in the role of H until we reach G itself, obtaining in succession a doubly transitive group on $m + 1$ letters, a triply transitive group on $m + 2$ letters, and ultimately G as an $n - m + 1$-ply transitive group.

In case H is imprimitive, this argument does not apply, but we note that we can increase the number s of letters in common between H and H' unless $b_1, \cdots, b_r$ are a set of imprimitivity for H' and $a_1, \cdots, a_r$ are a set of imprimitivity for H. Moreover, $H \cup H'$ is a transitive group on $s + 2r = m + r$ letters. Thus, if m is at most $n/2$, $m + r$ will be less than n. We may continue to form transitive subgroups on more and more letters until we have a transitive subgroup H on a number m of letters greater than $n/2$ but less than n. In this case any conjugate H' of H displaces some letters in common with H. Here, suppose H is transitive on the largest possible number of letters less than n. If $s + 2r = n$ and $r = 1$, then H is transitive on $n - 1$ letters and so G is doubly transitive. If this does not happen, we reach a group H where $s + 2r = n$ with $r \neq 1$. In this case the a's, b's, and c's are all the letters of G. But since G is primitive, there is an element g taking b_1 into some b_i but not all b's into themselves, and so at least one a or c into a b. Here H and $g^{-1}Hg$ both fix b_i, and their union is transitive on more letters than H. Thus we must ultimately reach a subgroup transitive on $n - 1$ letters and so G is doubly transitive.

The second alternative in the theorem can actually arise. Example 4 in Chap. 1 illustrates this, where the group is transitive on seven letters and thus primitive. It has a transitive subgroup on the four letters C, E, F, G, and it is doubly but not triply transitive.

5.7. Multiply Transitive Groups.

The symmetric group on n letters is, of course, n-ply transitive, and the alternating group A_n (as we remarked in §5.4) is $(n - 2)$-ply transitive. We shall exclude these in further discussion of multiple transitivity. There are infinitely many groups which are triply transitive. But apart from the alternating and symmetric groups, there are only four groups known which are quadruply transitive. These are the Mathieu groups on 11, 12, 23, and 24 letters, respectively, of which the groups on 12 and 24 letters are quintuply transitive and contain as subgroups fixing a letter the groups on 11 and 23 letters, respectively. These somewhat mysterious groups have been the subject of considerable investigation, but it is not known whether these groups are truly exceptional or whether they are part of an infinite family of groups which are quadruply transitive.

Theorem 5.7.1, due to G. A. Miller [1], gives a limit on the transitivity of groups of degree n. This theorem combined with "Bertrand's postulate," proves that for $n > 12$ a group of degree n cannot be t-fold transitive for $t \geq 3\sqrt{n} - 2$. Bertrand's postulate (proved correct by Chebyshev in 1850) states that for any real number $x \geqq 7$, there exists a prime number p in the interval $x/2 < p \leq x - 2$. Miller's theorem gives a considerably better limit for most specific values of n. Still better restrictions are known,* but their proofs are too complicated to include here.

THEOREM 5.7.1. *Let G be a t-ply transitive group on n letters. Let H be a subgroup fixing t letters, and let P be a Sylow p-subgroup of H, where P fixes $w \geq t$ letters. Then the normalizer in G of P is t-ply transitive on the w letters fixed by P.*

Proof: Let $a_1, \cdots, a_t$ and $b_1, \cdots, b_t$ be two ordered sets of t letters, both sets being from the w letters fixed by P. Then, since G is t-ply transitive, there is an element x of G taking a_i into b_i for $i = 1, \cdots, t$. Then $x^{-1}Px$ fixes $b_1, \cdots, b_t$, and thus both P and $x^{-1}Px$ are Sylow subgroups of the group fixing $b_1, \cdots, b_t$. By the second Sylow theorem these groups must be conjugate in the group fixing $b_1, \cdots, b_t$. Thus, for some y fixing $b_1, \cdots, b_t$, we have $y^{-1}(x^{-1}Px)y = P$. But here,

---

* E. Parker has obtained a limit with t of the order of magnitude $\sqrt[3]{n}$ for reasonable values of n. The best asymptotic value is due to Wielandt [1] which gives $t < 3 \log n$.

with $z = xy$, z takes $a_1, \cdots, a_t$ into $b_1, \cdots, b_t$ and $z^{-1}Pz = P$. Hence there is an element in the normalizer of P taking any ordered set of t of the w letters fixed by P into any other ordered set of t of these letters. Thus the normalizer in G of P is t-fold transitive on the w letters fixed by P, proving the theorem.

THEOREM 5.7.2. *Let the integer $n = kp + r$, where p is a prime and $p > k$, $r > k$. Except for $k = 1$, $r = 2$, a group of degree n cannot be as much as $(r + 1)$-fold transitive unless it is S_n or A_n.*

Proof: Suppose that G of degree n is $(r + 1)$-fold transitive. The subgroup H fixing the first r letters, $1, 2, \cdots, r$, is transitive on the remaining kp letters. Thus the order of H is divisible by the prime p, and contains a Sylow p-subgroup P. A subgroup of H fixing a letter is of index kp in H, and so its order is not divisible by the highest power of p dividing the order of H. Thus P must displace every one of the kp letters on which H is transitive. Furthermore since $kp < p^2$, by the hypothesis P cannot contain a transitive constituent on p^2 letters. As the number of letters in a transitive constituent of P is a divisor of the order of P, the group P must have on the kp letters of H exactly k transitive constituents of p letters each. (We have already excluded the possibility that any constituent is a single letter.) On each of these constituents P must be the cyclic group of order p. Thus P is a subdirect product of k cyclic groups of order p on p letters each. Thus every element of P is of order p, and P is an Abelian group. But for the most part we shall not have to concern ourselves with the manner in which P is a subdirect product.

Let N be the normalizer in G of P. By Theorem 5.7.1, N is the symmetric group S_r on the first r letters of G. Let us first consider cases with $r \geq 5$, and let N_1 be the subgroup of N which is the alternating group A_r on the first r letters. By Theorem 5.4.3, A_r is a simple group of order $r!/2$, and being of composite order, is not Abelian. Let $T_1, \cdots, T_k$ be the k transitive constituents of p letters of P. Then a homomorphic image of N_1 is given if we combine the permutations on the first r letters with the permutations on the transitive constituents T_i themselves, which are permuted among themselves in N_1. This image is the subdirect product of A_r on the first r letters and a group permuting the k T's among themselves in some manner. But a group on k symbols is of order at most $k!$ and so, since $k! < r!/2$, it can have no factor group isomorphic to A_r, which is a simple group; thus the only factor group of this group isomorphic to a factor group of A_r is the identity. Hence this group involving A_r and the constituents T_i is, by the results of §5.5, the direct product of A_r and the other group. Here A_r and the identity in the other group has as its inverse image in N_1 a group N_2, which is A_r on the first r letters, and takes the letters of each transitive constituent T_i into themselves. To analyze N_2 we must pause to consider the nature of the normalizer on p letters of the cyclic group generated by $a = (x_1, \cdots, x_p)$ on these letters. Since $a^p = 1$, if $b^{-1}ab = a^i$ and $c^{-1}ac = a^j$, we see that both bc and cb transform a into a^{ij}. Thus the automorphisms induced on a cyclic group by transformation themselves form an Abelian group. (We shall see in the next chapter that the automorphisms of a cyclic group of order p are themselves a cyclic group of order $p - 1$.)

Now an element u on $x_1, \cdots, x_p$ permuting with a, when multiplied by an appropriate power a^i of a, will be an element $v = ua^i$ which permutes with a and fixes the letter x_1. But with $a^{-1}va = v$ and v fixing x_1, we can readily show that v fixes $x_2, \cdots, x_p$, and so $v = 1$, whence $u = a^{-i}$. Thus N_2 on any one of the transitive constituents T_i of the k transitive constituents of p letters of P will have a normal subgroup of order p, consisting of the powers of a cycle of p letters and a factor group of elements inducing different automorphisms on the group of order p; this factor group is Abelian. Thus any factor group of this group is either Abelian or has an Abelian factor group. Therefore the only factor group isomorphic with a factor group of A_r is the identity. Thus, neglecting $T_2, \cdots, T_k$ momentarily, N_2, applying the results of §5.5 to the first r letters and T_1, has a subgroup which is A_r on the first r letters and the identity on the letters of T_1. This subgroup N_3 of N_2 in turn has a subgroup N_4 which is A_r on the first r letters and is the identity on both T_1 and T_2.

Continuing, we have a subgroup which is A_r on the first r letters and the identity on the remaining letters. But A_r contains a cycle (a, b, c) on three letters, and since G is at least 5-ply transitive on all n letters, this may be transformed into any cycle of three letters of the n letters. By Lemma 5.4.1 these three-cycles generate A_n, and since G contains A_n, G is either A_n or S_n.

The preceding argument required $r \geq 5$ and leaves to be considered the cases $r = 3$, $k = 1$ or 2, and $r = 4$, $k = 1$, 2, or 3. Let us first consider cases in which P is cyclic, generated by an element a, and $k = 1$ or 2. As we have remarked above, if $u = (12)(3) \cdots$ and $v = (1)(23) \cdots$ are elements of the normalizer N of P (which will be the symmetric group on the first three or four letters fixed by P), then since P is cyclic, uv and vu will both transform a into the same power of itself. Thus $u^{-1}v^{-1}uv = (1, 2, 3) \cdots$ will permute with a. This element $w = u^{-1}v^{-1}uv$ either interchanges the two constituents T_1, T_2 or takes both into themselves. In either event $w^2 = (1, 3, 2) \cdots$ fixes both constituents if there are two of them. This element will have order divisible by 3, and so some power of it will be of order 3^s and will still be a three-cycle on the first three letters, taking the constituents into themselves, and permuting with a. But for each cycle of a, the only permuting elements are the power of the cycle and are of order p if they are not the identity. Thus, unless $p = 3$, an element of order 3^s which permutes with a in this manner will be the three-cycle $(1, 2, 3)$ or $(1, 3, 2)$ on the first three letters and the identity on the remaining. Here G contains a three-cycle and is triply transitive, and so must be either A_n or S_n. We have excluded only $p = 3$, and this corresponds with $k = 1$ or 2 and $r = 3$ or 4 to $n = 6$, 7, 9, 10. Actually, with $p = 3$, $k = 1$, P itself is a three-cycle and the conclusion follows. This settles $n = 6$, 7, leaving $n = 9$, 10 to be treated as special cases. Here all cases with $k = 1$ are covered, since P is surely cyclic in these cases. Now if $k = 2$ and P is not cyclic, then P is the direct product of two p-cycles, and Theorem 5.6.2 applies with G primitive and H a cycle of order p, and thus a primitive group. Here G must be $(p + 4)$- or $(p + 5)$-fold transitive, and we may use the argument with $r = p + 3$ or $p + 4$ and $k = 1$ to conclude that $G = A_n$ or S_n.

There remain to be considered cases with $k = 3$, $r = 4$. First, if P is cyclic, we may argue as before that there are elements which are $(1, 2, 3)(4) \cdots$ $(1)(2, 3, 4) \cdots$, and, indeed, all eight possible three-cycles on the first four letters which permute with a generator a of P. But they may permute the three transitive constituents of a cyclically in either of the ways (T_1, T_2, T_3) or (T_1, T_3, T_2), and at least two of the eight must permute the T's in the same way. Combining these, we get an element either of the type $(1, 2, 3)(4) \cdots$ or $(1, 2)(3, 4) \cdots$ which takes T_1, T_2, T_3 into themselves. Here p is at least 5, and an element permuting with a and taking the cycles of a into themselves of one of these forms leads to an element of the same form fixing the $3p$ letters of a. Thus there is in G either a three-cycle $(1, 2, 3)$ or an element $(1, 2)(3, 4)$, and by quadruple transitivity, also $(1, 2)(3, 5)$ and also the three-cycle $(3, 4, 5)$. Hence G contains A_n and is either A_n or S_n. On the other hand, if P contains a single cycle of p letters, then Theorem 5.6.2 applies, and G is $(2p + 4)$- or $(2p + 5)$-fold transitive, whence again G is A_n or S_n.

As a final case we must consider the possibility that P is not cyclic nor does it contain a p-cycle. In this case P must be of order p^2. We may take a basis for P of two elements, $a = (x_1, x_2, \cdots, x_p)(y_1, y_2, \cdots, y_p)$ and $b = (y_1, y_2, \cdots, y_p)(z_1, z_2, \cdots, z_p)$, where we have chosen a and b with the same cycle on the y's. Thus ab^{-1} and its powers are the only elements of P fixing the constituent T_2 of the y's. Now an element normalizing P of the form $(1, 2, 3, 4) \cdots$ can be found so that it either fixes the three constituents or permutes two of them and fixes the third. Thus its square, $u = (1, 3)(2, 4) \cdots$, fixes all three constituents. Hence u transforms each of a, b, and ab^{-1} into some power of itself and thus transforms both a and b into the same power of themselves (say, the ith power, and so every element of P into its ith power). Such an automorphism must permute with any other automorphism of P, and in particular with an automorphism induced by an element $w = (1)(2, 3)(4) \cdots$. Therefore $v = w^{-1}uw = (1, 2)(3, 4) \cdots$ also transforms every element of P into its ith power and so naturally fixes the constituents of P. But now $uv^{-1} = (1, 4)(2, 3) \cdots$ permutes with every element of P, fixing the constituents. This leads to an element $(1, 4)(2, 3)$ in G, and since G is quadruply transitive on more than four letters, G will again be A_n or S_n.

5.8. On a Theorem of Jordan.

In 1872 Jordan [2] showed that a finite quadruply transitive group in which only the identity fixes four letters must be one of the following groups: the symmetric group on four or five letters; the alternating group on six letters, or the Mathieu group on eleven letters.

Jordan's theorem on quadruply transitive groups is generalized here in two ways. The number of letters is not assumed to be finite; instead of assuming that the subgroup fixing four letters consists of the identity alone, we assume only that it is a finite group of odd order. The conclusion is essentially the same as that of Jordan's theorem, the only other group satisfying the hypotheses being the alternating group on seven letters.

The theorem is the following:

THEOREM 5.8.1. *A group G quadruply transitive on a set of letters, finite or infinite, in which a subgroup H fixing four letters is of finite odd order, must be one of the following groups: S_4, S_5, A_6, A_7, or the Mathieu group on* 11 *letters.*

CASE 1. *G on not more than seven letters.* A quadruply transitive group on four or five letters must be the symmetric group. On six letters its order must be at least $6 \cdot 5 \cdot 4 \cdot 3$, and hence it is A_6 or S_6. On seven letters, its index is, at most, 6 in S_7. Since S_7 does not have a subgroup of index 3 or 6, the only possibilities are A_7 and S_7. In both S_6 and S_7 there are elements of order 2 fixing at least four letters, and so these groups do not satisfy our hypothesis.

To treat the case in which G is on more than seven letters, we begin with a lemma.

LEMMA 5.8.1. *Elements a and b in a group, satisfying the relations*

$$a^2 = 1, \quad b^2 = 1, \quad (ab)^s = 1,$$

generate the dihedral group of order $2s$. If $s = 2t - 1$ is odd, then a power of $y = ab$ transforms a into b. If $s = 2r$ is even, then a and b permute with y^r.

Proof: With $y = ab$, we have

$$a^2 = 1, \quad y^s = 1, \quad b = ay = y^{-1}a.$$

If $s = 2t - 1$, then

$$y^{-t}ay^t = ay^{2t} = b.$$

If $s = 2r$, then

$$ay^r = y^{-r}a = y^ra.$$

From here on, G will denote (as in Theorem 5.8.1) a group quadruply transitive on more than seven letters, and H will denote a subgroup of odd order m fixing four letters.

LEMMA 5.8.2. *The group G contains elements of order* 2, *and all elements of order* 2 *are conjugate. Either* (1) *every element of order* 2 *fixes two letters, or* (2) *every element of order* 2 *fixes three letters.*

Proof: By quadruple transitivity G contains an element

$$g = (12)\ (34)\ \cdots.$$

Here g^2 fixes 1, 2, 3, 4, and so belongs to H and will be of finite odd order m_1. Thus

$$x = g^{m_1} = (12)\ (34)\ \cdots,$$

with $x^2 = 1$. Since H is of odd order, any element u of order 2 will fix, at most, three letters and hence will displace at least four letters. With

$$u = (ab)\ (cd)\ \cdots,$$

there is a conjugate of u,

$$v = w^{-1}uw = (12)\ (34)\ \cdots.$$

Either $v = x$, or vx fixes four letters and is of odd order, whence, by Lemma 5.8.1, v and x are conjugate. Thus all elements of order 2 are conjugate. On the other hand, there is in G an element $z = (1)(2)$-$(34)\ \cdots$, and either z or an odd power of z is an element of order 2 fixing at least two letters. Hence every element or order 2 fixes either two or three letters, since they fix at least two and not as many as four.

CASE 2. *G on more than seven letters.*

Let

$$a_1 = (1)(2)(34)\ \cdots$$

be an element of order 2 and

$$b = (12)(34)\ \cdots,$$

another element of order 2. Then $f = a_1b = (12)(3)(4)\ \cdots$ will be of even order, and f^2 will be of odd order m_1. Hence $f^{m_1} = a_3$ is of order 2, and by Lemma 5.8.1, will permute with a_1. Hence in G we have permuting elements of order 2, with $a_2 = a_1a_3$.

$$\begin{aligned} a_1 &= (1)(2)(34)\ \cdots, \\ a_2 &= (12)(34)\ \cdots, \\ a_3 &= (12)(3)(4)\ \cdots. \end{aligned} \tag{5.8.1}$$

Now a_2 as an element of order 2 fixes either two letters 5 and 6, or three letters 5, 6, and 7. As a_1 permutes with the element a_2, it takes these letters into themselves. But a_1 fixes 1 and 2 and, at most, one other letter. Hence we have

$$(5.8.2)\quad \begin{aligned} a_1 &= (1)(2)(34)(56)\cdots, \\ a_2 &= (12)(34)(5)(6)\cdots, \\ a_3 &= (12)(3)(4)(56)\cdots; \end{aligned} \quad \text{or} \quad \begin{aligned} a_1 &= (1)(2)(34)(56)(7)\cdots, \\ a_2 &= (12)(34)(5)(6)(7)\cdots, \\ a_3 &= (12)(3)(4)(56)(7)\cdots. \end{aligned}$$

The first case arises if elements of order 2 all fix two letters; the second, if all fix three letters. The elements $a_1\, a_2\, a_3$ of (5.8.2) and the identity form a four-group, V. Further letters will occur in sets of four which will be sets of transitivity for V:

$$(5.8.3)\quad \begin{aligned} a_1 &= (1)(2)(34)(56)(7)(hi)(jk)\cdots, \\ a_2 &= (12)(34)(5)(6)(7)(hj)(ik)\cdots, \\ a_3 &= (12)(3)(4)(56)(7)(hk)(ij)\cdots. \end{aligned}$$

Here it is understood that the 7 may not be present.

The order of the subgroup K taking h, i, j, k into themselves will be $24m$, and $H = H(h, i, j, k)$, fixing these letters, of order m will be normal in K. There will be a subgroup U, $K \supset U \supset H$, in which h, i, j, k are permuted in the following way:

$$(5.8.4)\quad \begin{gathered} (h) \\ (hi)(jk) \\ (hj)(ik) \\ (hk)(ij) \\ (hjik) \\ (hkij) \\ (hi)(j)(k) \\ (h)(i)(jk). \end{gathered}$$

Now U is of order $8m$, and so a Sylow subgroup of U will be of order 8. The elements taking h, i, j, k into themselves in a particular way will be a coset of H in U. Since H is normal in U, a group of order 8 in U will have one element from each coset and will be isomorphic to U/H, and hence will be faithfully represented by the permutations on these letters. V will be contained in a Sylow subgroup of order 8 in U. This yields

$$\begin{aligned} a_1 &= (1)(2)(34)(56)(7)(hi)(jk)\cdots, \\ a_2 &= (12)(34)(5)(6)(7)(hj)(ik)\cdots, \\ a_3 &= (12)(3)(4)(56)(7)(hk)(ij)\cdots, \end{aligned}$$

(5.8.5)
$$\begin{aligned} u &= (1)(2)(3546)(7)(hjik)\cdots, \\ a_1u &= (1)(2)(3645)(7)(hkij)\cdots, \\ a_2u &= (12)(36)(45)(7)(hi)(j)(k)\cdots, \\ a_3u &= (12)(35)(46)(7)(h)(i)(jk)\cdots, \end{aligned}$$

or the same permutations with 5 and 6 interchanged. The way in which the last four elements permute the letters $1, \cdots, 7$ is determined by the relations

$$u^2 = a_1, \quad u^{-1}a_2u = a_3, \quad (a_2u)^2 = 1.$$

Here u normalizes V and so fixes the only letter, 7, fixed by V (if the 7 occurs). Also, u must take the fixed letters of a_3 into those of a_2, whence

$$u = \begin{pmatrix} 3, & 4, & \cdots \\ 5, & 6, & \cdots \end{pmatrix} \quad \text{or} \quad u = \begin{pmatrix} 3, & 4, & \cdots \\ 6, & 5, & \cdots \end{pmatrix};$$

but also $u^2 = a_1$, whence

$$u = (3546)\cdots \quad \text{or} \quad u = (3645)\cdots.$$

Finally, u must fix 1 and 2 or interchange them. But if u interchanges 1 and 2, then a_2u is of order 2 and fixes the letters $1, 2, j, k$. Thus

$$u = (1)(2)(3546)\cdots \quad \text{or} \quad u = (1)(2)(3645)\cdots,$$

and the rest follows.

Each further transitive constituent of V, such as h, i, j, k, yields a group S such as that in (5.8.5). The elements

$$(12)(36)(45)\cdots \quad \text{and} \quad (12)(35)(46)\cdots$$

in each of these groups fix two letters of the constituent. Since an element of order 2 cannot fix four letters, each constituent yields a different element, permuting the first six letters in the way (12)(36)(45). But there are, at most, m elements with this effect on the first six letters. Thus if there are t such constituents, $t \leq m$ is finite and G is a group on $n = 4t + 6$, or $4t + 7$ letters. If G is on 10 or 11 letters we have $t = 1$.

There is no quadruply transitive group on ten letters (except, of course, A_{10} and S_{10}), for the normalizer of a cycle of length 7 by Theorem 5.7.2 is S_3 on the remaining three letters; and so this normalizer, which is the subdirect product of S_3 and the normalizer on the letters of the seven-cycle, will pair a three-cycle with the identity. Hence G contains a three-cycle and, being quadruply transitive, all three-cycles; thus G contains A_{10}.

On 11 letters G is of order $11 \cdot 10 \cdot 9 \cdot 8m$, and even without assuming m odd, consideration of normalizers of Sylow subgroups fixing four letters shows that we must have $m = 1$. The group of order 8 fixing three letters contains a single element of order 2, and so it is the cyclic or quaternion group. The cyclic group, having only four automorphisms, could not have a normalizer triply transitive on the remaining three letters, for then G would contain a three-cycle. Hence the subgroup fixing three letters must be the quaternion group Q. Then G will be a transitive extension of Q, and the methods of T.C. Holyoke [1] will readily enable us to construct from Q not only the quadruply transitive Mathieu group on 11 letters, but also the quintuply transitive group on 12 letters.

We shall now show that $t > 1$ conflicts with the hypothesis that H is of odd order, and thus complete the proof of our theorem. If w, x, y, z is another transitive constituent of V, we have

$$a_2u = (12)(36)(45)(7)(hi)(j)(k) \cdots$$

from (5.8.5), and we will have another element

$$a_2u' = (12)(36)(45)(7)(wx)(y)(z) \cdots.$$

Each of these elements permutes with a_1 and transforms a_2 into a_3 and a_3 into a_2. Their product is an element q fixing the first six (or seven) letters and, so, is of odd order. Also, q centralizes V. By Lemma 5.8.1, a power of q transforms a_2u into a_2u', and so takes the fixed letters j, k, of a_2u into the fixed letters y, z, of a_2u'. Centralizing V, this element must take the entire constituent $h\ i\ j\ k$ into $w\ x\ y\ z$. Hence there is a group C in G which fixes the first six (or seven) letters, centralizes V, and is transitive on the t remaining constituents of V. An element of C taking a constituent of V into itself, being of odd order, must fix all four letters. Thus the transitive constituents of C are $(1)(2)(3)(4)(5)(6)(7)T_h$, T_i, T_j, T_k, the last four sets of t each, the letters h, i, j, k being in different constituents of C.

Let p be a prime dividing t. (Here we use the assumption $t > 1$.) Let P be the corresponding Sylow subgroup of C. Then P displaces all $4t$ letters which C displaces, since a subgroup of C fixing a letter is of index $t \equiv 0 \pmod{p}$ and cannot contain such a Sylow subgroup.

Now let P_1 be a Sylow subgroup of H, the subgroup fixing 1, 2, 3, 4, which contains P. Then P_1 displaces the $4t$ letters of C and no others, unless possibly we have the case

$$p = 3, \quad t = 3^w, \quad n = 4t + 7,$$

where P_1 might be on $4t + 3$ letters. This possibility will be considered later. With P_1 on $4t$ letters, by Theorem 5.7.2, the group $N_G(P_1)$ is quadruply transitive on the first six or seven letters and so contains A_6 or A_7 on these letters. But the subgroup taking the first six (or seven) letters into themselves also contains the element u of (5.8.5), which is not in the alternating group on these letters. Thus in G we have the full symmetric group on the first six or seven letters, and hence some element fixing the first four letters and interchanging the fifth and sixth. This conflicts with the hypothesis that H is of odd order. Finally, consider the possibility that

$$t = 3^w, \quad n = 4t + 7,$$

and that P_1 displaces 5, 6, 7, as well as the $4t$ letters of P. If $w > 1$, then surely (5, 6, 7) is a transitive constituent of P_1 and there is an element

$$z = (1)(2)(3)(4)(567) \cdots$$

in G. If $w = 1$, then P is of order 3, and (even though in P_1, 5, 6, 7 are in a constituent with 8, 9, 10, and 11, 12, 13 of P) since there is an element (5)(6)(7)(8, 9, 10)(11, 12, 13), there will also be one such as z fixing 8, 9, 10. But with $z = (1)(2)(3)(4)(567) \cdots$, and u of (5.8.5), we have

$$(zu)^3 = (1)(2)(35)(4)(6)(7) \cdots,$$

contradicting the assumption that a subgroup H fixing four letters is of odd order.

Let G be a quadruply transitive group on 11 letters, excluding S_{11} and A_{11}. If G contains an element of one of the forms (a, b), (a, b) (c, d), or (a, b, c), then by quadruple transitivity, G contains all such elements and must be A_{11} or S_{11}. If G contains a five-cycle or seven-cycle, such an element generates a group transitive and primitive on the letters it displaces. In this case, by Theorems 5.6.2 and 5.7.1, G must be S_{11} or A_{11}. With these exclusions a subgroup $V = V_{1234}$ fixing four letters is of order dividing $2^4 \cdot 3^2$. If V is not the identity, then V must have a Sylow 2-group or a Sylow 3-group. In either case because of our exclusion such a Sylow subgroup P must displace exactly six letters. By Theorem 5.7.2 the normalizer of P is quadruply transitive on the remaining five letters, and because P has transitive constituents of three and three letters, or four and two letters, or two, two, and two letters, it will follow that G contains a five-cycle, a possibility already excluded. Thus the only possibility remaining is that a subgroup V fixing four letters is the identity and G is of order $11 \cdot 10 \cdot 9 \cdot 8$.

The subgroup W fixing three letters, say, 9, 10, 11, is regular and transitive on the remaining eight letters, and so it is the regular representation of one of the five distinct groups of order 8. W will contain an element of order 2, say, $x = (1, 2)(3, 4)(5, 6)(7, 8)(9)(10)(11)$. In the subgroup H fixing the two letters 10 and 11, there are nine conjugates of W, each fixing one letter. If two different elements of order 2 contained the same transposition, say (i, j), their product would be an element different from the identity displacing, at most, seven letters. This cannot be. But each element of order 2 contains four transpositions and there are only $9 \cdot 8/2 = 36$ possible transpositions of $1, \cdots, 9$. Hence W contains only one element of order 2 and must be the cyclic group of order 8 or the quaternion group. But if W is the cyclic group, its normalizer contains an element of order 3, and this can only be the cycle $(9, 10, 11)$, which is not possible. Hence W must be the quaternion group Q.

The subgroup H fixing 10 and 11 is of order 72 and contains nine quaternion subgroups, any two of which intersect in the identity. The identity and the remaining eight elements form a subgroup U of order 9 which is normal on H. The eight elements of U different from the identity are conjugate under Q, and so U must be the elementary Abelian group.

From this information we can easily construct H, which is unique to within permutation isomorphism. U may be generated by

$$u = (123)\ (456)\ (789)\ (10)\ (11),$$
$$v = (147)\ (258)\ (369)\ (10)\ (11).$$

$H = QU$, where Q is the quaternion group generated by

$$a = (1)\ (2437)\ (5698)\ (10)\ (11),$$
$$b = (1)\ (2539)\ (4876)\ (10)\ (11),$$

and
$$a^2 = b^2$$
$$= (1)\,(23)\,(47)\,(59)\,(68)\,(10)\,(11).$$

The subgroup K fixing 11 will be generated by H and a conjugate x of a^2 fixing 2 and 11 and interchanging 1 and 10. Such an element must exist, since G is quadruply transitive. Clearly, x normalizes Q. Adjoining x to H must not yield an element different from the identity fixing four letters. The only possibilities are

$$x_1 = (1, 10)\,(2)\,(3)\,(11)\,(4, 5)\,(6, 8)\,(7, 9),$$
$$x_2 = (1, 10)\,(2)\,(3)\,(11)\,(4, 6)\,(5, 9)\,(7, 8),$$
$$x_3 = (1, 10)\,(2)\,(3)\,(11)\,(4, 7)\,(5, 6)\,(8, 9).$$

The element $(4, 5, 6)\,(7, 9, 8)$ transforms H into itself and permutes these three elements among themselves, therefore, to within permutation isomorphism, we may adjoin any one of these three. Let K be obtained by adjoining x_1 to H. Then G is obtained by adjoining to H a conjugate y of a^2 which interchanges 1 and 11 and fixes 2 and 10. Here y normalizes Q and also the subgroup fixing 1 and 11. The only possibilities for y are

$$y_1 = (1, 11)\,(2)\,(3)\,(10)\,(4, 6)\,(5, 9)\,(7, 8),$$
$$y_2 = (1, 11)\,(2)\,(3)\,(10)\,(4, 7)\,(5, 6)\,(8, 9).$$

Here the element $(4, 9)\,(5, 7)\,(6, 8)$ normalizes K and interchanges y_1 and y_2. Hence, to within permutation isomorphism, we may suppose G obtained by adjoining y_i to K. $G = \{H, x_1, y_1\}$. Strictly speaking, what we have shown so far is that if there is a quadruply transitive group on 11 letters, not A_{11} or S_{11}, then it is permutation isomorphic to G. Verification that G has these properties is given in Ex. 4. G is known as the Mathieu group on 11 letters, M_{11}. As a remarkable fact, if we regard M_{11} as a permutation group on 12 letters, fixing 12, and we take the group $M_{12} = \{M_{11}, z\}$, where

$$z = (1, 12)\,(2)\,(3)\,(10)\,(11)\,(4, 7)\,(5, 6)\,(8, 9),$$

we find that M_{12} is quintuply transitive of order $12 \cdot 11 \cdot 10 \cdot 9 \cdot 8$, and M_{11} is the subgroup fixing 12.

By arguments similar to those used in constructing M_{11}, we may show that the only quadruply transitive (not alternating or symmetic) groups on less than 35 letters are M_{11}, M_{12}, and the Mathieu groups on 23 and 24 letters, M_{23} and M_{24}, where if

$$\begin{aligned} A &= (0, 1, 2, 3, \cdots, 22) \\ B &= (2, 16, 9, 6, 8)\ (4, 3, 12, 13, 18) \cdot \\ &\quad (10, 11, 22, 7, 17)\ (20, 15, 14, 19, 21) \\ C &= (0, 23)\ (1, 22)\ (2, 11)\ (3, 15)\ (4, 17) \cdot \\ &\quad (5, 9)\ (6, 19)\ (7, 13)\ (8, 20)\ (10, 16) \cdot \\ &\quad (12, 21)\ (18, 14). \end{aligned}$$

Thus, $M_{23} = \{A, B\}$ and $M_{24} = \{A, B, C\} \cdot M_{23}$ is quadruply transitive of degree 23 and order $23 \cdot 22 \cdot 21 \cdot 19 \cdot 16 \cdot 3$, and M_{24} is quintuply transitive, M_{23} being the subgroup of M_{24} fixing 24.

5.9. The Wreath Product. Sylow Subgroups of Symmetric Groups.

Let G and H be permutation groups on sets A and B, respectively. We define the *wreath product* of G by H, written $G \wr H$ in the following way: $G \wr H$ is the group of all permutations θ on $A \times B$ of the following kind:

$$(5.9.1) \qquad (a, b)\theta = (a\gamma_b, b\eta), \quad a \in A,\ b \in B,$$

where for each $b \in B$, γ_b is a permutation of G on A, but for different b's the choices of the permutations γ_b are independent. The permutation η is a permutation of H on B. The permutations θ with $\eta = 1$ form a normal subgroup G^* isomorphic to the direct product of n copies of G, where n is the number of letters in the set B. The factor group $G \wr H/G^*$ is isomorphic to H, and the permutations θ with all $\gamma_b = 1$ form a subgroup isomorphic to H, whose elements may be taken as coset representatives of G^* in G.

The wreath product is associative in the sense that if K is a third permutation group on a set C, then $(G \wr H) \wr K$ and $G \wr (H \wr K)$ are isomorphic, and if we identify the sets $(A \times B) \times C$ and $A \times (B \times C)$ with $A \times B \times C$, then they are identical.

The Sylow subgroups of symmetric group S_n are easily constructed by means of the wreath product. What is the highest power of p dividing $n!$? The factors of $n!$ divisible by p are $p, 2p, 3p, \cdots, kp$, where $k = [n/p]$ is the largest integer not exceeding n/p. Hence $n!$ is divisible by p^k and the further powers of p dividing $k!$. We note that $[k/p] = [n/p^2]$ and continue, finding that the power of p dividing $n!$ is p^M, where

$$M = \left[\frac{n}{p}\right] + \left[\frac{n}{p^2}\right] + \left[\frac{n}{p^3}\right] + \cdots.$$

If we express n in the scale of p,

$$n = a_0p^u + a_1p^{u-1} + \cdots + a_{u-1}p + a_u, \tag{5.9.2}$$

where each a_i is in the range $0 \leq a_i \leq p - 1$, we find that

$$M = a_0(p^{u-1} + p^{u-2} + \cdots + p + 1) + a_1(p^{u-2} + \cdots + p + 1) + \cdots + a_{u-1}. \tag{5.9.3}$$

In particular, a Sylow subgroup of the symmetric group on p^r elements will be of order p^{N_r}, where $N_r = p^{r-1} + p^{r-2} + \cdots + 1$. Thus we see that, having constructed Sylow subgroups for symmetric groups on $p, p^2, \cdots, p^u$ letters, we can easily construct a Sylow subgroup for the symmetric group on n letters, where n is given by (5.9.2). We divide the n letters into a_0 blocks of p^u letters, a_1 or p^{u-1} letters, $\cdots$, a_{u-1} of p letters, and a_u single letters. Then, if in each block we construct the appropriate Sylow subgroup and take the direct product of these, we shall have a group P of order p^M, where M is given by (5.9.3). Hence P will be a Sylow subgroup of S_n.

A Sylow subgroup of S_p on $1, 2, \cdots, p$ will be of order p, and so a Sylow subgroup will be the cyclic group of order p generated by $a_1 = (1, 2, \cdots, p)$. S_{p^2} on $1, 2, \cdots, p^2$ will have a subgroup which is the direct product of the cyclic groups generated by $a_1 = (1, 2, \cdots, p)$, $a_2 = (p + 1, p + 2, \cdots, 2p)$, $\cdots$, $a_p = [p^2 - p + 1, \cdots, p^2]$. If we take a further element of order p, $b = [1, p + 1, 2p + 1, \cdots, p^2 - p + 1]\ (2, p + 2, \cdots), \cdots, (p, 2p, \cdots, p^2)$, then $b^{-1}a_ib = a_{i+1}$, where the subscripts are taken modulo p. Thus b and the a's generate a group P_2 of order p^{p+1}, which is the wreath product of the first cycle of b and the cyclic group $\{a_1\}$. Here P_2 is a Sylow subgroup of S_{p2}. In general let P_r be a Sylow subgroup of S_{p^r} on $1, \cdots, p^r$. Take letters $1, \cdots, p^r, p^r + 1, \cdots, 2p^r, \cdots, p^{r+1}$ as the letters of $S_{p^{r+1}}$. Then, choosing an element

$$c = [1, p^r + 1, 2p^r + 1, \cdots, (p - 1)p^r + 1] \cdots [j, p^r + j, \cdots, (p - 1)p^r + j] \cdots,$$

where j runs 1 to $-p^r$, we have $P_r^{(i)} = c^{-i}P_rc^i$ as a group of order p^{N_r} on the letters $ip^r + 1, \cdots (i + 1)p^r$. As each of $P_r^{(i)}$, $i = 0, 1, \cdots, p - 1$ displaces a distinct set of letters, the group which they generate is their direct product. Here c and P_r generate a group which is of order p^{pN_r+1}. But $pN_r + 1 = p[p^{r-1} + \cdots + (p + 1)] + 1 = N_{r+1}$, and so c and P_r generate P_{r+1}, a Sylow subgroup of the symmetric group on p^{r+1} letters. With P_r acting on letters $1, \cdots, p^r$, and taking c as the cycle $c = (u_0, u_1, \cdots, u_{p-1})$, then the wreath product $P_r \wr \{c\}$ permutes symbols (i, u_j), $i = 1, \cdots, p^r$, $j = 0, \cdots, p - 1$. If we identify (i, u_j) with $i + jp^r$ we see that P_{r+1} as defined before is precisely the wreath product $P_r \wr \{c\}$. Incidentally we note that P_r is generated by r elements of order p.

As an illustration, a Sylow 2-subgroup of S_8 is of order 2^7 and is generated by

$$\begin{aligned} a_1 &= (1, 2), \\ b_1 &= (1, 3)(2, 4), \\ c_1 &= (1, 5)(2, 6)(3, 7)(4, 8). \end{aligned}$$

EXERCISES

1. If an infinite group G has a subgroup H of finite index, show that there is a subgroup $K \subset H$, where K is normal and of finite index in G. (Hint: Represent G as a permutation group on the cosets of H.)
2. Show that there is only one simple group of order 60, namely, the alternating group on five letters.
3. Show that S_4 has two transitive representations on six letters which are both faithful but are not permutation isomorphic.
4. Given the permutations

$$\begin{aligned} u &= (1, 2, 3)\ (4, 5, 6)\ (7, 8, 9), \\ a &= (2, 4, 3, 7)\ (5, 6, 9, 8), \\ b &= (2, 5, 3, 9)\ (4, 8, 7, 6), \\ x &= (1, 10)\ (4, 5)\ (6, 8)\ (7, 9), \\ y &= (1, 11)\ (4, 6)\ (5, 9)\ (7, 8), \\ z &= (1, 12)\ (4, 7)\ (5, 6)\ (8, 9). \end{aligned}$$

Show that $\{u, a, b, x, y\}$ is the Mathieu group M_{11} quadruply transitive of degree 11 and of order $11 \cdot 10 \cdot 9 \cdot 8$ and M_{11}, and that $\{M_{11}, z\}$ is the quintuply transitive Mathieu group M_{12} in which M_{11} is the subgroup fixing 12.

5. Given the permutations

$$\begin{aligned} a &= (0, 1, 2, 3, 4, 5, 6, 7, 8, 9, 10, 11, 12, 13, 14, 15, 16, 17, 18, 19, 20, 21, 22), \\ b &= (2, 16, 9, 6, 8)\ (3, 12, 13, 18, 4)\ (7, 17, 10, 11, 22)\ (14, 19, 21, 20, 15), \\ c &= (0, 23)\ (1, 22)\ (2, 11)\ (3, 15)\ (4, 17)\ (5, 9)\ (6, 19)\ (7, 13)\ (8, 20) \\ &\quad (10, 16)\ (12, 21)\ (14, 18). \end{aligned}$$

Show that $\{a, b\}$ is the quadruply transitive Mathieu group of degree 23, M_{23}, of order $23 \cdot 22 \cdot 21 \cdot 20 \cdot 16 \cdot 3$, and that $M_{24} = \{a, b, c\}$ is the quintuply transitive Mathieu group in which M_{23} is the subgroup fixing 23.

6. AUTOMORPHISMS

6.1. Automorphisms of Algebraic Systems.

In §1.2 we saw that all the 1–1 mappings of any set onto itself form a group. In general those 1–1 mappings of a set S onto itself, which preserve certain properties P, will also form a group.

Let A be a general algebraic system with elements $X = \{x\}$ and operations f_μ such that $f_\mu\ (x_1, \cdots, x_n) = y$ is an element of A whenever $x_1, \cdots, x_n$ are elements of A. There may be arbitrarily many operations, but each operation is a single valued function of a finite number n of elements. The "laws" or "axioms" of A will be relations involving the operations. Then a 1–1 mapping α of X onto itself, $X \leftrightarrows X^\alpha$, is an *automorphism* of A if

$$(6.1.1) \qquad f_\mu(x_1, \cdots, x_n) = y \quad \text{implies} \quad f_\mu(x_1{}^\alpha, \cdots, x_n{}^\alpha) = y^\alpha$$

for every operation f_μ and for each f_μ for all $x_1, \cdots, x_n$. The mapping that is a product of two automorphisms will itself be an automorphism, and with respect to this product, the automorphisms will form a group. In particular the automorphisms of a group form a group. In a group there is a single binary operation, the product operation, and we require that $ab = c$ imply $a^\alpha b^\alpha = c^\alpha$, or more briefly, $(ab)^\alpha = a^\alpha b^\alpha$ for a 1–1 mapping α to be an automorphism.

The automorphisms of algebraic systems are a natural source of groups. Historically the development of group theory arose from the study of the automorphisms of algebraic fields.

6.2. Automorphisms of Groups. Inner Automorphisms.

If $\alpha : x \leftrightarrows x^\alpha$ is a 1–1 mapping of a group G onto itself, α will be an automorphism if, and only if, $ab = c$ implies $a^\alpha b^\alpha = c^\alpha$, or more briefly,

$$(6.2.1) \qquad (ab)^\alpha = a^\alpha b^\alpha.$$

The relation (6.2.1) alone defines an endomorphism, and in §2.4 we have already defined an automorphism as a 1–1 endomorphism. Thus the two definitions of a group automorphism agree.

For a fixed $a \in G$, the mapping A_a in which

$$A_a: x \rightleftarrows a^{-1}xa, \quad \text{all} \quad x \in G, \tag{6.2.2}$$

is in fact one to one, since $axa^{-1} \to a^{-1}(axa^{-1})a = x$. It is an automorphism, since $a^{-1}xya = a^{-1}xa \cdot a^{-1}ya$. The automorphism A_a of G in (6.2.2) is called an *inner automorphism*. Automorphisms of G not of this type are called *outer automorphisms*. Since $b^{-1}(a^{-1}xa)b = (ab)^{-1}x(ab)$, and $a(a^{-1}xa)a^{-1} = x$, we have

$$A_aA_b = A_{ab}; \quad A_{a^{-1}} = A_a^{-1}. \tag{6.2.3}$$

THEOREM 6.2.1. *The inner automorphisms of a group G are a normal subgroup $I(G)$ of the group $A(G)$ of all automorphisms of G. The mapping $a \to A_a$ is a homomorphism of G onto $I(G)$ whose kernel is the center of G.*

Proof: From (6.2.3) the inner automorphisms form a subgroup $I(G)$ of $A(G)$. Let α be any automorphism of G. Then $(a^{-1}xa)^\alpha = (a^\alpha)^{-1}x^\alpha a^\alpha$. Hence $\alpha^{-1}A_a\alpha$ maps x into $(a^\alpha)^{-1}xa^\alpha$, whence $\alpha^{-1}(A_a)\alpha = A_{a^\alpha}$ and so $I(G)$ is a normal subgroup of $A(G)$. From (6.2.3) the mapping $a \to A_a$ is a homomorphism of G onto $I(G)$. Now $A_a = 1$ if, and only if, $xa = ax$ for every $x \in G$. Thus $A_a = 1$ if, and only if, a belongs to the center of G. Thus the kernel of the homomorphism $G \to I(G)$ is the center of G.

A finite Abelian group X is the direct product of its Sylow subgroups (Theorem 3.2.3).

$$X = S(p_1) \times S(p_2) \times \cdots \times S(p_r). \tag{6.2.4}$$

$A(X)$, the group of automorphisms of X, must include the direct product of the automorphism groups $A[S(p_i)]$. But since an automorphism of X must map each of $S(p_i)$, $i = 1, \cdots, r$ onto itself, there can be no further automorphisms, and so

$$A(X) = A[S(p_1)] \times \cdots \times A[S(p_r)]. \tag{6.2.5}$$

More generally, the group of automorphisms of a periodic Abelian group is the Cartesian product of groups of automorphisms of the Sylow subgroups.

The problem of finding the automorphisms of a periodic Abelian group has thus been reduced to finding the automorphisms of an Abelian p-group. Any automorphism of a finite Abelian p-group, A_p, maps a basis onto another basis. Conversely, let $a_1, \cdots, a_s$ and

$b_1, = \cdots, b_s$ be two bases for A_p, arranged as they may be by Theorem 3.3.2 so that a_i is of the same order as b_i, $i = 1, \cdots, s$. Since

(6.2.6)
$$\begin{aligned} A_p &= \{a_1\} \times \{a_2\} \times \cdots \times \{a_s\} \\ &= \{b_1\} \times \{b_2\} \times \cdots \times \{b_s\}, \end{aligned}$$

it follows that the mapping

(6.2.7)
$$a_i \to (a_i)\alpha = b_i, \quad i = 1, \cdots, s$$

determines an automorphism α of A_p.

In the cyclic group C of order p, $C = \{a\}$, $a^p = 1$, every element a^i, $i = 1, \cdots, p - 1$ is a generator. Hence there are $p - 1$ automorphisms determined by $a \to (a)\alpha_i = a^i$. If r is a primitive root† modulo p, then $a \to (a)\beta = a^r$ determines an automorphism β. Here $a \to (a)\beta^j = a^{r^j}$. With r a primitive root, the first power of r such that $r^j \equiv 1 \pmod{p}$ is $j = p - 1$. Hence the automorphism β is of order $p - 1$, and the automorphism group $A(C)$ is cyclic of order $p - 1$ and generated by β.

6.3. The Holomorph of a Group.

Both the right and left regular representations of G are subgroups of the group S_G of all permutations of the elements of G (§1.4). In addition, if α is an automorphism of G, then $\alpha: x \rightleftarrows x^\alpha$ is an element of S_G fixing the identity 1 of G.

Since $(g_1x)g_2 = g_1(xg_2)$, we have $L(g_1)R(g_2) = R(g_2)L(g_1)$. Thus the right and left representations of G permute with each other elementwise.

THEOREM 6.3.1. *Each of the right and left regular representations of G is the centralizer of the other in S_G.*

Proof: Let π be a permutation in S_G belonging to the centralizer of $L(G)$. Let $(1)\pi = g$. Then $\pi R(g)^{-1} = \pi^*$ belongs to the centralizer of $L(G)$ and fixes the identity $(1)\pi^* = 1$. Here $(1)\pi^*L(g') = g'$. Hence also $(1)L(g')\pi^* = g'$, and so $(g')\pi^* = g'$. But g' may be any arbitrary element of G, whence $\pi^* = 1$, and so $\pi \in R(G)$. Hence the centralizer of $L(G)$ is $R(G)$. Similarly, $L(G)$ is the centralizer of $R(G)$.

This disposes of the centralizer of $R(G)$ in S_G. We shall call the normalizer of $R(G)$ in S_G the *holomorph* of G.

---

† For a treatment of primitive roots see Birkhoff and MacLane [1] p. 446, or Hardy and Wright [1] p. 236.

THEOREM 6.3.2. *Let H be the holomorph of G, the normalizer of $R(G)$ in S_G. The subgroup of H fixing the identity of G is the group $A(G)$ of automorphisms of G.*

Proof: Let H be the normalizer of $R(G)$ and let α be an element of H fixing 1. Here $R(g) \leftrightarrows \alpha^{-1}R(g)\alpha$ is surely an automorphism of $R(G)$, since $R(G)$ is a normal subgroup of H. Hence $\alpha^{-1}R(g)\alpha = R(g^\alpha)$ defines a 1–1 mapping $g \leftrightarrows g^\alpha$ of G onto itself. But since $(g_1g_2)^\alpha = g_1{}^\alpha g_2{}^\alpha$ under this mapping, $g \leftrightarrows g^\alpha$ is an automorphism of G. But α is in fact the permutation $g \leftrightarrows g^\alpha$. Since $(1)\alpha = 1$ and $\alpha^{-1}R(g)\alpha = R(g^\alpha)$, we have $(1)\alpha R(g^\alpha) = g^\alpha$ and also $(1)R(g)\alpha = g^\alpha$, whence $(g)\alpha = g^\alpha$. Thus, if α belongs to H and fixes 1, then α is an automorphism of G. Conversely, let $\alpha: g \leftrightarrows g^\alpha$ be an automorphism of G. Then α is an element of S_G fixing the identity 1 of G. We may now verify that $\alpha^{-1}R(g)\alpha = R(g^\alpha)$, whence α belongs to the normalizer of $R(G)$. For $(x)R(g)\alpha = x^\alpha g^\alpha$ and also $(x)\alpha R(g^\alpha) = x^\alpha g^\alpha$. Thus the subgroup of H fixing 1 consists entirely of automorphisms and contains every automorphism. In the proof of Theorem 6.3.1 we showed that only the identity of S_G fixes 1 and permutes with every element of $R(G)$. Hence every automorphism of G occurs exactly once in the subgroup of H fixing 1 whence this subgroup is $A(G)$. Since the normalizer of a group includes its centralizer it follows that $H \supset L(G)$.

6.4. Complete Groups.

DEFINITION: *A complete group is a group whose center is the identity and all of whose automorphisms are inner automorphisms.*

THEOREM 6.4.1. *Let G be a complete group which is a normal subgroup of a group T. Then T is the direct product $G \times K$ of G and the centralizer K of G in T.*

Proof: Let

$$T = G + Gx_2 + \cdots + Gx_i + \cdots. \tag{6.4.1}$$

Here $x_i^{-1}Gx_i = G$ since G is normal in T. Thus $g \leftrightarrows x_i^{-1}gx_i = g^\alpha$ is an automorphism of G. Since every automorphism of G is an inner automorphism, $g^\alpha = a^{-1}ga$ for some $a \in G$ and all g. Hence $x_i^{-1}gx_i = a^{-1}ga$ for all g. Here $y_i = x_ia^{-1}$ belongs to the centralizer K of G in T. But $Gx_i = x_iG = x_ia^{-1}G = y_iG = Gy_i$, and we may take y_i as the coset representative of G. Thus every coset of G in T contains an element of K. Hence $T = G \cup K = GK = KG$ as G is normal. But $K \cap G = 1$, since the center of G is the identity. Hence $T = G \times K$, since every element of K permutes with every element of G.

Corollary 6.4.1. *The holomorph H of a complete group G is the direct product $R(G) \times L(G)$.*

This follows since $L(G)$ is the centralizer of $R(G)$ in H.

6.5. Normal or Semi-direct Products.

Theorem 6.5.1. *Given two groups H and K and for every element $h \in H$ an automorphism of K,*

$$k \leftrightarrows k^h \quad \text{all} \quad k \in K, \tag{6.5.1}$$

such that

$$(k^{h_1})^{h_2} = k^{h_1h_2}, \ h_1, h_2 \in H. \tag{6.5.2}$$

Then the symbols $[h, k]$, $h \in H$, $k \in K$ form a group under the product rule

$$[h_1, k_1]\cdot[h_2, k_2] = [h_1h_2, k_1^{h_2}k_2], \tag{6.5.3}$$

called the normal product of K by H or the semi-direct product of K by H.

Proof: Since for every k and h, $k^h \in K$, the product rule (6.5.3) is well defined.

1) The product rule (6.5.3) is associative, since

$$\begin{aligned} &([h_1, k_1]\cdot[h_2, k_2])\cdot[h_3, k_3] \\ &= [h_1h_2, k_1^{h_2}k_2]\cdot[h_3, k_3] \\ &= [(h_1h_2)h_3, (k_1^{h_2}k_2)^{h_3}k_3] \\ &= [h_1h_2h_3, k_1^{h_2h_3}k_2^{h_3}k_3], \end{aligned} \tag{6.5.4}$$

using (6.5.1) and (6.5.2).

$$\begin{aligned} &[h_1, k_1]\cdot([h_2, k_2]\cdot[h_3, k_3]) \\ &= [h_1, k_1]\cdot[h_2h_3, k_2^{h_3}k_3] \\ &= [h_1h_2h_3, k_1^{h_2h_3}k_2^{h_3}k_3]. \end{aligned} \tag{6.5.5}$$

2) The element $[1, 1]$ is the identity, since

$$[1, 1][h, k] = [1h, 1^hk] = [h, k],$$
$$[h, k][1, 1] = [h1, k^11] = [h, k].$$

Here $k^1 = k$ because of (6.5.2).

3) An arbitrary $[h, k]$ has a left inverse $[h^{-1}, (k^{-1})^{h^{-1}}]$

$$[h^{-1}, (k^{-1})^{h^{-1}}] \cdot [h, k] = [h^{-1}h, k^{-1}k] = [1, 1]. \tag{6.5.6}$$

Hence the symbols $[h, k]$ with the product rule (6.5.3) form a group G.

THEOREM 6.5.2. *If G is the normal product of K by H, then the elements $[h, 1]$ of G form a subgroup isomorphic to H and the elements $[1, k]$ form a normal subgroup isomorphic to K. Moreover, the automorphism* (6.5.1) *of K as a subgroup of G is induced by transformation by the element $h = [h, 1]$ of H as a subgroup of G, since*

$$[h, 1]^{-1}[1, k][h, 1] = [1, k^h]. \tag{6.5.7}$$

Moreover, $G = H \cup K$, since

$$[h, 1][1, k] = [h, k]. \tag{6.5.8}$$

Proof: We have only to observe that $h \leftrightarrows [h, 1]$ and $k \leftrightarrows [1, k]$ are isomorphisms between H and K and subgroups of G, using the rule (6.5.3) and noting that $k^1 = k$. Also, (6.5.7) and (6.5.8) follow directly from the rule (6.5.3). Here (6.5.7) shows that K is a normal subgroup and that the automorphism (6.5.1) is induced by transformation by the element $h = [h, 1]$. Here $H \cap K = [1, 1] = 1$, and (6.5.8) shows that the elements of H may be taken as coset representatives of K.

THEOREM 6.5.3. *G is the normal product of K by H if, and only if, K is a normal subgroup of G and H is a subgroup of G whose elements may be taken as the coset representatives of K. Otherwise expressed*

1) *K is a normal subgroup of G.*
2) *H is a subgroup of G.*
3) *$K \cap H = 1$.*
4) *$H \cup K = G$.*

Proof: We have already observed that these properties hold if G is the normal product of K by H. Conversely, suppose these properties hold. Then from $K \cap H = 1$, $H \cup K = G$, with K normal in G, it follows (Theorem 2.3.3) that every element of G has a unique representation of the form

$$g = hk. \tag{6.5.9}$$

Since K is normal,

$$h^{-1}kh = k^h \in K, \tag{6.5.10}$$

and clearly $k \leftrightarrows k^h$ is an automorphism of K. Moreover, from (6.5.10) we have

$$(k^{h_1})^{h_2} = k^{h_1h_2}. \tag{6.5.11}$$

For the product of two elements of G,

$$\begin{aligned} g_1 &= h_1k_1, \quad g_2 = h_2k_2, \\ g_1g_2 &= h_1k_1h_2k_2 \\ &= h_1h_2(h_2^{-1}k^1h_2)k_2 \\ &= h_1h_2 \cdot k_1^{h_2}k_2, \end{aligned} \tag{6.5.12}$$

and so the rule for the product in G is precisely the same as (6.5.3), and G is the normal product of K by H.

We observe that the association of an automorphism of K with an element of H is a homomorphism of H into the group of automorphisms of K. If H is mapped into the identical automorphism of K, i.e., $k^h = k$ for every h, k, then the rule (6.5.3) is that for the direct product of H and K.

EXERCISES

1. Show that the dihedral group of order 8 is isomorphic to its group of automorphisms.
2. Show that the group of automorphisms of the elementary Abelian group of order p^r is of order $(p^r - 1)(p^r - p) \cdots (p^r - p^{r-1})$.
3. Find an outer automorphism of the symmetric group on six letters, S_6. This will interchange the two classes of elements of order 3.
4. Show that if the order of a group is divisible by p^2, the square of a prime, then the order of its group of automorphisms is divisible by p. (Hint: If there is no inner automorphism of order p, show that a Sylow p-subgroup is Abelian and a direct factor of G.)
5. An automorphism α of a group G is called a central automorphism if for every x of G, $x^{-1}(x)\alpha \in Z$, where Z is the center of G. Show that the group of central automorphisms, which are inner automorphisms of G, is isomorphic to the center of G/Z.
6. Let G be the group generated by elements a, b, c, with defining relations $a^8 = b^8 = c^4 = 1$, $b^{-1}ab = a^5$, $c^{-1}ac = a^5$, $c^{-1}bc = a^6b$. Show that $\{a, b\}$ is the normal product of $\{a\}$ by $\{b\}$, and that G is the normal product of $\{a, b\}$ by $\{c\}$. Hence conclude that these relations define a group of order 256 whose elements are of the form $a^ib^jc^k$.
7. Let G be the group of Ex. 6. Show that α: $a \to a^5$, $b \to b$, $c \to c$ is an outer automorphism of G which takes each conjugate class of G into itself.

7. FREE GROUPS

7.1. Definition of Free Group.

Suppose we are given a set of elements $S = s_1, \cdots s_n$, where it is not assumed that the elements $s_1 \cdots s_n$ are finite in number or even countable. But whenever it is desirable, we shall assume that the indices i of the s_i are well ordered. We now define symbols s_i^1, s_i^{-1} where $s_i^1 = s_i$ and s_i^{-1} is a new symbol.

A *word* or *string* is either void (written 1) or a finite succession $a_1a_2 \cdots a_t$, where each a_i is one of the s_j^ϵ, $\epsilon = \pm 1$.

A word is a *reduced word* if it is void or if in $a_1 \cdots a_t$ no pair a_ia_{i+1}, $i = 1 \cdots t - 1$ is of the form $s_j^\epsilon s_j^{-\epsilon}$, $\epsilon = \pm 1$.

Two words f_1 and f_2 are *adjacent* if they are of the form $f_1 = gs_j^\epsilon s_j^{-\epsilon}h$, $f_2, = gh$. Each of f_1 and f_2 is *adjacent* to the other.

Two words f and g are *equivalent* written $f \sim g$ if $f_1 = f$, $f_2 \cdots f_m = g$ exist such that f_i and f_{i+1} are adjacent for $i = 1 \cdots m - 1$. Clearly $f \sim g$ is a true equivalence relation. All words equivalent to f form a class which we may designate as $[f]$.

LEMMA 7.1.1. *Any class contains one, and only one, reduced word.*

Proof: If $f = a_1 \cdots a_t$ contains any $a_ia_{i+1} = s_j^\epsilon s_j^{-\epsilon}$, then there is a word adjacent to f, $a_1 \cdots a_{i-1}a_{i+2} \cdots a_t$ involving fewer symbols. After successive reductions we shall find in, at most, $t/2$ steps a reduced word equivalent to f. This shows that $[f]$ contains at least one reduced word.

Now, for $f = a_1a_2 \cdots a_t$ we define the W-process

$$\begin{aligned} W_0 &= 1 && \text{the void word} \\ W_1 &= a_1 \\ W_{i+1} &= W_ia_{i+1} && \text{if } W_i \text{ is } \textit{not} \text{ of the reduced form } Xa_{i+1}^{-1} \\ &= X && \text{if } W_i \text{ is of the reduced form } Xa_{i+1}^{-1}. \end{aligned}$$

Now, by induction, it is seen that $W_0, W_1, \cdots, W_t$ are all of reduced form and that $W_t = f$ if f is in reduced form. Now, if

$$\begin{aligned} f_1 &= a_1 \cdots a_ra_{r+1} \cdots a_t, \\ f_2 &= a_1 \cdots a_rs_j^\epsilon s_j^{-\epsilon}a_{r+1} \cdots a_t, \end{aligned}$$

let $W_0^1, W_1^1, \cdots, W_t^1$ be the words of the W-process for f_1 and $W_0^2, \cdots, W_{t+2}^2$ be the words for f_2. We want to show that $W_t^1 = W_{t+2}^2$. Now $W_0^1 = W_0^2 \cdots W_r^1 = W_r^2$, since the processes are identical. Consider two cases:

1) $W_r^1 = W_r^2$ is of the reduced form $Xs_j^{-\epsilon}$. Since $Xs_j^{-\epsilon}$ is in reduced form, X is not of reduced form Ys_j^{ϵ}. Here, for f_2, $W_{r+1}^2 = X$, $W_{r+2}^2 = Xs_j^{-\epsilon} = W_r^2 = W_r^1$.

2) $W_r^1 = W_r^2$ is not of the reduced form $Xs_j^{-\epsilon}$. Here $W_{r+1}^2 = W_r^2 s_j^{\epsilon}$, $W_{r+2}^2 = W_r^2 = W_r^1$.

Hence in both cases $W_{r+2}^2 = W_r^1$, and so inductively, $W_{r+2+i}^2 = W_{r+i}^1$, since the processes are identical. Thus the W-process yields the same reduced word for any two adjacent words, and hence for any two equivalent words. But also the W-process leaves a reduced word unchanged. Hence there cannot be two distinct reduced words in the same class.

We may define a multiplication for these classes of words, and under this definition these classes form a group which we shall call the free group F generated by S.

THEOREM 7.1.1. *For any two classes,* $[f_1]$, $[f_2]$ *of words on* S *define their product* $[f_1][f_2] = [f_1f_2]$. *This product is well defined, and with respect to this product, all the classes of words on* S *form a group, the free group* F *generated by* S.

Proof: Suppose $f_1 \sim f_1'$, $f_2 \sim f_2'$. Then $f_1f_2 \sim f_1'f_2'$ since we may first show $f_1f_2 \sim f_1'f_2$ by replacing f_1 in turn by the words adjacent to it which lead to f_1'. Similarly, $f_1'f_2 \sim f_1'f_2'$, whence $f_1f_2 \sim f_1'f_2'$, whence $[f_1'f_2'] = [f_1f_2]$, and so the product $[f_1f_2] = [f_1][f_2]$ depends only on the class of f_1 and f_2 and not on the particular representatives. The void word is the identity for this product, as $[1][f] = [f][1] = [f]$. Moreover, if $f = a_1 \cdots a_t$ and $h = a_t^{-1} \cdots a_1^{-1}$, then $[f][h] = [fh] = [1]$ and $[h][f] = [hf] = [1]$. Hence $[a_t^{-1} \cdots a_1^{-1}]$ is the inverse of the class $[a_1 \cdots a_t]$. We find moreover that $([f_1][f_2])[f_3] = [f_1f_2f_3] = [f_1]([f_2][f_3])$, whence the associative law holds. Thus the classes of words form a group, called the free group F generated by S. We may write F_S to indicate the generators.

It is convenient to write $f_1 = f_2$ if the two words are equivalent and hence represent the same element of F. We shall write $f_1 \equiv f_2$ to indicate that f_1 and f_2 are the same word. Naturally it is usually convenient to represent an element of F in its reduced form. Thus if $f = a_1 \cdots a_t$ is in reduced form, we say f is reduced as written.

In any group G a set of elements $X: x_1, \cdots, x_n$ generate a subgroup H consisting of all finite products $b_1 b_2 \cdots b_t$, each b_j being some x_j^ϵ, $\epsilon = \pm 1$. There is no difficulty in verifying that these finite products do form a subgroup. In general an element of H may be written in many ways as such a finite product. Moreover, it is trivial that all elements of G generate G. Hence every group G may be regarded as generated by a set of elements X, and we write $G = \{X\}$. The following theorem shows why free groups are interesting not merely in their own right, but also as a tool in the study of all groups.

THEOREM 7.1.2. *Let the group G be generated by a set of elements $X: x_1, \cdots, x_n$. Then if F is the free group generated by $S: s_1, \cdots, s_n$, there is a homomorphism $F \rightarrow G$ determined by $s_i \rightarrow x_i$ all i.*

Proof: Let $f = a_1 \cdots a_t$ be any word of S. Consider the element $g = b_1 \cdots b_t \in G$, where $b_i = x_j^\epsilon$ if $a_i = s_j^\epsilon$. Then $f \rightarrow g$ maps every word of S onto an element of G. Clearly adjacent, and therefore equivalent, words of S are mapped onto the same element of G. Hence the mapping $f \rightarrow g$ is in fact a mapping of elements of F onto elements of G. Moreover, if $f_1 \rightarrow g_1$, $f_2 \rightarrow g_2$, then $f_1 f_2 \rightarrow g_1 g_2$. Hence the mapping $s_i \rightarrow x_i$ determines a homomorphism of F onto G. From the theorems on homomorphisms we have the corollary:

COROLLARY 7.1.1. *Every group G given as generated by a set X is the factor group of a free group F with the same number of generators.*

As an alternate definition of a free group we may take the following:

DEFINITION: *The free group F generated by a set S of elements is the group with the following properties:*

1) *F is generated by S.*

2) *If G is any group generated by a set of elements X and if there is a one-to-one correspondence between S and X, $S \leftrightarrows X$, then there is a homomorphism of F onto G, $F \rightarrow G$ taking S onto X.*

In light of Theorem 7.1.2, this is a valid definition. From the previous definition the free group F_S satisfies these requirements. Moreover, if F' is a group generated by S and $F' \rightarrow F_S$, this homomorphism must be an isomorphism, since the only element of F' which can be mapped onto the identity is the identity.

However, there seem to be several disadvantages in this definition. It is not a constructive definition and it does not make clear, without the constructive process given above, that any group with properties 1 and 2 exists, nor does it make clear that if such a group exists, no nontrivial relations hold. Moreover, on broader grounds, the concept of a "free" system, a system in which no relations hold save those implied by the axioms, is tenable even though no theorem analogous to Theorem 7.1.2 may hold.

7.2. Subgroups of Free Groups. The Schreier Method.

The nature of subgroups is always fundamental in the study of groups, and for free groups, from Theorem 7.1.2, the normal subgroups are of particular interest. It was proved by Nielsen [1] and Schreier [3] that subgroups of free groups are themselves free groups. Nielsen's proof held only for finitely generated subgroups, but Nielsen's proof has been extended by Levi [1] and others so as to avoid this restriction. Nielsen's method works directly with the elements of the subgroup, Schreier's with the cosets of the subgroup. The first proof given here* is a simplification of the Schreier proof.

A set G of elements of a free group F is said to be a *Schreier system* if for each $g \in G$:

1) $g = a_1a_2 \cdots a_t$ is reduced as written.
2) $a_1a_2 \cdots a_{t-1}$ is also a $g \in G$.

We say that G is a *two-sided Schreier system* if in addition to 1 and 2 the following also holds:

3) $a_2 \cdots a_t$ is also a $g \in G$.

Note that a Schreier system always contains the identity.

Let F be the free group generated by S and let U be a subgroup of F. Consider the decomposition of F into left cosets of U:

$$F = U\cdot 1 + Ug_2 + \cdots + Ug_i + \cdots. \tag{7.2.1}$$

We shall always choose the identity as the representative of U itself. We find that it is advantageous to choose the representatives of the remaining cosets so that they will satisfy certain relations.

* See Hall and Rado [1]. For further results see M. Hall [4, 5].

LEMMA 7.2.1 (EXTENDED LEMMA OF SCHREIER). *If U is a subgroup of the free group F, it is possible to choose the representatives of the left cosets of U as a Schreier system. If U is a normal subgroup of F, it is possible to choose the representatives as a two-sided Schreier system.*

Proof: Let the generators of F, $S: s_1 \cdots s_n$ and their inverses be well ordered in any way; for example, $s_1 < s_1^{-1} < s_2 < s_2^{-1} \cdots < s_n < s_n^{-1}$ if the number n is finite. But it is not to be assumed that the set S is finite or even countable, merely that the set $S \cup S^{-1}$ may be well ordered.

This ordering of $S \cup S^{-1}$ may be extended to yield an *alphabetical ordering* for all the elements of F. If we have two elements of F, f and g, then we define $f < g$ in the alphabetical ordering if the reduced forms of f and g are

$$f = a_1 \cdots a_t,$$
$$g = b_1 \cdots b_u,$$

where the a_i and b_j belong to $S \cup S^{-1}$, and one of the following holds:

1) $t < u$.
2) $t = u$, $a_1 < b_1$.
3) $t = u$; $a_1 = b_1, \cdots, a_i = b_i$; $a_{i+1} < b_{i+1}$.

The alphabetical ordering so defined is clearly a simple order, indeed a well ordering, and the following useful properties hold:

If $f < g$ and gh is reduced as written, then $fh < gh$. If $f < g$ and hg is reduced as written, then $hf < hg$. This may be verified from the definition of the ordering.

To prove the lemma, let us choose the representative g_i of the coset Ug_i as that element of the coset earliest in the alphabetical ordering of F. Then we assert that the g_i form a Schreier system, and in fact a two-sided Schreier system, if U is a normal subgroup. Since the identity is the first element of F, the identity is chosen as the representative of the subgroup U. Let $g = a_1 \cdots a_{t-1}a_t$ be the representative of the coset Ug, being the earliest element in this coset. Let h be the earliest element in the coset containing $h^* = a_1 \cdots a_{t-1}$. If $h = b_1 \cdots b_u$, then $h \leq a_1 \cdots a_{t-1}$. But $ha_t \in Ug$, and so $g \leq ha_t$. But also $ha_t \leq a_1 \cdots a_{t-1}a_t = g$. Thus $g = ha_t$ and, so, $h = h^* = a_1 \cdots a_{t-1}$ is also a coset representative. Thus the g's form a Schreier system. If U is a normal subgroup, let $a_2 \cdots a_t$ be in the coset $Uf = fU$ whose earliest element is f. Then $f \leq a_2 \cdots$

a_t, and $a_1 f$ belongs to the same coset as $a_1 \cdots a_t U = gU = Ug$. Then $g \leq a_1 f$. But also $a_1 f \leq a_1 a_2 \cdots a_t = g$. Thus $g = a_1 f$ and $f = a_2 \cdots a_t$. Hence the g's form a two-sided Schreier system. Note that the lemma merely guarantees the existence of a Schreier system of left coset representatives. But the same subgroup may possess more than one set of coset representatives which is a Schreier system.

THE MAIN THEOREM: THEOREM 7.2.1. *Every subgroup of a free group is a free group.*

Let F be the free group generated by the set S, and let U be a given subgroup of F. Then by the Schreier lemma we may suppose the left coset representatives G to be a Schreier system.

$$(7.2.2) \qquad F = U \cdot 1 + Ug_2 + \cdots + Ug_i + \cdots.$$

We begin with a lemma which is true for any group F whether a free group or not. Let F be generated by a set of elements S, let U be a subgroup of F, and let (7.2.2) be the decomposition of F into left cosets of U.

If an element f of F belongs to the coset Ug_i in (7.2.2), let us define a function $\Phi(f)$ by putting $\Phi(f) = g_i$. Note that $\Phi(uf) = \Phi(f)$ if $u \in U$. $\Phi(f) = 1$ if, and only if, $f \in U$.

Suppose $f = a_1 a_2 \cdots a_t$ with each $a_i \in S \cup S^{-1}$. Write $f_0 = 1$, $f_1 = a_1, f_2 = a_1 a_2, \cdots, f_t = a_1 a_2 \cdots a_t = f$. Then write $h_0 = \Phi(f_0) = 1$, $h_1 = \Phi(f_1), \cdots, h_t = \Phi(f)$. Then, identically,

$$(7.2.3) \qquad \begin{aligned} fh_t^{-1} &= h_0 a_1 h_1^{-1} \cdot h_1 a_2 h_2^{-1} \cdot h_2 \cdots h_{t-1}^{-1} \cdot h_{t-1} a_t h_t^{-1} \\ &= f, \quad \text{if } f \in U, \end{aligned}$$

since then $h_t = 1$.

Now, since $h_i = \Phi(h_i) = \Phi(f_i) = \Phi(f_{i-1} a_i) = \Phi(h_{i-1} a_i) = \Phi(h_{i-1} s^\epsilon{}_\alpha)$, $a_i = s_\alpha{}^\epsilon$, $\epsilon = \pm 1$, $h_i \in G$, it is clear that in (7.2.3) we need only the function Φ for arguments of the form gs^ϵ, $g \in G$, $s^\epsilon \in S \cup S^{-1}$. Let us then write $\phi(gs^\epsilon) = \Phi(gs^\epsilon)$ so that $\phi(f)$ is defined only for arguments $f = gs^\epsilon$.

LEMMA 7.2.2. *In any group F the elements $gs\phi(gs)^{-1}$ are generators of the subgroup U, where g runs over the left coset representatives of U in* (7.2.2); *s over the generators of F and $\phi(gs^\epsilon)$ is the representative of the coset containing gs^ϵ.*

Proof: If $f \in U$, then $h_t = 1$ and (7.2.3) expresses f as a product of elements $h_{i-1}a_ih_i^{-1}$, and since $h_i = \Phi(h_{i-1}a_i)$, then $h_{i-1}a_ih_i^{-1}$ is of the form $gs^\epsilon\phi(gs^\epsilon)^{-1}$, with $h_{i-1} = g$ and $a_i = s^\epsilon$, since then $h_i = \phi(gs^\epsilon)$. But $gs^\epsilon \in U\phi(gs^\epsilon)$, whence for any gs^ϵ the element $gs^\epsilon\phi(gs^\epsilon)^{-1} \in U$. Note that if $\phi(g_js^\epsilon) = g_k$, then $\phi(g_ks^{-\epsilon}) = g_j$. Hence, if $g_js^\epsilon\phi(g_js^\epsilon)^{-1} = g_js^\epsilon g_k^{-1}$, its inverse is $g_ks^{-\epsilon}g_j^{-1} = g_ks^{-\epsilon}\phi(g_ks^{-\epsilon})^{-1}$, which is of the same form with the opposite sign for the exponent of s. Hence the elements $gs\phi(gs)^{-1}$ generate U.

Corollary 7.2.1. *If F is a finitely generated group and U is of finite index in F, then U is finitely generated.*

This follows since there are only a finite number of choices for g and s in $gs\phi(gs)^{-1}$.

From here on we shall assume that F is a free group and that the coset representatives G form a Schreier system.

We shall use the following properties of the function $\phi(gs^\epsilon)$:

1) $\phi(gs^\epsilon) \in G$.

2) If $gs^\epsilon \in G$, then $\phi(gs^\epsilon) = gs^\epsilon$.

3) $\phi[\phi(gs^\epsilon)s^{-\epsilon}] = g$.

As a generic notation let us write $v = gs^\epsilon\phi(gs^\epsilon)^{-1}$ and $u = gs\phi(gs)^{-1}$. Thus a u is a v with exponent $+1$ and a v is either a u or the inverse of a u, for if $v = g_is^{-1}\phi(g_is^{-1})^{-1}$, put $\phi(g_is^{-1}) = g_j$. Then, by the third property, $v^{-1} = g_jsg_i^{-1} = g_js\phi(g_js)^{-1}$ is a u, and also, similarly, the inverse of a u is a v.

Lemma 7.2.3. *A $v = gs^\epsilon\phi(gs^\epsilon)^{-1}$ is either reduced as written or equal to* 1.

Proof: Let $v = g_js_a^\epsilon\phi(g_js_a^\epsilon)^{-1} = g_js_a^\epsilon g_k^{-1}$, where $g_k = \phi(g_js_a^\epsilon)$. Both g_j and g_k^{-1} are reduced as written. Hence, if there is any cancellation in v, either (1) g_j ends in $s_a^{-\epsilon}$, or (2) g_k^{-1} begins with $s_a^{-\epsilon}$. If (1) holds, $g_j = a_1 \cdots a_{t-1}s_a^{-\epsilon}$ is the reduced form of g_j, whence $g_js_a^\epsilon = a_1 \cdots a_{t-1}$ is a g, and by property 2 of the ϕ function, $g_k = \phi(g_js_a^\epsilon) = g_js_a^\epsilon$; so, $v = g_js_a^\epsilon g_k^{-1} = 1$. If (2) holds, then similarly $g_j = \phi(g_ks_a^{-\epsilon}) = g_ks_a^{-\epsilon}$, and again $v = 1$.

For a $v = gs^\epsilon\phi(gs^\epsilon)^{-1} \neq 1$, let us call the factor s^ϵ the *significant factor* of v. Suppose $v = g_js_a^\epsilon\phi(g_js_a^\epsilon)^{-1} = g_ks_b^\eta\phi(g_ks_b^\eta)^{-1} \neq 1$. If g_j and g_k are of the same length, then since v is reduced as written, $g_j = g_k$, $s_a^\epsilon = s_b^\eta$. If g_j and g_k are of different lengths, say, g_k longer, then $g_js_a^\epsilon$ as a beginning section of g_k is itself a g; so, $\phi(g_js_a^\epsilon) = g_js_a^\epsilon$, and so $v = 1$ contrary to assumption. Thus a $v \neq 1$ has a unique expression of the form $gs^{\epsilon}\phi(gs^{\epsilon})^{-1}$ and in particular has a unique significant factor.

LEMMA 7.2.4. *In a product* v_1v_2, $v_1 \neq 1$, $v_2 \neq 1$, $v_2 \neq v_1^{-1}$, *the cancellation does not reach the significant factor of either* v.

Proof: Let $v_1 = g_i s_a^{\epsilon} g_j^{-1}$, $g_j = \phi(g_i s_a^{\epsilon})$, $v_2 = g_k s_b^{\eta} g_l^{-1}$, $g_l = \phi(g_k s_b^{\eta})$. v_1 and v_2 are both reduced as written, and since $v_2 \neq v_1^{-1}$, we cannot have both $g_k = g_j$ and $s_b^{-\eta} = s_a^{\epsilon}$ holding. Let us deny the lemma and assume that the cancellation reaches a significant factor. If the cancellation reaches s_b^{n} first, then gs_b^{n} is a beginning section of g_j, whence $\phi(gs_b^{n}) = gs_b^{n}$ and $v_2 = 1$, contrary to assumption. Similarly, if the cancellation reaches s_a^{ϵ} first, then $g_j s_a^{-\epsilon}$ is a beginning section of g_k and $v_1 = 1$, contrary to assumption. If the cancellation includes s_a^{ϵ} and s_b simultaneously, then $g_k = g_j$, $s_b^{n} = s_a^{-\epsilon}$ and $v_2 = v_1^{-1}$, contrary to assumption.

We are now close to the proof of the main theorem.

LEMMA 7.2.5. *A product of v's,* $v_1v_2 \cdots v_m$, $v_i \neq 1$, $i = 1 \cdots m$, $v_{i+1} \neq v_i^{-1}$, $i = 1, \cdots, m - 1$ *cannot be the identity.*

Proof: By the repeated application of Lemma 7.2.4, the cancellation between v_i and v_{i+1} does not reach either significant factor. Hence, when put in its reduced form in terms of the s's, the product $v_1 \cdots v_m$ contains all the original significant factors and cannot be the identity.

Now consider the elements u, $u = gs\phi(gs)^{-1} \neq 1$. From Lemma 7.2.2, all the u's and so the u's $\neq 1$ generate U. The u's will be *free* generators of U if no product of u's which is a reduced word in the u's is equal to the identity, i.e., reduces to 1 when expressed in terms of the s's. But every $v \neq 1$ is either a u or u^{-1} and in just one way. Hence a reduced word in the u's $\neq 1$ will be of the form $v_1v_2 \cdots v_m$, $v_i \neq 1$, $v_{i+1} \neq v_i^{-1}$ treated in Lemma 7.2.5, and therefore will not be the identity. Hence Lemma 7.2.6 will hold.

LEMMA 7.2.6. *The elements* $u = gs\phi(gs)^{-1} \neq 1$ *are free generators of* U.

Thus we have found free generators for U, and so U is a free group.

The role of the significant factor in Lemma 7.2.4 is the key to this proof of Theorem 7.2.1. We may generalize this idea by an independent definition of significant factor.

A set Y of elements such that $Y \cap Y^{-1} = 0$ is said to possess significant factors if for each $y \in Y$ we may select a factor from the reduced form of y and y^{-1}:

$$\begin{aligned} y &= a_1 \cdots a_i \cdots a_t, \\ y^{-1} &= a_t^{-1} \cdots a_i^{-1} \cdots a_1^{-1}, \end{aligned}$$

selecting a_i from y and a_i^{-1} from y^{-1} in such a way that in any product

$$zw,\ z \neq w^{-1},\ z,\ w \in Y \cup Y^{-1},$$

the cancellation does not reach the significant factor in z or w. In other words Y possesses significant factors if Lemma 7.2.4 is valid in $Y \cup Y^{-1}$ for these factors. The significant factors for a set Y are said to be *central significant factors* if for a y of odd length the significant factor is its central term and for a y of even length the significant factor is one of the two central terms.

THEOREM 7.2.2. *If a set Y possesses significant factors, then Y consists of free generators for the subgroup generated by its elements. If G is a Schreier system with each g a shortest element in its coset Ug, then for the u's with $u = gs\phi(gs)^{-1} \neq 1$, the s's form a set of central significant factors.*

Proof: By definition of the significant factor, Lemma 7.2.4 holds for $v_1, v_2 \in Y \cup Y^{-1}$. But then Lemma 7.2.5 holds also. Hence no word in the y's and their inverses is the identity unless its reduced form in the y's is the identity; thus the y's are free generators of the group which they generate.

If G is a Schreier system of coset representatives g for a subgroup U such that a coset Ug contains no element shorter than g, then since

$$\begin{aligned} gs &\in U\phi(gs), \\ \phi(gs)s^{-1} &\in Ug, \end{aligned}$$

we see that g and $\phi(gs)$ can differ in length by, at most, one. Thus s, which in Lemma 7.2.4 has already been shown to be a significant factor, is a central significant factor, since in $u = gs\phi(gs)^{-1}$ it is between two words of length differing, at most, by one.

We may prove a converse to Lemma 7.2.6 and the main theorem.

THEOREM 7.2.3. *Let G be a Schreier system in a free group F generated* *by a set S of free generators. Let $\phi(h)$ be a function defined for arguments $h = gs^\epsilon$, $\epsilon = \pm 1$, $g \epsilon G$, $s \epsilon S$ such that*

1) $\phi(gs^\epsilon) \epsilon G$.
2) *If* $gs^\epsilon \epsilon G$, *then* $\phi(gs^\epsilon) = gs^\epsilon$.
3) $\phi[\phi(gs^\epsilon)s^{-\epsilon}] = g$.

Then the elements $u = gs\phi(gs)^{-1} \neq 1$ are free generators of a subgroup U of F, and the Schreier system is a set of representatives of left cosets of U in F.

Proof: Let us write $v = gs^\epsilon\phi(gs^\epsilon)^{-1}$ as a generic notation. The proofs of Lemmas 7.2.3, 7.2.4, and 7.2.5 are valid under the hypotheses given here, since only the preceding properties of the ϕ function were used in the proof of these lemmas. From Lemma 7.2.5 it follows that the elements $u = gs\phi(gs)^{-1} \neq 1$ are free generators of some subgroup U of F.

In order to show that the Schreier system of G is a set of representatives of the left cosets of U, we define a function $\Phi(f)$ for every *word* f in $S \cup S^{-1}$. Suppose

$$f = a_1a_2 \cdots a_t, \quad a_i \epsilon S \cup S^{-1}, \quad i = 1 \cdots t.$$

Put

$$\begin{aligned} h_0 &= 1, \\ h_i &= \phi(h_{i-1}a_i),\ i = 1 \cdots t, \\ h_t &= \Phi(f). \end{aligned}$$

The essential properties of $\Phi(f)$ are easily proved:

1) $\Phi(a_1 \cdots a_ia_{i+1} \cdots a_k) = \Phi(a_1 \cdots a_is^\epsilon s^{-\epsilon}a_{i+1} \cdots a_t)$.

By definition every h_i, $i = 1 \cdots t$ is a $g \epsilon G$. Hence in evaluating the right hand side we have successively h_i, $\phi(h_is^\epsilon)$, and $\phi[\phi(h_is^\epsilon)s^{-\epsilon}] = h_i$ by property (3). Otherwise the process is identical in evaluating both sides. Thus $\Phi(f)$ is the same for any two words representing the same element of F.

2) $\Phi(g) = g$.

Here if $g = a_1 \cdots a_t$ is the reduced form of a $g \epsilon G$, every beginning section is also a g and, by property (2), $h_i = a_1 \cdots a_i$, $i = 1 \cdots t$.

3) $\Phi(f_1f_2) = \Phi[\Phi(f_1)f_2]$.

Write $f = f_1f_2$, $f_1 = a_1 \cdots a_i$, $f_2 = a_{i+1} \cdots a_t$. Then $h_i = \Phi(f_1)$ is a g, and so $\Phi(h_i) = h_i$. Thus, in evaluating $\Phi(h_if_2)$, we have a term equal to h_i and then further terms equal to $h_{i+1}, \cdots, h_t = \Phi(f_1f_2)$

4) $\Phi(gs^\epsilon) = \phi(gs^\epsilon)$.

Here $\Phi(gs^\epsilon) = \phi(\Phi(g)s^\epsilon)$ by the definition of the Φ function. Here $\Phi(g) = g$, and so $\Phi(gs^\epsilon) = \phi(gs^\epsilon)$.

5) $\Phi[gs^\epsilon\phi(gs^\epsilon)^{-1}] = 1$.

Here $\Phi[gs^\epsilon\phi(gs^\epsilon)^{-1}] = \Phi[\Phi(gs^\epsilon)\phi(gs^\epsilon)^{-1}] = \Phi[\phi(gs^\epsilon)\cdot\phi(gs^\epsilon)^{-1}] = \Phi(1) = 1$.

6) If $f \in U$, then $\Phi(f) = 1$.

This comes from a repeated application of (3) and (5).

7) If $\Phi(f) = g$, then $f \in Ug$.

Here
$$\begin{aligned} f &= a_1a_2 \cdots a_t \\ &= (1\cdot a_1\cdot h_1^{-1})(h_1a_2h_2^{-1}) \cdots (h_{t-1}a_th_t^{-1})h_t, \end{aligned}$$
and each of $h_{i-1}a_ih_i^{-1} = gs^\epsilon\phi(gs^\epsilon)^{-1} \in U$, $i = 1 \cdots t$, whereas $h_t = \Phi(f) = g$. In particular, if $\Phi(f) = 1$ then $f \in U$.

8) If $g_i \neq g_j$, then g_i and g_j are in different cosets of U.

Otherwise, $g_i = wg_j$ with $w \in U$ and $g_i = \Phi(g_i) = \Phi[\Phi(w)g_j] = \Phi(g_j) = g_j$, a contradiction.

Thus we have shown that the cosets Ug are all different and include all the free group F. In proving Theorem 7.2.3 we have shown even more than was required. We state this as a theorem.

THEOREM 7.2.4. *Given the Schreier system G and the function $\phi(gs^\epsilon)$ of Theorem 7.2.3. From these alone we may decide whether or not an arbitrary element f belongs to the subgroup U determined by G and ϕ.*

Proof: We may compute $\Phi(f)$ from ϕ and G, and $\Phi(f) = 1$ if, and only if, $f \in U$. Since ϕ and G determine U in this unambiguous fashion, we shall regard ϕ and G as *representing* U, and speak of $U = U[G, \phi(gs^\epsilon)]$ as a *standard representation* of U.

Two questions raise themselves naturally:

1) How are different standard representations of the same subgroup related to each other?

2) How many subgroups, if any, can be represented in terms of a given Schreier system?

We shall answer both these questions in turn.

THEOREM 7.2.5. *$U_1 = U_1[G_1, \phi_1(gs^\epsilon)]$ and $U_2 = U_2[G_2, \phi_2(gs^\epsilon)]$ are the same subgroup if, and only if, there is a one-to-one correspondence. $g^1 \leftrightarrows g^2$, including $1 \leftrightarrows 1$ between the Schreier system G_1 and G_2 such that whenever $g^1 \leftrightarrows g^2$ then $\phi_1(g^1s^\epsilon) \leftrightarrows \phi_2(g^2s^\epsilon)$ for any s^ϵ.*

Proof: If $U_1 = U_2 = U$, then each coset of U has a representative from G_1 and also from G_2. Thus, if $Ug^1 = Ug^2$, the correspondence $g^1 \leftrightarrows g^2$ is clearly one-to-one and includes $1 \leftrightarrows 1$. Since g^1s^ϵ and g^2s^ϵ are in the same coset, then $\phi_1(g^1s^\epsilon) \leftrightarrows \phi_2(g^2s^\epsilon)$.

Conversely, suppose a one-to-one correspondence $g^1 \leftrightarrows g^2$ given, including $1 \leftrightarrows 1$ such that $\phi_1(g^1s^\epsilon) \leftrightarrows \phi_2(g^2s^\epsilon)$ in all cases. We find that $\Phi_1(f) \leftrightarrows \Phi_2(f)$ for every f, and in particular, $\Phi_1(f) = 1$ if, and only if, $\Phi_2(f) = 1$. But this says that U_1 and U_2 contain the same elements f, and therefore are of the same subgroup U. Moreover, by induction on the length of g^1 we may show that $Ug^1 = Ug^2$.

Before answering the second question we shall note some properties of the function ϕ. The mappings $\pi(s):g \to \phi(gs)$, $\pi(s^{-1}):g \to \phi(gs^{-1})$ for all $g \in G$ and a fixed generator s map all of G into itself. From property (3) of the ϕ function the products $\pi(s)\pi(s^{-1})$ and $\pi(s^{-1})\pi(s)$ are both the identity. Hence $\pi(s)$ and $\pi(s^{-1})$ are permutations (one-to-one mappings) and are inverses of each other. In addition, from property (2), certain values of ϕ are compulsory in that they depend entirely on the nature of G and not on the subgroup U. Again consider a fixed s and all $g \in G$. The g's may be divided into classes $C(s)$ and $C^*(s)$ such that

$$\begin{aligned} g \in C(s) &\quad \text{if, and only if, } gs \in G, \\ g \in C^*(s) &\quad \text{if, and only if, } gs \notin G. \end{aligned}$$

Let $N(s)$ be the cardinal number of the class $C(s)$ and $M(s)$ the cardinal number of $C^*(s)$. Here

$$N(s) + M(s) = N,$$

where N is the cardinal number of G. Similarly,

$$\begin{aligned} g \in C(s^{-1}) &\quad \text{if, and only if, } gs^{-1} \in G, \\ g \notin C^*(s^{-1}) &\quad \text{if, and only if, } gs^{-1} \notin G, \end{aligned}$$

and again with $N(s^{-1})$ the cardinal of $C(s^{-1})$ and $M(s^{-1})$ the cardinal of $C^*(s^{-1})$ we have

$$N(s^{-1}) + M(s^{-1}) = N.$$

Now if g_i and g_j are g's such that $g_is = g_j$, whence $g_js^{-1} = g_i$, then $g_i \in C(s)$ and $g_j \in C(s^{-1})$. This relation establishes a one-to-one correspondence between $C(s)$ and $C(s^{-1})$, whence

$$N(s) = N(s^{-1}).$$

Now if N is finite, it will also follow that

$$M(s) = M(s^{-1}).$$

But if N is infinite, it need not follow that $M(s) = M(s^{-1})$ for an arbitrary Schreier system. In particular, take $1, s, s^2 \cdots s^i \cdots$. Here $M(s) = 0$, $M(s^{-1}) = 1$. On the other hand if a ϕ exists for a given G, $\pi(s)$ maps $C(s)$ onto $C(s^{-1})$, and being a permutation, also maps $C^*(s)$ onto $C^*(s^{-1})$. Hence $M(s) = M(s^{-1})$ is a necessary condition for the existence of ϕ.

THEOREM 7.2.6. *Given a Schreier system G such that $M(s) = M(s^{-1})$ for every generator s. Then it is possible to find a function $\phi(gs^\epsilon)$ satisfying the three properties:*

1) *$\phi(gs^\epsilon)$ is a $g \in G$.*
2) *If $gs^\epsilon \in G$ then $\phi(gs^\epsilon) = gs^\epsilon$.*
3) *$\phi[\phi(gs^\epsilon)s^{-\epsilon}] = g$.*

The most general choice for ϕ is given by taking for each s:

i) *$\phi(gs) = gs$ if gs is a g.*
ii) *For gs not a g choose the set of $\phi(gs)$ in any way such that $\pi(s): g \to \phi(gs)$ is a permutation of G.*
iii) *Having defined $\phi(gs)$ for all g, define $\phi(gs^{-1})$ so that $\pi(s^{-1})$: $g \to \phi(gs^{-1})$ is the inverse of $\pi(s)$.*

Proof: Given the condition $M(s) = M(s^{-1})$ on G for all generators s, the theorem not only asserts that $\phi(gs^\epsilon)$ exists but also describes what is clearly the most general construction if the construction is valid. Hence we must prove the validity of this construction. For a given s, clearly:

1) $\phi(gs)$ is a g.
2) If gs is a g, then $\phi(gs) = gs$.

If for some g_i we have $g_is = g_j \in G$, we have put $\phi(g_is) = g_j$. Here $g_js^{-1} = g_i$. Thus in $g \to \phi(gs)$ we have mapped the class $C(s)$ onto the class $C(s^{-1})$. There are $M(s)$ g's remaining to be mapped into the remaining $M(s^{-1})$ g's. Since $M(s) = M(s^{-1})$, a one-to-one correspondence is possible, mapping $C^*(s)$ onto $C^*(s^{-1})$ by $g \in C^*(s)$, $g \leftrightarrows g' \in C^*(s^{-1})$. We put $g' = \phi(gs)$. Here $\pi(s): g \leftrightarrows \phi(gs)$ is a one-to-one correspondence taking $C(s)$ onto $C(s^{-1})$ and $C^*(s)$ onto $C^*(s^{-1})$. Now $\pi(s)$ is a permutation, and so, if we take $\pi(s^{-1}): g \leftrightarrows$

$\phi(gs^{-1})$ as the inverse of $\pi(s)$, we have defined values for $\phi(gs^{-1})$. Here, clearly, $\phi(gs^{-1})$ is a g. Moreover, since $\pi(s)$ mapped $C(s)$ onto $C(s^{-1})$, it will follow that:

3) If gs^{-1} is a g, then $\phi(gs^{-1}) = gs^{-1}$. Thus properties (1) and (2) hold for all $g \in G$ and both s and s^{-1}. Finally property (3), $\phi[\phi(gs^{\epsilon})s^{-\epsilon}] = g$, holds since $\pi(s)$ and $\pi(s^{-1})$ are inverse permutations.

In both Theorems 7.2.5 and 7.2.6 the permutations $\pi(s)$ played a central role. If $g = a_1a_2 \cdots a_t$, we observe that the permutation $\pi(a_1)\pi(a_2) \cdots \pi(a_t)$ takes 1 into g and hence that the permutations $\pi(s)$ generate a group transitive on the g's. These permutations alone determine the subgroup U uniquely, as we shall now show.

THEOREM 7.2.7. *Let F be the free group on a set S of free generators. Let a set of permutations $\pi(s)$ be given, one for each $s \in S$, the permutations $\pi(s)$ being on symbols $1, y_2, \cdots y_i, \cdots$, and let the group generated by the $\pi(s)$ be transitive on the symbols. With each element f of F where $f = a_1a_2 \cdots a_t$, associate the permutation $\pi(f) = \pi(a_1)\pi(a_2) \cdots \pi(a_t)$. Then those elements f such that $\pi(f)$ fixes 1 will form a subgroup U. If $g_1 = 1, g_2, \cdots, g_i, \cdots$ is any Schreier system of left coset representatives for U, we may associate the g's with the symbols y_i, putting $g_i \leftrightarrows y_i$ if $\pi(g_i)$ takes 1 into y_i. In this way the $\pi(s)$ on the y_i are permutation isomorphic to the $\pi(s)$ of Theorems 7.2.5 and 7.2.6 on the g_i.*

Proof: Clearly, those f's with $\pi(f)$ fixing 1 form a subgroup U of F. By Theorem 5.3.1 we may regard the permutations $\pi(f)$ as a representation of F on cosets of U, replacing 1 by U and the y's by other left cosets of U. Hence each y_i corresponds uniquely to some left coset Ug_i, where $\pi(g_i)$ takes 1 into y_i. In this representation $\pi(s)$ takes the coset Ug into Ugs, which is the same as $U\phi(gs)$. Thus, if we replace a coset Ug_i by its representative g_i, the permutation $\pi(s)$ now becomes the permutation $\pi(s)$ of Theorems 7.2.5 and 7.2.6, and so we have fully established the permutation isomorphism of the original permutations on the y's with those on the Schreier system G.

For a subgroup U of finite index in a finitely generated free group, we may give some explicit values for the number of generators of U and for their total length.

THEOREM 7.2.8. *Let $U = U[G, \phi(gs^{\epsilon})]$ be a subgroup of finite index n in a free group F_r with r free generators $s_1, s_2, \cdots, s_r$. Then*

1) *U is a free group on $1 + n(r - 1)$ free generators.*

2) *If L is the total length of the Schreier system G, then the total length of the free generators of U, $u = gs\phi(gs^{-1}) \neq 1$, is $K = (2L + n)r - 2L$.*

Proof: We have shown that a set of free generators of U is given by the elements

$$u_{ia} = g_i s_a \phi(g_i s_a)^{-1}, \quad i = 1, \cdots, n; \quad a = 1, \cdots, r,$$

which are not equal to the identity. Moreover, by Lemma 7.2.3, u_{ia} is either reduced as written or equal to the identity. Now

$$\sum_{i=1}^{n} L(g_i) + L(s_a) + L[\phi(g_i s_a)] = 2L + n,$$

since for s_a fixed, $\phi(g_i s_a)$ is a permutation of the g's. Hence *before* cancellation we have $(nr)u$'s of total length $r(2L + n)$. Thus we must subtract from these totals, respectively, the number u_{ia} equal to the identity and the lengths $L(g_i) + L(s_a) + L(g_i s_a)$ counted for these u's. When is u_{ia} equal to the identity? Now g_i, s_a, and $\phi(g_i s_a)^{-1}$ are reduced as written. Hence there will be cancellation, and by Lemma 7.2.3 then $u_{ia} = 1$, if, and only if, s_a cancels with g_i or $\phi(g_i s_a)^{-1}$. In the first case g_i ends in s_a^{-1}, $g_i = g_j s_a^{-1}$, with $g_j \in G$ reduced as written. In the second case $\phi(g_i s_a) = g_k$ ends in s_a, and in fact, $g_k = g_i s_a$. Thus for s_a the number of u's equal to the identity is equal to the number of g's ending in s_a or s_a^{-1}. But every g except $g = 1$ ends in some s_a or s_a^{-1} and so is counted exactly once in this process. Hence there are $(n - 1)$ u's equal to the identity in all, and consequently there remain $nr - (n - 1) = n(r - 1) + 1$ free generators for U. What about the lengths? First, if $g_i = g_j s_a^{-1}$, then $\phi(g_i s_a) = g_j$, and so $L(g_i) + L(s_a) + L[\phi(g_i s_a)] = 2L(g_i) = 2L(g_j s_a^{-1})$. Secondly, if $g_i s_a = g_k$, then $L(g_i) + L(s_a) + L[\phi(g_i s_a)] = 2L(g_i s_a)$. Thus for s_a we have included for $u_{ia} = 1$ twice the length of every g ending in s_a or s_a^{-1}. Hence for all s_a we have included for u's equal to the identity twice the length of every g except $g = 1$. But $L(1) = 0$, and so we must subtract exactly $2L$, leaving $(2L + n)r - 2L$ as the total length of the free generators of U.

Finally, using Theorem 7.2.7, we may enumerate recursively the number of subgroups of index n in F_r.

THEOREM 7.2.9. *The number $N_{n,r}$ of subgroups of index n in F_r is given recursively by $N_{1r} = 1$*

$$N_{n,r} = n(n!)^{r-1} - \sum_{i=1}^{n-1} (n - i)!^{r-1} N_{i,r}.$$

Proof: $N_{1r} = 1$ asserts merely that F_r is its own unique subgroup of index 1.

Choose r permutations $P_1, \cdots, P_r$ on symbols $1, x_2, \cdots, x_n$. In general $P_1, \cdots, P_r$ need not generate a group transitive on all of $1, x_2, \cdots, x_n$. Let the transitive constituent including 1 be $1, b_2, \cdots, b_t$. Disregarding the remaining letters, we may take as $\pi(s_1) \cdots, \pi(s_r)$ the permutations on $1, b_2, \cdots, b_t$, and by Theorem 7.2.7, these will determine a unique subgroup of index t. The remaining $n - t$ letters could occur in $P_1, \cdots, P_r$ in $[(n - t)!]^r$ ways. In addition the same subgroup will be determined if we replace $1, b_2, \cdots, b_t$ by any other combination $1, c_2, \cdots, c_t$, and let the remaining $n - t$ letters occur in an arbitrary way. Thus a total of

$$(n-1)(n-2) \cdots (n-t+1)[(n-t)!]^r = (n-1)![(n-t)!]^{r-1}$$

different permutations $P_1, \cdots, P_r$ may be associated with the same subgroup of index t. Hence

$$\sum_{t=1}^{n} (n-1)! \, [(n-t)!]^{r-1} N_{t,r} = (n!)^r,$$

counting the $(n!)^r$ possible choices of $P_1, \cdots, P_r$ according to the index of the subgroup with which they are associated. Dividing by $(n - 1)!$ and transposing the sum from 1 to $n - 1$, we have the formula of the theorem.

7.3. Free Generators of Subgroups of Free Groups. The Nielsen Method.

In the Sec. 7.2 the properties of a subgroup U of a free group F were studied in terms of the cosets of U in F. In this section we shall be concerned more directly with the elements of U.

Let $A = \{a_i\}$ be a set of elements in a free group F indexed by a set I of indices i, and let us suppose that the set A consists of free generators of the group which they generate, which we shall designate as $[A]$. For an element $f \in [A]$ we write $L_A(f)$ for the length of f written as a reduced word in the a's and their inverses.

Let the set X be a free set of generators for the free group F.

Then a set A of elements of F will be said to have the *Nielsen property* with respect to the generators X if, and only if,

1) $A \cap A^{-1} = 0$ (A^{-1} is the set of inverses of elements of A).
2) If $a, b \in A \cup A^{-1}$, $L_X(ab) < L_X(a)$ implies that $b = a^{-1}$.
3) If $a, b, c \in A \cup A^{-1}$, $L_X(abc) \leq L_X(a) - L_X(b) + L_X(c)$ implies that either $b = a^{-1}$ or $b = c^{-1}$.

THEOREM 7.3.1. *If the set A has the Nielsen property with respect to a set of free generators X of F, then A consists of free generators of the subgroup $[A]$ which it generates. The Nielsen property is equivalent to the existence of central significant factors in the A's.*

Proof: It is sufficient to show that the Nielsen property is equivalent to the existence of a central significant factor, since by Theorem 7.2.2 this will imply that A consists of free generators of $[A]$.

Assume that A has the Nielsen property. Then from property (2), if $b \neq a^{-1}$, $L_X(ab) \geq L_X(a)$ and $L_X(b^{-1}a^{-1}) \geq L_X(b^{-1})$, whence $L_X(ab) \geq L_X(b)$. If more than one-half of one factor, say b, canceled with a in the reduced form of ab, we would have $a = uv^{-1}$, $b = vw$, $L_X(v) > L_X(w)$, and $L_X(ab) = L_X(uw) < L_X(u) + L_X(v) = L_X(a)$. Hence this cannot happen, and at most one-half of a or b is canceled in the reduced form of ab. Thus for an element of odd length its central term may be taken as a significant factor. If b is of even length, conceivably the first half v of b may be canceled in a product ab with $b \neq a^{-1}$. If also the second half w of b may be canceled in the reduced form of a product bc, $b \neq c^{-1}$, then we have $a = uv^{-1}$, $b = vw$, $c = w^{-1}z$, and $L_X(abc) = L_X(uz) \leq L_X(u) + L_X(z) = L_X(a) - L_X(b) + L(c)$, contrary to the third requirement for the Nielsen property. Since this cannot happen, one-half of b, either v or w, cannot be canceled in any product, and so, that one of the two central terms of b belonging to this half may be taken as its central factor. Thus the Nielsen property implies the existence of central significant factors. Conversely, if central significant factors exist for a set A with $A \cap A^{-1} = 0$, then if $b \neq a^{-1}$, half of b, at most, is canceled in ab against an equal number of terms in a; so, $L_X(ab) \geq L_X(a) + L_X(b) - 2 \cdot 1/2 L_X(b) = L_X(a)$, yielding the second requirement. Moreover in a product abc, with $b \neq a^{-1}$, $b \neq c^{-1}$, the cancellation between a and b and between b and c stops short of the significant factor of b; so, $L_X(abc) > L_X(a) + L_X(b) + L_X(c) - 2L_X(b)$, which is the third requirement. It is not difficult to show that the third requirement alone is equivalent to the existence of significant factors. For given b, take as $a \neq b^{-1}$ an element which cancels the greatest number of terms on the left of b, and take as $c \neq b^{-1}$ an element which cancels the greatest number of terms on the right of b. The requirement asserts that not all of b is canceled out, and any remaining term may be taken as the significant factor for b.

THEOREM 7.3.2. *Given a finite set B of elements $\beta_1, \cdots, \beta_m$ in a free group F on a given set X of free generators. In a finite number of changes of the following types:*

Type 1: *Delete a $\beta_i = 1$,*

Type 2: *Replace a β_i by β_i^{-1},*

Type 3: *Replace a β_j by $\beta_i\beta_j$, $i \neq j$,*

we may replace the set B by another set A: $\alpha_1, \cdots, \alpha_n$, $n \leq m$ such that A generates the same subgroup as B and A has the Nielsen property with respect to X. Hence A is a set of free generators for $[A] = [B]$.

Proof: Clearly, each change replaces a set by another set generating the same group. The first type reduces the number of elements, the second and third leave the number of elements unchanged. We note that a combination of changes of types 2 and 3 will replace β_j by $\beta_i^\epsilon\beta_j^\eta$ or $\beta_j^\eta\beta_i^\epsilon$, $\epsilon = \pm 1$, $\eta = \pm 1$, and leave the remaining β's unchanged.

If two β's are equal or inverses, we may make changes to replace a β by 1 and then delete this 1. This reduces the number of β's and so could happen, at most, m times. If for $a, b \in B \cup B^{-1}$, $b \neq a^{-1}$, we have $L_X(ab) < L_X(a)$, we cannot have $b = a$ since always $L_X(a^2) > L_X(a)$. Hence we may replace the $\beta = a^\epsilon$ by ab and so reduce the total length of all the β's. Thus there can be only a finite number of changes of this kind, and so requirements (1) and (2) for the Nielsen property can be satisfied in a finite number of steps. Satisfying the third requirement is more difficult.

Whether or not the set X is infinite, the set Y of generators in X which occur in the β's is certainly finite. Let us list the elements of F generated by Y according to length, the order for a given length being arbitrary but fixed. There are only a finite number of each length, and so every element has only a finite number of predecessors in this list.

If a β is of even length $2k$ write β in the form $\beta = \gamma\delta^{-1}$, where each of γ and δ is of length k. If $\beta \neq 1$, then $\delta \neq \gamma$. Since $\beta^{-1} = \delta\gamma^{-1}$, we may replace β by β^{-1} if necessary so that its first half is earlier than its second half in the list. If $\beta_i = \gamma\delta^{-1}$ and a β_j begins with the terms of δ, $\beta_j = \delta z$, we replace β_j by $\beta_i\beta_j = \gamma z$. Similarly, if β_k ends in δ^{-1}, we replace $\beta_k = w\delta^{-1}$ by $\beta_k\beta_i^{-1} = w\gamma^{-1}$. Hence we may change the β's so that if $\beta_i = \gamma\delta^{-1}$ no other β begins with δ or ends with δ^{-1}. Since we are replacing a series of terms δ by another γ of the same length but earlier in the list, this process will terminate in a finite number of steps. It is important to note that if we begin with the shortest β of even length and then continue with longer β's of even length, the process will terminate in a finite number of steps. For working with β's of the same length, we continually replace a half word by an earlier half word, and so we come to an end in a finite number of steps. In working with β's of greater length than $\beta_i = \gamma\delta^{-1}$, there will be no beginning section δ or end δ^{-1} in any of them. Naturally, if at any point either condition $A \cap A^{-1} = 0$ or $L_X(ab) \geq L_X(a)$ is violated, we make an appropriate change, either by reducing the number of β's or by reducing their total length and then starting over in replacement of half words, which leaves both the number and length of the β's unchanged. Hence after a finite number of changes this process will terminate, yielding a set A of elements $\alpha_1, \cdots, \alpha_n$, $n \leq m$. We assert that the set A has the Nielsen property with respect to X. Both $A \cap A^{-1} = 0$ and $L_X(ab) \geq L_X(a)$, $b \neq a^{-1}$, $a, b \in A \cup A^{-1}$ will surely hold, since otherwise we could reduce either the number or total length of the β's. Now consider a product abc, $b \neq a^{-1}$, $b \neq c^{-1}$. If b is of odd length, $2k + 1$ at most, the first k terms of b cancel with a and, at most, the last k terms cancel with c; so, $L_X(abc) > L_X(a) - L_X(b) + L_X(c)$ holds. If b is of even length, then b is of the form $\gamma\delta^{-1}$ or $\delta\gamma^{-1}$, with γ earlier than δ. Since the second property holds, half of b, at most, is canceled by a and at most half by c. But a cannot end with δ^{-1} or c begin with δ, and so the half of b which is either δ or δ^{-1} is not entirely canceled; thus b itself is not entirely canceled and $L_X(abc) > L_X(a) - L_X(b) + L_X(c)$, proving the third requirement for the Nielsen property for A.

THEOREM 7.3.3. *Two free groups are isomorphic if, and only if, they have the same cardinal number of free generators. A free group F_r with a finite number r of generators is freely generated by any set of r elements which generate it.*

Proof: Let F_X and F_Y be free groups on sets of free generators X and Y, respectively.

If X and Y have the same cardinality, there is a one-to-one correspondence between X and Y which can be extended to a one-to-one correspondence between F_X and F_Y, which is clearly an isomorphism.

Conversely, suppose that F_X and F_Y are isomorphic. Then F_X and F_Y have the same number of subgroups of index 2. A subgroup of index 2 is the kernel of a homomorphism onto the group of order 2. Such a homomorphism is uniquely determined by the set of generators mapped onto the identity. Thus the number of subgroups of index 2 of a free group F_Z on a set of generators Z is the number of nonvacuous subsets of Z. This number is uncountable if Z is infinite and is $2^r - 1$ if Z is finite with r elements. Thus, if F_X and F_Y are isomorphic, it follows that X and Y are either both infinite or both finite, and that in the latter case X and Y have the same number of elements. If X and Y are infinite, F_X and F_Y have the same cardinal number as X and Y, respectively. Since F_X and F_Y have the same cardinal number, so do X and Y.

Now suppose that F_r, the free group on $X: x_1, x_2, \cdots, x_r$, is also generated by $\beta_1, \beta_2, \cdots, \beta_r$. Then, by Theorem 7.3.2, after a certain number of changes of types 1, 2, 3 (from $\beta_1, \cdots, \beta_r$), we shall have F_r freely generated by $\alpha_1, \cdots, \alpha_s$, with $s \leq r$. But then we must have $s = r$, and so no changes of type 1 have been used. We may verify directly that if a change of type 2 or 3 is made from a set B to a set B', then if either B or B' consists of free generators so does the other. Hence, since $\alpha_1, \cdots, \alpha_r$ are free generators of F_r, so will $\beta_1, \cdots, \beta_r$ be free generators of F_r.

This proves the theorem, but we may obtain even more explicit information on $\alpha_1, \cdots, \alpha_r$. The α_i have the Nielsen property and thus possess central significant factors (Theorem 7.3.1). Moreover for each x_i, $i = 1, \cdots, r$, $x_i = \gamma_1 \cdots \gamma_m$, with each γ an α or its inverse and $\gamma_i\gamma_{i+1} \neq 1$. Now the product of the γ's in its reduced form includes every central factor. Hence there can be only one and this must be equal to x_i. Thus every x_i is an α_j or α_j^{-1}. Hence, if we further apply changes of the second type, the α's are precisely $x_1, \cdots, x_r$ in some order. Thus, apart from order, we know how any set of free generators $\beta_1, \cdots, \beta_r$ of F_r may be obtained from $x_1, \cdots, x_r$. But this is to say that we have a knowledge of the automorphisms of F_r.

THEOREM 7.3.4. *All automorphisms of a free group F_r on a finite number r of generators X are generated by the automorphisms:*

1) P_{ij}: $x_i \to x_j$, $x_j \to x_i$, $x_k \to x_k$, $k \neq i, j$,
2) V_i: $x_i \to x_i^{-1}$, $x_j \to x_j$, $j \neq i$,
3) W_{ij}: $x_j \to x_i x_j$, $i \neq j$, $x_k \to x_k$, $k \neq j$,

where $i \neq j$ are any of $1, \cdots, r$.

Proof: Each of these is surely an automorphism of F_r, since it replaces X by a set of r elements which generate F_r. We must show that an arbitrary automorphism of F_r is expressible as a product of these. We have just shown above that the most general automorphism of F_r is obtained by replacing X: $x_1, \cdots, x_r$ by a set of generators B: $\beta_1, \cdots, \beta_r$ and that the set B is related to X by a finite succession of replacements,

$$B = B_1, B_2, \cdots, B_{N-1}, B_N = X$$

where B_i is B_{i+1} changed by a type 2 or 3 change of Theorem 7.2.3 for $i = 1, \cdots, N-2$, and the change from B_{N-1} to B_N is a permutation of the set X and hence a product of transpositions P_{ij} (§5.4). Thus each of the replacements of B_{i+1} by B_i, $i = 1, \cdots, N-2$ is an automorphism V_i or W_{ij} in terms of the elements B_{i+1}. We must show that these may be expressed in terms of automorphisms V_i and W_{ij} in terms of the elements of X.

Now let

$$Y: y_1, \cdots, y_r,$$
$$Z: z_1, \cdots, z_r,$$
$$W: w_1, \cdots, w_r,$$

be three sets of free generators for F_r where

1) $z_i = y_i^{-1}$, $z_j = y_j$, $j \neq i$, or
2) $z_j = z_i z_j$, $z_k = y_k$, $k \neq j$, and
3) $w_m = z_m^{-1}$, $w_n = z_n$, $n \neq m$, or
4) $w_n = z_m z_n$, $w_t = z_t$, $t \neq n$.

Here the replacement of Y by Z is a V or W automorphism in Y; the replacement of Z by W, an automorphism of type V or W in Z. We must show that (3) or (4) can be expressed by V and W automorphisms on the Y's. This involves several cases, all relatively simple. Only the two most difficult will be given here. Suppose we have (2) $z_j = y_i y_j$ and (3) $w_m = z_m^{-1}$ with $m = j$. We must express the automorphism (3) which here replaces $y_i y_j$ by $y_j^{-1} y_i^{-1}$ and leaves

y_k fixed for $k \neq j$. This is equivalent to the replacement of y_j by $y_i^{-1}y_j^{-1}y_i^{-1}$, leaving all other y's fixed. But this is the product $y_j \to y_i^{-1}y_j \to y_i^{-1}y_j^{-1} \to y_i^{-1}y_j^{-1}y_i^{-1}$, which is $W_{ij}^{-1}(y)V_j(y)W_{ij}(y)$. Next, suppose we have (2) $z_j = y_iy_j$ and (4) $w_n = z_mz_n$ with $m = j$, $n = i$. Here the automorphism (4) replaces $z_j = y_iy_j$ by y_iy_j and $z_i = y_i$ by $y_iy_jy_i$ and leaves all other $z_k = y_k$ fixed. This is the same as the replacement

$$\begin{aligned} y_i &\to y_iy_jy_i, \\ y_j &\to y_i^{-1}. \end{aligned}$$

But this is the product $W_{ij}^{-1}(y)W_{ji}(y)W_{ij}(y)$; thus

$$\begin{aligned} y_i &\to y_i \to y_jy_i \to y_iy_jy_i, \\ y_j &\to y_i^{-1}y_j \to y_i^{-1} \to y_i^{-1}. \end{aligned}$$

Hence every V or W automorphism on the z's may be expressed in terms of V and W automorphisms on the y's. We may now proceed to the proof of the theorem, using induction on N. The replacement of B_{N-2} by B_1 may by induction be assumed to be a product of V's and W's on the generators of B_{N-2}. But with B_{N-1} as the set Y, and B_{N-2} as the set Z, we may express the replacement of B_{N-2} by B_1 in terms of V's and W's on the set B_{N-1}. The replacement of B_{N-1} by B_{N-2} is a V or W on B_{N-1}. Hence the replacement of B_{N-1} by B_1 is a product of V's and W's on B_{N-1}, and these are also V's and W's on X since B_{N-1} is merely a permutation of X. This proves the theorem.

If A is a set with the Nielsen property, with respect to the free generators X of F_X, then A in various ways may be regarded as the "shortest" set of generators for $[A]$.

THEOREM 7.3.5. *If A has the Nielsen property with respect to X and if*

$$f = a_1a_2 \cdots a_t, \quad a_i \in A \cup A^{-1}, \quad a_ia_{i+1} \neq 1,$$

then

$$L_X(f) \geq \tfrac{1}{2}L_X(a_1) + t - 2 + \tfrac{1}{2}L_X(a_t),$$

and

$$L_X(f) \geq L_X(a_i \cdots a_j), \quad 1 \leq i \leq j \leq t.$$

Moreover, if X is finite and the elements of A are listed in order of increasing length,

$$\alpha_1, \alpha_2 \cdots \alpha_r \cdots,$$

and if

$$\beta_1, \beta_2 \cdots \beta_i \cdots,$$

is any other set of free generators for A, also listed in order of increasing length, then

$$L_X(\beta_n) \geq L_X(\alpha_n),\ n = 1, 2, \cdots.$$

Proof: In $f = a_1a_2 \cdots a_t$ each a_i has a central factor which is not canceled in the reduced form for f. Hence, in the reduced form for f, at least the first half of a_1, the central factors of $a_2 \cdots a_{t-1}$, and the last half of a_t remain, yielding $L_X(f) \geq \frac{1}{2}L_X(a_1) + t - 2 + \frac{1}{2}L_X(a_t)$. In the reduced form of $a_1 \cdots a_{t-1}a_t$ the cancellation between the reduced form of $a_1 \cdots a_{t-1}$ and a_t involves k terms in a_{t-1} and k terms in a_t, where $k \leq \frac{1}{2}L_X(a_{t-1})$, $k \leq L_X(a_t)$, since neither central factor is canceled. Thus $L_X(f) = L_X(a_1 \cdots a_{t-1}) + L_X(a_t) - 2k$. But $2k \leq L_X(a_t)$, whence $L_X(a_1 \cdots a_t) \geq L_X(a_1 \cdots a_{t-1})$. Similarly, $L_X(a_1 \cdots a_t) \geq L_X(a_i \cdots a_j)$. By repeating this argument, dropping an a at one end or the other, we have $L_X(a_1 \cdots a_t) \geq L_X(a_i \cdots a_j)$.

If X is finite, there are only a finite number of elements of any given length, and therefore a listing of A in order of increasing length will exhaust all the set. This is given as $\alpha_1, \alpha_2, \cdots, \alpha_i, \cdots$. For a second set of generators this list is $\beta_1, \beta_2, \cdots, \beta_i, \cdots$. Let $\beta_1(\alpha), \cdots, \beta_n(\alpha)$ be the expressions for the first (n) β's in terms of the free generators A, and let α_r be the last α occurring in these expressions. We assert that $r \geq n$. Let us deny the assertion and assume $r < n$. Then modulo the commutator group K of $[A]$, we have

$$\begin{aligned} \beta_1 &\equiv \alpha_1^{e_{11}} \cdots \alpha_r^{e_{1r}} \pmod{K}, \\ &\cdots\cdots\cdots, \\ \beta_n &\equiv \alpha_1^{e_{n1}} \cdots \alpha_r^{e_{nr}} \pmod{K}. \end{aligned}$$

With $r < n$ there surely exist* integers $u_1, \cdots, u_n$, not all zero, such that

$$\begin{aligned} e_{11}u_1 + \cdots + e_{n1}u_n &= 0, \\ \cdots\cdots\cdots&, \\ e_{1r}u_1 + \cdots + e_{nr}u_n &= 0. \end{aligned}$$

But then $\beta_1^{u_1} \cdots \beta_n^{u_n} \in K$ with $u_1 \cdots u_n$ not all zero, contrary to the

---

* Birkhoff and MacLane [1], p. 48. See § 9.2 for properties of the commutator subgroup.

assertion that the β's are free generators of $[A]$. Hence $r \geq n$. Let α_r actually occur in $\beta_j(\alpha)$ for some $j \leq n$. Then, by the first part of the theorem, $L_X(\beta_j) \geq L_X(\alpha_r)$. But $L_X(\beta_n) \geq L_X(\beta_j)$ and $L_X(\alpha_r) \geq L_X(\alpha_n)$, since $r \geq n$, and so, $L_X(\beta_n) \geq L_x(\alpha_n)$.

EXERCISES

1. Let F be the free group generated by x and y. Show that a fully invariant subgroup of F containing x^2yxy^{-1} is either F itself or is of index 9 in F.
2. Let F be the free group with two generators. Find all its subgroups of index 3.
3. Let F be the free group generated by three elements a, b, c. Find a set of free generators of the subgroup of index 8 generated by the squares of all elements of F.
4. Let $A_1, A_2, \cdots, A_m$ be elements of a free group given in reduced form, no one of them the identity, such that $A_1A_2, \cdots, A_m = 1$. Show that for some i, A_i is completely canceled in the product $A_{i-1}A_iA_{i+1}$.
5. Given a reduced word $g = a_1a_2 \cdots a_t \neq 1$ in a free group F. Show that F has a subgroup H of index $t + 1$, such that $g \notin H$. (Hint: Take coset representatives of H to be $1, a_1, a_1a_2, \cdots, a_1a_2 \cdots a_t$.)
6. Show that if $g = g(x_1, \cdots, x_r)$ is a word in generators $x_1, \cdots, x_r$ which is not the identity in the free group generated by $x_1, \cdots, x_r$ as free generators, then there is a finite group G generated by elements $x_1, \cdots, x_r$ in which g is not the identity. (Use Ex. 5 of this chapter and Ex. 1 of Chap. 5.)

8. LATTICES AND COMPOSITION SERIES

8.1. Partially Ordered Sets.

Definition: *A partially ordered set is a system S of elements in which a relation $a \supseteq b$ (read "a contains b") is defined for some pairs of elements of S such that*

P1. $a \supseteq a$.
P2. If $a \supseteq b$ and $b \supseteq c$, then $a \supseteq c$.
P3. If $a \supseteq b$ and $b \supseteq a$, then $a = b$.

Definition: *An upper bound of a subset T of a partially ordered S is an element x of S such that $x \supseteq t$ for every t of T. Similarly, a lower bound of a subset T is a y such that $t \supseteq y$ for every t of T.*

Definition: *A least upper bound (l.u.b.) of a subset T of S is an element x such that*

1) x is an upper bound of T.
2) If z is any upper bound of T, then $z \supseteq x$.

Similarly, a greatest lower bound (g.l.b.) of a subset T is a y such that

a) y is a lower bound of T.
b) If z is any lower bound of T, then $y \supseteq z$.

In general a subset T need not possess either a least upper bound or a greatest lower bound. But if T does have a least upper bound x, then this is unique, for by the definition, two least upper bounds must contain each other and by P_3 they must be equal. The same applies to greatest lower bounds.

If a partially ordered set S also satisfies:

P4. For any pair a, b, either $a \supseteq b$ or $b \supseteq a$.

We say that S is a *simply ordered set* or a *chain.*

We write $b \subseteq a$ as meaning $a \supseteq b$. We also write $a \supset b$ if $a \supseteq b$ and $a \neq b$. Similarly, $b \subset a$ means $a \supset b$. A further useful notation is $a > b$ (read "a covers b"), which means $a \supset b$ and $a \supseteq x \supseteq b$ implies $x = a$ or $x = b$. Also $b < a$ means $a > b$.

Example: Let S be the set of elements a, b, c, d, e, f, where the inclusion relation is given by the diagram. $x \supseteq y$ if x is above y and

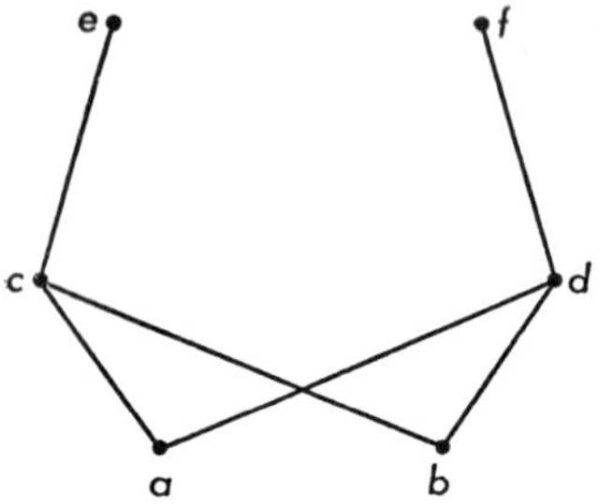

Fig. 3. A partially ordered set.

connected to it, or if $x = y$. Here the subset consisting of c and d has no upper bound and has two lower bounds but no greatest lower bound.

8.2. Lattices.

DEFINITION: *A lattice is a partially ordered set any two of whose elements* a, b *have a l.u.b. or union* $a \cup b$ *and a g.l.b. or intersection* $a \cap b$.

Since each of $a \cup b$ and $a \cap b$ is unique, union and intersection are well-defined binary operations in a lattice.

THEOREM 8.2.1. *In a lattice the following laws hold:*

$L1$. Idempotent laws. $x \cap x = x$ and $x \cup x = x$.

$L2$. Commutative laws. $x \cap y = y \cap x$ and $x \cup y = y \cup x$.

$L3$. Associative laws. $x \cap (y \cap z) = (x \cap y) \cap z$ and $x \cup (y \cup z) = (x \cup y) \cup z$.

$L4$. Absorption laws. $x \cap (x \cup y) = x$ and $x \cup (x \cap y) = x$.

Proof: $L1$, $L2$, and $L4$ are immediate consequences of the definition of l.u.b. and g.l.b. For $L3$ put $y \cap z = u$ and $x \cap u = w$. Here w is a lower bound of x and u, and hence of x, y, and z. But any lower bound of x, y, and z is contained in u, and so in $x \cap u = w$. Thus w is the g.l.b. of x, y, and z. But, similarly, $(x \cap y) \cap z$ is the g.l.b. of x, y, and z, whence $x \cap (y \cap z) = (x \cap y) \cap z$. In like manner each of $x \cup (y \cup z)$ and $(x \cup y) \cup z$ is the l.u.b. of x, y, and z.

THEOREM 8.2.2. *The laws* $L1$,-2,-3,-4 *completely characterize lattices.*

Proof: In any system satisfying $L1$,-2,-3,-4, $x \cap y = y$ if, and only if, $x \cup y = x$. If we define $x \supseteq y$ to mean $x \cap y = y$ in such a system, then the system is a partially ordered set with respect to this relation. Thus $a \cap a = a$ implies $P1$. If $a \cap b = b$ and $b \cap c = c$, then $a \cap c = a \cap (b \cap c) = (a \cap b) \cap c = b \cap c = c$, which proves $P2$. If $a \cap b = b$ and $b \cap a = a$, since $a \cap b = b \cap a$, we have $P3$. Thus, under this definition of inclusion the system is a partially ordered set. In addition $a \cap (a \cap b) = (a \cap a) \cap b = a \cap b$ and $b \cap (a \cap b) = a \cap b$, whence $a \cap b$ is a lower bound of a and b. But if $a \supseteq x$ and $b \supseteq x$, then $a \cap x = x$, $b \cap x = x$, whence $(a \cap b) \cap x = a \cap (b \cap x) = a \cap x = x$, and so $a \cap b$ is the g.l.b. of a and b. Similarly, if $y \supseteq a$ and $y \supseteq b$, then $a \cup y = y$ and $b \cup y = y$, whence $y = (a \cup b) \cup y$; it follows that not only is $a \cup b$ an upper bound of a and b but it is also the l.u.b.

Certain lattices possess further properties. The following are of some interest for our purposes.

DEFINITION: *A lattice L_1 is said to be isomorphic to a lattice L_2 if there is a one-to-one correspondence $x_i \leftrightarrows y_i$ between the elements x_i of L_1 and y_i of L_2 such that $x_i \cap x_j \leftrightarrows y_i \cap y_j$ and $x_i \cup x_j \leftrightarrows y_i \cup y_j$.*

DEFINITION: *A lattice L is said to be complete if every subset of L possesses a g.l.b. and a l.u.b.*

If the set of all elements of L possesses a l.u.b., this is called the *all element*; if a g.l.b., this is called the *zero element.*

DEFINITION: *A lattice L is said to be distributive if it satisfies the law:*

$$D_1 \cdot a \cap (b \cup c) = (a \cap b) \cup (a \cap c).$$

DEFINITION: *A lattice L is said to be modular if it satisfies the law:*

$$(M) \textit{ If } a \supseteq b, \textit{ then } a \cap (b \cup c) = b \cup (a \cap c).$$

A lattice, or more generally a partially ordered set, is said to satisfy the *minimal condition* if any chain $a_1 \supset a_2 \supset a_3 \supset \cdots$ is necessarily finite, and the *maximal condition* if any chain $a_1 \subset a_2 \subset a_3 \subset \cdots$ is necessarily finite.

DEFINITION: *In a lattice L, a finite chain $x = x_0 \supseteq x_1 \supseteq \cdots \supseteq x_d = y$ is maximal if x_i covers x_{i+1} for $i = 0, 1, \cdots, d - 1$; that is, $x = x_0 > x_1 > \cdots > x_d = y$. The chain is said to have length d.*

DEFINITION: *An element x of a lattice L has finite dimension d [written $d(x)$] if L has a zero element 0, providing that every chain from x to 0 is finite and that d is the length of the longest maximal chain from x to 0.*

8.3. Modular and Semi-modular Lattices.

In any lattice the set of x's such that $a \supseteq x \supseteq b$ form a sublattice, which we call the *quotient* a/b. Two quotients that may be put in the forms $a \cup b/b$ and $a/a \cap b$ are said to be *perspective* to each other, and if a_i/b_i is perspective to a_{i+1}/b_{i+1} for $i = 1, \cdots, n-1$, we say that a_1/b_1 is *projective* to a_n/b_n.

THEOREM 8.3.1. *In a modular lattice perspective quotients are isomorphic.*

Proof: Given the quotients $a \cup b/b$ and $a/a \cap b$ in a modular lattice. For any x in $a/a \cap b$ define

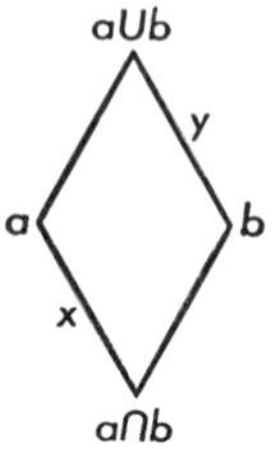

Fig. 4. Perspective quotients.

$$y(x) = x \cup b.$$

For any y in $a \cup b/b$ define

$$x(y) = y \cap a.$$

The first mapping takes elements of $a/a \cap b$ into elements of $a \cup b/b$, and the second takes elements of $a \cup b/b$ into elements of $a/a \cap b$. For x in $a/a \cap b$, $x[y(x)] = (x \cup b) \cap a$. Since $a \supseteq x$, we may apply the modular law and $a \cap (x \cup b) = x \cup (a \cap b) = x$, since $x \supseteq a \cap b$. Hence $x[y(x)] = x$. Similarly for y in $a \cup b/b$, by application of the modular law, $y[x(y)] = y$. Thus $x \to y(x)$ and $y \to x(y)$ yield a one-to-one correspondence between the two quotients. In addition this correspondence preserves the lattice operations. Thus for x_1, x_2 *in* $a/a \cap b$, $y(x_1 \cup x_2) = (x_1 \cup x_2) \cup b = (x_1 \cup b) \cup (x_2 \cup b) = y(x_1) \cup y(x_2)$. Also $x_1 = x(y_1)$, $x_2 = x(y_2)$, and $x_1 \cap x_2 = x(y_1) \cap x(y_2) = (y_1 \cap a) \cap (y_2 \cap a) = y_1 \cap y_2 \cap a = x(y_1 \cap y_2)$. Here $y(x_1 \cap x_2) = y[x(y_1 \cap y_2)] = y_1 \cap y_2 = y(x_1) \cap y(x_2)$; therefore both operations are preserved by the mapping $x \to y(x)$. From the fact that the correspondence is one to one, it therefore follows that the operations are also preserved by $y \to x(y)$. A similar proof would show that $y \to x(y)$ preserves both operations.

COROLLARY 8.3.1. *In a modular lattice projective quotients are isomorphic.*

THEOREM 8.3.2. *In a modular lattice if x is an element with finite dimension $d(x)$ then every maximal chain from x to zero has the same length.*

Proof: The proof will be by induction on the dimension of x. If $d(x) = 1$, then $x > 0$ is the only chain from x to 0. Let $x = x_0 > x_1 > \cdots > x_d = 0$ be one maximal chain from x to 0, and let $x = y_0 > y_1 \cdots > y_s = 0$ be another. If $x_1 = y_1$, then by induction the maximal chains from x_1 and y_1 have the same length $d - 1$, whence $s - 1 = d - 1$ and $s = d$. If $x_1 \neq y_1$, then write $z_2 = x_1 \cap y_1$. Here the quotients x/x_1 and y_1/z_2 are perspective and also x/y_1 and x_1/z_2. Both x/x_1 and x/y_1 contain no intermediate elements, and so $y_1 > z_2$ and $x_1 > z_2$. Since all maximal chains from x_1 to 0 are of length $d - 1$, all maximal chains from z_2 to 0 are of length $d - 2$. Hence, from y_1 to z_2 to 0 is of length $d - 1$, and so by induction, the chain $y_1 > y_2 \cdots > 0$. Hence $x = y_0 > y_2 > \cdots > y_s = 0$ is also of length $d = s$.

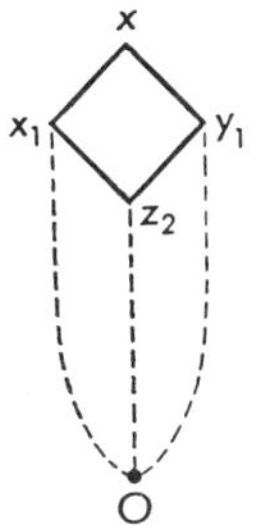

Fig. 5. Jordan-Dedekind condition.

As a consequence of this theorem we have the Jordan-Dedekind chain condition holding in modular lattices.

Jordan-Dedekind Chain Condition: *All finite maximal chains between two elements have the same length.*

If $a \supset b$ we may take b as the zero element in the quotient lattice a/b and apply the theorem.

In a modular lattice the dimension satisfies an important relation.

Theorem 8.3.3. *In a lattice whose elements are of finite dimension, the law*

$$d(x) + d(y) = d(x \cup y) + d(x \cap y)$$

holds if, and only if, the lattice is modular.

Proof: In a modular lattice the quotients $x \cup y/x$ and $y/x \cap y$ are isomorphic. The length of a maximal finite chain in each of these is respectively $d(x \cup y) - d(x)$ and $d(y) - d(x \cap y)$. From the isomorphism these two maximal lengths are equal and we have

$$\text{(M)} \qquad d(x) + d(y) = d(x \cup y) + d(x \cap y).$$

Conversely, suppose the law (M) holds in a lattice. Suppose $A \supseteq B$; consider the two expressions $A \cap (B \cup C)$ and $B \cup (A \cap C)$. Here

$$\begin{aligned} B &\subseteq A, \\ B &\subseteq B \cup C, \\ B &\subseteq A \cap (B \cup C), \\ A \cap C &\subseteq A, \\ A \cap C &\subseteq C \subseteq B \cup C, \\ A \cap C &\subseteq A \cap (B \cup C), \\ B \cup (A \cap C) &\subseteq A \cap (B \cup C). \end{aligned}$$

Hence these two expressions will be equal if their dimensions are equal. Using (M)

$$\begin{aligned} d[B \cup (A \cap C)] &= d(B) + d(A \cap C) - d(B \cap A \cap C) \\ &= d(B) + d(A \cap C) - d(B \cap C) \\ &= d(B \cup C) - d(C) + d(A \cap C) \\ &= d(B \cup C) + d(A) - d(A \cup C) \\ &= d(A) + d(B \cup C) - d(A \cup B \cup C) \\ &= d[A \cap (B \cup C)]. \end{aligned}$$

Hence $A \cap (B \cup C) = B \cup (A \cap C)$ and the modular law holds.

In terms of the covering relation $A > B$, we define two properties of *semi-modularity* which may hold in a lattice.

DEFINITIONS. Lower semi-modularity: *A lattice is lower semi-modular if whenever $A > B$ and $A > C$, $B \neq C$, then $B > B \cap C$ and $C > B \cap C$.*

Upper semi-modularity: *A lattice is upper semi-modular if whenever $A < B$ and $A < C$, $B \neq C$, then $B < B \cup C$ and $C < B \cup C$.*

Clearly, the two kinds of semi-modularity are dual to each other, and by Theorem 8.3.1, are both consequences of modularity. We shall show that in a finite dimensional lattice, both kinds of semi-modularity taken together imply modularity.

THEOREM 8.3.4. *In a semi-modular lattice L, if $A \supseteq B$, and if there is a finite maximal chain between A and B, then all finite maximal chains between A and B are of the same length.*

Proof: The proof is essentially the same as that of Theorem 8.3.2. Suppose that L is lower semi-modular. If there is a maximal chain of length one from A to B, then $A > B$, and there is no other chain from A to B. We proceed by induction on the length of a maximal chain from A to B. Suppose that

$$A = A_0 > A_1 > A_2 > \cdots > A_r = B$$

is a maximal chain of length r from A to B, the theorem being true for chain lengths less than r. Now let

$$A = U_0 > U_1 > U_2 > \cdots > U_s = B$$

be a second maximal chain from A to B. Then, if $U_1 = A_1$, maximal chains from $A_1 = U_1$ to B must by induction be of length $r - 1$, and the theorem follows. If, however, $U_1 \neq A_1$, then by lower semi-modularity,

$$A_1 > U_1 \cap A_1, \quad U_1 > U_1 \cap A_1.$$

Writing $U_1 \cap A_1 = V_2$, we shall have chains

$$\begin{aligned} A &= A_0 > A_1 > A_2 > \cdots > A_r = B, \\ A &= A_0 > A_1 > V_2 > \cdots > V_m = B, \\ A &= U_0 > U_1 > V_2 > \cdots > V_m = B, \\ A &= U_0 > U_1 > U_2 > \cdots > U_s = B. \end{aligned}$$

By induction on the chains from A_1 to B we have $m = r$, and the first two chains have the same length. The second and third have the same length m, and by induction on the chains from U_1 to B, we have $m = s$. Hence all four chains have the same length, and the theorem is proved for lower semi-modular lattices. A dual argument proves the same result for upper semi-modular lattices.

From this theorem, we see that in a semi-modular lattice, the dimension of an element $d(A)$ is the length of all maximal chains between A and the null element 0. In finite dimensional semi-modular lattices, we have inequalities relating the dimension functions of elements.

THEOREM 8.3.5. *Let L be a finite dimensional lattice. If L is upper semi-modular, then,* (1) $d(X \cup Y) + d(X \cap Y) \leq d(X) + d(Y)$. *If L is lower semi-modular, then* (2) $d(X \cup Y) + d(X \cap Y) \geqq d(X) + d(Y)$. *Conversely,* (1) *implies upper semi-modularity, but* (2) *does not imply lower semi-modularity.*

Proof: By Theorem 8.3.4, if L is semi-modular and if $R \supset S$, then $d(R) - d(S)$ is the length of a maximal chain between R and S, since all maximal chains from the zero element to R have the same length; therefore the dimension of R is the length of a maximal chain from R to zero including S. We shall use this fact in our proof.

Suppose L is upper semi-modular. Let us write $A \geqq B$ to mean $A = B$ or $A > B$; read this "A at most covers B." Then, if

$$\begin{aligned} X \cap Y &= U_0 < U_1 < U_2 < \cdots < U_m = X, \\ X \cap Y &= V_0 < V_1 < V_2 < \cdots < V_n = Y, \end{aligned}$$

we assert that $U_i \cup V_j \geqq U_{i-1} \cup V_j$ and $U_i \cup V_j \geqq U_i \cup V_{j-1}$ for all $i = 1, \cdots, m$ and $j = 1, \cdots, n$. This we prove by induction on $i + j$, the smallest significant value for $i + j$ being 2, and for this value the upper semi-modularity asserts that

$U_1 \cup V_1 \geqq U_1 = U_1 \cup V_0$ and $U_1 \cup V_1 \geqq V_1 = U_0 \cup V_1$.
Then $$U_i \cup V_j = (U_i \cup V_{j-1}) \cup (U_{i-1} \cup V_j),$$

and by induction $U_i \cup V_{j-1} \geqq U_{i-1} \cup V_{j-1}$ and $U_{i-1} \cup V_j \geqq U_{i-1} \cup V_{j-1}$, whence by upper semi-modularity, $U_i \cup V_j \geqq U_i \cup V_{j-1}$ and $U_i \cup V_j \geqq U_{i-1} \cup V_j$, as we wished to prove. From this for $j = n$, since $V_n = Y$,

$$Y \leq U_1 \cup Y \leq U_2 \cup Y \leq \cdots \leq U_m \cup Y = X \cup Y.$$

Thus the length of a maximal chain from Y to $X \cup Y$ is at most m. But as we have stated before, this means

$$d(X \cup Y) - d(Y) \leq m = d(X) - d(X \cap Y),$$

whence the inequality (1) holds in an upper semi-modular lattice. By a dual argument, the inequality (2) holds in a lower semi-modular lattice. This proves the direct part of the theorem.

LEMMA 8.3.1. *If inequality* (1) *holds in L, then* $U > V$ *implies* $d(U) = d(V) + 1$.

Proof: Let $0 = U_0 < U_1 < U_2 < \cdots < U_{t-1} < U_t = U$ be a longest chain from 0 to U. There cannot be a chain longer than i from 0 to U_i because if there were we could construct a longer chain from 0 to U. Hence $d(U) = t$ and $d(U_i) = i$, for $i = 0, \cdots, t - 1$. Also, since $U > V$, we have $d(U) \geqq d(V) + 1$, and so $t - 1 \geqq d(V)$. Let us select U_j so that $U_j \subseteq V$, $U_{j+1} \nsubseteq V$. There must be such a j in the range $0, 1, \cdots, t - 1$. Then $U_{j+1} \cup V = U$, $U_{j+1} \cap V = U_j$. By inequality (1), $d(U_{j+1} \cup V) + d(U_{j+1} \cap V) \leq d(V) + d(U_{j+1})$, whence $t + j \leq d(V) + j + 1$ or $t - 1 \leq d(V)$, and so, $d(V) = t - 1$, $d(U) = t = d(V) + 1$.

Now using the lemma and inequality (1), suppose $A < B$, $A < C$ and $B \neq C$. Then $A = B \cap C$, $d(B) = d(A) + 1$, $d(C) = d(A) + 1$. By inequality (1) $d(B \cup C) + d(B \cap C) \leq d(B) + d(C)$, which gives $d(B \cup C) \leq d(A) + 2$. But $B \cup C \neq B, C$, and so $d(B \cup C) = d(B) + 1 = d(C) + 1$, giving $B \cup C > B$, $B \cup C > C$, the conclusion that L is upper semi-modular. Because dimension (dX) is defined as the length of the longest chain from 0 to X and is not of a dual nature, the inequality (2) does not imply that L is lower semi-modular. The five-element lattice with elements 0, T, A_1, $B_1 \subset B_2$, such that $A_1 \cap B_1 = A_1 \cap B_2 = 0$, $A_1 \cup B_1 = A_1 \cup B_2 = T$ satisfies inequality (2) but is not lower semi-modular.

THEOREM 8.3.6. *A finite dimensional lattice is modular if, and only if, it is both upper and lower semi-modular.*

Proof: We have already observed that modularity implies both kinds of semi-modularity. But if both kinds hold, then by Theorem 8.3.5 we have $d(X \cup Y) + d(X \cap Y) = d(X) + d(Y)$, and by Theorem 8.3.3 this implies modularity.

THEOREM 8.3.7. *The subgroups of a finite p-group form a lower semi-modular lattice.*

Proof: Union and intersection of subgroups as defined in §1.4 do indeed satisfy the axioms for a lattice, the subgroups of a group being partially ordered by inclusion. If $A > B$, $A > C$, where A, B, C are subgroups of a finite p-group, then B and C are maximal subgroups of A and by Theorem 4.3.2 are of index p. By Theorem 1.5.5 on the inequality of indices $[B:B \cap C]$ and $[C:B \cap C]$ are, at most, p, and so either 1 or p. Thus, if $B \neq C$, we have $B > B \cap C$ and $C > B \cap C$.

8.4. Principal Series and Composition Series.

We shall now combine the results of the preceding sections and apply them to a study of the structure of the subgroups of groups. We shall consider a chain of subgroups of a group G, each a normal subgroup of the preceding group.

$$G = A_0 \supseteq A_1 \supseteq A_2 \supseteq \cdots \supseteq A_n \tag{8.4.1}$$

where each A_i is a normal subgroup of A_{i-1}, for which we write

$$A_i \triangleleft A_{i-1}, \quad i = 1 \cdots n. \tag{8.4.2}$$

The groups A_i are called *subinvariant groups* of G.

There will be associated with this chain the sequence of factor groups

$$A_{i-1}/A_i, \quad i = 1 \cdots n. \tag{8.4.3}$$

If every A_i is a normal subgroup of G, we shall call (8.4.1) a *normal* *chain* or *normal series*. We may also use the term *invariant series*. If $A_i \lhd A_{i-1}$, $i = 1 \cdots n$, it does not in general follow that $A_i \lhd G$, and so the requirements for a normal series are stronger than (8.4.2). If we assume only (8.4.2), we shall call the series a *subinvariant series*.* A normal series in which every A_i is a maximal normal subgroup contained in A_{i-1} will be called a *principal series* or *chief series*. A subinvariant series in which each A_i is a maximal normal subgroup of A_{i-1} will be called a *composition series*. In lattice terminology, if the inclusions in (8.4.1) are coverings, a normal series is called a *principal series*; a subinvariant series, a *composition series*. We may in addition require that the groups A_i be admissible subgroups with respect to a set of operators Ω.

We shall be able to interpret general theorems on modular lattices as theorems on subgroups, or as theorems on congruence relations on loops, or more generally, as theorems on congruence relations on any algebraic system whose congruence relations permute. The main theorem which will enable us to get the strongest result on groups is Theorem 2.4.1. The lattice theorems depend on the modular law, and this arises in different ways in the algebras. Thus by altering the hypotheses on the algebras, different theorems come from the same theorem on lattices. An auxiliary theorem on modularity in groups is needed. We shall say that subgroups A and B of a group G are *permutable* if the complexes AB and BA are equal. In this case it is readily verified that $A \cup B = AB = BA$, and the complex $AB = BA$ is in fact a subgroup. From theorem 2.3.3, subgroups A and B are permutable if either of them is a normal subgroup, and clearly, normality in $A \cup B$ is all that is required.

THEOREM 8.4.1. *Let A, B, C be subgroups of a group G such that $A \supseteq B$. Then a sufficient condition for*

$$A \cap (B \cup C) = B \cup (A \cap C)$$

to hold is that B and C be permutable.

Proof: As in the proof of Theorem 8.3.3 we note that always, if $A \supseteq B$,

$$B \cup (A \cap C) \subseteq A \cap (B \cup C).$$

---

* The more colorful term *subnormal series* has been urged on the writer by Irving Kaplansky, but this seems unnecessarily distracting.

It is necessary only to prove the opposite inclusion. An element of $A \cap (B \cup C)$ is of the form $a = bc$, $a \in A$, $b \in B$, $c \in C$, being simultaneously an element of A and also of $B \cup C$, and since B and C permute, the elements of $B \cup C$ are of the form bc. Here $c = b^{-1}a \in A$ since $B \subseteq A$. Hence this $c \in A \cap C$, and therefore $bc \in B \cup (A \cap C)$. Thus $A \cap (B \cup C) \subseteq B \cup (A \cap C)$, and the theorem is proved. This also holds for subloops of inverse loops where the permutability of B and C means $B \cup C = BC$. The conclusion $b^{-1}a = b^{-1}(bc) = c$ requires only the inverse law.

THEOREM 8.4.2. REFINEMENT THEOREM.* *Let* $U = A_0 \supseteq A_1 \supseteq \cdots \supseteq A_n = V$ *and* $U = B_0 \supseteq B_1 \supseteq \cdots \supseteq B_m = V$ *be two finite chains from U to V in a modular lattice. Then it is possible to refine both chains by inserting additional elements* $A_{i-1} = A_{i,0} \supseteq A_{i,1} \supseteq \cdots \supseteq A_{i,m} = A_i$, $i = 1, \cdots, n$, *and* $B_{j-1} = B_{j,0} \supseteq B_{j,1} \supseteq \cdots \supseteq B_{j,n} = B_j$, $j = 1, \cdots, m$ *in such a way that the quotients* $A_{i,j-1}/A_{i,j}$ *and* $B_{j,i-1}/B_{j,i}$ *are projective.*

Proof: Put $A_{i,j} = A_i \cup (A_{i-1} \cap B_j)$, $B_{j,i} = B_j \cup (B_{j-1} \cap A_i)$ $i = 1, \cdots, n$, $j = 1, \cdots, m$. Here $A_{i,j-1}/A_{i,j}$ is perspective to

(8.4.4) $$A_{i-1} \cap B_{j-1}/(A_{i-1} \cap B_j) \cup (A_i \cap B_{j-1}),$$

since from $B_j \subseteq B_{j-1}$ we have

(8.4.5)
$$(A_{i-1} \cap B_{j-1}) \cup A_i \cup (A_{i-1} \cap B_j) = A_i \cup (A_{i-1} \cap B_{j-1}).$$

Also

(8.4.6)
$$\begin{aligned}(A_{i-1} \cap B_{j-1}) &\cap [A_i \cup (A_{i-1} \cap B_j)] \\ &= (A_{i-1} \cap B_j) \cup (A_{i-1} \cap B_{j-1} \cap A_i) \\ &= (A_{i-1} \cap B_j) \cup (A_i \cap B_{j-1}),\end{aligned}$$

using modularity in (8.4.6). Similarly, $B_{j,i-1}/B_{j,i}$ is perspective to the quotient in (8.4.4) and our theorem is proved.

This theorem and its proof also holds for subinvariant series in a

* The original Jordan-Hölder theorem has been extended and generalized by a long series of authors. The original theorem is due to C. Jordan [1] and to O. Hölder [1]. Generalization to groups with operators is due to E. Noether [1] and W. Krull [1, 2]. The refinement theorem is due to O. Schreier [4] and H. Zassenhaus [1]. The lattice theoretical formulation given here is a modification of that given by O. Ore [2]. Generalization to partially ordered sets has been made by O. Ore [1] and S. MacLane [3].

group G where, if we take G to be a group with operators Ω, the subgroups are all admissible subgroups. This naturally includes groups without operators if we take Ω to be trivially the identical operator.

THEOREM 8.4.3 (REFINEMENT THEOREM FOR GROUPS). *Let G be a group with operators Ω, and let $G = A_0 \supseteq A_1 \supseteq \cdots \supseteq A_n = H$ and $G = B_0 \supseteq B_1 \supseteq \cdots \supseteq B_m = H$ be two subinvariant series from G to H of admissible subgroups. Then it is possible to refine both series by inserting additional admissible subinvariant groups*

$$A_{i-1} = A_{i,0} \supseteq A_{i,1} \supseteq \cdots \supseteq A_{i,m} = A_i, \quad i = 1, \cdots, n$$

and

$$B_{j-1} = B_{j,0} \supseteq B_{j,1} \supseteq \cdots \supseteq B_{j,n} = B_j, \quad j = 1, \cdots, m$$

in such a way that the quotient groups

$$A_{i,j-1}/A_{i,j} \quad \text{and} \quad B_{j,i-1}/B_{j,i}$$

are operator isomorphic.

Proof: By Theorem 2.4.1 perspective (and hence projective) quotient groups of admissible subgroups are operator isomorphic. Hence, to show that the proof of Theorem 8.4.2 gives this theorem, we must show that in the quotients X/Y occurring in the proof that $Y \triangleleft X$ and that the use of the modular law in (8.4.6) is valid. As the union and intersection of admissible subgroups are again admissible, all subgroups used in the proof are admissible. Now $A_{i,j} = A_i \cup (A_{i-1} \cap B_j)$ is a normal subgroup of $A_{i,j-1} = A_i \cup (A_{i-1} \cap B_{j-1})$, since both A_i and $A_{i-1} \cap B_j$ are transformed into themselves by $A_{i-1} \cap B_{j-i}$. Similarly, $B_{j,i} \triangleleft B_{j,i-1}$. Both $A_{i-1} \cap B_j$ and $A_i \cap B_{j-1}$, and so also their union, are normal subgroups of $A_{i-1} \cap B_{j-1}$, whence (8.4.4) is a quotient group. In (8.4.6), since A_i is normal in A_{i-1}, A_i permutes with any subgroup of A_{i-1} and in particular with $A_{i-1} \cap B_j$. Hence by Theorem 8.4.1 the modular law may be applied as was done in (8.4.6). Thus our theorem is proved.

In a principal series or composition series (with or without operators), no further refinement is possible, and so as a direct consequence of the refinement theorem we have the following:

THEOREM 8.4.4. THEOREM OF JORDAN-HÖLDER. *If $G = A_0 \supset A_1 \supset \cdots \supset A_n = H$ and $G = B_0 \supset B_1 \supset \cdots \supset B_m = H$ are two principal series (or two composition series), with operators Ω, then*

$m = n$ and the factor groups A_{i-1}/A_i are operator isomorphic to the factor groups B_{j-1}/B_j in some order.

The fact $m = n$ is of course a consequence of the one-to-one correspondence between the factor groups of the refinement theorem which are not the identity.

In the case of normal series all the subgroups, being normal subgroups, are admissible under all inner automorphisms $x \to a^{-1}xa$, and we may include all inner automorphisms in the set of operators Ω. An isomorphism preserved under all inner automorphisms is called a *central isomorphism.* Thus a consequence of the refinement theorem is the following:

THEOREM 8.4.5. *In the refinement of normal series, corresponding factor groups are centrally isomorphic.*

Now if $x \to (x)\alpha$ is a central automorphism of a group, then

$$a^{-1}\,(x)\alpha a = (a^{-1}xa)\alpha = (a)\alpha^{-1}\,(x)\alpha\,(a)\alpha,$$

whence $(a)\alpha a^{-1}$ permutes with every $(x)\alpha$ and must be an element of the center of the group, say, z. Hence, for a central automorphism, $(a)\alpha = az$ for every element a of the group and an appropriate z of the center, where z depends on a. Conversely, an automorphism with this form is readily seen to be a central automorphism.

8.5. Direct Decompositions.

Suppose that in a modular lattice we have m elements $A_1, \cdots, A_m$ such that if we write $\bar{A}_i = A_1 \cup \cdots \cup A_{i-1} \cup A_{i+1} \cdots \cup A_m$, $i = 1 \cdots m$, then $A_i \cap \bar{A}_i = 0$, the zero element for $i = 1 \cdots m$. We then say that $A = A_1 \cup \cdots \cup A_m$ is the *direct union* of $A_1, \cdots, A_m$ and write

$$A = A_1 \times A_2 \times \cdots \times A_m. \tag{8.5.1}$$

This will arise in groups when A is the direct product of $A_1, \cdots, A_m$.

THEOREM 8.5.1 (THEOREM OF ORE). *Let L be any modular lattice of finite dimension. If the all element T of L has two decompositions $T = A_1 \times \cdots \times A_m$, $T = B_1 \times \cdots \times B_n$, where the A_i and B_j are not further decomposable as direct unions, then $m = n$ and the A_i and B_j are projective in pairs.*

Proof: We shall show that any given A (say, A_1) may be replaced by some B_j projective to it, where $T = A_1 \times A_2 \times \cdots \times A_m = B_j \times A_2 \times \cdots \times A_m$. This is the main part of our proof. Having replaced A_1 by B_j, we proceed to replace A_2 in the second decomposition by some B_k' and so on. In the process of replacement we cannot possibly use the same B_j twice, since this would be in conflict with the requirement that any factor intersect the union of the remaining ones in zero. We must have enough B's to replace all the A's, and clearly, since every $B \subseteq T$, we cannot have any remaining when all A's have been replaced. Thus $m = n$. We write $\bar{A}_i = A_1 \cup \cdots \cup A_{i-1} \cup A_{i+1} \cup \cdots \cup A_m$, $i = 1, \cdots, m$, and $\overline{B}_j = B_1 \cup \cdots \cup B_{j-1} \cup B_{j+1} \cdots \cup B_n$, $j = 1, \cdots, n$, and base our proof on induction on the dimension of T, the theorem being trivial for dimension one.

Case 1. $A_1 \cup \overline{B}_j = \bar{A}_1 \cup B_j = T$ for some j. Here

$$\begin{aligned} d(A_1) &= d(T) - d(\overline{B}_j) + d(A_1 \cap \overline{B}_j) \\ &= d(B_j) + d(A_1 \cap \overline{B}_j) \geq d(B_j), \end{aligned}$$

and similarly, $d(B_j) \geq d(A_1)$, giving $d(A_1) = d(B_j)$. Thus $d(A_1 \cap \overline{B}_j) = d(\bar{A}_1 \cap B_j) = 0$, and so $T = A_1 \times \overline{B}_j = \bar{A}_1 \times B_j$ and A_1 and B_j are mutually replaceable.

Case 2. Suppose $A_1 \cup \overline{B}_j \subset T$ for some j (say, $j = 1$).

Write $D_h = A_1 \cup \overline{B}_h$, $Q_h = D_h \cap B_h$, $h = 1, \cdots, n$. If $D_1 = A_1 \cup \overline{B}_1 \supseteq B_1$, then $D_1 \supseteq \overline{B}_1 \cup B_1 = T$, contrary to hypothesis. Hence $Q_1 = D_1 \cap B_1 \subset B_1$ and $d(Q_1) < d(B_1)$. T is the direct union of the B's, and therefore the union of the Q's will be their direct union, since $Q_h \subseteq B_h$, $h = 1, \cdots n$.

Define

$$C = \bigcup_{h=1}^{n} Q_h.$$

Thus both T and C, being direct unions and $Q_1 \subset B_1$, $d(T) = d(B_1) + \cdots + d(B_n)$,

$$d(C) = d(Q_1) + \cdots + d(Q_n) < d(T). \tag{8.5.2}$$

Since C is properly contained in T, we may by induction on dimension assume the theorem true for C.

Let us write $U_r = Q_1 \cup \cdots \cup Q_r$. We wish to prove $U_r = M_r \cap N_r$, where $M_r = B_1 \cup \cdots \cup B_r$, $N_r = D_1 \cap \cdots \cap D_r$. For $r = 1$ this reduces to $U_1 = B_1 \cap D_1$, the definition of $U_1 = Q_1$. The proof is by induction. We assume $U_j = M_j \cap N_j$. Then $U_{j+1} = U_j \cup Q_{j+1} = (M_j \cap N_j) \cup (B_{j+1} \cap D_{j+1})$. Here $D_{j+1} \supseteq \overline{B}_{j+1} \supseteq M_j \supseteq M_j \cap N_j$. By modularity $U_{j+1} = D_{j+1} \cap [(M_j \cap N_j) \cup B_{j+1}]$. Here $B_{j+1} \subseteq \overline{B}_h \subseteq D_h$, $h = 1, \cdots, j$, whence $B_{j+1} \subseteq N_j$. Finally, $U_{j+1} = D_{j+1} \cap [N_j \cap (B_{j+1} \cup M_j)] = N_{j+1} \cap M_{j+1}$, proving the induction. For $r = n$, $M_r = T$, whence

$$C = Q_1 \cup \cdots \cup Q_n = D_1 \cap D_2 \cap \cdots \cap D_n \supseteq A_1, \tag{8.5.3}$$

the last relation holding since $D_h \supseteq A_1$, $h = 1, \cdots, n$. Since $C \supseteq A_1$ we may apply modularity to find $(C \cap \bar{A}_1) \cup A_1 = C \cap (\bar{A}_1 \cup A_1) = C \cap T = C$. Since, trivially, $C \cap \bar{A}_1 \cap A_1 = 0$, we have

$$C = A_1 \times (C \cap \bar{A}_1) = Q_1 \times \cdots \times Q_n. \tag{8.5.4}$$

Hence, by induction on dimension, the theorem is valid for C, and so A_1 is replaceable by some indecomposable factor of some Q (say, $E \subseteq Q_h$). By replaceability in C, $d(E) = d(A_1)$. Also, since $C = E \times (C \cap \bar{A}_1)$, we have $0 = E \cap C \cap \bar{A}_1 = E \cap \bar{A}_1$. Hence, $d(E \cup \bar{A}_1) = d(E) + d(\bar{A}_1) = d(A) + d(\bar{A}_1) = d(T)$, and so $T = \bar{A}_1 \cup E = E \times \bar{A}_1$. Moreover, $E \subseteq Q_h = (A_1 \cup \overline{B}_h) \cap B_h \subseteq B_h$ and $E \cap (\bar{A}_1 \cap B_h) = E \cap \bar{A}_1 = 0$. $E \cup (\bar{A}_1 \cap B_h) = B_h \cap (E \cup \bar{A}_1) = B_h \cap T = B_h$, whence

$$B_h = E \times (\bar{A}_1 \cap B_h). \tag{8.5.5}$$

But by assumption B_h was indecomposable and $d(E) = d(A_1) > 0$. Hence $B_h = E$ and $\bar{A}_1 \cap B_h = 0$. This yields

$$T = A_1 \times \bar{A}_1 = B_h \times \bar{A}_1. \tag{8.5.6}$$

Also $B_h = E \subseteq Q_h \subseteq B_h$, and so, $B_h = Q_h = (A_1 \cup \overline{B}_h) \cap B_h$. Thus $B_h \subseteq A_1 \cup \overline{B}_h$, and so, $A_1 \cup \overline{B}_h \supseteq B_h \cup \overline{B}_h = T$. Since $d(A_1) = d(B_h)$, we must also have $d(A_1 \cap \overline{B}_h) = d(A_1) + d(\overline{B}_h) - d(A_1 \cup \overline{B}_h) = d(B_h) + d(\overline{B}_h) - d(T) = 0$. Hence

$$T = A_1 \times \overline{B}_h \tag{8.5.7}$$

and A_1 and B_h are mutually replaceable. Here $h \neq 1$, since $A_1 \cup \overline{B}_h = T$, while $A_1 \cup \overline{B}_1 \neq T$.

Case 3. $A_1 \cup \overline{B}_j = T$ for all j but $\bar{A}_1 \cup B_j \subset T$ for all j, the only possibility not covered by Cases 1 or 2.

Reversing the roles of the A's and B's, we may apply Case 2 and then any specified B (say, B_n) is mutually replaceable with some A not A_1, which by renumbering we may take as A_m. Then

$$(8.5.8) \qquad \begin{aligned} T &= A_1 \times \cdots \times A_{m-1} \times A_m \\ &= B_1 \times \cdots \times B_{n-1} \times A_m = \overline{B}_n \times A_m. \end{aligned}$$

Here $z \to (z \cup A_m) \cap \bar{A}_m$ is a projectivity of the quotient $\overline{B}_n/0$ onto $\bar{A}_m/0$ and by the corollary to Theorem 8.3.1, is a lattice isomorphism. Hence, if we put $B_j^* = (B_j \cup A_m) \cap \bar{A}_m$, $j = 1, \cdots, n-1$, we find from the isomorphism

$$(8.5.9) \qquad \bar{A}_m = A_1 \times \cdots \times A_{m-1} = B_1^* \times \cdots \times B_{n-1}^*.$$

By induction on dimension the theorem is true for $\bar{A}_m$, and so, A_1 is replaceable by some B_j^* (say, B_1^*) in $\bar{A}_m$. Here $B_1 \cup A_m = (B_1 \cup A_m) \cap (\bar{A}_m \cup A_m) = [(B_1 \cup A_m) \cap \bar{A}_m] \cup A_m = B_1^* \cup A_m$. Hence $B_1 \cup \bar{A}_1 = (B_1^* \cup A_2 \cup \cdots \cup A_{m-1}) \cup A_m = (A_1 \cup A_2 \cdots \cup A_{m-1}) \cup A_m = T$, since B_1^* replaces A_1 in $\bar{A}_m$. But here $A_1 \cup \overline{B}_1 = T = B_1 \cup \bar{A}_1$, and therefore Case 1 applies, and A_1 and B_1 are mutually replaceable. Note that Case 3 does not actually arise and that in every case for a given A_1, there is a B_j such that A_1 and B_j are mutually replaceable.

For groups the theorem is:

Theorem 8.5.2 (Theorem of Wedderburn-Remak-Schmidt†). *Let G be a group whose normal subgroups form a finite dimensional lattice. Then if G has two representations as direct products of indecomposable subgroups*

$$G = A_1 \times \cdots \times A_m,$$
$$G = B_1 \times \cdots \times B_n,$$

then $m = n$, any A_i is mutually replaceable by some B_j, and the A's and B's are pairwise centrally isomorphic. The theorem is valid for G as a group with operators or for the congruence relations on inverse loops.

Proof: Since we have already established that normal subgroups form a modular lattice, we need only observe that the definitions of

† The original proof of this theorem is due to J. H. M. Wedderburn [1]. R. Remak corrected an omission [1, 2]. P. Schmidt extended this to groups with operators. The lattice theorem (8.5.1) was proved by Ore [1], but the form given here is taken from G. Birkhoff [1] with a few changes.

direct product agree. In $G = A_i \times \bar{A}_i = B_j \times \bar{A}_i$ we have both A_i and B_j perspective to $G/\bar{A}_i$ and hence projective. Thus there is a central isomorphism established between A_i and B_j which becomes a central automorphism of G if we map $\bar{A}_i$ into itself. Hence corresponding elements of A_i and B_j differ by a factor in the center of G.

8.6. Composition Series in Groups.

Suppose $G = A_0 \supset A_1 \supset \cdots \supset A_n = H$ is a composition series from G to a subgroup H. By definition A_{i+1} is a maximal normal subgroup of A_i. Hence A_i/A_{i+1} is a simple group, since a normal subgroup of A_i/A_{i+1} would correspond to a normal subgroup of A_i containing A_{i+1} (Theorem 2.3.4). Hence if A_i/A_{i+1} is Abelian it can contain no proper subgroup and must be finite of prime order. There is a relation between chief series and composition series given in the following theorem:

THEOREM 8.6.1. *Let H be a normal subgroup of G such that there is a composition series from G to H. Then there is a chief series from G to H,*

$$G = B_0 \supset B_1 \supset \cdots \supset B_m = H,$$

and each factor group B_i/B_{i+1} is the direct product of a finite number of isomorphic simple groups. Conversely, if such a series exists with B_i/B_{i+1} a direct product of a finite number of isomorphic simple groups, then there is a series of composition from G to H.

Proof: Any normal series from G to H can be refined to a composition series by inserting further terms. Hence any normal series from G to H is necessarily shorter than a composition series and therefore is of finite length. Hence there must be a chief series from G to H,

$$G = B_0 \supset B_1 \supset \cdots \supset B_m = H.$$

If $m = 1$, G/H is a simple group and the theorem is true. Let us use induction on m, whence each of $B_0/B_1, \cdots, B_{m-2}/B_{m-1}$ is the direct product of a finite number of isomorphic simple groups. It remains to be proved that B_{m-1}/B_m is the direct product of a finite number of isomorphic simple groups.

Any normal subgroup of B_{m-1}/B_m corresponds to a group normal in B_{m-1} containing B_m. Hence there exists a minimal normal subgroup

K/B_m where $K \supset B_m$ and K is normal in B_{m-1}. If $K = B_{m-1}$, then B_{m-1}/B_m is simple and there is nothing further to prove. Now consider the conjugates K_j of K under G. $K_j \subseteq B_{m-1}$, since B_{m-1} is normal in G. Moreover, since transformation by an element of G induces an automorphism in B_{m-1}, every K_j is a normal subgroup of B_{m-1}. Also, $\bigcup_j K_j$ is a normal subgroup of G, since transformation by an element of G merely permutes the K_j among themselves. Hence $\bigcup_j K_j = B_{m-1}$ since there is no normal subgroup of G between B_{m-1} and B_m. Take $K = K_1$, $K_2 \nsubseteq K_1$, $K_3 \nsubseteq K_1 \cup K_2$, and $K_j \nsubseteq K_1 \cup \cdots \cup K_{j-1}$. Each of $U_j = K_1 \cup \cdots \cup K_j$ is a normal subgroup of B_{m-1} and contains the preceding U_{j-1}. Since there is a composition series from G to B_m including B_{m-1}, there can be only a finite number of U_j's, whence for some finite j, $B_{m-1} = K_1 \cup \cdots \cup K_j$. Now a K_i not contained in the union of the remaining K's must intersect the union of the remaining ones in B_m, since every K is a minimal normal subgroup of B_{m-1} containing B_m. Hence, deleting the K's contained in the union of the remainder, $B_{m-1}/B_m = K_1/B_m \cup \cdots \cup K_s/B_m$, where each K_i/B_m is a normal subgroup of B_{m-1}/B_m intersecting the union of the remainder in the identity. But by Theorem 3.2.2, B_{m-1}/B_m is the direct product of $K_1/B_m, \cdots, K_s/B_m$. Now if K_1/B_m had a proper normal subgroup, this would be a normal subgroup of B_{m-1}/B_m, since it would be normal in K_1/B_m and surely normalized by the remaining direct factors. But K_1/B_m was assumed to be a minimal normal subgroup; therefore K_1/B_m is a simple group and B_{m-1}/B_m is the direct product of the s isomorphic simple groups.

For the converse part of the theorem we observe that $B_m \subset K \subset U_2 \subset U_3 \cdots \subset B_{m-1}$ is part of a composition series since each factor group is simple.

THEOREM 8.6.2.* *The intersection of two subinvariant subgroups of G is a subinvariant group of G. Both the union and intersection of two subgroups occurring in composition series will occur in a composition series.*

Proof: Suppose A and B are two subinvariant groups of G. Then by definition we have two chains:

$$A = A_r \triangleleft A_{r-1} \triangleleft \cdots \triangleleft A_1 \triangleleft G,$$
$$B = B_s \triangleleft B_{s-1} \triangleleft \cdots \triangleleft B_1 \triangleleft G.$$

* These results are due to H. Wielandt [2].

Here, in the chain $A = A_r \supseteq A_r \cap B_1 \supseteq \cdots A_r \cap B_s = A \cap B$, each subgroup is either equal to or normal in its predecessor (Theorem 2.4.1). Hence

$$A \cap B \lhd C_u \lhd C_{u-1} \cdots \lhd C_1 \lhd A_r \lhd \cdots \lhd A_1 \lhd G,$$

where the C_i are the distinct subgroups of the set above and $A \cap B$ is subinvariant.

Now suppose the preceding two chains are composition series. Then if $B_1 \neq A_1$, $G = A_1 \cup B_1$, since both B_1 and A_1 were maximal normal subgroups of G. Here $A_1 \cap B_1$ is a normal subgroup of G, and $A_1/A_1 \cap B_1 \cong G/B_1$ and is therefore simple; therefore $A_1 \cap B_1$ is a maximal normal subgroup of A_1. Here either $A_1 \cap B_1 = A_2$ or $A_1 \cap B_1$ and A_2 are both maximal normal subgroups of A_1, whence $A_1 = A_2 \cup (A_1 \cap B_1)$ and $A_2 \cap B_1 = A_2 \cap (A_1 \cap B_1)$ and so $A_2/A_2 \cap B_1 \cong A_1/A_1 \cap B_1 \cong G/B_1$ is simple. Also $A_1 \cap B_1/A_2 \cap B_1 \cong A_1/A_2$. Continuing in this way, either $A = A_r = A_r \cap B_1$ or $A_r \cap B_1 \lhd A_r$ and $A_r/A_r \cap B_1 = G/B_1$ is simple. Here we have series of composition,

$$A_r \cap B_1 \lhd A_{r-1} \cap B_1 \lhd \cdots \lhd A_1 \cap B_1 \lhd B_1 \lhd G,$$
$$B_s \lhd B_{s-1} \lhd \cdots \lhd B_2 \lhd B_1 \lhd G,$$

similar to those above but involving fewer terms below B. Now repeat with B_2 in the role of B_1, etc., and we shall ultimately find a composition series from G to $A \cap B$.

To show that the union of two composition groups (as we shall refer to subgroups occurring in composition series) is again a composition group is more difficult. We use induction on the lengths r and s of the two composition series from $A = A_r$ and $B = B_s$ to G. Specifically, we shall use induction on $r + s$, the theorem being true for $r + s = 2$, since $A_1 \cup B_1$ is a normal subgroup of G. For this we need a lemma.

LEMMA 8.6.1. *If C is a composition group of G which properly contains the composition group A, then there is a composition series from G to A which includes C, and in particular, the length of a composition series from G to C is less than the length of a composition series from G to A.*

This follows since if

$$C = C_t \lhd C_{t-1} \lhd \cdots \lhd C_1 \lhd G$$

and

$$A = A_r \lhd A_{r-1} \lhd \cdots \lhd A_1 \lhd G$$

are composition series for A and C, then as before,

$$A_r = A_r \cap C_t \lhd A_{r-1} \cap C_t \cdots \lhd A_1 \cap C_t \lhd C_t \cdots \lhd C_1 \lhd G,$$

and the distinct groups from C_t to A_r will complete a composition series from G to A_r which is therefore of length r, and hence $r > t$.

By induction both $A_{r-1} \cup B_s$ and $A_r \cup B_{s-1}$ are composition groups of G. If $A_{r-1} \cup B_s$ is a proper subgroup of G then A_r and B_s are composition groups in $A_{r-1} \cup B_s$ with lengths $r' < r$ and $s' < s$ (by the lemma) as composition groups in $A_{r-1} \cup B_s$. Then by induction $A_r \cup B_s$ is a composition group of $A_{r-1} \cup B_s$, and hence of G. Hence assume $A_{r-1} \cup B_s = G$. Similarly, we may apply induction unless we also assume $A_r \cup B_{s-1} = G$. Now suppose by symmetry that $r < s$. Here, if $b \in B_s$,

$$b^{-1}A_r b \lhd b^{-1}A_{r-1}b \lhd \cdots \lhd b^{-1}A_2 b \lhd A_1 \lhd G,$$

where $b^{-1}A_1 b = A_1$, since A_1 is normal. Now if $b^{-1}A_r b \neq A_r$, then in A_1, A_r, and $b^{-1}A_r b$ are composition groups and the length of the series is $r - 1$ in both cases. Hence by induction $A^* = A_r \cup b^{-1}A_r b$ is a composition group in A_1, where the length of a chain from A_1 to A^* is less than $r - 1$ Hence by induction $B_s \cup A^*$ is a composition group. But $B_s \cup A^* = B_s \cup A_r = B \cup A$. Thus we may suppose that A_r is transformed into itself by every element of B_s. But A_r is also transformed into itself by every element of A_{r-1}. Hence A_r is normal in $A_{r-1} \cup B_s = G$. As a normal subgroup of G we may take A_r as A_1. But then $B \cup A = B_s \cup A_1 \lhd B_{s-1} \cup A_1 \lhd \cdots B_1 \cup A_1 \lhd G$. This holds since B_i and A_1 are transformed into themselves by B_{i-1}, and clearly, A_1 as a subgroup of $B_i \cup A_1$ transforms it into itself. Hence $B_i \cup A_1 \lhd B_{i-1} \cup A_1$. Thus $B \cup A$, as a subinvariant group of G containing a group A in a composition series, is also a composition group.

EXERCISES

1. Let the group G be of order $p^r q^s$. If G has two composition series $1 \subset A_1 \subset A_2 \subset \cdots \subset A_r \subset A_{r+1} \subset \cdots \subset A_{r+s} = G$ and $1 \subset B_1 \subset B_2 \subset \cdots \subset B_s \subset B_{s+1} \subset \cdots \subset B_{r+s} = G$, where A_r is of order p^r and B_s is of order q^s, show that G is the direct product of A_r and B_s.
2. Generalize the result of Ex. 1 to show that if G is a finite group and if for every prime p dividing the order of G there is a composition series of G, one of whose terms is a Sylow subgroup $S(p)$, then G is the direct product of its Sylow subgroups.

3. Show that an automorphism of the direct product of a finite number of non-Abelian simple groups permutes the factors.
4. If a finitely generated group G has exactly one maximal subgroup A, show that G is generated by any element not in A. Prove that G is cyclic of prime power order.
5. If a finitely generated group G has exactly two maximal subgroups A and B and $[G:A] = p$, $[G:B] = q$, where p and q are different primes, show that G is cyclic of order $p^i q^j$. (Hint: Show that $A \cap B$ is normal and that $G/A \cap B$ is cyclic.)
6. Suppose that G is a finite group such that $L(G)$ is of dimension 2. Show that if the order of G is not divisible by a square, then at least one Sylow subgroup is normal. Hence conclude that G is of order p^2 or pq, p and q being primes.

9. A THEOREM OF FROBENIUS; SOLVABLE GROUPS

9.1. A Theorem of Frobenius.

Theorem 9.1.1. in its original form due to Frobenius [2], is of an entirely different nature from most of the other results in group theory. It does not deal with subgroups, homomorphisms, or permutation representation but with the number of solutions of an equation in a finite group. It has been greatly generalized by Philip Hall [3], who has generalized both the equation studied and the information on the solutions. But here we shall give only a mild generalization of the original theorem.

THEOREM 9.1.1. *Let G be a group of order g and let C be a class of h conjugate elements. The number of solutions of $x^n = c$, where c ranges over C is a multiple of (hn, g).*

Proof: Let $A(K, n)$ designate the complex of those elements of G whose nth powers lie in the complex K, and let $a(K, n)$ designate the number of elements in $A(K, n)$. For $g = 1$, $(hn, 1) = 1$, and the result is trivial, while for $n = 1$ the number of solutions is $h = (h, g)$. We shall use induction on g and n, assuming the theorem for any $g' < g$ or $g' = g$ and $n' < n$.

If $c' = u^{-1}cu$ and $x^n = c$, then $(u^{-1}xu)^n = c'$, giving a one-to-one correspondence between the solutions for an element c and any of its conjugates. Thus $a(C, n) = h \cdot a(c, n)$. If $x^n = c$, then $x^{-1}cx = x^{-1}(x^n)x = x^n = c$, and the solutions of $x^n = c$ lies in the normalizer N_c of c, which by Theorem 1.6.1 is of order g/h. Hence if $h > 1$, the theorem being true in N_c, $a(c, n)$ is a multiple of $(n, g/h)$, and so, $a(C, n) = h \cdot a(c, n)$ is a multiple of $h(n, g/h) = (hn, g)$, proving the theorem.

Hence, suppose $h = 1$. If $n = n_1n_2$, $(n_1, n_2) = 1$, $n_1 > 1$, $n_2 > 1$, and if $D = A(C, n_2)$, then $A(C, n) = A(D, n_1)$. D consists of complete classes. By induction (n_1, g) is a divisor of $a(C, n)$ and, similarly, (n_2, g) is a divisor of $a(C, n)$. But then, since (n_1, g) and (n_2, g) are relatively prime, their product $(n_1, g)(n_2, g) = (n_1n_2, g) = (n, g)$

divides $a(C, n)$, proving the theorem. We may now suppose $n = p^e$ is the eth power of a prime. If p divides the order u of c, then an element x in $A(c, n)$ has order nu. Then exactly n elements in the cyclic subgroup generated by x belong to $A(c, n)$, and all these generate the same subgroup. Hence $A(c, n)$ is divisible by n.

Finally, we suppose that $n = p^e$ is relatively prime to the order u of c. Since $h = 1$, c is in the center of G. The elements in the center of G whose orders are not divisible by p form an Abelian group B whose order b is not divisible by p.

Now let c_1 and c_2 be two elements of B. Since $p \nmid b$, the equation $c_2 = c_1 y^n$ has a unique solution y in B. But then if $x^n = c_1$, we have $(xy)^n = c_2$ and so, $a(c, n)$ has the same value for every $c \in B$. Finally, the equation

$$g = \sum_{C \notin B} a(C, n) + ba(c, n)$$

counts the g elements of G according to the class in which their nth powers lie, counting first for those classes not in B, and last for B, b times the number for one of them. Now (n, g) divides every term $a(C, n)$ in the first sum, each term of this being covered by induction or a previous part of the proof. Also, since (n, g) divides g and is prime to b, it must follow that (n, g) divides $a(c, n)$, completing the proof of the theorem in all cases.

If c is the identity then $h = 1$ and we have the original form of the Theorem of Frobenius. Here $x^g = 1$ for all elements, and so, if $(n, g) = m$, from $x^n = 1$ follows $x^m = 1$.

THEOREM 9.1.2. *If n is a divisor of the order of a group G, then the number of solutions of $x^n = 1$ in G is a multiple of n.*

Note that since the identity satisfies the equations, the number of solutions is not zero and hence must be at least n.

In connection with this theorem there is an interesting conjecture: *If n divides the order of G and there are exactly n solutions of $x^n = 1$, then these solutions form a normal subgroup of G.*

Note that if G contains a subgroup H of order n, then the elements of H will be the solutions. Moreover, if $x^n = 1$, then for an arbitrary z, $(z^{-1}xz)^n = 1$, whence H will be a normal subgroup. The problem then consists in showing that the n solutions form a subgroup H. The assumption that n divide the order of G is essential, since by the Theorem of Lagrange the order of a subgroup divides the order of the group. Also, $x^4 = 1$ has exactly four solutions in the symmetric group on three letters, which is of order 6, but these do not form a subgroup.

9.2. Solvable Groups.

The element $x^{-1}y^{-1}xy$ of a group G is called the *commutator* of x and y, and we write $x^{-1}y^{-1}xy = (x, y)$. We also define commutators of higher order by the recursive rule $(x_1, \cdots, x_{n-1}, x_n) = ((x_1, \cdots, x_{n-1}), x_n)$. These are the *simple commutators.* More generally, the set of all elements which can be obtained by successive commutation are called *complex commutators*; for example, $((a, b), (c, d, e))$. We define the weight ω of a commutator recursively by saying that elements g of G are of weight one, $\omega(g) = 1$, and putting $\omega(x, y) = \omega(x) + \omega(y)$. Thus the weight of an element which is a commutator depends on the form of the commutator by which it is expressed and not on the element itself.

From its definition $(x, y) = 1$ if, and only if, $yx = xy$. Thus all commutators in G are 1 if, and only if, G is an Abelian group, and the commutators may be regarded as measuring the extent to which a group departs from being Abelian. The subgroup G' of G generated by all commutators $x^{-1}y^{-1}xy$ is called the *commutator subgroup* or *derived group.* Clearly, G' is a fully invariant subgroup of G.

THEOREM 9.2.1. *The factor group G/G' is Abelian. If K is a normal subgroup of G such that G/K is Abelian, then $K \supseteq G'$.*

Proof: In the mapping $G \to G/G' = H$, let u, v be arbitrary elements of H, and suppose $x \to u$, $y \to v$. Then $x^{-1}y^{-1}xy \to u^{-1}v^{-1}uv$. But $x^{-1}y^{-1}xy \in G'$, whence $x^{-1}y^{-1}xy \to 1 = u^{-1}v^{-1}uv$, and hence $vu = uv$ and G/G' is Abelian. Now suppose that G/K is Abelian. For $x, y \in G$, and $x \to u$, $y \to v$ in $G \to G/K$, we have $x^{-1}y^{-1}xy \to u^{-1}v^{-1}uv = 1$. Thus every commutator $x^{-1}y^{-1}xy$ belongs to K, and therefore $K \supseteq G'$.

DEFINITION: *A group G is said to be solvable if the sequence $G \supseteq G' \supseteq G'' \cdots \supseteq \cdots \supseteq G^{(i)} \cdots$, where each group $^{(i)}$ is the derived group of the preceding, terminates in the identity in a finite number of steps, say, $G^{(e)} = 1$.*

By Theorem 9.2.1 each factor group $G^{(i)}/G^{(i+1)}$ is Abelian. Note that if $G^{(i)} = G^{(i+1)}$, then $G^{(i)} = G^{(j)}$ all $j \geq i$. Hence the inclusions of Theorem 9.2.1 are proper until $G^{(i)} = 1$.

THEOREM 9.2.2. *Every subgroup and factor group of a solvable group is solvable.*

Proof: Let G be solvable and H a subgroup of G. Then by definition $H' \subseteq G'$, since H' is generated by all commutators of elements in H and G' by all commutators in G. Hence $H'' \subseteq G''$, etc., and so if $G^{(e)} = 1$, then $H^{(e)} = 1$ and H is solvable. Here $H^{(i)}$ may be the identity for some $i < e$. If $Q = G/K$ is a factor group of G, consider the homomorphism $G \to Q$. Here every commutator in Q is the image of a commutator in G, whence $G' \to Q'$. Continuing, $G^{(e)} \to Q^{(e)}$, whence $Q^{(e)} = 1$ if $G^{(e)} = 1$. Again $Q^{(i)}$ may be the identity for some $i < e$.

THEOREM 9.2.3.* *A group of finite order is solvable if, and only if, the factor groups in a series of composition from G to 1 are cyclic of prime order.*

Proof: Suppose $G = A_0 \supset A_1 \supset \cdots \supset A_r = 1$, where each A_{i-1}/A_i, $i = 1 \cdots r$ is cyclic of some prime order. By Theorem 9.2.1, since G/A_1 is Abelian, $A_1 \supseteq G'$. Similarly, $A_2 \supseteq A_1' \supseteq G''$, and finally $A_r \supseteq G^{(r)}$, whence $G^{(r)} = 1$ and G is solvable. Conversely, suppose G is solvable and finite. Since G/G' is Abelian, in

$$G \supset G' \supset G'' \supset \cdots \supset G^{(e)} = 1,$$

a maximal normal subgroup $A_1 \supseteq G'$ will exist. Since G/A_1 is simple and Abelian, it is cyclic of prime order. Similarly, since A_1 is solvable, A_1 contains a maximal normal subgroup A_2 such that A_1/A_2 is cyclic of finite order. Continuing, we have $G = A_0 \supset A_1 \supset \cdots \supset A_r = 1$ with each A_{i-1}/A_i cyclic of prime order. By the Jordan-Hölder theorem the same is true of every composition series.

THEOREM 9.2.4. *In a chief series for a solvable finite group G*

$$G = C_0 \supset C_1 \supset \cdots \supset C_s = 1,$$

the factor groups C_{i-1}/C_i, $i = 1, \cdots, s$ are elementary Abelian groups.

---

* Historically, this property of a composition series was the original definition of solvability, but such a definition is inapplicable to infinite groups. The Galois theory shows that a polynomial equation $f(x) = 0$ is solvable by radicals if, and only if, its Galois group is solvable.

Proof: By Theorem 8.6.1, C_{i-1}/C_i is the direct product of isomorphic simple groups. By Theorem 9.2.2 these simple groups are solvable and hence cyclic of prime order. Thus C_{i-1}/C_i is the direct product of cyclic groups of the same prime order p and is an elementary Abelian group. Conversely, if G has such a chief series, since the factor groups are Abelian, G will be solvable. The numbers $c_1, \cdots, c_s$, which are the orders of $C_0/C_1, \cdots, C_{s-1}/C_s$, respectively, are called the *chief factors* of G and are prime powers as shown. Clearly, for a factor group G/K, the chief factors are a subset of those for G, since there will be a chief series of G including the normal subgroup K. For a subgroup H of G, the distinct members of

$$H \supseteq H \cap C_1 \supseteq H \cap C_2 \supseteq \cdots \supseteq H \cap C_s = 1$$

will be a normal series in H and either this or its refinement will be a chief series for H, whence the chief factors of H will be divisors of those for G, since $H \cap C_{i-1}/H \cap C_i$ is isomorphic to a subgroup of C_{i-1}/C_i.

THEOREM 9.2.5. *The following two properties of a group G are equivalent to solvability:*

1) *G has a finite normal series*

$$G = A_0 \supseteq A_1 \supseteq A_2 \supseteq \cdots \supseteq A_s = 1$$

in which every A_{i-1}/A_i, $i = 1, \cdots, s$ is Abelian.

2) *G has a finite subinvariant series*

$$G = B_0 \supseteq B_1 \supseteq B_2 \supseteq \cdots \supseteq B_t = 1$$

in which every B_{i-1}/B_i, $i = 1, \cdots, t$ is Abelian.

Proof: If G is solvable, then its derived series

$$G \supset G' \supset G'' \supset \cdots \supset G^{(r)} = 1$$

is a finite normal series in which $G^{(i-1)}/G^{(i)}$ is Abelian for $i = 1, \cdots, r$, whence property (1) holds and *a fortiori* property (2) holds. It remains to show that property (2) implies solvability. Here if $G = B_0 \supseteq B_1 \supseteq B_2 \supseteq \cdots \supseteq B_t = 1$ is a subinvariant series with B_{i-1}/B_i Abelian for $i = 1, \cdots, t$; then, as $G/B_1 = B_0/B_1$ is Abelian, $B_1 \supseteq G'$. Similarly, if $B_{i-1} \supseteq G^{(i-1)}$, then $B_i \supseteq B'_{i-1} \supseteq G^{(i)}$. Hence, ultimately, $1 = B_t \supseteq G^{(t)}$ and $G^{(t)} = 1$, whence G is solvable.

COROLLARY 9.2.1. *A group G is solvable if it has a normal subgroup H such that both H and G/H are solvable.*

If $G/H \supseteq A_1/H \supseteq \cdots \supseteq A_{r-1}/H \supseteq H/H$, and $H \supseteq B_1 \supseteq \cdots \supseteq B_{s-1} \supseteq 1$ are series satisfying the second property for G/H and H, respectively, then $G \supseteq A_1 \supseteq \cdots \supseteq A_{r-1} \supseteq H \supseteq B_1 \supseteq \cdots \supseteq B_{s-1} \supseteq 1$ is a series satisfying the second property for G.

9.3. Extended Sylow Theorems in Solvable Groups.

A Sylow subgroup of a finite group has the property that its order $m = p^a$ is prime to the order of its index n. Philip Hall [1] has shown that the Sylow theorems generalize for solvable groups in terms of subgroups whose order m is prime to their index n without the requirement that m be a prime power.

THEOREM 9.3.1. *Let G be a solvable group of order mn where $(m, n) = 1$. Then*

1) *G possesses at least one subgroup of order m.*

2) *Any two subgroups of order m are conjugate.*

3) *Any subgroup whose order m' divides m is contained in a subgroup of order m.*

4) *The number h_m of subgroups of order m may be expressed as a product of factors, each of which (a) is congruent to 1 modulo some prime factor of m, and (b) is a power of a prime and divides one of the chief factors of G.*

Proof: Note that for $m = p^a$, a prime power, properties (1) and (3) are given in the first Sylow theorem (Theorem 4.2.1), property (2) is the second Sylow theorem, and property (4) in a stronger statement than the third Sylow theorem.

The proof will be by induction on the order of G being trivially true if the order of G is a power of a prime. The proof will rest heavily on the structure of a chief series of G as given in Theorem 8.3.3 and the structure of factor groups (Theorem 2.3.4).

CASE 1. G has a proper normal subgroup H of order m_1n_1 and index m_2n_2, where $m = m_1m_2$, $n = n_1n_2$, and $n_1 < n$.

For property (1) G/H by induction contains a subgroup of order m_2 which corresponds to a subgroup D of G or order mn_1. D by induction contains a subgroup of order m.

For property (2), if M and M' are two subgroups of order m,

$M \cup H = MH$ and $M' \cup H = M'H$ are subgroups whose orders divide $m_1m_2 \cdot m_1n_1$, since $M \cup H/H \cong M/M \cap H$ (Theorem 2.4.1). Since the order also divides mn, it must divide mn_1. But it is also a multiple of m and a multiple of n_1. Hence both $M \cup H$ and $M' \cup H$ are of order $mn_1 = m_1n_1m_2$, and therefore $M \cup H/H$ and $M' \cup H/H$ are subgroups of G/H of order m_2 and are by induction conjugate. If a^* in G/H transforms $M' \cup H/H$ into $M \cup H/H$, and a in G is mapped into a^* by the homomorphism $G \to G/H$, then $a^{-1}(M' \cup H)a$ is mapped into $M \cup H/H$; in other words, $a^{-1}(M' \cup H)a = M \cup H$. Here $a^{-1}M'a$ and M are of order m in $M \cup H$ and are by induction conjugate. Hence M and M' are conjugate in G.

For property (3), if M_1 is a subgroup of order m', a divisor of m, then the order of $M_1 \cup H/H$ is a divisor of m_2 and hence it belongs to a subgroup of G/H of order m_2. Thus M_1 belongs to the corresponding subgroup of G order mn_1 and by induction on this group M_1 belongs to a subgroup of order m.

For property (4), following the proof of (2), the number h_m of conjugates of M of order m is the product of h_{m_2}, the number of subgroups of order m_2 in G/H and the number of conjugates of M in $M \cup H = D$. Here the chief factors of D divide those of G and the chief factors of G/H are a subset of those of G. Thus by induction h_m is a product of two factors, both of which satisfy property (4) and thus the property is proved.

Now the least normal subgroup K in a chief series is of order p^a, with p a prime. K will satisfy the requirements for the H of Case 1 unless $n = p^a$. Thus we may assume that every minimal normal subgroup is of order p^a. But as Sylow subgroups of order p^a there can be only one.

Case 2. G contains a unique minimal normal subgroup K of order $n = p^a$.

For property (1) let L be a minimal normal subgroup properly containing K. Then L/K is of order q^b with $q \neq p$. Let Q be a Sylow subgroup of L of order q^b, and let M be the normalizer of Q in G. Consider $M \cap K = T$. T is a normal subgroup of M and, as a subgroup of K, is elementary Abelian. Every element of T permutes with every element of Q, since a commutator of an element in Q and an element in T lies in $T \cap Q = 1$. Hence T belongs to the center C of L, which, as a characteristic subgroup of L, is a normal subgroup of G. Since K is minimal and unique, $C = K$ or $C = 1$. If $C = K$, then $L = K \times Q$, and Q is a normal subgroup of G contrary to the uniqueness of K. Hence $T = C = 1$. Thus Q is its own normalizer in L and has as many conjugates in L as its index in L; that is, Q has $n = p^a$ conjugates in L. Any conjugate of Q in G lies in L, since L is normal. Hence Q has $n = p^a$ conjugates in G, whence M is of index $n = p^a$ in G and hence of order m.

For properties (2) and (4), the normalizers of the p^a conjugates of Q are conjugate and distinct. Thus we have p^a conjugate subgroups of order m. Also $p^a \equiv 1 \pmod{q}$ as the number of Sylow subgroups of order q^b in L. Now, if M' is any subgroup of order m, the order of $M' \cup L$ is divisible by both m and n, whence $M' \cup L = G$. As $G/L = M'/M' \cap L$, we see that $M' \cap L$ is of order q^b and hence a conjugate of Q. Also, $M' \cap L$ is normal in M', whence M' is the normalizer of a conjugate of Q. Thus the p^a conjugate subgroups of order m already found constitute all subgroups of order m. This proves both (2) and (4).

For property (3), let M' be a subgroup of order $m' | m$. Then, if M is of order m, $M \cap (M' \cup K) = M^*$ is of order m', and by property (2) for $M' \cup K$, M^* is conjugate to M'. Hence M' is contained in a conjugate of M, proving (3).

The above properties of solvable groups are usually violated in simple groups. The simple group of order 60 (the alternating group on five letters) has no subgroup of order 15 and therefore violates (1); it contains a subgroup of order 6, generated by (123) and (12)(45) which is not contained in a subgroup of order 12 and therefore violates (3). Finally the number of Sylow subgroups of order 5 is six, and since $6 = 2 \cdot 3$, the property (4) is also violated. The group of automorphisms of the elementary Abelian group A of order 8 is a simple group G of order 168. G permutes the seven subgroups of A of order 2 transitively and also the seven subgroups of order 4 transitively. Hence G possesses two distinct conjugate sets of subgroups of index 7 and order 24 and therefore violates property (2).

The first property of Theorem 9.3.1 in fact characterizes solvable groups. For the proof of this we need a theorem, which will be proved in Chap. 16 as Theorem 16.8.7.

THEOREM 9.3.2 (BURNSIDE). *A group of order p^aq^b, where p and q are primes, is solvable.*

Assuming this theorem we may characterize solvable groups in terms of the first property. In a group G of order g, a p-complement is a subgroup S_p' whose index p^e is the highest power of p dividing its order g. Thus the first property asserts the existence of p-complements in solvable groups, and with the aid of the Burnside theorem we shall prove the converse.

THEOREM 9.3.3. *If a group G contains a p-complement for every prime p dividing its order, then G is solvable.*

Proof: Let the order of G be g and let $g = p_1^{e_1} \cdots p_r^{e_r}$, where the p_i are primes. If H_1 and H_2 are subgroups of indices, $p_i^{e_i}$ and $p_j^{e_j}$, respectively, then because the indices are relatively prime (Theorem 1.5.6), $H_{12} = H_1 \cap H_2$ is of index $p_i^{e_i}p_j^{e_j}$. The intersection of H_{12} with a p^k complement will again by Theorem 1.5.6 be of index $p_i^{e_i}p_j^{e_j}p_k^{e_k}$. Continuing in this way, if $g = mn$ with $(m, n) = 1$, we may find a subgroup of order m and index n, which will be the intersection of p-complements for primes p dividing n. Thus, the existence of p-complements is sufficient to prove the existence of a subgroup of order m prime to its index n and thus to prove the full first property.

We shall assume the theorem true for groups of order less than g and proceed by induction. In a group of order p^a every maximal subgroup is of index p and a normal subgroup (Corollary 4.1.2), and therefore a group of order p^a is solvable. We assume the Burnside theorem that a group of order p^aq^b is solvable, and hence we may now consider only cases in which the order of G is divisible by at least three distinct primes. G contains a subgroup H of order $p^aq^b = m$ prime to its index n, $mn = g$, where p and q are two different primes dividing g. Now H, as a solvable group, contains a least normal subgroup K which (Theorem 9.2.4) is elementary Abelian of prime power order, say, p^i. Now K will be contained in a Sylow subgroup $P \subseteq H \subseteq G$ of order p^a. Here a q-complement L^* in G will contain a Sylow subgroup P^* conjugate to P in G. Hence transforming by some element in G will take L^* into a q-complement L containing P. Here $L \supseteq P$ and $H \supseteq P$, and so by their orders, $L \cap H = P$, $L \cup H = G$, and in fact, $LH = G$, since LH contains g distinct elements. Thus every coset of L contains an element of H, and therefore all conjugates of L are obtained by transforming by elements $h \in H$. But $h^{-1}Lh \supseteq K$, since $h^{-1}Kh = K$, K being normal in H. Thus the intersection M of the conjugates of L is a proper subgroup of G, since

$K \subseteq M \subset L$ and being an intersection of a complete set of conjugates is a normal subgroup of G.

Hence G contains a proper normal subgroup M. If S'_p is a p-complement in G, then $S'_p \cap M$ is a p-complement in M and $S'_p \cup M/M$ is a p-complement in G/M. Hence both M and G/M possess p-complements and by induction are solvable. Thus G is solvable.

9.4. Further Results on Solvable Groups.

THEOREM 9.4.1. *If G is a solvable group of order g and if n is a divisor of g such that $x^n = 1$ has exactly n solutions, then these solutions form a normal subgroup in G.*

Proof: We shall assume the theorem true for solvable groups of order lower than G, the theorem being true if g is a prime. Now, as a solvable group, G contains a least normal subgroup K which is elementary Abelian of order p^i. We consider two cases, one in which p divides n and one in which it does not.

CASE 1. p divides n.

Here every element of K is of order p and hence is among the solutions of $x^n = 1$. Let $n = p^j n_1$, $g = p^s g_1$. Here G/K is of order $p^{s-i}g_1$ and has order divisible by $u = p^{j-i}n_1$ if $j \geq i$, $u = n_1$ if $j < i$. Hence in G/K there are ku elements z such that $z^u = 1$. Now if x is an element of G such that $x \rightarrow z$ in the homomorphism $G \rightarrow G/K$ with $z^u = 1$, then $x^u \in K$, whence $x^{up} = 1$, and since up divides n, $x^n = 1$ for any such x. But these x's are the elements of ku cosets of K in G/K. Hence there are at least these kup^i x's in G satisfying $x^n = 1$. Now if $j < i$, up^i is a proper multiple of n, yielding more than n solutions of $x^n = 1$ contrary to assumption. Hence $j \geq i$ and $up^i = n$ and there are at least kn solutions. Thus $k = 1$ and there are exactly u solutions of $z^u = 1$ in G/K. By induction these u solutions form a normal subgroup H/K of G/K, and then H, the corresponding group in G, is a normal subgroup of G of order $up^i = n$ whose elements are the n solution of $x^n = 1$.

CASE 2. p does not divide n.

Here n divides the order of G/K and there are kn solutions of $z^n = 1$ in G/K. Here, if $y \in G$ with $y \rightarrow z$ in the homomorphism, $y^n \in K$ and $y^{pn} = 1$. Hence in G there are kn cosets of K of elements y

satisfying $y^{pn} = 1$. We assert that each coset Ky yields a distinct solution of $x^n = 1$. For, let Ky_1 and Ky_2 be distinct cosets of K with $y_1 \to z_1$, $y_2 \to z_2$, $z_1 \neq z_2$. Here $y_1^{pn} = 1$, $y_2^{pn} = 1$, and therefore $y_1^p = x_1$, $y_2^p = x_2$ are solutions of $x^n = 1$ in G. If $y_1^p = y_2^p$, then $z_1^p = z_2^p$. But $z_1^n = 1$, $z_2^n = 1$, and since $(p, n) = 1$, from this we would have $z_1 = z_2$, contrary to assumption. Hence if $z^n = 1$ has kn solutions in G/K, then $x^n = 1$ has at least kn solutions in G. Thus $k = 1$, and by induction, G/K contains a normal subgroup U/K of order n. Here the corresponding group U in G is of order p^in. But as a solvable group, U contains a p-complement H of order n. Thus the n elements of H are the n solutions of $x^n = 1$, and since transformation by an arbitrary element of G permutes the solutions of $x^n = 1$ among themselves, H is a normal subgroup of G.

THEOREM 9.4.2. *If two consecutive factor groups of derived groups* $G' \supset G'' \supset G''' \cdots$ *of a group G are cyclic, then the latter is the identity.*

Proof: We may take $G''' = 1$, taking G'/G'' and G''/G''' as cyclic, and must show $G'' = 1$. Let b be a generator of G''. Now G is the normalizer of G'', and if Z_b is the centralizer of G'', G/Z_b is isomorphic to a group of automorphisms of a cyclic group, and so, Abelian. Hence $Z_b \supseteq G'$. But then G'' is in the center of G' and G' is given by adjoining a single element to G''. But then G' is Abelian, and so, $G'' = 1$, as was to be shown.

We say that G is *metacyclic* if G/G' and G' are both cyclic. Here $G'' = 1$ and we have a two-step metacyclic group. By Theorem 9.4.2 there could not be a three-step metacyclic group.

THEOREM 9.4.3. *If the Sylow subgroups of a finite group G of order g are all cyclic, then G is metacyclic and is generated by two elements a and b with defining relations:*

$$a^m = 1, \quad b^n = 1, \quad b^{-1}ab = a^r,$$
$$mn = g,$$
$$[(r-1)n, m] = 1,$$
$$r^n \equiv 1 \pmod{m}.$$

Conversely, a group given by such defining relations has all its Sylow subgroups cyclic.

Proof: We must first show that G is solvable. Let $g = p_1^{e_1} \cdots p_s^{e_s}$, $p_1 < p_2 < \cdots < p_s$ be the decomposition of g as a product of primes.

We show first that for $m = p_j^{f_j} p_{j+1}^{e_{j+1}} \cdots p_s^{e_s}$, $f_j \leq e_j$, the equation $x^m = 1$ has exactly m solutions. This is surely true for $m = g$. Hence it suffices to show that if $x^{mp} = 1$ has exactly mp solutions and p is the smallest prime dividing mp, then $x^m = 1$ has exactly m solutions. Since the Sylow subgroup belonging to p is cyclic, then if p^{f+1} is the highest power of p dividing pm, there are elements of order p^{f+1} in G; therefore not all solutions of $x^{mp} = 1$ are also solutions of $x^m = 1$. Hence the km solutions of $x^m = 1$ (Theorem 9.1.2) are a proper part of the solutions of $x^{mp} = 1$, and hence $1 \leq k < p$. An element satisfying $x^{mp} = 1$ but not $x^m = 1$ has order t exactly divisible by p^{f+1}. Here there will be $\phi(t)$ elements, all generating the same cyclic group, all of which have order exactly divisible by p^{f+1}. Here, since p^{f+1} divides t, $\phi(t)$ is divisible by $p-1$. Hence $pm - km = (p - k)m$, the number of elements satisfying $x^{pm} = 1$ but not $x^m = 1$, is divisible by $p - 1$. Since p was the smallest prime dividing m, $p-1$ has no factor in common with m. Thus $p-1$ divides $p-k$, and since $1 \leq k < p$, this is possible only if $k = 1$; that is, if $x^m = 1$ has exactly m solutions. In particular for $m = p_s^{e_s}$, $x^m = 1$ has exactly m solutions. But there is a Sylow subgroup of this order which must therefore be a normal subgroup of G. This is cyclic and so, of course, solvable.

We have shown that a group G with cyclic Sylow subgroups must have a normal subgroup H. Then both H and G/H also have cyclic Sylow subgroups. We may assume inductively that H and G/H are solvable and so, G is also solvable, since a group of prime order is solvable.

An Abelian group whose Sylow subgroups are cyclic is itself cyclic. Hence in $G \supset G' \supset G'' \cdots$ the factor groups are cyclic, and hence by Theorem 9.4.2, $G'' = 1$. If $G' = 1$, then G is cyclic, and this case is covered if we take $b = 1$, $r = 1$, $n = 1$, $m = g$. Hence, suppose $G' \neq 1$, and let a be a generator of G' with $a^m = 1$. Let b be an element from a coset $G'b$ which is a generator of the cyclic factor group G/G'. Here a and b generate G and $b^{-1}ab = a^r$ with $r \neq 1$, since G' is a normal subgroup; if $r = 1$, G would be Abelian and hence cyclic, contrary to assumption. If G/G' is of order n, then $b^{-n}ab^n = a^{r^n} = a$ and $r^n \equiv 1 \pmod{m}$. Now every element of G is of the form $b^j a^i$, whence the most general commutator $(b^u a^v, b^j a^i)$ may be expressed in terms of commutators of the form (a^k, b^t); these in turn are powers of $a^{-1}b^{-1}ab = a^{r-1}$. Hence a^{r-1} generates G' and therefore

$(r - 1, m) = 1$. Now $b^n \epsilon G'$ is a power a^j of a which permutes with b, whence $a^{rj} = a^j$, but since $(r - 1, m) = 1$, $j = 0$ and so $b^n = 1$. If m and n had a prime factor p in common, $a^{m/p}$ and $b^{n/p}$ would generate a noncyclic subgroup of order p^2, contrary to the fact that Sylow subgroups are cyclic. Hence $(m, n) = 1$. This completes the direct part of the proof.

Conversely, suppose m, n, r, and g satisfy the relations above. Then $a \to a^r$, since $r^n = 1 \pmod{m}$, is an automorphism of the cyclic group generated by a, whose nth power (and possibly a lower power) is the identity. Thus with mn elements $b^j a^i$, j modulo n, i modulo m, and the product law $b^j a^i \cdot b^k a^t = b^{j+k} a^h$, $h = ir^k + t$, we may verify the associative law and the existence of inverses whence we have a group of order $g = mn$ with relations $a^m = 1$, $b^n = 1$, $b^{-1}ab = a^r$ and observe that the product law is a consequence of these defining relations. In this group every commutator is a power of $a^{-1}b^{-1}ab = a^{r-1}$, whence since $(r - 1, m) = 1$, G' is generated by a. Since $(m, n) = 1$, every Sylow subgroup is a conjugate of the subgroup $\{a\}$ or the subgroup $\{b\}$ and hence cyclic.

COROLLARY 9.4.1. *Every group G of square free order is metacyclic of the type in Theorem* 9.4.3.

This follows, since the Sylow subgroups are all of prime order and necessarily cyclic.

EXERCISES

1. Show that if a group G is of finite order divisible by 12 and if $x^{12} = 1$ has exactly 12 solutions in G, then these solutions form a normal subgroup.
2. Show that if G is of order p^2q, where p and q are different primes, then one of the Sylow subgroups is normal and G is solvable.
3. Show that if G is of order p^2qr, where p, q, r are different primes, then either G is solvable or G is the alternating group A_5 of order 60. Use Theorem 14.3.1 and its corollary.
4. Show that if $x^n = 1$ has exactly m solutions, $x_1 = 1, x_2, \cdots, x_m$, in a group G, then $K = \{x_1, \cdots, x_m\}$ is a normal subgroup of G and its elements are of the form $x_2^{a}x_3^{a_3} \cdots x_m^{a_m}$ and K is of order at most (n^{m-1}).

10. SUPERSOLVABLE AND NILPOTENT GROUPS

10.1. Definitions.

There are two properties of groups, qualitatively stronger than solvability, which are of considerable importance. These are supersolvability and nilpotence.

DEFINITION: *A group G is supersolvable if it possesses a finite normal series $G = A_0 \supseteq A_1 \supseteq A_2 \supseteq \cdots \supseteq A_r = 1$, in which each factor group A_{i-1}/A_i is cyclic.*

DEFINITION: *A group G is nilpotent if it possesses a finite normal series $G = A_0 \supseteq A_1 \supseteq A_2 \supseteq \cdots \supseteq A_r = 1$, in which A_{i-1}/A_i is in the center of G/A_i for $i = 1, \cdots, r$.*

Since in both these cases A_{i-1}/A_i is Abelian, these properties do imply solvability of G. Note that in a supersolvable group G, $A_{i-1} = \{b_{i-1}, A_i\}$, where b_{i-1} is any element of A_{i-1} mapped onto a generator of the cyclic group A_{i-1}/A_i, and thus G is finitely generated. Since nilpotent groups include all Abelian groups, it is clear that a nilpotent group need not be finitely generated.

Baer [12] defines supersolvability in a more general way, saying that G is supersolvable if every homomorphic image of G contains a cyclic normal subgroup. He shows this definition to be equivalent to ours for finitely generated groups, but with the broader definition, the properties proved in this chapter do not hold.

10.2. The Lower and Upper Central Series.

We write the commutator $x^{-1}y^{-1}xy$ as (x, y). For subgroups A, B, the notation (A, B) will mean the group generated by all (a, b) with $a \in A$, $b \in B$. We have defined *simple commutators* by the rule

$$(x_1, \cdots, x_{n-1}, x_n) = ((x_1, \cdots, x_{n-1}), x_n),$$

and similarly for subgroups $A_1, \cdots, A_{n-1}, A_n$ we define

$$(A_1, \cdots, A_{n-1}, A_n) = ((A_1, \cdots, A_{n-1}), A_n).$$

Let us represent conjugation by an exponent; thus

$$a^x = x^{-1}ax.$$

There are a number of important identities on the higher commutators:

(10.2.1.1) $(y,x) = (x,y)^{-1}$.
(10.2.1.2) $(xy,z) = (x,z)^y(y,z) = (x,z)(x,z,y)(y,z)$.
(10.2.1.3) $(x,yz) = (x,z)(x,y)^z = (x,z)(x,y)(x,y,z)$.
(10.2.1.4) $(x,y^{-1},z)^y(y,z^{-1},x)^z(z,x^{-1},y)^x = 1$.
(10.2.1.5) $(x,y,z)(y,z,x)(z,x,y)$
$= (y,x)(z,x)(z,y)^x(x,y)(x,z)^y(y,z)^x(x,z)(z,x)^y$.

These may be verified by direct calculation from the definitions of the commutators.

We define a series of subgroups of a group G by the rules:

$$\Gamma_1(G) = G,$$
$$\Gamma_k(G) = \{(x_1, \cdots, x_k)\},$$

for arbitrary $x_i \in G$.

Since $(y_1, y_2, \cdots, y_{k+1}) = [(y_1, y_2), y_3, \cdots, y_{k+1}]$, we see that $\Gamma_{k+1}(G) \subseteq \Gamma_k(G)$ for all k. Clearly, the $\Gamma_k(G)$ are fully invariant subgroups of G. The series

$$G = \Gamma_1(G) \supseteq \Gamma_2(G) \supseteq \Gamma_3(G) \supseteq \cdots$$

is called the *lower central series* of G.

THEOREM 10.2.1. $\Gamma_{k+1}(G) = (\Gamma_k(G), G)$.

Proof: Since $(y_1, \cdots, y_k, y_{k+1}) = ((y_1, \cdots, y_k), y_{k+1})$, we have trivially $\Gamma_{k+1}(G) \subseteq (\Gamma_k(G), G)$. To prove the inclusion in the other direction, we need the identities (10.2.1). In (10.2.1.2) put $x = (a_1, \cdots, a_k)$, $y = (a_1, \cdots, a_k)^{-1}$, $z = a_{k+1}$. Then $1 = (1, a_{k+1}) = (a_1, \cdots, a_k, a_{k+1})^y((a_1, \cdots, a_k)^{-1}, a_{k+1})$. Thus we have $((a_1, \cdots, a_k)^{-1}, a_{k+1}) \in \Gamma_{k+1}(G)$, since the other term belongs to $\Gamma_{k+1}(G)$. Now $(\Gamma_k(G), G)$ is generated by elements $(u_1u_2 \cdots u_n, g)$, where $u_i = (a_1, \cdots, a_k)$ or $(a_1, \cdots, a_k)^{-1}$. We have shown that $(u_i, g) \in \Gamma_{k+1}(G)$. We show by induction on n that $(u_1u_2 \cdots u_n, g) \in \Gamma_{k+1}(G)$. This we do by putting $x = u_1u_2 \cdots u_{n-1}$, $y = u_n$, $z = g$ in (10.2.1.2) so that we have $(u_1 \cdots u_{n-1}u_n, g) = (u_1 \cdots u_{n-1}, g)^{u_n}(u_n, g)$; by induction the two expressions on the right are in $\Gamma_{k+1}(G)$. Hence we have shown $(\Gamma_k(G), G) \subseteq \Gamma_{k+1}(G)$ and have proved the theorem.

This theorem leads to an important corollary.

COROLLARY 10.2.1. *$\Gamma_k(G)/\Gamma_{k+1}(G)$ is in the center of $G/\Gamma_{k+1}(G)$.*

We may also define an *upper central series* for an arbitrary group G.

$$Z_0 = 1 \subseteq Z_1(G) \subseteq Z_2(G) \subseteq \cdots \subseteq Z_i(G) \subseteq Z_{i+1}(G) \subseteq \cdots,$$

where we define $Z_{i+1}(G)$ by the rule: *$Z_{i+1}(G)/Z_i(G)$ is the center of $G/Z_i(G)$*. Since the center of a group is a characteristic subgroup (but not in general fully invariant), each Z_i is a characteristic subgroup of G. The following theorem justifies the use of the terms *upper* and *lower* as applied to the central series we have defined.

A series $G = A_1 \supseteq A_2 \supseteq A_3 \supseteq \cdots \supseteq A_{r+1} = 1$, in which each A_i/A_{i+1} is in the center of G/A_{i+1}, is called a *central series.*

Theorem 10.2.2. *Let $G = A_1 \supseteq A_2 \supseteq A_3 \supseteq \cdots \supseteq A_{r+1} = 1$ be a central series for G. Then $A_i \supseteq \Gamma_i(G)$, $i = 1, \cdots, r+1$ and $A_{r+1-j} \subseteq Z_j(G)$, $j = 0, 1, \cdots, r$.*

Proof: We have $A_1 = G = \Gamma_1(G)$. Suppose that $A_i \supseteq \Gamma_i(G)$. Since A_i/A_{i+1} is in the center of G/A_{i+1}, we have $(A_i, G) \subseteq A_{i+1}$. But then $\Gamma_{i+1}(G) = (\Gamma_i(G), G) \subseteq (A_i, G) \subseteq A_{i+1}$, and this proves by induction that $A_i \supseteq \Gamma_i(G)$ for all i. Suppose for some i $A_{r+1-i} \subseteq Z_i(G)$. Then $T = G/Z_i(G)$ is a homomorphic image of $U = G/A_{r+1-i}$ with kernel $Z_i(G)/A_{r+1-i}$. Now A_{r-i}/A_{r+1-i} is in the center of U, whence its homomorphic image in T must lie in the center of T. But this image is $A_{r-i} \cup Z_i/Z_i$, while the center of T is Z_{i+1}/Z_i. Hence $A_{r-i} \subseteq A_{r-i} \cup Z_i \subseteq Z_{i+1}$, proving our theorem by induction.

As a consequence of this theorem we have the following corollary:

COROLLARY 10.2.2. *In a nilpotent group G, the upper and lower central series have finite length and both have the same length c.* For, if there is a finite central series of length r, the theorem shows that the upper and lower central series have at most length r. And if the two series are compared with each other, we conclude that neither one can be longer than the other. Hence they both have the same length c, and this number c is called the *class* of the nilpotent group. A nilpotent group of class 1 is simply an Abelian group.

THEOREM 10.2.3. *If a group G is generated by elements $x_1, \cdots, x_r$,* *then* $\Gamma_k(G)/\Gamma_{k+1}(G)$ *is generated by the simple commutators* $(y_1, y_2, \cdots, y_k)$ *mod* $\Gamma_{k+1}(G)$, *where the* y's *are chosen from* $x_1, \cdots, x_r$ *and are not necessarily distinct.*

COROLLARY 10.2.3. *If* G *is generated by* r *elements, then* $\Gamma_k(G)/\Gamma_{k+1}(G)$ *is generated by at most* r^k *elements.*

Proof: We proceed by induction on k, the theorem being immediate for $k = 1$. Assume the theorem true for $k - 1$. $\Gamma_k(G)$ is generated by all commutators $C = (a_1, \cdots, a_{k-1}, a_k)$ with $a_i \in G$. Here $C = ((a_1, \cdots, a_{k-1}), a_k)$ and $(a_1, \cdots, a_{k-1}) \in \Gamma_{k-1}(G)$, whence by induction $(a_1, \cdots, a_{k-1}) = u_1^{\epsilon_1}u_2^{\epsilon_2} \cdots u_n^{\epsilon_n}w$, $\epsilon_i = \pm 1$, with $u_1, \cdots, u_n$ being commutators of the form $(y_1, \cdots, y_{k-1})$ and the y's being x's and $w \in \Gamma_k(G)$. Then $C = (u_1^{\epsilon_1}u_2^{\epsilon_2} \cdots u_n^{\epsilon_n}w, a_k)$. Applying (10.2.1.2) we have $C = (u_1^{\epsilon_1}u_2^{\epsilon_2} \cdots u_n^{\epsilon_n}, a_k)(u_1^{\epsilon_1}u_2^{\epsilon_2} \cdots u_n^{\epsilon_n}, a_k, w)(w, a_k) \equiv (u_1^{\epsilon_1}u_2^{\epsilon_2} \cdots u_n^{\epsilon_n}, a_k) \pmod{\Gamma_{k+1}}$. Now $a_k = x_{i_1}^{\eta_2} \cdots x_{i_m}^{\eta_m}$, $\eta_j = \pm 1$, and since the u's $\in \Gamma_{k-1}$, we find by repeated application of (10.2.1.2) and (10.2.1.3) that modulo Γ_{k+1}, C is a product of commutators $(u_j^{\epsilon_j}, x_{i_s}^{\eta_s})$. Since also from these rules $(u^\epsilon, x^\eta) \equiv (u, x)^{\epsilon\eta}$ (mod $\Gamma_{k+1}(G))$], it follows that $\Gamma_k(G)/\Gamma_{k+1}(G)$ is generated by commutators (u, x) mod $\Gamma_{k+1}(G)$ or $(y_1, \cdots, y_{k-1}, x_{ik})$ mod $\Gamma_{k+1}(G)$, as we wished to prove. Note that we do not need finiteness of r for this theorem.

An almost immediate consequence of this is the following, which gives the relationship between nilpotent and supersolvable groups:

THEOREM 10.2.4. *A finitely generated nilpotent group is supersolvable.*

Proof: Let G be finitely generated and nilpotent. Let its lower central series be

$$G = \Gamma_1(G) \supset \Gamma_2(G) \supset \cdots \supset \Gamma_c(G) \supset \Gamma_{c+1}(G) = 1.$$

Since $\Gamma_c(G)$ is Abelian and finitely generated, it is the direct product of, say, m cyclic groups. Also since $\Gamma_c(G)$ is in the center of G, any subgroup of it is normal in G. Thus there is a chain $\Gamma_{c+1} = 1 \subset \{a_1\} \subset \{a_1, a_2\} \subset \cdots \subset \{a_1, a_2, \cdots, a_m\} = \Gamma_c(G)$, all being normal subgroups of G and having the property that the factor group of consecutive groups is cyclic. Similarly, we may insert normal subgroups between $\Gamma_{i+1}(G)$ and $\Gamma_i(G)$, with the property that the factor group of consecutive groups is cyclic. In this way we find a series for G which is the defining property for G to be supersolvable.

COROLLARY 10.2.4. *A finitely generated nilpotent group satisfies the maximal condition.*

A group G satisfies the maximal condition if there are no infinite ascending chains of subgroups. This is equivalent to the requirement that G and every subgroup of G be finitely generated. But we shall show in Theorem 10.5.1 that every subgroup of a supersolvable group is supersolvable and so finitely generated. The corresponding statement is false for solvable groups. Thus if F is the free group with two generators, a, b, then F/F'' is a solvable group, but F'/F'' has infinitely many generators $a^{-i}b^{-j}a^{i}b^{j}$.

10.3. Theory of Nilpotent Groups.

We note that if a group G is nilpotent of class c, then every commutator $(a_1, \cdots, a_{c+1})$ is the identity, and conversely, that if every $(a_1, \cdots, a_{c+1}) = 1$, then G is nilpotent of class c at most. We describe the property that $(a_1, \cdots, a_{c+1}) = 1$ for all $a_i \in G$ by saying that G has nil-c.

THEOREM 10.3.1. *If G has nil-c, then every subgroup and factor group of G has nil-c.*

Proof: If G has nil-c, then *a fortiori* for a subgroup H all commutators $(a_1, \cdots, a_{c+1})$ with $a_i \in H$ must be 1, and so H has nil-c. Also if T is a homomorphic image of G, then every commutator $(b_1, \cdots, b_{c+1})$ with $b_i \in T$ is the homomorphic image of some commutator $(a_1, \cdots, a_{c+1})$ in G and hence is the identity, whence T has nil-c.

The following theorem applies to nilpotent normal subgroups of a group G which may not itself be nilpotent.

THEOREM 10.3.2. *If H, K are normal subgroups of G, and if H has nil-c and K has nil-d, then $H \cup K = HK$ has nil-$(c + d)$.*

Proof: $\Gamma_m(HK)$ is generated by all commutators $(u_1, u_2, \cdots, u_m)$ with $u_i \in HK$, whence $u_i = h_ik_i$, $h_i \in H$, $k_i \in K$. We assert that $(u_1, u_2, \cdots, u_m)$ is a product of commutators of the form $w = (v_1, v_2, \cdots, v_m)$, where each v_i is an $h \in H$ or a $k \in K$. This is trivial for $m = 1$. Suppose this to be true for $m - 1$. Then

$$\begin{aligned}(u_1, \cdots, u_{m-1}, u_m) &= (w_1w_2 \cdots w_t, h_mk_m)\\ &= (w_1 \cdots w_t, k_m)(w_1 \cdots w_t, h_m)^{k_m}\\ &= (w_1 \cdots w_t, k_m)(w_1^{k_m} \cdots w_t^{k_m}, h_m^{k_m})\\ &= (w_1 \cdots w_t, k_m)(w_1' \cdots w_t', h_m')\end{aligned}$$

by applying (10.2.1.3) and the normality of H and K.

Similarly, applying (10.2.1.2),

$$\begin{aligned}(w_1 \cdots w_t, k_m) &= (w_1, k_m)^{w_2 \cdots w_t}(w_2 \cdots w_t, k_m)\\ &= (w_1'', k_m'')(w_2 \cdots w_t, k_m).\end{aligned}$$

Continuing, we finally express $(u_1, \cdots, u_{m-1}, u_m)$ as a product of terms $(w, h_m^{(i)})$ and $(w, k_m^{(i)})$, which will be of the form $(v_1, \cdots, v_m)$, with each v an h or a k. This proves our assertion by induction. We have now shown that $\Gamma_{c+d+1}(HK)$ is generated by commutators $(v_1, \cdots, v_{c+d+1})$, with each v_i an h or a k. We have in general $(v_1, \cdots, v_{t-1}, v_t) = (v_1, \cdots, v_{t-1})^{-1}v_t^{-1}(v_1, \cdots, v_{t-1})v_t$. By the normality of $\Gamma_i(H)$ in HK, if $(v_1, \cdots, v_{t-1}) \in \Gamma_i(H)$, then if v_t is a k, then $(v_1, \cdots, v_t) \in \Gamma_i(H)$, whereas if v_t is an h, then $(v_1, \cdots, v_t) \in \Gamma_{i+1}(H)$. Hence if there are as many as $(c + 1)$ h's in $(v_1, \cdots, v_{c+d+1})$, it will belong to $\Gamma_{c+1}(H) = 1$, and hence be the identity. If not, there must be at least $(d + 1)$ k's in $(v_1, \cdots, v_{c+d+1})$, and it will follow in the same way that it is in $\Gamma_{d+1}(K) = 1$. In all cases $(v_1, \cdots, v_{c+d+1}) = 1$, and therefore $H \cup K = HK$ has nil-$(c + d)$.

THEOREM 10.3.3. *If a group G has nil-c, $H = H_0$ is any subgroup and H_{i+1} is the normalizer of H_i in G, then $H_c = G$.*

Proof: $H_0 \supseteq Z_0 = 1$ trivially. We prove by induction that $H_m \supseteq Z_m$ for all m. Assume that $H_i \supseteq Z_i$. Then by definition of Z_{i+1}, we have for any $z_{i+1} \in Z_{i+1}$ and any $g \in G$, $z_{i+1}^{-1}g^{-1}z_{i+1}\,g = z_i \in Z_i$, whence with $g^{-1} = h_i \in H_i$, we have $z_{i+1}^{-1}h_iz_{i+1} = z_ih_i \in H_i$, and so Z_{i+1} normalizes H_i, whence $H_{i+1} \supseteq Z_{i+1}$, proving our assertion by induction. Since $Z_c = G$, we must have $H_c = G$.

COROLLARY 10.3.1. *Every proper subgroup of a nilpotent group is a proper subgroup of its normalizer.*

COROLLARY 10.3.2. *Every maximal subgroup of a nilpotent group is normal, is of prime index, and contains the derived group.*

Let M be a maximal subgroup of the nilpotent group G. Since $N_G(M)$ properly contains M, we must have $N_G(M) = G$, or $M \triangleleft G$. Then, by the maximality of M, G/M contains no proper subgroup, whence it must be a cyclic group of prime order. Thus M is of prime index, and as G/M is Abelian, M contains the derived group G'.

COROLLARY 10.3.3. *If G is nilpotent and H is a subgroup such that $G = G'H$, then $H = G$.*

Here if $H \neq G$, then by the theorem, with $H = H_0$ and $H_{i+1} = H_i Z_{i+1}$, we shall have each H_i normal in H_{i+1}. If $H_j \neq G$, but $H_{j+1} = G$, then H_j is a proper normal subgroup of G and G/H_j is Abelian, whence $H_j \supseteq G'$. But then $HG' \subseteq H_j G' = H_j \neq G$, contrary to our hypothesis. Hence we must have $H = G$, proving our theorem. Note that we have *not* assumed here that G possesses maximal subgroups.

THEOREM 10.3.4. *Finite p-groups are nilpotent. A finite group is nilpotent if, and only if, it is the direct product of its Sylow subgroups.*

Proof: By Theorem 4.3.1, every finite p-group P has a center different from the identity. Hence the upper central series for P terminates with the entire group, whence P is nilpotent. The same argument holds for a direct product of finite p-groups. Now suppose that G is any finite nilpotent group, and let P be a Sylow p-subgroup of G. Then $N_G(P)$ is its own normalizer by Theorem 4.2.4, and by Corollary 10.3.1, $N_G(P)$ cannot therefore be a proper subgroup of G. Hence $P \lhd G$. As every Sylow subgroup of G is normal, G must be the direct product of its Sylow subgroups.

COROLLARY 10.3.4 (WIELANDT): *A finite group is nilpotent if, and only if, its maximal subgroups are normal.*

For, by Corollary 10.3.2 of Theorem 10.3.3, the maximal subgroups of a nilpotent group are normal. On the other hand, by Theorem 4.2.4, $N_G(P)$ cannot be contained in a proper normal subgroup of G if P is a Sylow p-subgroup. Hence if maximal subgroups are normal, then $P \lhd G$, and G is the direct product of its Sylow subgroups.

THEOREM 10.3.5. *If X, Y, Z are subgroups of a group G, and if K is a normal subgroup of G containing (Y, Z, X) and (Z, X, Y), then K also contains (X, Y, Z).*

Proof: From (10.2.1.4) we have

$$(x, y, z) = ((z, x^{-1}, y^{-1})^{xy})^{-1}((y^{-1}, z^{-1}, x)^{zy})^{-1},$$

and the conclusion follows.

THEOREM 10.3.6. *If $H = H_0 \supseteq H_1 \supseteq \cdots$ are normal subgroups of a group G such that $(H_{i-1}, L) \subseteq H_i$ for all i and a subgroup L, then $(H_i, \Gamma_j(L)) \subseteq H_{i+j}$.*

COROLLARY 10.3.5. $(\Gamma_i(G), \Gamma_j(G)) \subseteq \Gamma_{i+j}(G)$.

Proof: We proceed by induction on j, the hypothesis including the case $j = 1$. Suppose that $(H_i, \Gamma_{j-1}(L)) \subseteq H_{i+j-1}$ for all i. Then by induction $(L, H_i, \Gamma_{j-1}(L)) \subseteq (H_{i+1}, \Gamma_{j-1}(L)) \subseteq H_{i+j}$ and $(H_i, \Gamma_{j-1}(L), L) \subseteq (H_{i+j-1}, L) \subseteq H_{i+j}$. Since $(\Gamma_{j-1}(L), L) = \Gamma_j(L)$, we may apply Theorem 10.3.5 to conclude that

$$(H_i, \Gamma_j(L)) = (H_i, (\Gamma_{j-1}(L), L)) = (\Gamma_{j-1}(L), L, H_i) \subseteq H_{i+j},$$

the conclusion of our theorem.

10.4. The Frattini Subgroup of a Group.

Let G be an arbitrary group. We define a subgroup Φ of G, called the Frattini subgroup, in the following way: $\Phi = G \bigcap_M M$, where M ranges over the maximal subgroups of G if G has any maximal subgroups. Thus $\Phi = G$ if, and only if, G has no maximal subgroups. Since any automorphism of G permutes the maximal subgroups among themselves, the Frattini subgroup is clearly a characteristic subgroup.

The Frattini subgroup has an interesting relation to the generation of G. It consists of the elements of G which are nongenerators of G in the following precise sense:

DEFINITION: *An element x of a group G is said to be a nongenerator of G if whenever $G = \{T, x\}$ for a subset T of G, then also $G = \{T\}$.*

Note that we require $\{T, x\} = \{T\}$ for *every* set T for which $\{T, x\} = G$. Here if $G \neq 1$, surely 1 is a nongenerator.

THEOREM 10.4.1. *If a group G is not the identity alone, then its Frattini subgroup Φ consists of the set of nongenerators of G.*

Proof: Let x be an element of G. If there is a maximal subgroup M which does not contain x, then the group $\{M, x\}$ properly contains M, and as M is maximal, we must have $\{M, x\} = G$. But here $\{M\} = M \neq G$. Thus x is an essential generator in $\{M, x\} = G$. Thus the nongenerators of G belong to every maximal subgroup, and so every nongenerator is an element of $\Phi = G \bigcap_M M$. We must show conversely that if $u \in \Phi$, then u is a nongenerator of G. By hypothesis $G \neq 1$, whence 1 is surely a nongenerator.

Now suppose that $G = \{T, u\}$ for a subset T of G. We show that if $\{T\} = H \neq G$, we reach a contradiction. Now if $H \neq G$, we cannot have $u \in H$, since in this case $H = \{H, u\} \supseteq \{T, u\} = G$. Hence $u \notin H$. Then, by Zorn's lemma, there exists a subgroup, $K \supseteq H$ maximal with respect to the property that $u \notin K$. Now $\{K, u\} \supseteq \{T, u\} = G$, whence $\{K, u\} = G$. But by our choice of K, any group containing K properly must contain u. Hence $K = M$ is a maximal subgroup not containing u, which conflicts with $u \in \Phi = G \bigcap_M M$. Hence we must have $\{T\} = G$, and so every $u \in \Phi$ is a nongenerator of G.

THEOREM 10.4.2. *The Frattini subgroup of a finite group is nilpotent.*

Proof: Let G be a finite group and Φ its Frattini subgroup. Let P be a Sylow p-subgroup of Φ. Now Φ as a characteristic subgroup of G is a normal subgroup. Thus every conjugate of P in G lies in Φ and so is conjugate to P in Φ, being a Sylow p-subgroup of Φ. Thus P has as many conjugates in Φ as it does in G, and so $[G:N_G(P)] = [\Phi:N_\Phi(P)]$. But $[G:N_\Phi(P)] = [G:\Phi][\Phi:N_\Phi(P)] = [G:N_G(P)][N_G(P):N_\Phi(P)]$, whence $[G:\Phi] = [N_G(P):N_\Phi(P)]$. We note that $N_\Phi(P) = N_G(P) \cap \Phi$ and apply the inequality on indices of Theorem 1.5.5 to find $[N_G(P) \cup \Phi:\Phi] \geqq [N_G(P):N_G(P) \cap \Phi] = [G:\Phi]$. From this we conclude that $N_G(P) \cup \Phi = G$. Now, since $G = \{N_G(P), \Phi\}$, we also have, removing the elements of Φ one at a time, since Φ is finite, $G = \{N_G(P)\} = N_G(P)$. Thus $P \triangleleft G$, and *a fortiori* $P \triangleleft \Phi$. Since every Sylow subgroup of Φ is normal, Φ must be the direct product of its Sylow subgroups and is therefore a nilpotent group.

Theorem 10.4.3. *The Frattini subgroup of a nilpotent group contains the derived group.*

Proof: From Corollary 10.3.3 if G is nilpotent and $G = HG'$, then $G = H$. This says that G' can be omitted from any set of generators for G, whence it follows that $\Phi \supseteq G'$. The converse holds for finite groups.

THEOREM 10.4.4 (WIELANDT). *If the Frattini subgroup of a finite group G contains the derived group G', then G is nilpotent.*

Proof: Let P be a Sylow subgroup of G. If $N_G(P) = H \neq G$, then H is contained in some maximal subgroup M of G. Now $M \supseteq \Phi$, and by hypothesis, $\Phi \supseteq G'$. As G/G' is Abelian, M is a normal subgroup of G. On the other hand, by Theorem 4.2.4, since $M \supseteq N_G(P)$, M is its own normalizer. This is a contradiction and we conclude that we must have $N_G(P) = G$. The Sylow subgroups of G all being normal, we conclude that G is their direct product and is nilpotent.

10.5. Supersolvable Groups.

Theorem 10.5.1. *Subgroups and factor groups of supersolvable groups are supersolvable.*

Proof: Let G be supersolvable and $G = A_0 \supset A_1 \supset A_2 \supset \cdots \supset A_r = 1$ be a normal series with every A_{i-1}/A_i a cyclic group. Then, for a factor group $G/K = T$, the homomorphic images B_i of the A_i will form a normal series $T = B_0 \supseteq B_1 \supseteq B_2 \supseteq \cdots \supseteq B_r = 1$, where, if we delete repetitions of the same group, consecutive terms B_{i-1}, B_i will have a cyclic factor group B_{i-1}/B_i, since every homomorphic image of a cyclic group is cyclic or the identity. For a subgroup H take

$$H = C_0 \supseteq C_1 \supseteq C_2 \supseteq \cdots \supseteq C_r = 1,$$

where $C_i = H \cap A_i$. For every i, $H \cap A_i$ is normal in H, and by Theorem 2.4.1, we have $C_i/C_{i+1} = H \cap A_i/H \cap A_{i+1} \cong A_{i+1} \cup (H \cap A_i)/A_{i+1}$. But the right-hand side of this is a subgroup of A_i/A_{i+1}, and hence cyclic or the identity. Thus C_i/C_{i+1} is cyclic or the identity, and so H is supersolvable.

Corollary 10.5.1. *Supersolvable groups satisfy the maximal condition.*

A supersolvable group is finitely generated, and by Theorem 10.5.1, its subgroups are also finitely generated, whence the maximal condition will be satisfied.

Theorem 10.5.2. *A supersolvable group G has a normal series $G = B_0 \supset B_1 \supset B_2 \supset \cdots \supset B_k = 1$ in which every factor group B_{i-1}/B_i is either infinite cyclic or cyclic of prime order.*

Proof: Let $G = A_0 \supset A_1 \supset A_2 \supset \cdots \supset A_r = 1$ be a normal series, with each A_{i-1}/A_i cyclic. If A_{j-1}/A_j is of finite order $p_1 p_2 \cdots p_s$, where $p_1, p_2, \cdots, p_s$ are primes (not necessarily distinct), then

A_{j-1}/A_j has a unique cyclic subgroup of each of the orders p_1, p_1p_2, $\cdots$, $p_1 \cdots p_{s-1}$, and these are characteristic subgroups. Hence the $s - 1$ corresponding subgroups between A_{j-1} and A_j are normal in G, and the factor groups of consecutive groups are cyclic of prime order. Refining in this way every factor group A_{j-1}/A_j of finite order, we obtain the normal series of the theorem in which every factor group is either infinite cyclic or cyclic of prime order.

This theorem can be further improved since we can rearrange the prime factor groups according to the magnitude of the primes.

THEOREM 10.5.3. *A supersolvable group G has a normal series*

$$G = C_0 \supset C_1 \supset C_2 \supset \cdots \supset C_k = 1,$$

in which every C_{i-1}/C_i is either infinite cyclic or cyclic of prime order, and if C_{i-1}/C_i and C_i/C_{i+1} are of prime orders p_i and p_{i+1}, we have $p_i \leq p_{i+1}$.

Proof: Take a series $G = B_0 \supset B_1 \supset B_2 \supset \cdots \supset B_k = 1$ given by Theorem 10.5.2. If B_{i-1}/B_i and B_i/B_{i+1} are of prime orders q and p, respectively, with $q > p$, then B_{i-1}/B_{i+1} is of order pq, with $p < q$, and this has a characteristic subgroup of order q whose inverse image B_i^* will be normal in G. If we replace B_i by B_i^*, then B_{i-1}/B_i^* will be of order p and B_i^*/B_{i+1} will be of order q. Continuing this process, which does not alter the length of the normal series, we shall ultimately get a series in which the orders of consecutive factor groups of prime order do not increase in magnitude, as stated in the theorem.

COROLLARY 10.5.2. *If G is a finite supersolvable group of order $p_1p_2 \cdots p_r$, where $p_1 \leq p_2 \leq \cdots \leq p_r$ are primes, then G has a chief series $G = A_0 \supset A_1 \supset \cdots \supset A_r = 1$, where A_{i-1}/A_i is of order p_i.*

THEOREM 10.5.4. *The derived group of a supersolvable group is nilpotent.*

Proof: Suppose $G = A_0 \supset A_1 \supset \cdots \supset A_r = 1$ is a normal series for G, with A_{i-1}/A_i cyclic. Write $H_i = G' \cap A_i$. Then $G' = H_0 \supseteq H_1 \supseteq \cdots \supseteq H_r = 1$ is a normal series, and the distinct terms of this series K_i are such that $G' = K_0 \supset K_1 \supset \cdots \supset K_s = 1$, with K_{i-1}/K_i cyclic. We assert that the K's form a central series for G'. Every K_i is the intersection of normal subgroups of G, and hence normal in G. Thus, in G/K_i, K_{i-1}/K_i is a cyclic normal subgroup, and transformation by an element of G induces an automorphism in the cyclic group K_{i-1}/K_i. Now the automorphisms of a cyclic group form an Abelian group, and so two elements of G/K_i induce permuting automorphisms in K_{i-1}/K_i. But then the commutator of any two elements $x^{-1}y^{-1}xy$ induces the identical automorphism in K_{i-1}/K_i. But this is to say that in G'/K_i, K_{i-1}/K_i lies in the center, and therefore the K's form a central series for G', and so G' is nilpotent.

There is a very interesting property of chains of arbitrary subgroups in a supersolvable group. We shall say that H_2 is of index ∞^1 in H_1 if $H_1 = \sum_x H_2a^x$ for some element a and x running over all integers from $-\infty$ to $+\infty$. Thus if $A_j \triangleleft A_{j-1}$ and A_{j-1}/A_j is an infinite cyclic group, then A_j is of index ∞^1 in A_{j-1}, since for a we may take any element of the coset of A_j which is a generator of the cyclic group A_{j-1}/A_j. But H_2 may be of index ∞^1 in H_1 without being normal in H_1.

THEOREM 10.5.5. *In a supersolvable group G any chain of subgroups $G = M_0 \supset M_1 \supset M_2 \supset \cdots \supset M_s = 1$ may be refined by the insertion of further groups:*

$$M_{i-1} = M_{i,0} \supset M_{i,1} \supset \cdots \supset M_{i,t} = M_i, \quad t = t(i), \quad i = 1, \cdots, s,$$

so that $M_{i,j}$ is of prime index or index ∞^1 in $M_{i,j-1}$.

Proof: Since M_1 is supersolvable, it is sufficient to show that the series may be refined by inserting terms between $G = M_0$ and M_1 with the required properties. For repeating the argument with $M_1, \cdots, M_{s-1}$ in turn, we may refine the entire series.

Let $G = A_0 \supset A_1 \supset A_2 \supset \cdots \supset A_r = 1$ be a normal series for G, where each A_{i-1}/A_i is cyclic of prime or infinite order. Surely $M_1 \supseteq A_r = 1$ and $M_1 \not\supseteq A_0 = G$. Hence for some i in the range $1, \cdots, r$ we have $M_1 \supseteq A_i$, and $M_1 \not\supseteq A_{i-1}$. We consider two cases: (1) A_{i-1}/A_i of prime order, and (2) A_{i-1}/A_i infinite cyclic.

CASE 1. A_{i-1}/A_i OF PRIME ORDER. Here $A_{i-1} \supset M_1 \cap A_{i-1} \supseteq A_i$. Since A_i is of prime index in A_{i-1}, there can be no subgroup between A_i and A_{i-1}. Hence $M_1 \cap A_{i-1} = A_i$. If $A_{i-1} = \sum_x A_ia^x$, $x = 0, \cdots, p-1$, then $M_1 \cup A_{i-1} = M_1A_{i-1} = M_1^*$, and $M_1^* = \sum_x M_1a^x$,

$x = 0,1, \cdots, p-1$ since M_1 contains A_i and $a^p \in A_i$ but not the element a. Here M_1 is of prime index in M_1^* and $M_1^* \supseteq A_{i-1}$.

CASE 2. A_{i-1}/A_i INFINITE CYCLIC. Here $A_{i-1} \supset M \cap A_{i-1} \supseteq A_i$. Now every subgroup of A_{i-1}/A_i is characteristic, whence $M_1 \cap A_{i-1}$ is a normal subgroup of G. If $M_1 \cap A_{i-1} = A_i$ and $A_{i-1} = \sum_x A_i a^x$, then put $M_1^* = M_1 \cup A_{i-1} = M_1 A_{i-1} = \sum_x M_1 a^x$, and M_1 is of index ∞^1 in M_1^*, since M_1 contains A_i but no power of the element a. But if $M_1 \cap A_{i-1} \supset A_i$, then since every subgroup of an infinite cyclic group is of finite index, $M_1 \cap A_{i-1}$ is of finite index in A_{i-1}. Thus in our normal series we may insert terms between A_{i-1} and $M_1 \cap A_{i-1}$ each of prime index in the one above, and as in Case 1, find an M_1^* in which M_1 is of prime index. Repeating the construction, we find a chain $M_1 \subset M_1^* \subset M_1^{**} \cdots \subset M_1^{(u)}$ with each of prime index in the next and $M_1^{(u)} \supseteq A_{i-1}$.

Continuing the construction, we shall in a finite number of steps reach an $M_1^{(v)} \supseteq A_0 = G$ and thus have found the refinement between G and M_1 as required for the theorem. As already remarked, the same procedure will give the needed refinement for the entire chain.

For finite groups this theorem takes an interesting form.

THEOREM 10.5.6. *In a finite supersolvable group G, all maximal chains of subgroups have the same length, this being the number r if G is of order $p_1p_2 \cdots p_r$, the p's being primes, but not necessarily distinct.*

Proof: By the previous theorem, in a maximal chain every index of one subgroup in the next is a prime, and so the length of a maximal chain is r.

COROLLARY 10.5.1. *Every maximal subgroup of a finite supersolvable group is of prime index.*

It is a remarkable fact, first proved by Huppert [1], that the converse of this corollary is true. For this we must use some of the theorems on group representation which will be proved in Chap. 16. First we give an unpublished theorem of P. Hall.

THEOREM 10.5.7 (P. HALL). *Suppose G is a finite group with the property (M) that all its maximal subgroups are of index a prime, or the square of a prime. Then G is solvable.*

Proof: We proceed by induction on the order of G. Let p be the largest prime dividing this order, S a Sylow p-subgroup of G, N its normalizer in G. If $N = G$, S is normal in G and G/S has property (M), whence G/S is solvable by induction. S is a p-group and so G is solvable. If on the other hand, $N \subset G$, choose a maximal subgroup H of G containing N. N is the normalizer of S in H as well as in G, and so $[G:N] = 1 + k_1p$, $[H:N] = 1 + k_2p$ by the third Sylow theorem, these being the number of Sylow p-subgroups in G and H, respectively. Hence $[G:H] = 1 + kp$. But by hypothesis $[G:H] = q$ or q^2 for some prime q; and clearly, $q < p$, $k > 0$. Hence $kp = q^2 - 1 = (q-1)(q+1)$. Since $p \geq q + 1$, we must have $p = q + 1$. This is possible only if $p = 3$, $q = 2$, and the order of G is of the form 2^a3^b. By Theorem 16.8.7, G is solvable.

THEOREM 10.5.8 (HUPPERT). *Suppose G is a finite group with the property (M^1) that all its maximal subgroups are of index a prime. Then G is supersolvable.*

Proof: If the theorem is not true, choose G to have property (M^1), but to be not supersolvable, and to have the smallest possible order subject to these two conditions. Then G is solvable by Theorem 10.5.7. Let N be a minimal normal subgroup of G, and let its order be p^α, p prime. By the minimal property of G, G/N is supersolvable so that, of the chief factors of G, only N is noncyclic. We conclude that N is the only minimal normal subgroup of G. Let H/N be a minimal normal subgroup of G/N. There are two cases according to whether (1) $[H:N] = p$, (2) $[H:N] = q$, a prime different from p. In case (1) H must be Abelian, since otherwise we would have $1 \subset H' \subset N$ and H' normal in G. Since $\alpha > 1$, H cannot have elements of order p^2, for this would make N contain a characteristic subgroup of H of order p—the same argument again. Thus H is elementary Abelian.

We now have G represented in a natural way by automorphisms of H, i.e., effectively by linear transformations modulo p of degree $\alpha + 1$ (since H is of order $p^{\alpha+1}$). Let K be the set of all elements $a \in G$ such that for $x \in H$, we have $a^{-1}xa = x^m$, where $m = m(a)$ is independent of x, and let L be the centralizer of H in G. Then K/L is contained in the center of G/L. Also, $K \subset G$, since N is the only minimal normal subgroup of G, and every subgroup of H is normal in K. Let M/K be a minimal normal subgroup of G/K. If $[M:K] = p$, we shall have M/L as a direct product of K/L, which is of order prime to p and in the center of G/L, with a group $\{L, a\}/L$, say, of order p, and $M_1 = \{L, a\}$ will be normal in G. Since the group of commutators $(N, L) = 1$ and $[M_1:L] = p$, N must contain elements $\neq 1$ in the center of M_1. The center of M_1 is a normal subgroup of G, and so by the minimality of N, N is in the center of M_1 and $(M_1, N) = 1$. If $H = \{N, b\}$, the group (H, M_1) will be of order p and generated by $(a, b) = c \in N$, $c \neq 1$. This group is, however, normal in G, and hence N could not be a minimal normal subgroup of G.

Therefore $[M:K] = q$, some prime different from p, and so M/L is of order prime to p. By the Theorem of Complete Reducibility, Theorem 16.3.1, it follows that $H = N \times P$, where P is normal in M and of order p. The conjugates of P in G are normal subgroups of order p in M and their union Q is a normal subgroup of G. Since N is the only minimal normal subgroup of G and $Q \neq N$, it follows that $Q = H$. Since P does not lie in N, no conjugate of P lies in N. Let $P = \{b\}$, where $b^p = 1$, and if P_i is any conjugate of P except itself, then $PP_i \cap N = R$, where, since $[H:N] = p$, R is of order p. We may take a generator c of P_i such that $P_i = \{c\}$, $R = \{bc\}$. Since P, P_i, and R are normal subgroups of M, it follows that for any a of M, $a^{-1}ba = b^m$, $a^{-1}ca = c^n$, $a^{-1}(bc)a = (bc)^t$. But then $(bc)^t = b^m c^n$ and $t = n = m$. But P_i was any conjugate of P, and it follows that for any x of H we have $a^{-1}xa = x^m$, where $m = m(a)$ is independent of x. Hence $M \subseteq K$ is a contradiction. Thus case (1) cannot arise.

Case (2) can be dismissed at once. If $[H:N] = q$ different from p, let Q be a Sylow q-subgroup of H; T the normalizer of Q in G. Any conjugate of Q in G lies in H, and hence is a conjugate of Q by an element of N. Hence $G = NT$. Then $N \cap T$ is normal in G. But $T \not\supseteq N$, since this would make $T = G$ and Q normal in G. Hence $N \cap T = 1$, $[G:T] = p^\alpha$. But T is a maximal subgroup of G, since if $T \subset T_1 \subset G$ we should have $1 \subset T_1 \cap N \subset N$ and $T_1 \cap N$ normal in G. Thus G has a maximal subgroup of index not a prime, contrary to hypothesis.

EXERCISES

1. Let $I^{(1)} = I^{(1)}(G)$ be the group of inner automorphisms of a group G and $I^{(n)}$ the group of inner automorphisms of $I^{(n-1)}$. If any group of the sequence $G, I^{(1)}, I^{(2)}, \cdots$ is the identity, show that G is nilpotent.

2. Let G be a group satisfying the maximal condition. If $A(G)$, the group of automorphisms of G, is supersolvable, show that G is supersolvable.
3. Let a and b be elements of a nilpotent group G, where $a^m = b^n = 1$ and $(m, n) = 1$. Put $w = a^{-1}b^{-1}ab$. Show that if $w \in \Gamma_i(G)$, then $w^m \in \Gamma_{i+1}(G)$, $w^n \in \Gamma_{i+1}(G)$, whence $w \in \Gamma_{i+1}(G)$. Hence conclude $w = 1$, $ba = ab$.
4. Prove the converse of Ex. 2 of Chap. 8, i.e.: If G is a finite nilpotent group and if $p_1, p_2, \cdots, p_s$ is any arrangement of the primes whose product is the order of G, then G has a composition series $G = A_0 \supset A \supset \cdots \supset A_s = 1$, where A_{i-1}/A_i is of order p_i.
5. Let G be a p-group with $\Gamma_3(G) = 1$. Show that if p^m is the highest order of an element of $G/\Gamma_2(G)$, then no element of $\Gamma_2(G)$ has an order higher than p^m.

11. BASIC COMMUTATORS

11.1. The Collecting Process.

We consider formal *words* or *strings* $b_1 \, b_2 \, \cdots \, b_n$ where each b is one of the letters $x_1, x_2, \cdots, x_r$. We also introduce *formal commutators* c_j and weights $\omega(c_j)$ by the rules:

1) $c_i = x_i$, $i = 1, \cdots, r$ are the commutators of weight 1; i.e., $\omega(x_i) = 1$.

2) If c_i and c_j are commutators, then $c_k = (c_i, c_j)$ is a commutator and $\omega(c_k) = \omega(c_i) + \omega(c_j)$.

Note that these definitions yield only a finite number of commutators of any given weight. We shall order the commutators by their subscripts, numbering $c_i = x_i$, $i = 1, \cdots, r$, and listing in order of weight, but giving an arbitrary ordering to commutators of the same weight.

A string $c_{i_1} \cdots c_{i_m}$ of commutators is said to be in collected form if $i_1 \leq i_2 \leq \cdots \leq i_m$, i.e., if the commutators are in order read from left to right. An arbitrary string of commutators,

$$c_{i_1} \cdots c_{i_m} c_{i_{m+1}} \cdots c_{i_n}, \tag{11.1.1}$$

will in general have a *collected part* $c_{i_1} \cdots c_{i_m}$ if $i_1 \leq \cdots \leq i_m$ and if $i_m \leq i_j$, $j = m + 1, \cdots, n$, and will have an uncollected part $c_{i_{m+1}} \cdots c_{i_n}$, where i_{m+1} is not the least of i_j, $j = m + 1, \cdots, n$. The collected part of a string $c_{i_1} \cdots c_{i_n}$ will be void unless i_1 is the least of the subscripts.

We define a collecting process for strings of commutators. If c_u is the earliest commutator in the uncollected part and if $c_{i_j} = c_u$ is the leftmost uncollected c_u, we replace

$$c_{i_1} \cdots c_{i_m} \cdots c_{i_{j-1}} c_{i_j} \cdots c_{i_n}$$

by

$$c_{i_1} \cdots c_{i_m} \cdots c_{i_j} c_{i_{j-1}} (c_{i_{j-1}}, c_{i_j}) \cdots c_{i_n}.$$

This has the effect of moving c_{i_j} to the left and introducing the new commutator $(c_{i_{j-1}}, c_{i_j})$ which by its weight is surely later than c_{i_j}. Thus c_{i_j} is still the earliest commutator in the uncollected part.

After enough steps c_{i_j} will be moved to the $(m + 1)$st position and will become part of the collected part. Since at each step a new commutator is introduced, the process will not in general terminate.

If $x_1, \cdots, x_r$ are generators of a group F (and we shall be concerned chiefly with the case in which F is the free group with these generators), and if a commutator $(u, v) = u^{-1}v^{-1}uv$, then we note that

$$c_{i_{j-1}}c_{i_j} = c_{i_j}c_{i_{j-1}}(c_{i_{j-1}}, c_{i_j}), \tag{11.1.2}$$

and that the collecting process does not alter the group element represented by a word. As it stands, the collecting process has not been defined for all elements of F but only for the *positive words*, those elements which can be expressed as a product of the generators without using any inverses of generators. This defect will be remedied below.

In applying the collecting process to a positive word, not all commutators will arise. Thus (x_2, x_1) may arise but not (x_1, x_2), since x_1 is collected before x_2. The commutators that may actually arise are called *basic commutators.* We give a formal definition of the basic commutators for a group F generated by $x_1, \cdots, x_r$.

Definition of Basic Commutators:

1) $c_i = x_i$, $i = 1, \cdots, r$ are the basic commutators of weight one, $\omega(x_i) = 1$.

2) Having defined the basic commutators of weight less than n, the basic commutators of weight n are $c_k = (c_i, c_j)$, where

(a) c_i and c_j are basic and $\omega(c_i) + \omega(c_j) = n$, and

(b) $c_i > c_j$, and if $c_i = (c_s, c_t)$, then $c_j \geq c_t$.

3) The commutators of weight n follow those of weight less than n and are ordered arbitrarily with respect to each other. Basic commutators will always be numbered so that they are ordered by their subscripts.

We note that if commutators are ordered according to weight, but arbitrarily otherwise, the collection process when applied to a positive word will yield only basic commutators. For, in replacing

$$c_u c_v = c_v c_u (c_u, c_v), \tag{11.1.3}$$

we collect c_v before c_u, whence $c_u > c_v$, and if $c_u = (c_s, c_t)$, we have collected c_t before collecting this c_v, whence $c_v \geq c_t$.

We shall now show that modulo $\Gamma_{k+1}(F)$, the $(k + 1)$st term of the lower central series of F (k being arbitrary), which we shall write F_{k+1}, an arbitrary element, can be written in the form

(11.1.4) $$f = c_1^{e_1}c_2^{e_2} \cdots c_t^{e_t} \bmod F_{k+1},$$

where $c_1, \cdots, c_t$ are the basic commutators of weights $1, 2, \cdots, k$. In the collection process we have

(11.1.5) $$vu = uv\,(v, u),$$

where u, v, and (v, u) are basic commutators. We must also consider collecting u or u^{-1} in expressions vu^{-1}, $v^{-1}u^{-1}$, and $v^{-1}u$. Now $vu^{-1} = u^{-1}v(v, u^{-1})$, and from (10.2.1.3) we have

(11.1.6) $$1 = (v, uu^{-1}) = (v, u^{-1})(v, u)(v, u, u^{-1}),$$

whence $(v, u^{-1}) = (v, u, u^{-1})^{-1}(v, u)^{-1}$. Similarly, $(v, u, u^{-1}) = (v, u, u, u^{-1})^{-1}(v, u, u)^{-1}$. Writing $v_0 = v$, $v_{t+1} = (v_t, u)$, we have

(11.1.7) $$\begin{aligned}(v, u^{-1}) &= (v_1, u^{-1})^{-1}v_1^{-1}\\ &= v_2(v_2, u^{-1})v_1^{-1}\\ &= v_2v_4 \cdots v_5^{-1}v_3^{-1}v_1^{-1} \pmod{F_{k+1}},\end{aligned}$$

and we note that if $v_1 = (v, u)$ is basic, then also $v_2, v_3, \cdots$ are basic. Modulo F_{k+1} we may ignore (v_s, u^{-1}) if s is so large that this is of weight $k + 1$ or higher. Hence as a step in collection we have

(11.1.8) $$vu^{-1} = u^{-1}v \cdot v_2v_4 \cdots v_5^{-1}v_3^{-1}v_1^{-1} \pmod{F_{k+1}}.$$

Similarly,

(11.1.9) $$v^{-1}u = u(v,u)^{-1}v^{-1}.$$

Also, $v^{-1}u^{-1} = u^{-1}(uvu^{-1})^{-1}$, and from (11.1.8),

(11.1.10) $$uvu^{-1} = v \cdot v_2v_4 \cdots v_5^{-1}v_3^{-1}v_1^{-1} \pmod{F_{k+1}},$$

whence

(11.1.11) $$v^{-1}u^{-1} = u^{-1}v_1v_3v_5 \cdots v_4^{-1}v_2^{-1}v^{-1} \pmod{F_{k+1}}.$$

Repeated applications of (11.1.5, -8, -9, -11) will lead to the expression (11.1.4) for an arbitrary element f in terms of a sequence of basic commutators.

If F is the free group generated by $x_1, x_2, \cdots, x_r$, then for a given sequence of basic commutators we shall show in §11.2 that the expression (11.1.4) is unique. In particular the basic commutators of weight k are a free basis for F_k/F_{k+1}, which is consequently a free Abelian group. This is, of course, the justification for the term *basic* as applied to these commutators.

11.2. The Witt Formulae. The Basis Theorem.

Suppose we are given a sequence of basic commutators $c_1, c_2, \cdots$ formed from the generators $x_1, x_2, \cdots, x_r$. We call a product of basic commutators,

$$c_{i_1}c_{i_2} \cdots c_{i_n}, \tag{11.2.1}$$

a *basic product* if it is in collected order, i.e., $i_1 \leq i_2 \leq \cdots \leq i_n$. For an arbitrary product of commutators $p = a_1a_2 \cdots a_n$, we define the weight $\omega(p)$ as $\omega(p) = \omega(a_1) + \cdots + \omega(a_n)$. The collecting process alters the weight of a product. We define here a *bracketing process* similar to the collecting process which leaves weights unchanged. In this if u, v, and (u, v) are basic commutators, we replace

$$\cdots uv \cdots \quad \text{by} \quad \cdots (u, v) \cdots,$$

rather than the $\cdots vu(u, v) \cdots$ of the collecting process.

THEOREM 11.2.1. *The number of basic products of weight n formed from generators $x_1, \cdots, x_r$ is r^n.*

Proof: For each $k = 1, 2, \cdots$ we define the family $P_k = P_k^{(n)}$ of all products of weight n, $a_1a_2 \cdots a_t$, the a's being basic commutators which are of the form

$$c_1^{e_1}c_2^{e_2} \cdots c_k^{e_k}c_{i_1} \cdots c_{i_s}, \tag{11.2.2}$$

where $e_i \geq 0$, $i_1 > k$, $i_2, \cdots, i_s \geq k$, and for each c_{i_j} which is a commutator, $c_{i_j} = (c_u, c_v)$, c_v precedes c_k. Thus P_k may be regarded as the family in which $c_1, \cdots, c_{k-1}$ have been collected but not c_k. We denote the number of products in P_k by $|P_k|$. Clearly, P_1 is the family of all products of n generators, and so $|P_1| = r^n$. But we may set up a one-to-one correspondence between the members of P_k and P_{k+1}. For, if $c_1^{e_1} \cdots c_k^{e_k}c_{i_1} \cdots c_{i_s}$ is a member of P_k, c_{i_1} is later than c_k and so, though there may be a succession of c_k's in the uncollected part, each such string is immediately preceded by a c_y with $y > k$. For each string

$$c_y c_k \cdots c_k c_w \quad y > k, \quad w > k$$

we bracket $(((c_y, c_k), c_k) \cdots, c_k)c_w$, and since if $c_y = (c_u, c_v)$, $k > v$, the new commutator introduced is basic and later than c_k. This gives a unique member of P_{k+1}. Conversely, if in a member of P_{k+1} we remove all brackets involving c_k, we have a unique member of P_k. Hence $|P_k| = |P_{k+1}|$, and so for every k, $|P_k| = |P_1| = r^n$. But for k sufficiently large, P_k consists of all basic products of weight n. This proves the theorem.

We may use Theorem 11.2.1 to find the number of basic commutators of weight n, and even more, we may find the number of basic commutators whose weights in each generator are specified. We define weights $\omega_i(c)$, $i = 1, \cdots, r$ by the rules $\omega_i(x_i) = 1$, $\omega_i(x_j) = 0$, $i \neq j$, and recursively by $\omega_i[(c_u, c_v)] = \omega_i(c_u) + \omega_i(c_v)$. Let $M_r(n)$ be the number of commutators of weight n in r generators $x_1, x_2, \cdots, x_r$, and let $M(n_1, n_2, \cdots, n_r)$ be the number of commutators c such that $\omega_i(c) = n_i$, $i = 1, \cdots, r$, with $n = n_1 + n_2 + \cdots + n_r$.

THEOREM 11.2.2 (THEOREM OF WITT).

$$M_r(n) = \frac{1}{n} \sum_{d|n} \mu(d) r^{n/d} \tag{11.2.3}$$

$$M(n_1, n_2, \cdots, n_r) = \frac{1}{n} \sum_{d|n_i} \mu(d) \left(\frac{n}{d}\right)! \Big/ \left(\frac{n_1}{d}\right)! \cdots \left(\frac{n_r}{d}\right)! \tag{11.2.4}$$

Here $\mu(m)$ is the Möbius function which is defined for positive integers by the rules $\mu(1) = +1$, and for $n = p_1^{e_1} \cdots p_s^{e_s}$; $p_1, \cdots, p_s$ being distinct primes, $\mu(n) = 0$ if any $e_i > 1$, and $\mu(p_1 p_2 \cdots p_s) = (-1)^s$.

Proof: From Theorem 11.2.1 the number of basic products is r^n. This leads to the formal identity in a power series for a variable z,

$$\frac{1}{1 - rz} = \prod_{n=1}^{\infty} (1 - z^n)^{-M_r(n)}. \tag{11.2.5}$$

The bracketing process leaves all the weights ω_i, $i = 1, \cdots, r$ unchanged. The number of words W in the x's with $\omega_i(W) = n_i$ is, of course, the multinomial coefficient

$$\frac{n!}{n_1! \cdots, n_r!}.$$

This leads to the formal identity in variables $z_1, \cdots, z_r$

$$(11.2.6)\quad \frac{1}{1 - z_1 - \cdots - z_r} = \prod_{n_1, \ldots, n_r = 0}^{\infty} (1 - z_1^{n_1} \cdots z_r^{n_r})^{-M(n_1, \ldots, n_r)}.$$

Witt [2] used these identities, taking logarithms, and applied Möbius inversion to find the formulae of the theorem. Here we shall modify a result of Meier-Wunderli [1], proving it along lines similar to those of the proof of Theorem 11.2.1 to obtain the Witt formulae.

We call a word $a_1 \cdots a_n$ *circular* if a_1 is regarded as following a_n, where $a_1a_2 \cdots a_n$, $a_2 \cdots a_na_1$, $\cdots$, $a_na_1 \cdots a_{n-1}$ are all regarded as the same word. A circular word C of length n may conceivably be given by repeating a segment of d letters n/d times, where d is some divisor of n. We say that C is of period d if this is the case. Each circular word belongs to a unique smallest period, and this smallest period d corresponds to a unique circular word of length d.

LEMMA 11.2.1. *There is a one-to-one correspondence between basic commutators of weight n and circular words of length and period n. This is given by an appropriate bracketing of the circular word.*

Proof: Let $a_1a_2 \cdots a_n$ be a circular word of length n. The circular words of weight n form a family $C_k{}^n = C_k$ if they are of the form $c_{i_1}c_{i_2} \cdots c_{i_s}$, where the c's are basic commutators and for any $c_{i_j} = c_w$ which is a commutator $c_w = (c_u, c_v)$, we have $v < k$ and either (1) $i_1 = i_2 \cdots = i_s$ (including the case $s = 1$) or (2) $i_1, \cdots, i_s \geq k$ and some $i_j > k$. If (1) holds, the word is as it stands a word of C_{k+1}. If (2) holds, we take every circular subsequence (if any) of the form

$$c_w, c_k, \cdots, c_k, c_t, \quad w > k, \ t > k,$$

and bracket thus:

$$((\cdots((c_w, c_k), \cdots, c_k)c_t,$$

obtaining a unique circular word of C_{k+1}. By removing the brackets involving c_k from a word of C_{k+1}, we obtain a unique word of C_k. Thus there is a unique correspondence between words of C_1 and words of C_k for arbitrary k. If k is large enough, the commutator c_k is of weight greater than n and (2) cannot hold. Hence, ultimately, our bracketing ceases and (1) holds. Here our word is either a basic commutator of weight n or a succession of $s = n/d$ identical basic commutators of weight d. A bracketing by which we pass from C_k to C_{k+1} involves one c_w and a number of c_k's. Hence each such bracketing lies in a single period and will be exactly duplicated in every other period. Thus at every stage the number of periods in a word is the same. Hence bracketing all circular words of length n yields all basic commutators of weight n and for $d|n$ all basic commutators of weight d repeated identically n/d times, for these are the members of C_k if k is sufficiently large. This proves the lemma and somewhat more.

How many circular words of length and period n are there? A circular word of length n and period d, $d|n$ yields exactly d ordinary words of length n:

$$\begin{array}{l} a_1 \cdots a_d a_1 \cdots a_d \cdots a_1 \cdots a_d \\ a_2 \cdots a_d a_1 \cdots a_1 \cdots a_d a_1 \\ \cdot \quad \cdot \quad \cdot \quad \cdot \quad \cdot \quad \cdot \quad \cdot \quad \cdot \quad \cdot \\ a_d a_1 \cdots a_d \cdots a_1 \cdots a_{d-1}. \end{array}$$

Thus

$$r^n = \sum_{d|n} dM_r(d),$$

since the number of circular words of length and period d is $M_r(d)$ and every one of the r^n ordinary words corresponds to a unique period d. From

(11.2.7)
$$r^n = \sum_{d|n} dM_r(d)$$

we may find $M_r(n)$, since the Möbius inversion formula* says that if

(11.2.8)
$$f(n) = \sum_{d|n} g(d),$$

then

(11.2.9)
$$g(n) = \sum_{d|n} \mu\left(\frac{n}{d}\right) f(d).$$

Hence

$$nM_r(n) = \sum_{d|n} \mu\left(\frac{n}{d}\right) r^d$$

or

(11.2.10)
$$M_r(n) = \frac{1}{n} \sum_{d|n} \mu\left(\frac{n}{d}\right) r^d,$$

the Witt formula.

---

* Hardy and Wright [1] p. 235.

The number of ordinary words W such that $\omega_i(W) = n_i$, $n_1 + \cdots + n_r = n$ is the multinomial coefficient

$$\frac{n!}{n_1! \cdots n_r!}.$$

This leads to the formula

$$(11.2.11) \qquad \frac{n!}{n_1! \cdots n_r!} = \sum_{d|n_1,\cdots n_r} dM\left(\frac{n_1}{d}, \frac{n_2}{d}, \cdots, \frac{n_r}{d}\right).$$

Here d ranges over the divisors of $(n_1, \cdots, n_r) = n_0$. Applying the Möbius inversion we have

$$(11.2.12) \qquad M(n_1, n_2, \cdots, n_r) = \frac{1}{n} \sum_{d|n_1,\ldots,n_r} \mu(d) \frac{\left(\frac{n}{d}\right)!}{\left(\frac{n_1}{d}\right)! \cdots \left(\frac{n_r}{d}\right)!},$$

the second of the Witt formulae.

Consider the free associative ring R with integer coefficients having r generators, $x_1, x_2, \cdots, x_r$. The elements R_m of degree m form an additively free Abelian group with a basis of the r^m products $x_{i1} \cdots x_{i_m}$. In a ring we define a commutator $[u, v]$ by the rule

$$(11.2.13) \qquad [u, v] = uv - vu.$$

The formal properties of bracketing will apply to the ring commutators quite as well as to the group commutators. Indeed we shall show that there is a very close relation between group and ring commutators, originally established by Magnus [1].

THEOREM 11.2.3. *The basic products of degree m form an additive basis for R_m.*

COROLLARY 11.2.1. *The basic commutators of degree m are linearly independent.*

Proof: Since by Theorem 11.2.1 the number of basic products of degree m is r^m, which is the right number for a basis of R_m, it is sufficient to show that every element of R_m can be expressed as a linear combination with integral coefficients of basic products. Since $P_1^{(m)} = P_1$ is the basis of the r^m products $x_{i1} \cdots x_{i_m}$, and since P_k consists of the basic products for k sufficiently large, it will be enough to express the elements of P_k as linear combinations with integral coefficients of elements of P_{k+1}. For this we need an identity. For simplicity of notation write $[\cdots [u, v], v \cdots], v] = [u, \overbrace{v, \cdots, v}^{s}] = [{}^s u, v^s]$ if there are s v's. The identity is

$$(11.2.14) \qquad uv^s = v^s u + \sum_{j=1}^{s} \binom{s}{j} v^{s-j} [{}^j u, v^j].$$

For $s = 1$ this reduces to

$$uv = vu + [u, v].$$

We note the identity

$$(11.2.15) \qquad [{}^j u, v^j]v = [{}^{j+1}u, v^{j+1}] + v[{}^j u, v^j].$$

Hence (11.2.14) is proved by induction on s by multiplying (11.2.14) by v on the right, making replacements throughout by means of (11.2.15) and combining similar terms.

If a term in P_k has a sub-sequence $\cdots uc_k \cdots c_k w \cdots u$, $w \neq c_k$, where u is later than c_k and there are s c_k's, we apply (11.2.14) with $u = n$, $v = c_k$. This gives terms either belonging to P_{k+1} or terms of P_k with fewer c_k's, or having the c_k's nearer the beginning. Repeated application of (11.2.14) will ultimately express an element of P_k as a linear combination with integer coefficients of the terms of P_{k+1}. This proves the theorem.

Let us adjoin a unit 1 to R, making the rational integers the elements R_0 of degree zero, and take this ring modulo the two-sided ideal generated by all terms of degree $n + 1$ or higher. Call the resulting ring $\overline{R}$. Then

$$(11.2.16) \qquad \overline{R} = R_0 + R_1 + \cdots + R_n.$$

In $\overline{R}$ the elements with constant term 1 of the form $1 + z$, $z \in R_1 + \cdots + R_n$ form a group G, for, since $z^{n+1} = 0$,

$$(11.2.17) \qquad (1 + z)^{-1} = 1 - z + z^2 + \cdots + (-1)^n z^n.$$

If $1 + z = 1 + u_m + u_{m+1} + \cdots + u_n$, with $u_j \in R_j$ for $j = m, \cdots, n$ and $u_m \neq 0$, we say that u_m is the *leading term* of $1 + z$. The leading term of 1 is 0.

LEMMA 11.2.2. *Let $u, v \neq 1$ be elements of G with leading terms u_s, v_t of degree s and t, respectively. The leading terms of u^{-1} and v^{-1} are*

$-u_s$ and $-v_t$. If $s < t$, the leading term of uv is u_s. If $t < s$, the leading term of uv is u_t. If $t = s$ and $u_s + v_t \neq 0$, the leading term of uv is $u_s + v_t$. If the ring commutator $[u_r, v_s]$ is not zero, it is the leading term of the group commutator (u, v).

Proof: Let $u = 1 + a$, $v = 1 + b$, $u^{-1} = 1 + a'$, $v^{-1} = 1 + b'$. Then

$$\begin{gathered} a + a' + aa' = 0 \quad aa' = a'a \\ b + b' + bb' = 0 \quad bb' = b'b \\ a = u_s + \cdots + u_n, \quad b = v_t + \cdots + v_n \\ uv = 1 + a + b + ab. \end{gathered}$$

From these relations we get immediately the statements of the lemma about the leading terms of u^{-1}, v^{-1}, and uv. Using these relations we find

$$\begin{aligned} (u, v) &= u^{-1}v^{-1}uv \\ &= (1 + a')(1 + b')(1 + a)(1 + b) \\ &= 1 + ab - ba \\ &\qquad + aa'b - bb'a + b'ab + a'b'a + a'b'ab, \end{aligned}$$

whence

$$(u, v) = 1 + [u_s, v_t] + \text{higher terms}, \tag{11.2.18}$$

giving the final statement of the lemma.

Let $c_1, c_2, \cdots$ be a sequence of basic commutators in the free group F generated by elements $y_1, \cdots, y_r$, and $d_1, d_2, \cdots$ be the ring commutators in R obtained by replacing $y_1, \cdots, y_r$ by $x_1, \cdots, x_r$. Also let c_t be the last commutator of weight n. Then there is a correspondence between the c's and the d's in $\overline{R}$ given by the following lemma:

Lemma 11.2.3. *If we make $y_i \to 1 + x_i$, $i = 1, \cdots, r$, mapping F onto G we map $c_i \to g_i \in G$, and for $i = 1, \cdots, t$, the leading term of g_i is d_i.*

Proof: Since $y_i \to 1 + x_i$, $i = 1, \cdots, r$, the leading term of $g_i = 1 + x_i$ is x_i for $i = 1, \cdots, r$. We proceed by induction. If $c_w = (c_u, c_v)$, $w \leq t$, then by induction the leading term of g_u is d_u and of g_v is d_v. Hence, by Lemma 11.2.2, the leading term of (g_u, g_v) is $[d_u, d_v]$ if this is not zero; as a basic commutator, it is not zero from the corollary to Theorem 11.2.3. Hence the leading term of $g_w = (g_u, g_v)$ is $[d_u, d_v] = d_w$ as the lemma asserts.

THEOREM 11.2.4 (BASIS THEOREM).* *If F is the free group with free generators $y_1, \cdots, y_r$ and if in a sequence of basic commutators $c_1, \cdots, c_t$ are those of weights $1, 2, \cdots, n$, then an arbitrary element f of F has a unique representation,*

$$f = c_1^{e_1}c_2^{e_2} \cdots c_t^{e_t} \bmod F_{n+1}. \tag{11.2.19}$$

The basic commutators of weight n form a basis for the free Abelian group F_n/F_{n+1}.

Proof: We prove the second statement first. Suppose $c_s, \cdots, c_t$ are the basic commutators of weight n. By Lemma 11.2.3, if we take the mapping of F into G determined by

$$y_i \to 1 + x_i = g_i, \quad i = 1, \cdots, r, \tag{11.2.20}$$

then the leading terms of $c_s, \cdots, c_t$ are the corresponding ring commutators $d_s, \cdots, d_t$, which are the basic ring commutators of degree n. By the corollary to Theorem 11.2.3, $d_s, \cdots, d_t$ are linearly independent, and by Lemma 11.2.2, the leading term of $c_s^{e_s} \cdots c_t^{e_t}$ is $e_sd_s + \cdots e_td_t$ and so is not zero unless $e_s = \cdots = e_t = 0$. Hence $c_s, \cdots, c_t$ are independent elements of F_n/F_{n+1}, and hence a basis, since we already know from (11.1.4) that every element of F_n/F_{n+1} can be expressed in terms of $c_s, \cdots, c_t$. The existence of at least one expression for f in the form (11.2.19) was given by (11.1.4). We must show its uniqueness. But if

$$c_1^{e_1} \cdots c_t^{e_t} = c_1^{h_1} \cdots c_t^{h_t} \pmod{F_{n+1}}, \tag{11.2.21}$$

and $h_i = e_i$, $i = 1 \cdots j - 1$ but $h_j \neq e_j$, if c_j is of weight k this would lead to a dependence between the basic commutators of weight k modulo F_{k+1}. Since this cannot be the case, the expression (11.2.18) is unique. This completes our proof.

---

* See Marshall Hall, Jr. [6].

12. THE THEORY OF p-GROUPS; REGULAR p-GROUPS

12.1. Elementary Results.

In Chaps. 4 and 10, some elementary properties of finite p-groups P were established. We list them here:

1) P has a center Z greater than the identity. (Theorem 4.3.1.)

2) A proper subgroup H of P is not its own normalizer. (Theorem 4.2.1.)

3) If P is of order p^n, then every maximal subgroup M is of order p^{n-1} and is normal. (Theorem 4.3.2.)

4) A normal subgroup of order p in P is contained in the center of P. (Theorem 4.3.4.)

5) P is supersolvable. (Theorem 10.3.4 and Theorem 10.2.4.)

6) P is nilpotent. (Theorem 10.3.4.)

12.2. The Burnside Basis Theorem. Automorphisms of p-Groups.

Let P be of order p^n. The intersection of all its maximal subgroups will be a characteristic subgroup D, the Frattini subgroup of P. Then, in the homomorphism $P \rightarrow P/D$, elements generating P will be mapped onto elements generating P/D. The converse of this is true in a strong sense, which is the subject of the Burnside basis theorem.

THEOREM 12.2.1 (THE BURNSIDE BASIS THEOREM). *Let D be the intersection of the maximal subgroups of the p-group P. The factor group $P/D = A$ is an elementary Abelian group. If A is of order p^r, then every set of elements $z_1, \cdots, z_s$ which generates P contains a subset of r elements $x_1, \cdots, x_r$ which generate P. In the mapping $P \rightarrow A$, the elements $x_1, \cdots, x_r$ are mapped onto a basis $a_1, \cdots, a_r$ of A. Conversely, any set of r elements of P which, in $P \rightarrow A$ is mapped onto a set of generators of A, will generate P.*

Proof: If M is a maximal subgroup of P, then M is of index p and is normal. Thus P/M is the cyclic group of order p. Hence the pth power of every element of P and every commutator are contained in M. Hence D, the intersection of all the maximal subgroups, contains every pth power and every commutator. Thus P/D is an elementary Abelian group A. If A is of order p^r, then every basis of A consists of r elements, say, $a_1, \cdots, a_r$. If $b_1, \cdots, b_s$ are elements generating A, we may find a basis for A by deleting from them the b's equal to 1 and those b_i's belonging to the subgroup generated by $b_1, \cdots, b_{i-1}$. Hence $s \geqq r$ and $b_1, \cdots, b_s$ contains a subset which is a basis for A.

Now suppose that $z_1, \cdots, z_s$ generate P. In the mapping $P \to P/D = A$, let $z_i \to b_i$, $i = 1, \cdots, s$. Then $b_1, \cdots, b_s$ generate A and so contain a subset $a_1, \cdots, a_r$, which is a basis for A. Let $x_1, \cdots, x_r$ be the subset of $z_1, \cdots, z_s$ mapped onto $a_1, \cdots, a_r$. The theorem will be proved if we can show that any set $x_1, \cdots, x_r$ of elements of P mapped onto a basis $a_1, \cdots, a_r$ of A will generate P. Let $H = \{x_1, \cdots, x_r\}$. If $H \neq P$, then H is contained in some maximal subgroup M of P. But then in $P \to P/D = A$ we have $H \to HD/D \subseteq M/D = B$, where B is a subgroup of A of order p^{r-1}. This is in conflict with $H = \{x_1, \cdots, x_r\} \to \{a_1, \cdots, a_r\} = A$. Hence $H = P$ and $x_1, \cdots, x_r$ generate P.

As an application of this theorem we may obtain some information on the group $A(P)$ of automorphisms of P. We may choose a basis $a_1, \cdots, a_r$ of P/D in $\theta(p^r) = (p^r - 1)(p^r - p) \cdots (p^r - p^{r-1})$ ways. This is easily seen since a_1 may be taken as any of the $p^r - 1$ elements different from the identity, and having chosen $a_1, \cdots, a_i$, we may take a_{i+1} as any one of the $p^r - p^i$ elements not in the subgroup generated by $a_1, \cdots, a_i$. Thus there are $\theta(p^r)$ choices for a basis of A, and every mapping of a fixed basis $a_1, \cdots, a_r$ onto another $b_1, \cdots, b_r$ yields an automorphism of A. But since every automorphism of A must map $a_1, \cdots, a_r$ onto a basis, there are exactly $\theta(p^r)$ automorphisms of A.

There will be exactly $p^{r(n-r)}\theta(p^r)$ ordered sets $X = (x_1, \cdots, x_r)$ which generate P, since in a mapping $x_i \to a_i$, $i = 1, \cdots, r$ of X onto a basis of A, the basis of A may be chosen in $\theta(p^r)$ ways, and for a single a_i, any of the p^{n-r} elements in the coset of D mapped onto a_i will be a permissible choice for x_i. Every automorphism of P will map a set X onto another. Hence the group $A(P)$ of automorphisms of A may be regarded as a permutation group on the X's. But $A(P)$ is a regular group on the X's, since an automorphism fixing any set X fixes every product of these x's and hence the entire group P, and so is the identical automorphism. Hence the sets X are permuted among themselves in transitive constituents each of which has k sets in it if k is the order of $A(P)$. Hence $p^{r(n-r)}\theta(p^r) = kt$. Here the number t may be interpreted as the number of essentially different ways of generating P by r elements. Two sets $X = (x_1, \cdots, x_r)$ and $Y = (y_1, \cdots, y_r)$ are said to generate P in essentially the same way if every relation $w(x_1, \cdots, x_r) = 1$ is such that $w(y_1, \cdots, y_r) = 1$, and conversely.

In the same way, let $A_1(P)$ be the normal subgroup of $A(P)$ which leaves A/D fixed elementwise. These automorphisms permute regularly the $p^{r(n-r)}$ generating sets $X = (x_1, \cdots, x_r)$ which are mapped onto the same basis $a_1, \cdots, a_r$ of A in the homomorphism $P \to P/D = A$. Thus the order of $A_1(P)$ divides $p^{r(n-r)}$. These results, due to P. Hall [2], we state as a theorem.

THEOREM 12.2.2. *If P is a p-group of order p^n, D the intersection of the maximal subgroups of P, and $[P:D] = p^r$, then the order of $A(P)$, the group of automorphisms of P, divides $p^{r(n-r)}\theta(p^r)$. The order of $A_1(P)$, the group of automorphisms fixing P/D elementwise, is a divisor of $p^{r(n-r)}$.*

12.3. The Collection Formula.

Let G be a group generated by elements $a_1, a_2, \cdots, a_r$. We shall develop a formula for $(a_1a_2 \cdots a_r)^n$ in terms of the higher commutators of $a_1, \cdots, a_r$. We may take G to be the free group generated by $a_1, \cdots, a_r$, for the formula will then hold *a fortiori* in any group generated by r elements.

We repeat the definition of basic commutators, given in §11.1, but make the ordering more precise.

1) $a_1, \cdots, a_r$ are the commutators of weight one, and are simply ordered by the rule $a_1 < a_2 < \cdots < a_r$.

2) If basic commutators of weights less than n have been defined and simply ordered, then (x, y) is a basic commutator of weight n if, and only if,

(a) x and y are basic commutators with $\omega(x) + \omega(y) = n$.

(b) $x > y$.

(c) If $x = (u, v)$, then $y \geq v$.

3) Commutators of weight n follow all commutators of weight less

than n, and for weight n, $(x_1, y_1) < (x_2, y_2)$ if $y_1 < y_2$ or if $y_1 = y_2$ and $x_1 < x_2$.

Consider

$$(12.3.1) \qquad (a_1a_2 \cdots a_r)^n = a_1(1)a_2(1) \cdots a_r(1)a_1(2) \cdots a_r(2) \cdots a_r(n),$$

where we have labeled the individual generators a_i as $a_i(1)$, $a_i(2)$, $\cdots$, $a_i(n)$ from left to right so as to be able to distinguish each letter in the formula. Since $SR = RS(S, R)$ by definition of the commutator, we may replace the right-hand side of (12.3.1) by another expression equal to it in which a pair of consecutive elements SR is replaced by $RS(S, R)$. This replacement puts R nearer the beginning of the expression and introduces a commutator (S, R). By a succession of such replacements we may move any letter as near to the beginning as we choose. We shall alter (12.3.1) in a specific way. We begin by moving $a_1(2)$ to the left until it is next to $a_1(1)$, then move $a_1(3)$ to the left until it is next to $a_1(2)$, and continue until we have collected all a_1's at the beginning. This completes the first stage of collection. Next we collect in order the a_2's immediately to the right of the a_1's.

Let us describe the collection process precisely. At the end of the ith stage we have

$$(12.3.2) \qquad (a_1a_2 \cdots a_r)^n = c_1^{e_1}c_2^{e_2} \cdots c_i^{e_i}R_1R_2 \cdots R_t,$$

where c_1, c_2, $\cdots$, c_i are the first i basic commutators and R_1, $\cdots$, R_t are basic commutators later than c_i. If R_{j_1}, R_{j_2}, $\cdots$, R_{j_s} are in order, the basic commutators among R_1, $\cdots$, R_t which are c_{i+1}, we first move R_{j_1} to the position immediately following $c_i^{e_i}$, then R_{j_2}, R_{j_3}, $\cdots$, and finally R_{j_s}, so that with $e_{i+1} = s$, (12.3.2) takes the form

$$(12.3.3) \qquad (a_1a_2 \cdots a_r)^n = c_1^{e_1}c_2^{e_2} \cdots c_{i+1}^{e_{i+1}}R^*_1 \cdots R_k^*,$$

which is the $(i + 1)$st stage. In (12.3.2) we call $c_1^{e_1} \cdots c_i^{e_i}$ the collected part and $R_1 \cdots R_t$ the uncollected part. But to validate this description we must show that only basic commutators appear in any formula. The initial formula (12.3.1) is stage zero and contains only generators a_i which are basic commutators of weight one. Let us assume by induction that at stage i the uncollected part $R_1 \cdots R_t$ contains only basic commutators later than c_i. In collecting R's equal to c_{i+1}, we introduce only further commutators $(c_j, c_{i+1}, \cdots, c_{i+1})$ where $j \geq i + 2$. Such a commutator is basic, since if $c_j =$

(c_r, c_s), then c_j arose at stage s when c_s was collected, whence $s < i + 1$ and so $c_s < c_{i+1}$. Thus (c_j, c_{i+1}) is basic and so also is $(c_j, c_{i+1}, \cdots, c_{i+1})$.

We have already in (12.3.1) labeled the generators a_i with labels j, $a_i^{(j)}$, $j = 1, \cdots, n$. If a commutator R of weight w_1 has a label $(\lambda_1, \cdots, \lambda_{w_1})$ and S of weight w_2 has a label $(\mu_1, \cdots, \mu_{w_2})$, we assign to (R, S) the label $(\lambda_1, \cdots, \lambda_{w_1}, \mu_1, \cdots, \mu_{w_2})$. The calculation of the exponents $e_1, \cdots, e_i, e_{i+1}$ in (12.3.3) may be made to depend on these labels. Here $e_{i+1} = s$ is the number of uncollected commutators at stage i equal to c_{i+1}. Thus it is the number E_{i+1} of commutators c_{i+1} existing at this stage. Also if $c_{i+1} = (c_r, c_s)$, then c_{i+1} arose when c_s was collected and this particular c_r preceded this particular c_s in the uncollected part. Hence we must also consider precedence conditions for a commutator c_r to precede c_s when they both exist in an uncollected part.

At stage zero the commutators of weight one (and no others) exist, and $a_k^{(\lambda)}$ exists for any label $\lambda = 1, \cdots, n$. Moreover, $a_k^{(\lambda)}$ precedes $a_s^{(\mu)}$ at stage zero when $k > s$ if $\lambda < \mu$ and when $k < s$ if $\lambda \leq \mu$. More formally at stage zero we have existence and precedence conditions on the uncollected part in terms of labels:

$$E_k^0[a_k^{(\lambda)}] \text{ is that } \lambda \text{ exists (a vacuous condition),}$$
$$P^0_{rs}[a_r^{(\lambda)} \text{ precedes } a_s(\mu)] \qquad \begin{array}{l} \lambda < \mu \text{ if } r \geq s \\ \lambda \leq \mu \text{ if } r < s. \end{array}$$

Let $\lambda_1, \cdots, \lambda_m$ be a set of integers and consider conditions of the type $\lambda_t < \lambda_u$, $\lambda_t \leq \lambda_u$. Any logical sum and product of such conditions we shall call conditions (L). We shall show that conditions E_k^i for existence of a commutator c_k with label $(\lambda_1, \cdots, \lambda_m)$ at stage i are conditions (L) on $\lambda_1, \cdots, \lambda_m$, and the precedence conditions $p_{rs}{}^i$ for the precedence of a commutator c_r before a commutator c_s in the uncollected part of the ith stage are conditions (L) on $\lambda_1, \cdots, \lambda_m$, $\mu_1, \cdots, \mu_q$ if $(\lambda_1, \cdots, \lambda_m)$ is the label of c_r and $(\mu_1, \cdots, \mu_q)$ is the label of c_s. We have observed that at stage zero, existence and precedence conditions were conditions (L) as above. We prove this true in general by induction on the stage. Suppose this to be true at the ith stage. To show this to be true at the $(i + 1\text{st})$ stage, we compare (12.3.2) and (12.3.3). With $R_{j_1} = R_{j_2} = \cdots = R_{j_s} = c_{i+1}$, we collected first R_{j_1}, then R_{j_2}, and finally R_{j_s}. Each step in the collection was a replacement $SR = RS(S, R)$. Here any commutators existing at stage i different from c_{i+1} also exist at stage $i + 1$ and are in the same order. Thus

$$E_k^{i+1} = E_k^i \quad \text{and} \quad P_{rs}^{i+1} = P_{rs}^i$$

for such commutators. Hence we need consider only the existence of commutators c_k arising in the $(i + 1)$st stage and precedence P_{rs} where one or both of c_r, c_s arose at this stage. A commutator arising at this stage will be of the form $c_k = (c_j, R_{u_1}, \cdots, R_{u_m})$, obtained by moving R_{u_1} past c_j, then R_{u_2} past this commutator, and so on until we move R_{u_m} past $(c_j, R_{u_1}, \cdots, R_{u_{m-1}})$. Here all of $R_{u_1}, \cdots, R_{u_m}$ are equal to c_{i+1}. Here E_k^{i+1} is the logical product of the conditions for existence of $c_j, R_{u_1}, \cdots, R_{u_m}$ at stage i together with the precedence conditions that $c_j, R_{u_1}, \cdots, R_{u_m}$ are in this order at stage i. Thus E_k^{i+1} is a condition (L) on the label of c_k. In the collecting for the $(i + 1)$st stage, a commutator (S, R) arises in $SR = RS(S, R)$ immediately to the right of S and to the left of all commutators following S. We must find the precedence condition P_{rs}^{i+1} where $c_r = c_{j_1}$ or $(c_{j_1}, R_{u_1}, \cdots, R_{u_m})$ and $c_s = c_{j_2}$ or $(c_{j_2}, R_{v_1}, \cdots, R_{v_w})$. Here $P_{rs}^{i+1} = P_{j_1 j_2}^i$ if $c_{j_1} \neq c_{j_2}$. If, however, $c_{j_1} = c_{j_2}$, P_{rs}^{i+1} involves the Rs'. Suppose e is the largest integer such that $R_{u_1} = R_{v_1}, \cdots, R_{u_e} = R_{v_e}$. Then c_r precedes c_s if either (1) $m = e$ and there is no $R_{u_{e+1}}$, in which case c_s is a commutator of c_r, or (2) $R_{v_{e+1}}$ precedes $R_{u_{e+1}}$. Here P_{rs}^{i+1} is a logical sum of precedence conditions, and so, are conditions (L) on the labels of c_r and c_s are combined.

LEMMA 12.3.1. *The number of sets $\lambda_1, \cdots, \lambda_m$ with $1 \leq \lambda_i \leq n$ satisfying given conditions (L) is an integer valued polynomial in n $b_1 n + b_2 n^{(2)} + \cdots + b_m n^{(m)}$, where $n^{(i)} = n(n-1) \cdots (n-1+i)/i!$ and the b's are integers determined by the conditions (L) but not depending on n.*

Proof: Let us divide the indices $1, \cdots, m$ into disjoint sets $S_1, S_2, \cdots, S_t$. Then an ordering of $\lambda_1, \cdots, \lambda_m$ is given by $\lambda_j = v_i$, $j \in S_i$, $i = 1, \cdots, t$, where $v_1 < v_2 < \cdots < v_t$. Every possible choice of the λ's belongs to a unique ordering of this type, and there are $n^{(t)}$ choices for the v's, this being merely the number of combinations of n things t at a time. For this ordering either all λ's satisfy the conditions (L) or none. Hence the number of sets of λ's satisfying given conditions (L) is the polynomial $b_1 n + b_2 n^{(2)} + \cdots + b_m n^{(m)}$, where b_t is the number of orderings with t distinct values which satisfy the conditions (L), and clearly, b_t depends on the conditions but not on n.

For example, if $\lambda_1, \lambda_2, \lambda_3$ satisfy conditions (L) $\lambda_1 < \lambda_2, \lambda_3 \leq \lambda_2$, the orderings satisfying (L) are

1) $\lambda_1 = v_1, \quad \lambda_2 = \lambda_3 = v_2, \quad v_1 < v_2,$
2) $\lambda_1 = \lambda_3 = v_1, \quad \lambda_2 = v_2, \quad v_1 < v_2,$
3) $\lambda_1 = v_1, \quad \lambda_3 = v_2, \quad \lambda_2 = v_3, \quad v_1 < v_2 < v_3,$
4) $\lambda_3 = v_1, \quad \lambda_1 = v_2, \quad \lambda_2 = v_3, \quad v_1 < v_2 < v_3,$

and the number of sets satisfying the conditions (L) is $2n^{(2)} + 2n^{(3)}$.

We have shown that the exponent e_i in (12.3.2) of the commutator c_i is the number of commutators in the uncollected part at stage $i - 1$ equal to c_i and that this number is given as the number of sets $\lambda_1, \cdots, \lambda_m$ satisfying certain conditions (L), where m is the weight of c_i. Thus Lemma 12.3.1 gives us information on these exponents. We state our results in a theorem.

Theorem 12.3.1. *We may collect the product* $(a_1a_2 \cdots a_r)^n$ *in the form* $(a_1a_2 \cdots a_r)^n = a_1^n a_2^n \cdots a_r^n c_{r+1}^{e_{r+1}} \cdots c_i^{e_i} R_1 \cdots R_t$, *where* $c_{r+1}, \cdots, c_i$ *are the basic commutators on* $a_1, \cdots, a_r$ *in order, and* $R_1, \cdots, R_t$ *are basic commutators later than* c_i *in the ordering. For* $1 \leq j \leq i$, *the exponent* e_j *is of the form* $e_j = b_1 n + b_2 n^{(2)} + \cdots + b_m n^{(m)}$, *where* m *is the weight of* c_j, *the* b'*s are non-negative integers and do not depend on* n *but only on* c_j. *Here* $n^{(k)} = n(n-1) \cdots (n-k+1)/k!$

We may prove immediately an important corollary if G is a p-group whose class is less than p. Collecting all commutators of weight less than p, the uncollected part reduces to the identity. Moreover, with $n = p^\alpha$ all exponents are multiples of p^α, since an $n^{(i)}$, $i \leq p - 1$ is a binomial coefficient with n as a factor of the numerator and denominator with factors not exceeding $p - 1$.

Corollary 12.3.1. *If* P *is a* p-*group of class less than* p, *then with* $n = p^\alpha$

$$(a_1a_2 \cdots a_r)^n = a_1^n a_2^n \cdots a_r^n S_1^n S_2^n \cdots S_t^n$$

where $S_1, S_2, \cdots, S_t$ *belong to the commutator subgroup of the group generated by* $a_1, a_2, \cdots, a_r$.

12.4. Regular p-Groups.

We define a regular p-group as a group P in which for any two elements a, b, and any $n = p^\alpha$ satisfy

$$(ab)^n = a^n b^n S_1^n \cdots S_t^n, \tag{12.4.1}$$

with $S_1, \cdots, S_t$ appropriate elements from the commutator subgroup of the group generated by a and b. Immediate consequences of the definition and the corollary to Theorem 12.3.1 are

1) Every p-group of class less than p is regular.
2) Every p-group of order at most p^p is regular.
3) P is regular if every subgroup generated by two elements is regular.
4) Every subgroup and factor group of a regular group is regular.

For every p there is an irregular group of order p^{p+1}, namely, the Sylow subgroup $S^{(p)}$ of the symmetric group S_{p^2} on p^2 letters. This group is generated by two elements of order p and yet it contains elements of order p^2. This will be shown impossible for a regular group.

THEOREM 12.4.1 *In a regular p group with $n = p^\alpha$, $a^n b^n = (ab)^n S_1^n = (ab\ S_2)^n$, with S_1, S_2 in the derived group $H_2\ (a, b)$ of the group $H(a, b)$, generated by a and b.*

By repeated application of the theorem we get the corollary

COROLLARY 12.4.1. *In a regular p-group with $n = p^\alpha$, $a_1^n a_2^n \cdots a_r^n = (a_1\, a_2 \cdots a_r\, S_2)^n = (a_1 \cdots a_r)^n S_1^n$ with S_1, S_2 in $H_2\ (a_1 \cdots a_r)$.*

Proof: The theorem and corollary both hold in an Abelian group with $S_1 = 1$, $S_2 = 1$. We shall use induction to prove the theorem for a group H, assuming the theorem and its corollary to be true for any proper subgroup of H. We note that if H is generated by $a_1 \cdots a_r$, then $H_2(a_1 \cdots a_r)$, the derived subgroup of H, is a proper subgroup of H. From (12.4.1)

$$a^n b^n = (ab)^n\, S_t^{-n} \cdots S_1^{-n}. \tag{12.4.2}$$

By induction $S_t^{-n} \cdots S_1^{-n} = S^n$ with $S \in H_2$. But if $H = H(a, b)$ is not Abelian, then H_2 and ab generate a proper subgroup of H whence by induction $(ab)^n S^n = (abS_2)^n$. For it follows from the Burnside basis theorem that if H/H_2 is cyclic, then H is cyclic. Thus the theorem holds in H if both the theorem and corollary hold in any proper subgroup of H. Applying the theorem $r - 1$ times to $a_1^n\, a_2^n \cdots a_r^n$, we get

$$a_1^n\, a_2^n \cdots a_r^n = (a_1 a_2 \cdots a_r)^n\, S_1^n \cdots S_{r-1}^n$$

with all of $S_1 \cdots S_{r-1}$ in H_2.
Thus $a_1^n \cdots a_r^n = (a_1\, a_2 \cdots a_r)^n\, S^n$, applying the corollary to H_2, and by the theorem, $(a_1\, a_2 \cdots a_r)^n\, S^n = (a_1\, a_2 \cdots a_r S^1)^n$.

THEOREM 12.4.2. *A finite p-group P is regular if, and only if, for any a, b in P we have*

$$a^p\, b^p = (ab)^p\, S^p, \tag{12.4.3}$$

with S in the derived group of the group generated by a and b.

The condition (12.4.3) is clearly necessary in a regular p-group since it is a special case of Theorem 12.4.1. We must show conversely that (12.4.3) implies

$$a^n b^n = (ab)^n S_1^n, \quad n = p^\alpha, \quad S_1 \in H_2\,(a, b). \tag{12.4.4}$$

Now the relations

$$a_1^p\, a_2^p \cdots a_r^p = (a_1\, a_2 \cdots a_r)^p\, S_1^p = (a_1\, a_2 \cdots a_r\, S_2)^p \tag{12.4.5}$$

with $S_1, S_2 \in H_2\,(a_1 \cdots a_r)$ are surely satisfied with $S_1 = S_2 = 1$ when H is Abelian. If (12.4.5) is satisfied for every proper subgroup of H, then applying (12.4.3) $r - 1$ times $a_1^p\, a_2^p \cdots a_r^p = (a_1 a_2 \cdots a_r)^p$ $u_1^p \cdots u_{r-1}^p$,with $u_1, \cdots, u_{r-1}$ in H_2. By induction $u_1^p \cdots u_r^p{}_1 = S_1^p$. But $b = a_1\, a_2 \cdots a_r$ and S_1 generate a proper subgroup of H, whence $(a_1 a_2 \cdots a_r)^p S_1^p = (a_1 \cdots a_r S_2)^p$, proving (12.4.5) in general.

LEMMA 12.4.1. *Assuming* (12.4.3), $x^{-p}\, y^{-p}\, x^p\, y^p = S^p$, *with S in the derived group of* $\{x, y\}$.

Proof:

$$x^p\, y^p = (x\, y)^p\, S_1^p,$$
$$y^p\, x^p = (y\, x)^p\, S_2^p,$$

whence

$$x^{-p}\, y^{-p}\, x^p\, y^p = S_2^{-p}\, (y\, x)^{-p}\, (x\, y)^p\, S_1^p,$$

and also

$$(y\ x)^{-p}\ (x\ y)^p = (x^{-1}\ y^{-1}\ x\ y)^p\ S_3{}^p$$
$$= (x,\ y)^p\ S_3{}^p,$$

and so
$$x^{-p}\ y^{-p}\ x^p\ y^p = S_2{}^{-p}\ (x,\ y)^p\ S_3{}^p\ S_1{}^p = S^p.$$

From this it follows that any commutator in $a_1{}^p, a_2{}^p, \cdots, a_r{}^p$ is the pth power of an element in the derived group of $\{a_1, \cdots, a_r\}$.

From (12.4.3) we have

(12.4.6)
$$a^{p^2}\ b^{p^2} = (a^p\ b^p)^p\ S_1{}^p$$
$$= [(a\ b)^p\ S_2{}^p]^p\ S_1{}^p$$
$$= (ab)^{p^2}\ S_2{}^{p^2}\ S_3{}^p\ S_1{}^p,$$

where S_1 is in the derived group of $\{a^p, b^p\}$ and S_3 in the derived group of $(ab)^p, S_2{}^p$. By the lemma these are pth powers of elements in the derived group of $\{a, b\}$, whence

(12.4.7)
$$a^{p^2}b^{p^2} = (ab)^{p^2}\ S_2{}^{p^2}\ S_4{}^{p^2}\ S_5{}^{p^2},$$

and applying induction (12.4.4) holds for $n = p^2$. The same procedure and use of lemma enables us to prove (12.4.4), going from $n = p^\alpha$ to $n = p^{\alpha+1}$.

Theorem 12.4.3. *If P is a regular p-group then with $n = p^\alpha$.*

1) *Each of $(a^n,b) = 1$ and $(a,b)^n = 1$ implies the other.*

2) *If $(a^n,b) = 1$, then $(a,b^n) = 1$.*

3) *A commutator S involving an element u has order at most that of u modulo the center of P.*

4) *The order of a product $a_1\ a_2 \cdots a_r$ cannot exceed the order of all of $a_1, a_2, \cdots, a_r$.*

Proof: In an Abelian group the first three properties are vacuously true and the fourth is true. We shall assume by induction that the theorem holds for all proper subgroups of P, and we also take P to be non-Abelian.

Let us apply (12.4.4) to

(12.4.8)
$$a^{-n}b^{-1}a^nb = (a^{-1})^n(b^{-1}ab)^n = (a^{-1}b^{-1}ab)^ns^n,$$

with s in the derived group of $K(a, b^{-1}ab) \subset H(a, b)$; this becomes

(12.4.9)
$$(a^n,\ b) = (a,\ b)^ns_1{}^n.$$

Now if $(a^n, b) = 1$, then the order of a modulo the center of $H(a, b)$

is n or less, whence by property (3) for the proper subgroup $K(a, b^{-1}ab)$ with $u = a$, every commutator in K involves a and is of order at most n. The element s_1 in (12.4.9) is a product of commutators in K, and by (4) for K, the order of s_1 is at most n. Thus $(a^n, b) = 1$ implies $s_1^n = 1$ in (12.4.9), and so, $(a, b)^n = 1$. Conversely, if $(a, b)^n = 1$, then in $K = K(a, a^{-1}b^{-1}ab) = K(a, u)$ with $u = (a, b)$, the order of u modulo the center is at most n and every commutator involves u. Thus by (3) for K, all commutators in K_2 are of order at most n, and so by (4) in K_2, the order of s_1 in (12.4.9) is at most n.

Thus $(a, b)^n = 1$ implies $S_1^n = 1$, and so, $(a^n, b) = 1$, This proves property (1) for P, and of course (2) follows immediately from (1). Property (3) in P follows from repeated application of (1). If the order of u modulo the center of P is n, then *a fortiori* $(u^n, v) = 1$, whence $(u, v)^n = 1$. Here, with $x = (u, v)$ and since $x^n = 1$, it follows that $(x, y)^n = 1$.

It remains to prove property (4) for P. If $a^n = 1$, $b^n = 1$, then by (3), any commutator involving a or b is of order at most n. Hence in (12.4.4), S_1 is a product of commutators of order at most n, and by property (4) for the proper subgroup P_2, S_1 itself is of order at most n. Hence $S_1^n = 1$, and so, $(ab)^n = 1$. Thus the product of two factors has order not exceeding that of both factors, and by repetition it follows that the product of r factors has an order not exceeding that of all the factors.

THEOREM 12.4.4. *If $a^n = b^n$ with $n = p^\alpha$, then $(a\, b^{-1})^n = 1$, and conversely.*

Proof: In $H(a,b)$ all commutators are of order at most n from property (3) in Theorem 12.4.3. Hence in $1 = a^nb^{-n} = (ab^{-1})^n s_1^n$ we have $s_1^n = 1$, and so, $(ab^{-1})^n = 1$. Conversely, with $a^nb^{-n} = (ab^{-1})^n s_1^n$ and $(ab^{-1})^n = 1$, we may write $H(a, b) = H(a, ab^{-1})$, and so by property (3) with $u = ab^{-1}$, we have $s_1^n = 1$ and hence $a^n = b^n$.

THEOREM 12.4.5. *In a regular p-group P the (p^α)th powers of the elements form a characteristic subgroup $C^\alpha(P)$, the elements of order at most p^α a characteristic subgroup $C_\alpha(P)$.*

Proof: With $n = p^\alpha$ the relation of Theorem 12.4.1, $a^nb^n = (abs_2)^n$, shows that the (p^α)th powers of elements form a subgroup $C^\alpha(P)$ which is necessarily characteristic and in fact fully invariant. Property (4) of Theorem 12.4.3 shows that elements whose orders are at most p^α form a subgroup which will be fully invariant.

12.5. Some Special p-Groups. Hamiltonian Groups.

THEOREM 12.5.1. *The groups of order p^n which contain a cyclic subgroup of index p are of the following types:*

Abelian,

$n \geqq 1$, *cyclic:*

1) $a^{p^n} = 1$.

$n \geqq 2$:

2) $a^{p^{n-1}} = 1, b^p = 1, ba = ab$.

Non-Abelian,

p *odd*, $n \geqq 3$:

3) $a^{p^{n-1}} = 1, b^p = 1, ba = a^{1+p^{n-2}}b$.

$p = 2, n \geqq 3$;

4) *Generalized quaternion group.*

$a^{2^{n-1}} = 1, b^2 = a^{2^{n-2}}, ba = a^{-1}b$.

$p = 2, n \geqq 3$

5) *Dihedral group.*

$a^{2^{n-1}} = 1, b^2 = 1, ba = a^{-1}b$

$p = 2, n \geqq 4$

6) $a^{2^{n-1}} = 1, b^2 = 1, ba = a^{1+2^{n-2}}b$.

$p = 2, n \geqq 4$

7) $a^{2^{n-1}} = 1, b^2 = 1, ba = a^{-1+2^{n-2}}b$.

Proof: An Abelian group of order p^n which contains an element of order p^{n-1} must have a basis element of order p^{n-1} or p^n. This settles the theorem for Abelian groups, giving the first two cases.

In considering non-Abelian groups of order p^n containing an element of order p^{n-1}, let us first suppose p odd. If $a^{p^{n-1}} = 1$, then $\{a\}$ as a subgroup of index p is a normal subgroup, and so for $b \notin \{a\}$, we have $bab^{-1} = a^r$, where $r \not\equiv 1 \pmod{p^{n-1}}$, since our group is not Abelian. We find that $b^iab^{-i} = a^{r^i}$ by induction on i, since $(bab^{-1})^j = ba^jb^{-1} = a^{rj}$ for any j, and in particular for $j = r$ we have $b(bab^{-1})b^{-1} = b^2ab^{-2} = ba^rb^{-1} = a^{r^2}$. The general case $b^iab^{-i} = a^{r^i}$ follows readily by induction. As $b^p \in \{a\}$, we have $b^pab^{-p} = a$, whence $r^p \equiv 1 \pmod{p^{n-1}}$. Since p is odd, we may conclude from this congruence that $r \equiv$

$1 + kp^{n-2} \pmod{p^{n-1}}$, where $k \not\equiv 0 \pmod p$, since $r \not\equiv 1 \pmod{p^{n-1}}$. Now take $b_1 = b^i$, where i is determined by the congruence $ik \equiv 1 \pmod p$. Then $r^i \equiv (1 + kp^{n-2})^i \equiv 1 + ikp^{n-2} \equiv 1 + p^{n-2} \pmod{p^{n-1}}$. Hence $b_1ab_1^{-1} = b^iab^{-i} = a^{r^i} = a^{1+p^{n-2}}$. Let us write $h = 1 + p^{n-2}$. Then $(a^jb_1)^2 = a^jb_1a^jb_1^{-1}b_1^2 = a^{j(1+h)}b_1^2$, and we find by induction that $(a^jb_1)^t = a^{jT}b^t$, where $T = 1 + h + \cdots + h^{t-1}$. For $t = p$ we have $1 + h + \cdots + h^{p-1} \equiv p + p^{n-2}[1 + 2 + \cdots + (p-1)] \equiv p + p^{n-1}(p-1)/2 \equiv p \pmod{p^{n-1}}$ since p is odd. Thus $(a^jb_1)^p = a^{jp}b_1^p$. This formula could also have been found by an appeal to the collection formula. Now $b_1^p = a^u \in \{a\}$, where $u = pv$ since b_1 is not of order p^n, and since the group is not cyclic. If we put $b_2 = a^{-v}b_1$, then $b_2^p = (a^{-v}b_1)^p = a^{-vp}b_1^p = a^{-pv}a^{pv} = 1$, and also $b_2ab_2^{-1} = a^{-v}b_1ab_1^{-1}a^v = a^{-v}a^{1+p^{n-2}}a^v = a^{1+p^{n-2}}$. Thus a and b_2 satisfy the relations given in the theorem as type 3 for non-Abelian groups with p odd.

Let us now take $p = 2$ and find the non-Abelian groups of order 2^n containing an element of order 2^{n-1}. Let $a^{2^{n-1}} = 1$, $b \notin \{a\}$. Then $bab^{-1} = a^r$, where $r^2 \equiv 1 \pmod{2^{n-1}}$, $r \not\equiv 1 \pmod{2^{n-1}}$. This gives three distinct choices of r modulo 2^{n-1}, $r = -1$, $r = 1 + 2^{n-2}$, $r = -1 + 2^{n-2}$. Also let $b^2 = a^w \in \{a\}$. Then, since $b(b^2)b^{-1} = b^2$, we have $a^{wr} = a^w$ or $wr \equiv w \pmod{2^{n-1}}$ as a condition on w. For $r = -1$ we find $-w \equiv w \pmod{2^{n-1}}$, whence $a^w = 1$ or $a^w = a^{2^{n-2}}$. Thus with $r = -1$ we find the generalized quaternion group or the dihedral group, types 4 and 5 in the theorem, respectively. For $n = 3$ these are the only groups, as we determined in §4.4.

Suppose now $n \geqq 4$ and $ba = a^rb$, with $r = 1 + 2^{n-2}$. With $b^2 = a^w$, the condition on w that $wr \equiv w \pmod{2^{n-2}}$ is merely that $2^{n-2}w \equiv 0 \pmod{2^{n-1}}$ or that w be an even number $w = 2w_1$. Determine j by the congruence $j(1 + 2^{n-3}) + w_1 \equiv 0 \pmod{2^{n-2}}$. Then with $b_1 = a^jb$, we have $b_1^2 = a^j(ba^j)b = a^{j(2+2^{n-2})}b^2 = a^{2[j(1+2^{n-3})+w_1]} = a^{2^{n-1}} = 1$. Here $b_1a = a^{1+2^{n-2}}b_1$, and a and b_1 satisfy the relations of type 6 in the theorem. Finally, if $n \geq 4$, $ba = a^rb$ with $r = -1 + 2^{n-2}$, the condition on w in $b^2 = a^w$ that $w \equiv rw \pmod{2^{n-1}}$ is that $(2 + 2^{n-2})w \equiv 0 \pmod{2^{n-1}}$ or $w \equiv 0 \pmod{2^{n-2}}$. Thus $b^2 = 1$ or $b^2 = a^{2^{n-2}}$. If $b^2 = a^{2^{n-2}}$, take $b_1 = ab$ and $b_1^2 = a(ba)b = a(a^{-1+2^{n-2}})\,b^2 = a^{2^{n-2}}a^{2^{n-2}} = 1$. Thus either a and b or a and b_1 satisfy the relations of type 7 in the theorem.

All the relations in Theorem 12.5.1 determine groups, as may be verified in every case, except that of the generalized quaternion groups, by means of Theorem 6.5.1. For the generalized quaternion groups, we may make a direct verification or refer ahead to Theorem 15.3.1.

THEOREM 12.5.2. *A p-group which contains only one subgroup of order p is cyclic or a generalized quaternion group.*

Proof: Let P be of order p^n and contain only one subgroup of order p. We prove by induction on n that P is cyclic or of generalized quaternion type. This is trivial for $n = 1$. First, suppose p odd. Then by induction a subgroup P_1 of index p is cyclic, and so by Theorem 12.5.1, P is of one of types 1, 2, or 3 for p odd, and of these types 2 and 3 contain more than one subgroup of order p. Hence P is cyclic. When $p = 2$, if P contains a cyclic subgroup P_1 of index 2, then by Theorem 12.5.1, P is one of types 1 through 7 for $p = 2$, and each of these contains more than one subgroup of order 2, except for the cyclic group and the generalized quaternion group. Thus P is cyclic or of generalized quaternion type.

There remains to be considered the case in which, by induction every subgroup P_1 of index 2 is generalized quaternion. We shall show that this situation cannot arise. Here $n \geqq 4$. First let $n = 4$ and a subgroup of index 2 be a quaternion group Q, and let c be an element not in Q. Q is given by $a^4 = b^4 = 1$, $a^2 = b^2$, $ba = a^{-1}b$, and $P = Q + Qc$. The element c, being of order a power of 2, must transform into itself at least one of the three subgroups of order 4 in Q, $\{a\}$, $\{b\}$, $\{ab\}$. Relabeling if necessary, we may take this to be $\{a\}$. Then $c^{-1}ac = a$ or $c^{-1}ac = a^{-1}$. If $c^{-1}ac = a$, then $\{a, c\}$ is an Abelian subgroup of index 2, contrary to assumption. If $c^{-1}ac = a^{-1}$, then $(cb)^{-1}a(cb) = a$, and $\{a, cb\}$ is an Abelian subgroup of index 2, contrary to assumption. This takes care of $n = 4$.

Finally, suppose $n \geqq 5$, and P_1 a generalized quaternion subgroup of index 2. Then P_1 is given by $a^{2^{n-2}} = 1$, $b^2 = a^{2^{n-3}}$, $ba = a^{-1}b$, and $P = P_1 + P_1c$. Here $\{a\}$ is the only subgroup of P_1 of order 2^{n-2}, all elements of P_1 not in $\{a\}$ being of order 4. Thus $c^{-1}ac = a^r$, and $c^2 = a^ib$ or $c^2 = a^i$. If $c^2 = a^ib$, then $c^{-2}ac^2 = a^{-1}$ and $r^2 \equiv -1 \pmod{2^{n-2}}$ which is impossible. If $c^2 = a^i$, then $\{a, c\}$ is a subgroup of index 2 and by assumption a generalized quaternion group. Then $c^{-1}ac = a^{-1}$, $(cb)^{-1}a(cb) = a$, and $\{cb, a\}$ is an Abelian subgroup of index 2, contrary to assumption. This completes the proof of the theorem in all cases.

THEOREM 12.5.3. *A group of order p^n which contains only one subgroup of order p^m, where $1 < m < n$ is cyclic.*

Proof: If $m = n - 1$, then a group P of order p^n with only one subgroup of order p^{n-1} is generated by any element x not in the subgroup, since $\{x\}$ is not contained in the unique maximal subgroup, and so $\{x\} = P$ and P is cyclic. This proves the theorem for $n = 3$, the first value of n to which the theorem applies, and for all cases with $m = n - 1$. We proceed by induction on n. We have proved the theorem when $m = n - 1$, and therefore we may assume $m < n - 1$.

Let P_1 be the unique subgroup of order p^m, and suppose P_1 contained in a maximal subgroup A of order p^{n-1}. Since $1 < m < n - 1$, by induction A is cyclic, and so, P_1 as a subgroup of A is also cyclic. Every subgroup of order p or p^2 is contained in a subgroup of order p^m since $m \geq 2$, and so, in P_1. But P_1, being cyclic, contains a unique subgroup of order p and a unique subgroup of order p^2. Thus P contains a unique subgroup of order p and a unique subgroup of order p^2. By Theorem 12.5.2 P is cyclic or generalized quaternion. But the generalized quaternion group contains more than one subgroup of order 4. Hence P must be cyclic.

It is trivial that every subgroup of an Abelian group is normal. But the quaternion group is an example of a non-Abelian group in which every subgroup is normal. We call a group H *Hamiltonian* if H is non-Abelian and every subgroup of H is normal.

THEOREM 12.5.4. *A Hamiltonian group is the direct product of a quaternion group with an Abelian group in which every element is of finite odd order and an Abelian group of exponent two.*

Proof: Let a and b be two elements of a Hamiltonian group H. Then the commutator $c = (a, b) = (a^{-1}b^{-1}a)b = b^s = a^{-1}(b^{-1}ab) = a^r$, since $\{a\}$ and $\{b\}$ are both normal subgroups. Note that this implies that c permutes with a and also with b. By (10.2.1).

$$(a^2, b) = (a, b)(a, b, a)(a, b) = (a, b)(c, a)(a, b) = (a, b)^2,$$

and we may prove similarly by induction that

$$(a^i, b) = (a, b)^i = c^i.$$

If a and b do not permute, then $c = a^r \neq 1$, and putting $i = r$ or $i = -r$, whichever is positive, then (a^i, b) is either (c, b) or (c^{-1}, b)

and is the identity in either event since c permutes with b. Then $(a^i, b) = 1 = (a, b)^i = c^i$. Hence $c^i = 1$ and $a^{ri} = 1$, $b^{si} = 1$. Hence two elements of H which do not permute are of finite order. If an element x of H permutes with both a and b, then xa does not permute with b, and it follows that xa, and so also x, is of finite order. Thus every element of H is of finite order.

Let a and b be elements of H which do not permute, and $a^N = 1$, $b^M = 1$, where we suppose N and M minimal. If p is any prime divisor of N, then by the minimality of N, a^p permutes with b, and so, $(a^p, b) = (a, b)^p = 1$. The same will hold for any prime dividing M. As $c = (a, b) \neq 1$, there can be only one prime dividing M and N, and $M = p^m$, $N = p^n$. Thus $a^{p^n} = 1$, $b^{p^m} = 1$, $c = (a, b)$, $c^p = 1$, where by symmetry we may assume $n \geq m$. Further, since $c \in \{a\}$ and $c \in \{b\}$, $c = a^{jp^{n-1}} = b^{kp^{m-1}}$, where $j, k \not\equiv 0 \pmod{p}$.

In $\{a, b\}$ the derived group is $\{c\}$ and is in its center. Thus in $\{a, b\}$ all commutators of weight three or more are the identity. We may establish the formula

$$(ab)^i = a^i b^i (b, a)^{i(i-1)/2}$$

by induction. It is true for $i = 1$, and we have

$$\begin{aligned}
(ab)^{i+1} &= (ab)^i ab = a^i b^i (b, a)^{i(i-1)/2} ab \\
&= a^i b^i ab (b, a)^{i(i-1)/2} \\
&= a^i a b^i (b^i, a) b (b, a)^{i(i-1)/2} \\
&= a^{i+1} b^i (b, a)^i b (b, a)^{i(i-1)/2} \\
&= a^{i+1} b^{i+1} (b, a)^{i(i+2)/2}.
\end{aligned}$$

This proves the formula by induction for any group $\{a, b\}$ in which (a, b) is in the center. This formula is also a consequence of the collection formula.

If $b_1 = a^u b^k$, where $u = -jp^{n-m}$, then $\{a, b_1\} = \{a, b\}$, whence b_1 does not permute with a, and therefore by assumption, the order of b_1 is at least as great as that of b. The formula just established yields

$$\begin{aligned}
b_1{}^p = (a^u b^k)^p &= a^{up} b^{kp} (b^k, a^u)^{p(p-1)/2} \\
&= a^{pu} b^{kp} c^{-ukp(p-1)/2},
\end{aligned}$$

whence

$$\begin{aligned}
b_1{}^{p^{m-1}} &= a^{-jp^{n-1}} b^{kp^{m-1}} c^{jkp^{n-1}(p-1)/2} \\
&= c^{jkp^{n-1}(p-1)/2}.
\end{aligned}$$

Here $b_1{}^{p^{m-1}} \neq 1$, but since $c^p = 1$, we must have $p = 2$, $n = 2$.

Thus the relations on a and b are $a^2 = b^2 = a^{-1}b^{-1}ab = c$, $c^2 = 1$, and $\{a, b\}$ is the quaternion group. This shows that any non-Abelian subgroup of H contains a quaternion group.

We next show that H is the union of the quaternion group Q, given by $a^4 = b^4 = 1$, $a^2 = b^2$, $ba = a^{-1}b$, and the group Z of elements centralizing Q. If an element x of H does not permute with a, then $x^{-1}ax = a^{-1}$ and xb permutes with a. Similarly, if x (or xb) does not permute with b, then xa (or xba) permutes with a. Hence one of the elements x, xb, xa, xba lies in Z. Hence $H = Q \cup Z = QZ$. We now show that Z cannot contain an element of order 4. For, if $x^4 = 1$, $x \in Z$, then $(a, bx) \neq 1$. Since $(bx)^4 = 1$, we have $a^{-1}(bx)a = (bx)^{-1}$, whence $a^{-1}bax = b^{-1}x^{-1}$, giving $x^2 = 1$. Since Z contains no element of order 4, Z cannot contain a quaternion group, and it follows that Z is Abelian. $Z \cap Q = a^2$. By use of Zorn's lemma we find a subgroup Z_1 of Z maximal with respect to the property of not containing a^2. Then we easily find that $Z = Z_1 + Z_1a^2$, $H = Q \times Z_1$. Z_1 is the direct product of an Abelian group U, whose elements are of odd order, and an Abelian group V of exponent 2, since Z_1 contains no element of order 4. Thus $H = Q \times U \times V$.

Conversely, a group of the form $Q \times U \times V$ is Hamiltonian, for Q is non-Abelian. It suffices to show that every cyclic subgroup $\{quv\}$ is normal. U and V are in the center of $Q \times U \times V$, and we need only show that a and b transform this group into itself. Here $a^{-1}(quv)a = q^iuv$, where $i = 1$ or 3. The order of u is an odd number n, and the order of v is 2. Hence the congruences $r \equiv i \pmod 4$, $r \equiv 1 \pmod n$ are solvable, and $a^{-1}(quv)a = (quv)^r$. This completes our theorem.

13. FURTHER THEORY OF ABELIAN GROUPS

13.1. Additive Groups. Groups Modulo One.

Any group may be written with the group operation designated as addition. It is a common practice to write Abelian groups additively, and it is particularly convenient to do so if there are operators. Also, certain groups arise naturally in the addition of familiar systems. Two (which we shall consider here) are the additive group of rational numbers which we shall designate as r_+ and the additive group of real numbers which we shall designate as R_+.

When we use the additive notation for groups, we shall change our terminology appropriately, speaking of the sum of elements, Cartesian sums, and direct sums.

A cyclic group in additive form consists of all the integral multiples na of a generator a. The groups r_+ and R_+ are both aperiodic, since $na = 0$ implies $a = 0$. In an infinite cyclic group generated by a, there is no element x such that $2x = a$. Since for any a in r_+ there is an x with $2x = a$, it is clear that r_+ is not a cyclic group. But r_+ is very nearly a cyclic group. Any finite set of elements in r_+ will generate a cyclic group. We describe this property by saying that r_+ is of *rank one*, or *locally cyclic*. More generally we shall say that an Abelian group is of *rank* k if a subgroup generated by any finite number of elements can be generated by at most k elements, although some finitely generated subgroup requires k generators.

THEOREM 13.1.1. *The additive group of rational numbers, r_+, is locally cyclic.*

Proof: Consider a subgroup of r_+ generated by the finite set of elements $a_1/b_1, \cdots, a_t/b_t$. Its elements will be the numbers $m_1a_1/b_1 + \cdots + m_ta_t/b_t$, the m's being arbitrary integers. These can be expressed in the form $(m_1a_1b_2 \cdots b_t + \cdots + m_ta_tb_1 \cdots b_{t-1})/b_1b_2 \cdots b_t$. Here we readily verify that the numerators form an additive subgroup of the additive group of integers, which is cyclic. Hence these form a cyclic group consisting of all integral multiples of some integer w.

Thus our group consists of the numbers $nw/b_1b_2 \cdots b_t$ and is a cyclic group.

In the group R_+ the integers form a subgroup, which like all subgroups of an Abelian group is a normal subgroup. In the factor group all numbers differing by integers are identified, and so we speak of the factor groups as the group R_+ modulo 1. Similarly, r_+ has a factor group r_+ modulo 1, which is, of course, a subgroup of R_+ modulo 1.

The group r_+ (mod 1) is a periodic group, since if a/b is any rational number (a, b being integers), we have $b(a/b) \equiv 0 \pmod 1$. By Theorem 3.2.3, r_+ (mod 1) is the direct sum of its Sylow subgroups $S(p)$. An $S(p)$ of r_+ (mod 1) we designate as $Z(p^\infty)$. $Z(p^\infty)$ is generated by the infinite set $1/p, 1/p^2, \cdots, 1/p^i \cdots \pmod 1$. An element of $Z(p^\infty)$ is of the form m/p^n, $(m, p) = 1$, and such an element generates the same cyclic group as $1/p^n$. Hence a subgroup of $Z(p^\infty)$ is either finite or contains infinitely many of the set $1/p, 1/p^2, \cdots, 1/p^i, \cdots \pmod 1$, and so it is the entire group $Z(p^\infty)$. Thus $Z(p^\infty)$ is an infinite group, all of whose proper subgroups are finite cyclic groups.

13.2. Characters of Abelian Groups. Duality of Abelian Groups.

Given an arbitrary Abelian group A. A character χ of A is a homomorphism of A into the group R_+ (mod 1). Thus our definition is

$$(13.2.1) \qquad \chi(a_1) + \chi(a_2) = \chi(a_1 + a_2) \quad \text{all } a_1, a_2 \in A.$$

Here the addition $a_1 + a_2$ is the addition in A, the addition of the values of the characters is, of course, in R_+ (mod 1). We shall also define an addition of characters. If χ_1 and χ_2 are two characters of A, we define

$$(13.2.2) \qquad \chi_3(a) = \chi_1(a) + \chi_2(a) \quad \text{all } a \in A.$$

Then χ_3 is also a character of A, since

$$(13.2.3) \qquad \begin{aligned} \chi_3(a_1 + a_2) &= \chi_1(a_1 + a_2) + \chi_2(a_1 + a_2) \\ &= \chi_1(a_1) + \chi_1(a_2) + \chi_2(a_1) + \chi_2(a_2) \\ &= \chi_1(a_1) + \chi_2(a_1) + \chi_1(a_2) + \chi_2(a_2) \\ &= \chi_3(a_1) + \chi_3(a_2). \end{aligned}$$

We readily verify that if we use (13.2.2) to define an addition

(13.2.4)
$$\chi_3 = \chi_1 + \chi_2,$$

then, with respect to the addition (13.2.4), the characters themselves form an additive group A^* whose zero element is the character which maps every element of A onto zero.

THEOREM 13.2.1. *The character group A^* of a finite Abelian group A is isomorphic to A.*

Proof: For any homomorphism we must have $\chi(0) = 0$. Hence for an element a of finite order m, we have $m\chi(a) = \chi(ma) = \chi(0) = 0$. Thus $\chi(a)$ must have one of the m values $0, 1/m, \cdots, (m-1)/m \pmod 1$. In a finite Abelian group it is clear that a character is completely determined if it is known for a basis. Let a_i, $i = 1, \cdots, r$ be a basis of A where a_i is of order n_i, and A is of order $n = n_1 n_2 \cdots n_r$. Since there are at most n_i choices for $\chi(a_i)$, we see that there are at most $n = n_1 n_2 \cdots n_r$ different characters for A. But we easily see that there are indeed this many. For if we put $\chi_i(a_i) = 1/n_i$, $\chi_i(a_j) = 0.\ j \neq i$, we can show that for each $i = 1, \cdots, r$ this defines a character and that the correspondence $a_i \leftrightarrows \chi_i$ determines an isomorphism between A and A^*. We note, however, that the isomorphism between A and A^* is not uniquely determined but depends upon a particular choice of a basis for A.

The following theorem is true for any Abelian group, finite or not:

THEOREM 13.2.2. *Let H be a subgroup of the Abelian group A. Then the characters of A for which $\chi(h) = 0$ for every $h \in H$ are precisely the characters of the factor group A/H.*

Proof: If a character assigns 0 to every element of H, then it assigns the same value to every element of a coset $H + x$. We may take this as assigning a value in $R_+ \pmod 1$ to the coset as an element of the factor group A/H. This is readily seen to be a character for A/H. Conversely, $A \to A/H$ is a homomorphism which when followed by a homomorphism into $R_+ \pmod 1$ yields a homomorphism of A into $R_+ \pmod 1$. This will be a character of A in which all elements of H go onto the zero of A/H, which is in turn mapped onto 0.

COROLLARY 13.2.1. *If $a \neq 0$ in a finite Abelian group A, then there is a character of A for which $\chi(a) \neq 0$.*

For if this were not so, then every character of A would be a character of the factor group $A/\{a\}$, and by Theorem 13.2.1, A^* would be isomorphic both to A and also to $A/\{a\}$, which is, of course, of lower order.

A *duality* between groups A and B is a one-to-one correspondence $H \leftrightarrows K$ between subgroups H of A and subgroups K of B which reverses inclusions; i.e., if $H_1 \leftrightarrows K_1$ and $H_2 \leftrightarrows K_2$ and if $H_1 \supset H_2$, then $K_1 \subset K_2$, and conversely, if $K_1 \subset K_2$, then $H_1 \supset H_2$. There is a natural duality between a finite Abelian group A and its character group A^*, which is given in the following theorem.

THEOREM 13.2.3. *There is a duality between a finite Abelian group A and its character group A^* given by the rule $H \leftrightarrows K$ where, given H a subgroup of A, K consists of all characters of A such that $\chi(h) = 0$ for every $h \in H$, and given K a subgroup of A^*, H consists of all elements of A such that $\chi(h) = 0$ for every $\chi \in K$. A is dual to itself.*

Proof: With every subgroup H of A, let us associate the subgroup H^* of A^* consisting of all those characters χ such that $\chi(h) = 0$ for every $h \in H$. If $H_1 \neq H_2$ are distinct subgroups of A, then one of H_1, H_2 (say, H_1) contains an element b not contained in the other. Then, by Theorem 13.2.2, H_2^* is the character group of A/H_2, and by Corollary 13.2.1, there exists a $\chi \in H_2^*$ such that $\chi(b) \neq 0$. Hence $H_1^* \neq H_2^*$. Since A and A^* are finite and isomorphic, it follows that the mapping $H \to H^*$ is a one-to-one correspondence between the subgroups of A and those of A^*, and in particular, that every subgroup K of A is of the form $K = H^*$ for a unique subgroup H of A. If $H_1 \leftrightarrows K_1 = H_1^*$ and $H_2 \leftrightarrows K_2 = H_2^*$, then $H_1 \supset H_2$ implies $K_1 \subset K_2$, since $\chi(h) = 0$ for every $h \in H_1$ implies *a fortiori* $\chi(h) = 0$ for every $h \in H_2 \subset H_1$. Similarly, $K_1 \subset K_2$ implies $H_1 \supset H_2$. Thus the correspondence of the theorem is a duality between A and A^*. The isomorphism between A and A^* then leads to a duality of A with itself.

THEOREM 13.2.4. *An Abelian group which is periodic and has all its Sylow subgroups finite is self dual.*

Proof: If A is a periodic Abelian group whose Sylow subgroups are finite, then an $S(p)$ as a finite Abelian group is self-dual. Let us write this $H_p \leftrightarrows H_p^d$ where, for H_p any subgroup of $S(p)$, the dual subgroup is H_p^d. Now if H is any subgroup of A, then H is the direct sum of its Sylow subgroups H_p. Then let us put $H^d = \sum_p H_p^d$. This is easily seen to be a duality of A. Note that this argument does not work if we take without restriction a direct sum of finite Abelian groups because in general such a direct sum will have many subgroups which are not the direct sum of subgroups of the summands. It has been shown by Baer [6] that the Abelian groups which possess duals are precisely those covered by this theorem.

13.3. Divisible Groups.

An additively written Abelian group A is said to be *divisible* if, for every $a \in A$ and integer n, there is an element $x \in A$ such that $nx = a$.

THEOREM 13.3.1. *A divisible group is a direct summand of every Abelian group A which contains it.*

Proof: Suppose we are given an Abelian group A and a divisible subgroup D. We wish to show the existence of a subgroup B such that

$$(13.3.1) \qquad A = D \oplus B \qquad \text{or} \qquad D \cap B = 0 \quad \text{and} \quad D \cup B = A.$$

For this proof it is convenient to make use of Zorn's lemma which we discussed in §1.8. If $U_1 \subset U_2 \subset U_3 \subset \cdots$ is an ascending chain of subgroups of A such that $D \cap U_i = 0$, then $U = \bigcup_i U_i$ also has the property that $D \cap U = 0$. Hence, by Zorn's lemma, A contains a subgroup K maximal with respect to the property that $K \cap D = 0$. We may take $B = K$ in (13.3.1) if it can be shown that $D \cup K = A$. Suppose that x is an element of A not in $K \cup D$. Then by the maximality of K, $\{x\} \cup K$ has a nonzero element in common with D. Hence for some non-negative integer n and $k \in K$, we have $nx + k = d \in D$, $d \neq 0$. Here $n \neq 0$, since $D \cap K = 0$. And if $n = 1$, then $x \in K \cup D$, contrary to assumption. Since D is divisible, $d = nd_1$ with $d_1 \in D$, and $n(x - d_1) = -k$. Putting $x_1 = x - d_1$, then if $x_1 \in K \cup D$, also $x \in K \cup D$, contrary to assumption. The elements of $K \cup \{x_1\}$ are of the form $mx_1 + k$, $0 \leq m < n$. By the maximality of K there must be an element common to $\{x_1\} \cup K$ and D, $n_1x_1 + k_1 = d = n_1d_2$, with $n_1 < n$, $d,d_2 \in D$. Here, with $x_2 = x_1 - d_2$, we have $n_1x_2 = -k_1 \in K$, and if $x_2 \in K \cup D$, then also $x_1 \in K \cup D$. This process leads to a contradiction because we ultimately find an $n_i = 1$, whence $x_i, x_{i-1}, \cdots, x_1$ and x all belong to $K \cup D$, contrary to assumption. Thus $K \cup D = A$ and our theorem is proved.

See Kaplansky [1] for a proof that every divisible group is a direct sum of groups isomorphic to r_+ or groups $Z(p^\infty)$.

13.4. Pure Subgroups.

We say that H is a pure subgroup of the Abelian group A if it is true that whenever $nx = h \in H$ for some $x \in A$, then there is an $h_1 \in H$ such that $nh_1 = h$. Thus the property of being pure is a sort of relative divisibility, division being possible in H if it is possible at all. A divisible group is certainly a pure subgroup of any Abelian group containing it. A direct summand is a pure subgroup. But although a divisible group is necessarily infinite, there can be pure subgroups in finite groups, and hence the concept is useful in the study of finite groups.

The periodic subgroup of an Abelian group is pure, since if $nx = h$ where h is of finite order, then x, if it exists, must also be of finite order. The union of an ascending chain of pure subgroups will be pure, for if h is any element of such a union, then h is an element of one of the groups in the chain and so $nx = h$ will have a solution in the chain.

Theorem 13.4.1 shows that in a great many cases, a pure subgroup is indeed a direct summand.

THEOREM 13.4.1. *Let A be an Abelian group, H a pure subgroup, and suppose that A/H is the direct sum of cyclic groups. Then H is a direct summand of A.*

Proof: We first prove a lemma.

LEMMA 13.4.1. *If H is a pure subgroup of A, and the element y is in A/H, then there is an element x of A mapping onto y in the homomorphism $A \to A/H$ of the same order as y.*

If y is of infinite order, then any x mapping onto y will do. If $ny = 0$ and $u \to y$, then $nu \to 0$, $nu = h \in H$. But then, by the purity of H, $h = nh_1$. Here put $x = u - h_1$. Then $x \to y$ and $nx = n(u - h_1) = nu - nh_1 = h - h = 0$, as we wished to show.

The proof of the theorem is now fairly simple. Let A/H be the direct sum of cyclic groups generated by basis elements y_i, $i \in I$. Choose in A elements $x_i \to y_i$ where in every case we have chosen x_i of the same order as y_i, this choice being possible by the lemma.

Let K be the subgroup generated by the x_i. If a relation $n_{i_1}x_{i_1} + \cdots + n_{i_s}x_{i_s} = h \in H$ holds in A, then in A/H we have $n_{i_1}y_{i_1} + \cdots + n_{i_s}y_{i_s} = 0$, and so as the y's are a basis for A/H ,we have $n_iy_i = 0$ for each of these terms. But since the x's are of the same order as the y's, then also $n_ix_i = 0$ for each of the terms and also the h is zero. Hence $K \cap H = 0$. Also $K \cup H = A$, since K contains one element from each coset of A. Thus $A = H \oplus K$, as was to be proved.

13.5. General Remarks.

For a more detailed study of Abelian groups the reader is referred to Kaplansky's monograph [1] and to Part II of Kurosch's book [2]. A particularly useful feature of Kaplansky's monograph is a section discussing the literature.

In general, Theorem 3.2.3 reduces the study of periodic groups to that of primary groups. One of the major results on primary groups is the Theorem of Ulm, which fully characterizes countable primary Abelian groups in terms of certain cardinal numbers, the "Ulm invariants" of the group.

The direct sum of infinite cyclic groups is called a *free* Abelian group. Every Abelian group with r generators is the homomorphic image of the free Abelian group with r generators. Every subgroup of a direct sum of cyclic groups is itself a direct sum of cyclic groups, and in particular, a subgroup of a free Abelian group is free Abelian.

As remarked in §13.3, every divisible group is the direct sum of groups isomorphic to r_+ and of groups isomorphic to various $Z(p^\infty)$'s. Every Abelian group can be embedded in a divisible group, and so in a certain sense, the study of all Abelian groups is the study of subgroups of divisible groups. Thus a torsion-free (i.e., aperiodic) group of rank one is a subgroup of r_+.

An Abelian group which contains both elements of finite and elements of infinite order is called *mixed.* Examples show that in general a mixed group is not the direct sum of its periodic subgroup and a torsion-free group. But since the periodic subgroup is pure, Theorem 13.4.1 will often give the decomposition of a mixed group as the direct sum of the periodic part and another group.

14. MONOMIAL REPRESENTATIONS AND THE TRANSFER

14.1. Monomial Permutations.

Let us consider a set S of indeterminates $u_1, \cdots, u_n$ which may be multiplied on the left by elements of a group H. We postulate the rules

$$1u_i = u_i, \tag{14.1.1}$$

1 the identity of H; and

$$h_1(h_2u_i) = (h_1h_2)u_i.$$

A monomial permutation M is a mapping $u_i \to h_{ij}u_j$, $i = 1 \cdots n$, $j = j(i)$, where $u_i \to u_j$ is a permutation of S. For the product of two mappings M_1 and M_2, if M_1 is $u_i \to h_{ij}u_j$ and M_2 is $u_j \to h_{jk}u_k$, we define M_1M_2 as $u_i \to (h_{ij}h_{jk})u_k$. Under this definition the mappings form a group whose identity is the mapping $u_i \to u_i$. If we associate with the mapping $M_1: u_i \to h_{ij}u_j$ the matrix (h_{ij}), which has for its ith row h_{ij} in the jth column and zeros elsewhere, then the rule for multiplying the mappings is the same as the ordinary matrix multiplication.

In the group M of all monomial permutations the multiplications $u_i \to h_{ii}u_i$ form a normal subgroup D, and the factor group M/D is the symmetric group of permutations of $u_1 \cdots u_n$. More generally, if G is a subgroup of M, then if $g \in G$ is $u_i \to h_{ij}u_j$, $g \to g^*: u_i \to u_j$ is a homomorphism of G onto a group of permutations whose kernel is $G \cap D$.

We shall say that a monomial permutation group G is transitive if the corresponding permutation group is transitive.

THEOREM 14.1. *Let G be a group with a subgroup K and $G = K + Kx_2 \cdots + Kx_n$. Also let $K \to H$ be a homomorphism of K onto a group H. Then a transitive monomial representation of G with H as multipliers is given in the following way: For $g \in G$ let $x_ig = k_{ij}x_j$, $i = 1, \cdots, n$, $j = j(i)$, $k_{ij} \in K$. Also let $k_{ij} \to h_{ij}$ in the homomorphism $K \to H$. Then $\pi\ (g): u_i \to h_{ij}u_j$ is a transitive monomial representation* *of G with H as multipliers. Conversely, every transitive monomial representation is of this type or is conjugate under the group of multiplications D to a representation of this type.*

Proof: Given G, the left coset representation $G = K + Kx_2 + \cdots + Kx_n$ and the homomorphism $K \to H$. Let g_1 and g_2 be any two elements of G. Then, if $x_i g_1 = k_{ij} x_j$ and $x_j g_2 = k_{js} x_s$, we have $x_i(g_1 g_2) = k_{ij} k_{js} x_s$, whence we see that $\pi(g_1 g_2) = \pi(g_1)\pi(g_2)$ for the corresponding monomial permutations, whence we have a representation of G (of course not necessarily faithful). The corresponding permutation group is the permutation group of left cosets discussed in §5.3 and is, of course, transitive.

Conversely, let us consider any transitive monomial representation R of G, $g \in G$, $g \to \pi(g): u_i \to h_{ij} u_j$. Let us select a particular letter u_1 and consider all elements k of G such that $\pi(k)$ maps u_1 onto $h_{11} u_1$ for some $h_{11} \in H$. These form a subgroup K. By the transitivity of R, for each $i = 2, \cdots, n$ there is an element x_i such that $\pi(x_i)$ takes u_1 into $h_{1i} u_i$. Then we see easily that

$$G = K + Kx_2 + \cdots + Kx_n. \tag{14.1.2}$$

If we transform R by the multiplication $d: u_1 \to u_1, \cdots, u_i \to h_{1i}^{-1} u_i$, then in $d^{-1}Rd$ we see that $d^{-1}\pi(x_i)d$ takes u_1 into u_i. Let us consider $R^* = d^{-1}Rd$. Here, if for $k \in K$, $\pi(k)$ takes u_1 into hu_1, $\pi(x_i^{-1} k x_j)$ takes u_i into hu_j, and conversely. Thus in R^* every h that occurs as a multiplier at all occurs in K. These may indeed be a proper subgroup H_1 of the group H originally used. But if $\pi(k)$ takes u_1 into hu_1, then $k \to h$ is a homomorphism of K onto H_1. Moreover, if $\pi(g)$ takes u_i into $h_{ij} u_j$, then $\pi(x_i g x_j^{-1})$ takes u_1 into $h_{ij} u_1$, whence $x_i g x_j^{-1} = k_{ij} \in K$ and $k_{ij} \to h_{ij}$ in the homomorphism of K onto H_1.

We note in passing that changing the representatives of the left cosets of K in G yields another monomial representation conjugate to the first under the group D of multiplications.

14.2. The Transfer.

Suppose we have a monomial representation R of a group G with multipliers from H:

$$\pi(g): u_i \to h_{ij} u_j, \quad i = 1 \cdots n, \quad j = j(i). \tag{14.2.1}$$

Suppose further that the number n of letters is finite. Then the mapping

$$g \to \prod_{i=1}^{n} h_{ij} \bmod H' \tag{14.2.2}$$

is easily seen to be a homomorphism of G onto the factor group H/H', where H' is the derived group of H. Let us take in particular the case where we have $H = K$:

$$G = K + Kx_2 + \cdots + Kx_n. \tag{14.2.3}$$

Here we have, if $\phi(z) = x_j$ for $z = kx_j$, $k \in K$,

$$V_{G\to K}(g) \equiv \prod_{i=1}^{n} x_i g \phi(x_i g)^{-1} \bmod K', \tag{14.2.4}$$

and $V_{G\to K}(g)$ is a homomorphism of G into K/K'. This homomorphism is called the *transfer* (in German: *Verlagerung*) of G into K. If H is a homomorphic image of K, then the mapping of (14.2.2) is a homomorphic image of the transfer, since if $K \to H$, K/K' is mapped onto H/H', K' being a fully invariant subgroup of K. The chief properties of the transfer are given by Theorem 14.2.1.

Theorem 14.2.1.

1) *The mapping* $g \to V_{G\to K}(g)$ *is a homomorphism of* G *into* K/K'.

2) *The transfer* $V_{G\to K}(g)$ *is independent of the choice of representatives* x_i.

3) *If* $G \supset K \supset T$, *then* $V_{G\to T}(g) = V_{K\to T}[V_{G\to K}(g)]$.

Proof: We have already observed the first property as a consequence of the theory of monomial representations. But we shall prove all three properties directly from the definition (14.2.4) of the transfer. For the first property we observe that if $x_i g_1 = k_{ij} x_j$, $i = 1 \cdots n$, $x_j g_2 = k_{js} x_s$, $j = 1 \cdots n$, then $V_{G\to K}(g_1) \equiv \prod_i k_{ij} \bmod K'$, $V_{G\to K}(g_2) = \prod_j k_{js} \bmod K'$, and $V_{G\to K}(g_1 g_2) \equiv \prod_i (k_{is}{}^*)$, where $k_{is}{}^* = k_{ij} k_{js}$. For the second property if $x_i{}^* = a_i x_i$ is the relation between the first and second choice of representatives and if $x_i g = k_{ij} x_j$, then $x_i{}^* g = a_i x_i g = a_i k_{ij} x_j = a_i k_{ij} a_j{}^{-1} x_j{}^*$; then in the first case $V(g)$ is $\prod_i k_{ij} \bmod K'$, and in the second case is $\prod_i (a_i k_{ij} a_j{}^{-1}) \equiv \prod_i a_i \cdot \prod_i k_{ij} \cdot \prod a_j{}^{-1} \equiv \prod_i k_{ij}$ mod K_j. For the third property let

(14.2.5)
$$G = K + Kx_2 + \cdots + Kx_n,$$
$$K = T + Ty_2 + \cdots + Ty_m.$$

Then

$$\begin{aligned} G = T &+ Ty_2 + \cdots + Ty_m \\ &+ \cdots \\ &+ Tx_i + Ty_2x_i \cdots + Ty_mx_i \qquad (14.2.6) \\ &+ \cdots \\ &+ Tx_n + Ty_2x_n \cdots + Ty_mx_n. \end{aligned}$$

Here, for $g \in G$, let $x_ig = k_{ij}x_j$ and $y_rk_{ij} = t_{ijrs}y_s$.

Thus
$$y_rx_ig = t_{ijrs}y_sx_j.$$

Then $V_{G\to T}(g) \equiv \prod_{i,r} t_{ijrs} \bmod T'$ and $V_{G\to K}(g) \equiv \prod_i k_{ij} \bmod K'$.

Now
$$V_{K\to T}(k_{ij}) = \prod_r t_{ijrs} \bmod T'.$$

Hence
$$\begin{aligned} V_{K\to T}(g) &\equiv \prod_i V_{K\to T}(k_{ij}) \bmod G' \\ &\equiv V_{K\to T}(\prod_i k_{ij}) \bmod T' \equiv V_{K\to T}[V_{G\to K}(g)]. \end{aligned}$$

We note here that, as the transfer of K onto T maps K' onto the identity, there is no ambiguity in the transfer of $V_{G\to K}(g)$ into T, although this is an element of K/K' rather than of K.

14.3. A Theorem of Burnside.

THEOREM 14.3.1. *If a Sylow subgroup P of a finite group G is in the center of its normalizer, then G has a normal subgroup H which has the elements of P as its coset representatives.*

Proof: We begin with a lemma.

LEMMA 14.3.1. *If two complexes K_1 and K_2 are normal in a Sylow subgroup P of G and are conjugate in G, then K_1 and K_2 are conjugate in $N_G(P)$.*

Proof of the lemma: Suppose $x^{-1}K_1x = K_2$ with $x \in G$. As K_1 is normal in P, then $K_2 = x^{-1}K_1x$ is normal in $x^{-1}Px = Q$. Thus both P and Q are contained in the normalizer of K_2, and hence as Sylow subgroups are conjugate in $N_G(K_2)$. Hence $y^{-1}Qy = P$ for some y with $y^{-1}K_2y = K_2$. Thus for $z = xy$, $z^{-1}Pz = P$, $z^{-1}K_1z = K_2$, proving the lemma. For the proof of the theorem, since P is in the center of $N_G(P)$, P is Abelian and $P' = 1$. Let us consider $V_{G\to P}$. Let $u \in P$. In calculating $V_{G\to P}(u)$, let us use as representatives of P in G, $x_i, x_iu \cdots x_iu^{r-1}$ if $x_iu^r \in Px_i$ but $x_iu^j \notin Px_i$ for $j < r$. Here $x_iu^{j-1}\cdot u\cdot\phi(x_iu^j)^{-1} = x_iu^ju^{-j}x_i^{-1} = 1$ for $j < r$ and $x_iu^{r-1}\cdot u\cdot\phi(x_iu_r)^{-1} = x_iu^rx_i^{-1}$. Hence, for each cycle of length r in representing u on the left cosets of P, there is a term $x_iu^rx_i^{-1}$ in the product for $V_{G\to P}(u)$ and the rest are the identity. Thus $V_{G\to P}(u) = \prod_i x_iu^rx_i^{-1}$. Now $x_iu^rx_i^{-1} \in P$ is conjugate to u^r in G, and as P is Abelian, both elements are normal in P. By the lemma $x_iu^rx_i^{-1} = y^{-1}u^ry$ with $y \in N_G(P)$. By hypothesis P is in the center of its normalizer, whence $y^{-1}u^ry = u^r$. Hence $V_{G\to P}(u) \equiv \prod u^r \equiv u^n$, where $n = [G:P]$ is the sum of the lengths of all the cycles. Since P is a Sylow subgroup of order, say, p^s, it follows that $p \nmid n = [G:P]$. Thus, in the transfer of G onto P, P is mapped isomorphically onto itself and $V_{G\to P}(G) = P$, since trivially the transfer can be no larger than P. The kernel of this homomorphism must be a group H of index p^s in G and of order $n = [G:P]$. Hence H is a normal subgroup of index p^s, and so the elements of P can be taken as the coset representatives of H.

COROLLARY 14.3.1. *The order of a finite simple group is either divisible by* 12 *or by the cube of the smallest prime dividing its order.*

Proof: Let p be the smallest prime dividing the order of the simple group G, and suppose that a Sylow p-group P is of order p or p^2 and hence Abelian. By the theorem, unless $N_G(P)$ induces a nontrivial automorphism in P, then G has P as a factor group. If P is of order p, its automorphisms are of order dividing $p - 1$, and so of orders less than p. If P is cyclic of order p^2, the automorphisms are of order dividing $p(p - 1)$, and if noncyclic of order p^2, of order $(p^2 - 1)(p^2 - p) = p(p - 1)^2(p + 1)$. No one of these numbers is divisible by a prime greater than p if p is odd, since $p + 1 = 2[(p + 1)/2]$, and hence no nontrivial automorphism can be induced by $N_G(P)$. If $p = 2$, then in the last case $p + 1 = 3$ and $N_G(P)$ can induce an automorphism of order 3 in P only if the order of $N_G(P)$ is divisible by 12.

14.4. Theorems of P. Hall, Grün, and Wielandt.

The following theorems have as their main content the relationship between the Sylow p-subgroups of a group G and the factor groups G/K of G which are p-groups.

To describe these relationships, we introduce the concepts of strong and weak closure.

Definition: *If H is a subgroup of G and B is a subgroup of H, we say that B is strongly closed in H (with respect to G) if $H \cap B^x \subseteq B$ for $B^x = x^{-1}Bx$, any $x \in G$, and that B is weakly closed in H if $B^x \subseteq H$ implies $B^x = B$.*

We say that a group G is *p-normal* if the center Z of a Sylow p-subgroup P is the center of every Sylow p-subgroup P_1 which contains it. This is a special case of weak closure, being equivalent to the assertion that the center Z of P is weakly closed in P with respect to G. For suppose that G is p-normal. Then let $x \in G$ be such that $Z^x \subseteq P$. Then Z is contained in $P_1 = P^{x^{-1}}$. By p-normality Z is the center of P_1. But then Z^x is the center of $P_1^x = P$, whence $Z^x = Z$, and so Z is weakly closed in P. Conversely, suppose that Z is weakly closed in P, and that $Z \subseteq P_1$, another Sylow subgroup. Then for some $x \in G$, $P_1^x = P$. Then $Z^x \subseteq P$. By weak closure $Z = Z^x$. But if Z_1 is the center of P_1, then Z_1^x is the center of $P_1^x = P$. Hence $Z_1^x = Z = Z^x$, and $Z = Z_1$ is the center of P_1, whence G is p-normal.

It is clear that strong closure implies weak closure. A weakly closed finite subgroup B of H must be normal in H. A subgroup of H generated by all x's satisfying some equation $x^k = 1$ will be weakly closed, and if these x's form a subgroup X, then X will be strongly closed in H. This will be the case if H is a regular p-group, and also under certain other circumstances.

We shall write the transfer $V_{G \to H}(g)$ as $V(g)$ when no ambiguity may arise. Here if

$$G = H + Hx_2 + \cdots + Hx_n,$$

$$V(g) \equiv \prod_{i=1}^{n} x_i g \phi(x_i g)^{-1} \bmod H'.$$

We may also replace congruences modulo H' by congruences modulo H_0, where H_0 is any subgroup of H containing H', so that H/H_0 is Abelian. All congruences we use will be modulo H_0.

For $g \in G$ and $i = 1, \cdots, n$ define ig as that one of $1, \cdots, n$ such that $x_i g x_{ig}^{-1} \in H$. Then for a fixed g, $i \to ig$ is the permutation $\pi(g)$ of the transitive permutation representation of G on the left cosets of H. Thus we may write

$$V(g) \equiv \prod_i x_i g x_{ig}^{-1}.$$

There will be a number of cycles in the permutation $\pi(g)$, including fixed letters as cycles of length one. Choose one value from each cycle and call this set $C_H(g)$. For $i \in C_H(g)$ let r_i be the order of the cycle in which i appears. Then

$$\sum_{i \in C_H(g)} r_i = n,$$

which merely says that the total length of the cycles is n.

LEMMA 14.4.1.

$$V(g) \equiv \prod_{i \in C_H(g)} x_i g^{r_i} x_i^{-1}.$$

Here $x_i g^{r_i} x_i^{-1}$ is the first power of $x_i g x_i^{-1}$ which lies in H.

Proof: In a cycle of $\pi(g)$ beginning with i we have $i, ig, \cdots, ig^{r_i-1}$ all different, and we may take $x_i, x_i g, \cdots, x_i g^{r_i-1}$ as representatives of the corresponding cosets of H. These cosets make the contribution

$$x_i g(x_i g)^{-1} \cdot (x_i g) g (x_i g^2)^{-1} \cdot (\cdots) \cdot (x_i g^{r_i-1})\, g x_i^{-1} = x_i g^{r_i} x_i^{-1}$$

to $V(g)$, since $\phi(x_i g^s) = x_i g^s$, $s = 1 \cdots r_i - 1$, $\phi(x_i g^{r_i}) = x_i$. Since $x_i g^s \notin H x_i$ for $s < r_i$, $x_i g^{r_i} x_i^{-1}$ is the first power of $x_i g x_i^{-1}$ which lies in H.

We shall call the contribution of the cycles of length one to $V(g)$ the *diagonal contribution* $d(g)$ and write

$$d(g) \equiv \prod_{i=ig} x_i g x_i^{-1} \bmod H_0.$$

Here, as with $V(g)$, $d(g)$ is independent modulo H_0 of the order of the factors and the choice of the coset representatives x_i.

LEMMA 14.4.2. *If u and v are conjugate in G, then $d(u) \equiv d(v)$. Also $d(u^{-1}) \equiv [d(u)]^{-1}$.*

Proof: Let $v = t^{-1}ut$. Then $iu = i$ is equivalent to $itv = it$, and so by definition,

$$\begin{aligned} d(v) &\equiv \prod_{i=iu} x_{it} v x_{it}^{-1} \\ &\equiv \prod_{i=iu} (x_{it} t^{-1} x_i^{-1})(x_i u x_i^{-1})(x_i t x_{it}^{-1}) \\ &\equiv \prod_{i=iu} (x_i u x_i^{-1}) \equiv d(u). \end{aligned}$$

This follows since $x_{it}t^{-1}x_i^{-1}$ and $x_itx_{it}^{-1}$ are in H and are inverses of each other. Also, since $i = iu$ is equivalent to $i = iu^{-1}$, we have

$$d(u^{-1}) \equiv \prod_{i=iu} x_iu^{-1}x_i^{-1} \equiv \left(\prod_{i=iu} x_iux_i^{-1}\right)^{-1}$$
$$\equiv [d(u)]^{-1}.$$

For $h \in H$ define $d^*(h) \equiv h^{-1}d(h)$. Then $h \equiv d(h)[d^*(h)]^{-1} \equiv d(h)d^*(h^{-1})$ by Lemma 14.4.2, which also gives $d(h^r) \equiv d(x_ih^rx_i^{-1})$, and so if $x_ih^rx_i^{-1} \in H$, we get

$$\begin{aligned} x_ih^rx_i^{-1} &\equiv d(h^r)d^*(x_ih^{-r}x_i^{-1}) \\ &\equiv h^rd^*(h^r)d^*(x_ih^{-r}x_i^{-1}) \end{aligned}$$

and so finally by Lemma 14.4.1, we have:

Lemma 14.4.3. *If $h \in H$, then*

$$V(h) \equiv h^n \prod_{i \in C_H(h)} d^*(h^{r_i})d^*(x_ih^{-r_i}x_i^{-1})$$

Corollary 14.4.1. *If $d^*(h) \in H_0$ for all $h \in H$, then for any $h \in H$, $V(h) = h^n$.*

Let p be a prime, G_1 any finite group, and define $G = u_p(G_1)$ to be the group generated by all elements of G_1 of order prime to p. Thus G_1/G is the maximal p-factor group of G. Let P_1 be a Sylow p-subgroup of G_1, N_1 its normalizer in G_1, and H_1 any subgroup of G_1 containing N_1. Let us put $P = P_1 \cap G, N = N_1 \cap G, H = H_1 \cap G$, so that $G_1 = GP_1 = GN_1 = GH_1$ and $P_1/P = N_1/N = H_1/H = G_1/G$. G is a fully invariant subgroup of G_1, and we note that P is a Sylow p-subgroup of G, and that N normalizes both P_1 and G, whence N normalizes $P_1 \cap G = P$. Now $u_p(G) = G$ since G is generated by elements of order prime to p, but it may happen that $u_p(H) \subset H$. Let us suppose that $u_p(H) \subset H$.

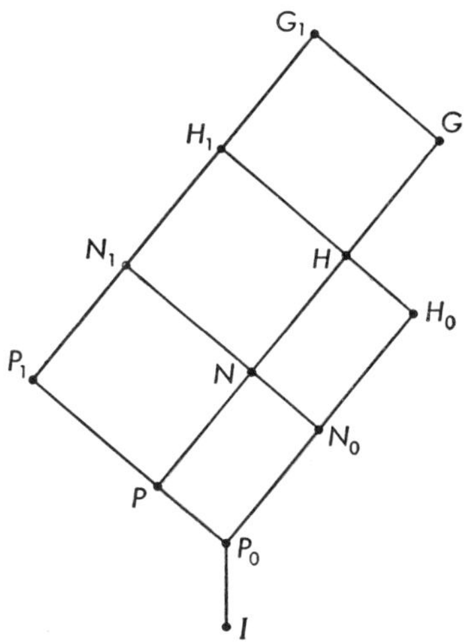

Fig. 6. A theorem of Philip Hall.

Here $u_p(H)$ is a fully invariant subgroup of H, and H is normal in H_1. Indeed, since H_1/H is a p-group, it is evident that $u_p(H) = u_p(H_1)$. Let us define

$$\begin{aligned} H_0 &= H^p \cup (H, H_1) \cup u_p(H) \\ &= H^p(H, H_1)u_p(H). \end{aligned}$$

Here H^p is the group generated by pth powers of elements of H, and (H, H_1) is the group generated by commutators (h, h_1), $h \in H$, $h_1 \in H_1$. Since $H_1/u_p(H)$ is a p-group, and these three groups are characteristic subgroups, their union is their product. Since $u_p(H) = u_p(H_1)$, $H_1/u_p(H)$ is a p-group and so nilpotent. Thus $(H, H_1)/u_p(H)$ is a proper subgroup of $H/u_p(H)$. Moreover, since H^p is contained in any subgroup T such that $H \supset T \supset (H, H_1)u_p(H)$ with $[H:T] = p$, it follows that if $u_p(H)$ is a proper subgroup of H, then also H_0 is a proper normal subgroup of H and H/H_0 is a p-group. We consider the problem: *What elements of P must be adjoined to H_0 to give H?*

LEMMA 14.4.4. *H is generated by H_0 together with the set of all elements $d^*(u)$ with $u \in P$.*

Proof: Here we have as before

$$G = H + Hx_2 + \cdots + Hx_n,$$

$$d(u) \equiv \prod_{i=iu} x_i u x_i^{-1} \bmod H_0,$$

$$d^*(u) \equiv u^{-1}d(u) \bmod H_0,$$

and we note that with our particular choice of H_0, $H_0 \supseteq (H, H_1) \supseteq (H, H) = H'$, surely H/H_0 is Abelian. Since $u \in P \subseteq H$, we surely have all $d^*(u) \in H$. Thus $K = \{d^*(u) \mid u \in P\} \cup H_0 \subseteq H$. To prove $H \subseteq K$, we use the fact that since H/H_0 is an Abelian p-group, $V(w) \equiv 1 \pmod{H_0}$ for every element w of G whose order is prime to p. But by our construction G is generated by such elements, whence $V(u) \equiv 1 \pmod{H_0}$ for every $u \in G$. Hence *a fortiori* $V(u) \in K$ for every $u \in P$. Now for $u \in P$, by Lemma 14.4.3,

$$V(u) \equiv u^n \prod_{i \in CH(u)} d^*(u^{r_i}) d^*(x_i u^{-r_i} x_i^{-1}).$$

Here $d^*(u^{r_i}) \in K$ by definition, and $v = x_i u^{-r_i} x_i^{-1}$ is a p-element of H, whence for some $y \in H$, $y^{-1}vy \in P$, and so $d^*(v) \equiv v^{-1}d(v) \equiv v^{-1}d(y^{-1}vy)$ by Lemma 14.4.2, and so

$$d^*(v) \equiv v^{-1}y^{-1}vy\, d^*(y^{-1}vy)$$
$$\equiv (v, y)d^*(y^{-1}vy).$$

But $(v, y) \in H' \subseteq H_0$ and $d^*(y^{-1}vy) \in K$ by definition, whence $d^*(v) = d^*(x_iu^{-r_i}x_i^{-1}) \in K$. It then follows that for $u \in P$,

$$u^n \equiv V(u) \in K.$$

But $(n, p) = 1$ and every element of P is an nth power of some other element of P. Thus $P \subseteq K$, and since H/H_0 is a p-group and P a Sylow p-subgroup of H, we have $H = H_0 \cup P \subseteq K$, proving $H = K$ and thus the lemma.

Since $G \triangleleft G_1$ and $H = H_1 \cap G$, $G \cup H_1 = G_1$, we may use the left coset representatives $1, x_2, \cdots, x_n$ of H in G as left coset representatives of H_1 in G_1. Thus $G_1 = H_1 + H_1x_2 + \cdots + H_1x_n$. Hence, writing G in terms of double cosets of H_1 and P_1, we have

$$G_1 = H_1 + H_1t_1P_1 + \cdots + H_1t_sP_1,$$

where $1, t_1, \cdots, t_s$ are a subset of $1, x_2, \cdots, x_n$. Let $\pi_i(i = 1, \cdots, s)$ be the transitive permutation representation of P_1 on the cosets of H_1 in $H_1t_iP_1$. Here π_i is of degree greater than one, since otherwise $H_1t_iP_1 = H_1t_i$ and then $t_iP_1t_i^{-1} \subseteq H_1$, but then by Sylow's theorem, $t_iP_1t_i^{-1} = y^{-1}P_1y$ for some $y \in H_1$, yielding $yt_i \in N_1 \subseteq H_1$, whence $t_i \in H_1$ which is not the case. Thus the representation π_i of P_1 is not the identity, and thus its kernel K_i is properly contained in P_1 and π_i faithfully represents P_1/K_i. As P_1/K_i is a p-group, its center is not the identity. Hence we may choose an element $z_i \in P_1$ such that $\pi_i(z_i)$ is of order p and in the center of $\pi_i(P_1)$. Here $\pi_i(z_i)$ permutes with every $\pi_i(u)$ for $u \in P_1$. Now an element in the center of a transitive permutation group cannot fix one letter without fixing all letters. Hence $\pi_i(z_i)$ fixes no one of the cosets of $H_1t_iP_1$ and consists exclusively of cycles of length p. For any $u \in P \subseteq P_1$, $\pi_i(u)$ permutes with $\pi_i(z_i)$ and so if $\pi_i(u)$ leaves fixed any coset, say, H_1x_{j+1} contained in $H_1t_iP_1$, then it must also fix all the cosets $H_1x_{j+1}, \cdots, H_1x_{j+p}$ in the cycle of $\pi_i(z_i)$ to which H_1x_{j+1} belongs. Hence we may write for $u \in P$,

$$d(u) \equiv u \cdot \prod d_j(u) \bmod H_0,$$

where

$$d_j(u) = h_1h_2 \cdots h_p$$

and

$$h_k = x_{j+k}ux_{j+k}^{-1} \quad k = 1 \cdots p,$$

where $x_{j+1}, \cdots, x_{j+p}$ are, as above, representatives of the cosets of a cycle of $\pi_i(z_i)$ for some i. Here the single factor $u = 1 \cdot u \cdot 1^{-1}$ is the contribution to $d(u)$ from H_1. We note also that with $x_{j+k}ux_{j+k}^{-1} = h_k \in H_1$, since $x_{j+k} \in G$, $u \in P$, we have $h_k \in G$, whence $h_k \in H_1 \cap G = H$, and so these are indeed the factors entering into $d(u)$. Now $d^*(u) = u^{-1}d(u)$, whence

$$d^*(u) \equiv \prod_j d_j(u) \bmod H_0.$$

With Lemma 14.4.4 this relation immediately establishes:

LEMMA 14.4.5. *H is generated by H_0 together with all the $d_j(u)$ for $u \in P$.*

Consider one of these $d_j(u)$ in more detail, writing $w_k = x_{j+k}$, $k = 1, \cdots, p$ for convenience of notation.

$$H_1w_kz_i = H_1w_{k+1},$$

with subscripts mod p. Thus

$$w_kz_i = y_kw_{k+1},$$

with $y_k \in H_1$. Also

$$w_kuw_k^{-1} = h_k, \quad d_j(u) = h_1 \cdots h_p.$$

Now

$$\begin{aligned} w_k(u, z_i)w_k^{-1} &= w_ku^{-1}w_k^{-1} \cdot w_kz_i^{-1}uz_i \cdot w_k^{-1} \\ &= h_k^{-1}y_{k-1}^{-1}w_{k-1}uw_{k-1}^{-1} \cdot w_{k-1}z_iw_k^{-1} \\ &= h_k^{-1}y_{k-1}^{-1}h_{k-1} \cdot y_{k-1} \\ &= h_k^{-1}h_{k-1}(h_{k-1}, y_{k-1}). \end{aligned}$$

But the y's $\in H_1$, the h's $\in H$, and since $(H_1, H) \subseteq H_0$, we have

$$w_k(u, z_i)w_k^{-1} \equiv h_k^{-1}h_{k-1} \bmod H_0.$$

Now P is normal in P_1, whence $(u, z_i) \in P$, and so for $u_1 = (u, z_i)$ the diagonal contribution to $d(u_1)$ from the cosets Hw_k, $k = 1, \cdots p$ will be $h_k^{-1}h_{k-1} \bmod H_0$, $k = 1, \cdots p$. Thus, from $w_kuw_k^{-1} \equiv h_k \pmod{H_0}$, $k = 1, \cdots p$, we conclude that $w_k(u, z_i)w_k^{-1} = h_k^{-1}h_{k-1} \pmod{H_0}$, $k = 1, \cdots, p$. Now with $u = u_0$, $u_1 = (u, z_i)$, $u_2 = (u_1, z_i)$, and recursively, $u_{s+1} = (u_s, z_i)$. We have seen that if

$$w_k u_s w_k^{-1} \equiv h_{k,s} \pmod{H_0} \quad k = 1, \cdots, p,$$

then

$$w_k u_{s+1} w_k^{-1} \equiv h_{k-1,s} h_{k,s}^{-1} \equiv h_{k,s+1} \pmod{H_0}.$$

Hence, by induction on s,

$$w_k u_s w_k^{-1} \equiv h_{k-s} h_{k-s+1}^{-s} h_{k-s+2}^{\binom{s}{2}} \cdots h_k^{(-1)^s},$$

the exponents being the binomial coefficients with alternating signs. From the properties of binomial coefficients and the fact that $H^p \subseteq H_0$, we have

$$w_k u_{p-1} w_k^{-1} \equiv h_1 h_2 \cdots h_p \equiv d_j(u) \bmod H_0.$$

Thus

$$d_j(u) \equiv w_k(u, \overbrace{z_i, \cdots, z_i}^{p-1}) w_k^{-1} \pmod{H_0}$$

with $u \in P$, $z_i \in P_1$. If we write

$$e_p(u, z_i) = (u, \overbrace{z_i, \cdots, z_i}^{p-1}),$$

then Lemma 14.4.5 tells us that H can be obtained by adjoining for all $u \in P$ certain elements of the form $x_{j+k} e_p(u, z_i) x_{j+k}^{-1}$ which belong to H; i.e., certain diagonal coefficients of the $e_p(x, z_i)$ for $i = 1, \cdots, s$ and $u \in P$. Since these coefficients are p-elements lying in H, and P is a Sylow p-subgroup of H, we may transform them by elements of H so that they lie in P. This will not affect them mod H_0 since H/H_0 is Abelian.

This proves our main theorem.

THEOREM 14.4.1 (P. HALL). *Let G_1 be any finite group, P_1 a Sylow p-subgroup, N_1 its normalizer, and H_1 a subgroup containing N_1. Let $G = u_p(G_1)$ be the subgroup generated by all elements of G_1 of orders prime to p, and put $H = G \cap H_1$, $N = G \cap N_1$, $P = G \cap P_1$. Then $u_p(H_1) = u_p(H)$, and if $u_p(H) \neq H$, $H_0 = H^p(H_1, H)u_p(H)$ is a proper subgroup of H, and H can be obtained by adjoining to H_0 certain conjugates lying in H of elements $e_p(u, z_i) = (u, \overbrace{z_i, \cdots, z_i}^{p-1})$ where $u \in P$ and z_i, $i = 1, \cdots s$ are elements in P_1. If*

$$G_1 = H_1 + H_1 t_1 P_1 + \cdots H_1 t_s P_1$$

is a decomposition of G_1 into double cosets of H_1 and P_1, let π_i, $i = 1, \cdots s$ be the transitive representation of P_1 on the cosets of H_1 in $H_1 t_i P_1$. Then π_i is not of degree one, and we choose z_i so that $\pi_i(z_i)$ is of order p in the center of $\pi_i(P_1)$.

COROLLARY 14.4.2. *If* $e_p(u, z) = 1$ *for all* $u, z \in P_1$, *then* $u_p(N_1) = N = u_p(G_1) \cap N$ *and* $G_1/u_p(G_1) = N_1/u_p(N_1)$. *This will happen in particular if the class of* P_1 *is less than* p.

Here we have taken $H_1 = N_1$, and so $H = N$.

Suppose that Q_1 is a weakly closed subgroup of P_1. Then, as we have already remarked, Q_1 is normal in the normalizer N_1 of P_1, and so we may take the normalizer of Q_1 as a subgroup $H_1 \supseteq N_1$. Then the preceding theorem will give a result which is an improvement of a theorem of Wielandt's [3].

THEOREM 14.4.2 (HALL-WIELANDT). *Let* P_1 *be a Sylow p-subgroup of* G_1 *and* Q_1 *be a weakly closed subgroup of* P_1. *Let* N_1 *be the normalizer of* P_1 *and* H_1 *the normalizer of* Q_1. *Then any one of the following conditions will ensure* $u_p(H_1) = H = u_p(G_1) \cap H_1$, *whence* $G_1/u_p(G_1) = H_1/u_p(H_1)$.

1) $e_p(u, z) = 1$ *for all* $u \in P_1$, *all* $z \in Q_1$.
2) $e_{p-1}(u, z) = 1$ *for all* $u, z \in Q_1$.
3) $Q_1 \subseteq Z_{p-1}(P_1)$ *the* $(p - 1)$*st term of the ascending central series for* P_1.

Proof: As in the proof of Theorem 14.4.1, let K_i be the kernel of the representation π_i of P_1 on the cosets of $H_1t_iP_1$. Suppose, if possible, that $Q_1 \subseteq K_i$. Then $H_1t_iQ_1 = H_1t_i$, and so $t_iQ_1t_i^{-1} \subseteq H_1$. Thus $t_iQ_1t_i^{-1}$ is a p-subgroup of H_1, and there exists a $y \in H_1$ such that $y^{-1}t_iQ_1t_i^{-1}y \subseteq P_1$, which is a Sylow p-subgroup of H_1. By the weak closure of Q_1 this means $y^{-1}t_iQ_1t_i^{-1}y = Q_1$, and so $y^{-1}t_i \in H_1$, the normalizer of Q_1 and also $t_i \in H_1$, which is not the case. Hence $Q_1 \nsubseteq K_i$. Now Q_1 is normal in P_1, and so the image of Q_1 in P_1/K_i is a normal subgroup and must therefore contain elements of its center. Hence we may choose our elements z_i in Q_1. This gives the first condition, where we note that it would be sufficient to take $u \in P = P_1 \cap G$, but *a priori* we do not know which subgroup of P_1 is P. The third condition implies the first, for if $Q_1 \subseteq Z_{p-1}(P_1)$, then $z \in Z_{p-1}(P_1)$ and $(u, z) \in Z_{p-2}(P_1)$, $(u, z, z) \in Z_{p-3}$, and continuing,

$$e_p(u, z) = (u, \overbrace{z, \cdots, z}^{p-1}) = 1.$$

As to the second condition, $e_p(u, z) = e_{p-1}(u_1, z)$, where $u_1 = (u, z)$, and for $u \in P$, $u_1 \in Q_1$, whence also the second condition implies the first.

Corollary 14.4.3. *Let Q_1 be a characteristic subgroup of P_1. If Q_1 is not weakly closed in P_1, then there is another Sylow p-subgroup P_2 which contains Q_1 but in which Q_1 is not normal. This must be the case if Q_1 satisfies the conditions* (1), (2), *or* (3) *of the theorem, but $G_1/u_p(G_1)$ and $H_1/u_p(H_1)$ are not isomorphic.*

Proof: As Q_1 is characteristic in P_1, then Q_1 is normal in N_1. Hence $N_1 \subseteq H_1$, the normalizer of Q_1. If Q_1 is not weakly closed in P_1, then for some x, $x^{-1}Q_1x \subseteq P_1$, but $x^{-1}Q_1x \neq Q_1$. If $x^{-1}Q_1x$ were normal in P_1, then by Lemma 14.3.1, Q_1 and $x^{-1}Q_1x$ would be conjugate to each other in N_1, which is not the case. Hence $x^{-1}Q_1x$ is in P_1 but not normal in P_1, and so Q_1 is in $P_2 = xP_1x^{-1}$ but not normal in P_2. If Q_1 satisfies the conditions (1), (2), or (3) of the theorem, then the conclusion of the theorem can fail only because Q_1 is not weakly closed in P_1.

The following theorems arc somewhat more elementary than the preceding theorems.

Theorem 14.4.3. *Let P be a Sylow p-subgroup of G, and G' the derived group of G. Then $V_{G\to P}(G) \cong P/P \cap G'$.*

Proof: Since $V_{G\to P}(G)$ is a homomorphism of G into P/P', a p-group, every element of order prime to p is mapped onto the identity. Since G is generated by P and Sylow subgroups belonging to other primes, $V(G) = V(P)$.

Suppose

$$G = P + Px_2 + \cdots Px_n.$$

By Lemma 14.4.1, for $u \in P$,

$$V(u) \equiv \prod_{i \in C_P(u)} x_i u^{r_i} x_i^{-1} \bmod P',$$

$$V(u) \equiv \prod_{i \in C_P(u)} u^{r_i}(u^{r_i}, x_i^{-1}) \bmod P',$$

and

$$V(u) \equiv \prod_{i \in C_P(u)} u^{r_i} \equiv u^n \bmod G'.$$

Hence, as $(n, p) = 1$, $V(u) \not\equiv 1 \bmod G'$ if $u \in P$, $u \notin G'$. But as $V(G)$ is Abelian, $V(G') \equiv 1$. Hence the kernel of $P \to V_{G\to P}(P)$ is exactly $P \cap G'$, and so $V_{G\to P}(G) \cong P/P \cap G'$.

THEOREM 14.4.4 (FIRST THEOREM OF GRÜN) [1]. *Let P be a Sylow p-subgroup of G. Then $V_{G\to P}(G) \cong P/P^*$, where*

$$P^* = [P \cap N_G'(P)] \bigcup_{z \in G} (P \cap z^{-1}P'z).$$

Proof: From Theorem 14.4.3 we know that $V_{G\to P}(G) \cong P/P \cap G'$. From its construction P^* is the union of subgroups contained in $P \cap G'$, and so $P^* \subseteq P \cap G'$. We must show that $P \cap G' \subseteq P^*$. We prove that every element u in $P \cap G'$ is also in P^*, using induction on the order of u. Here, trivially, $1 \in P^*$.

Let

$$G = P + Py_2P + \cdots + Py_sP$$

be the decomposition of G into double cosets of P. We suppose $u \in P \cap G'$. Then, by Lemma 14.4.1,

$$V(u) \equiv \prod_{i \in C_P(u)} x_i u^{r_i} x_i^{-1} \bmod P'.$$

Here the contribution to $V(u)$ from a double coset PyP is of the form

$$w = \prod_k y v_k u^{r_k} v_k^{-1} y^{-1},$$

with $v_1 = 1$ and $v_k \in P$. Also $\sum_k r_k = p^t$ if there are p^t left cosets of P in PyP. In considering the contribution w, we distinguish two cases: Case 1, $t \geq 1$ in p^t; Case 2, $t = 0$, $p^t = 1$.

CASE 1. We have

$$w \equiv y u^{p^t} y^{-1} \pmod{yP'y^{-1}}.$$

Also, for $v_1 = 1$, we have a factor $y u^{p^b} y^{-1} \in P$, and since $b \leq t$, we have $y u^{p^t} y^{-1} \in P$. But also $w \in P$, and so

$$w \equiv y u^{p^t} y^{-1} \pmod{P \cap yP'y^{-1}},$$

whence *a fortiori*

$$w \equiv y u^{p^t} y^{-1} \pmod{P^*}.$$

Since $u \in P \cap G'$, $V(u) \equiv 1 \pmod{P'}$, and thus $V(y u^{p^t} y^{-1}) \equiv 1 \pmod{P'}$. But then $y u^{p^t} y^{-1}$, since it belongs to P, is in the kernel $P \cap G'$, and since $t > 1$, it is of lower order than u, whence by our induction $yu^{p^t}y^{-1} \in P^*$. Since also by induction $u^{p^t} \in P^*$, we have

$$w = yu^{p^t}y^{-1} = 1 = u^{p^t} \pmod{P^*}.$$

CASE 2. Here $PyP = Py$, and therefore $Py \subseteq N_G(P)$. Also

$$w \equiv yuy^{-1} \equiv u[N_G'(P)],$$

and $$w \equiv u[\text{mod } P \cap N_G'(P)],$$

and *a fortiori* $w \equiv u \pmod{P^*}$. Hence in all cases

$$w_j \equiv u^{p^{t_j}} (\text{mod } P^*),$$

if w_j is the contribution from Py_jP which contains p^{t_j} left cosets of P. Hence

$$V(u) \equiv u^n \pmod{P^*},$$

where $n = [G:P]$ is prime to p. But $V(u) \equiv 1$ for $u \in P \cap G'$, and so $V(u) \in P' \subseteq P^*$. Thus $u^n \equiv 1 \pmod{P^*}$, and so $u \in P^*$, as we wished to show.

THEOREM 14.4.5 (SECOND THEOREM OF GRÜN). *If G is p-normal, then the greatest Abelian p-group which is a factor group of G is isomorphic to that for the normalizer of the center of a Sylow p-subgroup.*

Proof: Let P be a Sylow p-subgroup of G, Z its center. Let $G'(p) \supseteq G'$ be the smallest normal subgroup of G such that $G/G'(p)$ is an Abelian p-group. Then $G = G'(p) \cup P$, since the order of $G'(p)$ must contain every factor of the order of G except for powers of p. If $G^* = P \cup G'$, then $G'(p) \cup G^* = G$. Also $G^* \cap G'(p) = G'$, since G^*/G' contains only p-elements and $G'(p)/G'$ contains only elements of orders prime to p. By Theorem 2.4.1, $G/G'(p) = G^*/G' = P/P \cap G'$. Let N be the normalizer of P, and H the normalizer of Z. As Z is characteristic in P, $H \supseteq N$. Now if $H'(p)$ is the least normal subgroup of H such that $H/H'(p)$ is an Abelian p-group then, as with G, $H/H'(p) = P/P \cap H'$. Hence to prove our theorem, we must show $P \cap G' = P \cap H'$. Trivially, $G \supseteq H$, $G' \supseteq H'$, and $P \cap G' \supseteq P \cap H'$. Thus we need to show $P \cap H' \supseteq P \cap G'$. By the First Theorem of Grün,

$$P \cap G' = (P \cap N') \bigcup_{x \in G} (P \cap x^{-1}P'x).$$

Since $H \supseteq N$, $P \cap H' \supseteq P \cap N'$. We must also show for every $x \in G$ that

$$P \cap H' \supseteq P \cap x^{-1}P'x.$$

Write $M = P \cap x^{-1}P'x$. Then $Z \subseteq N_G(M)$ and $x^{-1}Zx \subseteq N_G(M)$, since $x^{-1}Zx$ is the center of $x^{-1}Px$. Here Z is in a Sylow subgroup R of $N_G(M)$ and $x^{-1}Zx$ is in a Sylow subgroup S of $N_G(M)$. Hence, for some $y \in N_G(M)$, both Z and $y^{-1}x^{-1}Zxy$ are in the same Sylow subgroup Q of G containing R. By p-normality both Z and $y^{-1}x^{-1}Zxy$ are the center of Q and so equal to each other. Thus $Z = y^{-1}x^{-1}Zxy$, and so $xy = h \in N_G(Z) = H$. But $y \in N_G(M)$, whence

$$\begin{aligned} M &= y^{-1}My = y^{-1}Py \cap y^{-1}x^{-1}P'xy \\ &= y^{-1}Py \cap h^{-1}P'h \subseteq H'. \end{aligned}$$

Thus $M = P \cap x^{-1}P'x \subseteq P \cap H'$ and our theorem is proved.

The Theorem of P. Hall also yields an improvement of the Second Theorem of Grün, dropping the requirement "Abelian."

THEOREM 14.4.6 (HALL-GRÜN). *If G is p-normal, then the greatest factor group of G which is a p-group is isomorphic to that for the normalizer of the center of a Sylow p-subgroup.*

Proof: In Theorem 14.4.2 take G_1 as G, P_1 a Sylow p-subgroup, Q_1 as the center of P_1, and H_1 the normalizer of Q_1. Then the p-normality of G_1, as we have observed, means that Q_1 is weakly closed in P_1. Here, since $Q_1 = Z(P_1)$, the third condition holds and we conclude $G_1/u_p(G_1) \cong H_1/u_p(H_1)$. These are the maximal factor p-groups and the theorem is proved.

We can also improve on the Theorem of Burnside. Under what circumstances is a Sylow p-subgroup P of a group G isomorphic to a factor group of G? That is to say, when is $G/u_p(G) = P$? Assume that this is the case, writing $B = u_p(G)$; then B consists of all the elements of G of orders prime to p. Here $B \cap P = 1$, $B \cup P = BP = G$. If Q is any subgroup of P, then $B \cup Q = BQ$ is a subgroup containing Q and all elements of orders prime to p. Here B is normal in BQ. Write $W = N_{BQ}(Q)$. Then $W \cap B$ consists of the elements of W of orders prime to p. Clearly, $W \cap B \triangleleft W$ and, of course, $Q \triangleleft W$. But then $W = (W \cap B) \times Q$. Hence *every element of order prime to p which normalizes Q also centralizes Q.* This condition, which is necessary for $G/u_p(G) \cong P$, we shall show is also sufficient and to this extent generalizes Theorem 14.3.1.

THEOREM 14.4.7. *A group G has a factor group $G/u_p(G)$ isomorphic to a Sylow p-subgroup P if, and only if, for every subgroup Q of P an element of order prime to p which normalizes Q also centralizes Q.*

Proof: We proceed by induction on the order of G, the result being trivially true if $G = P$. First we show that G is p-normal. Let Z be the center of P. By the corollary to Theorem 14.4.2, if G is not p-normal, then Z is contained in another Sylow p-group P_2, but is not normal in P_2. Then by Theorem 4.2.5, there is a subgroup Q of P which is normalized but not centralized by an element of order prime to p. By our hypothesis, this does not happen, and so G must be p-normal. By Theorem 14.4.6, $G/u_p(G) \cong H/u_p(H)$, where H is the normalizer of Z. If H is a proper subgroup of G, then by induction $H/u_p(H) \cong P$ and our theorem is proved.

Hence we may suppose that $G = H$, and so, that Z is normal in G. But if G/Z contains a p-group Q/Z, which is normalized but not centralized by an element of order prime to p, then the same holds for its inverse image Q. Thus our hypothesis holds for G/Z, and so, G/Z has a normal subgroup K/Z such that the factor group is isomorphic to P/Z. Since Z is normalized by K and K/Z is of order prime to p, then Z is centralized by K, and so, $K = Z \times K_1$, where K_1 is of order prime to p. But $K_1 = u_p(K) = u_p(G)$ consists exclusively of elements of orders prime to P. Hence $G/u_p(G) = P$, as was to be shown.

15. GROUP EXTENSIONS AND COHOMOLOGY OF GROUPS

15.1. Composition of Normal Subgroup and Factor Group.

Generally speaking, any group G which contains a given group U as a subgroup is called an *extension* of U. General group extensions have been studied in a broad way by Reinhold Baer [11]. Here, however, we shall consider only cases in which U is normal in G.

Otto Schreier [1, 2] first considered the problem of constructing all groups G such that G will have a given normal subgroup N and a given factor group $H \cong G/N$. There is always at least one such group, since the direct product of N and H has this property.

Let us first assume such a group G given, and examine it closely. Let the elements of the factor group $H \cong G/N$ be designated as 1, $u, v, \cdots, w$. Each element x of H corresponds to a coset of N in G. Let us choose a representative $\bar{x}$ in G of the coset $\bar{x}N$ corresponding to x, with the convention that the identity 1 of G shall be chosen as the representative of N. Then

$$G = N + \bar{u}N + \bar{v}N + \cdots + \bar{w}N, \tag{15.1.1}$$

and in every case the homomorphism $G \to H$ is such that

$$\bar{u} \to u, \quad \bar{u} \in G, \quad u \in H. \tag{15.1.2}$$

Then the mapping

$$a \leftrightarrows \bar{u}^{-1}a\bar{u} = a^u, \tag{15.1.3}$$

all $a \in N$, is an automorphism of N, since N is a normal subgroup. Also

$$\bar{u}\cdot\bar{v} = \overline{uv}(u, v), \tag{15.1.4}$$

with $(u, v) \in N$, since $\bar{u} \to u$, $\bar{v} \to v$ in the homomorphism from G onto H. The set of all elements (u, v) defined by (15.1.4) we call the *factor set*. Thus in the structure of G the four following structures enter:

1) The normal subgroup N.
2) The factor group H.
3) The automorphisms of N: $a \leftrightarrows a^u$, $a \in N$, $u \in H$.
4) The factor set of $(u, v) \in N$; $u, v \in H$.

It is to be emphasized that, in general, the automorphisms and the factor set as defined by (15.1.3), and (15.1.4) depend on the choice of representative $\bar{u}$ of the coset $\bar{u}N$ corresponding to u.

The automorphisms and factor set must satisfy certain conditions. Transforming an element $a \in N$ by both sides of (15.1.4), we have

$$(a^u)^v = (u, v)^{-1}(a^{uv})(u, v). \tag{15.1.5}$$

Also, since in G $(\bar{u}\bar{v})\bar{w} = \bar{u}(\bar{v}\bar{w})$ we have

$$(\bar{u}\bar{v})\bar{w} = [\overline{uv}(u, v)]\bar{w} = \overline{uv}\bar{w}(u, v)^w = \overline{uvw}(uv, w)(u, v)^w,$$

and also

$$\bar{u}(\bar{v}\bar{w}) = \bar{u}[\overline{vw}(v, w)] = \overline{uvw}(u, vw)(v, w),$$

whence it follows that

$$(uv, w)(u, v)^w = (u, vw)(v, w). \tag{15.1.6}$$

For the product of two elements $\bar{u}a$, $\bar{v}b$ of G we have

$$(\bar{u}a)(\bar{v}b) = \bar{u}\bar{v}a^vb = \overline{uv}(u, v)a^vb \qquad \text{or}$$

$$(\bar{u}a)(\bar{v}b) = \overline{uv}(u, v)a^vb. \tag{15.1.7}$$

The convention of taking 1 as the representative of N in G yields, from (15.1.4),

$$(u, 1) = 1 = (1, v), \tag{15.1.8}$$

for all $u, v \in H$.

Conversely, the conditions (15.1.5) and (15.1.6) on the automorphisms and factor set are sufficient for G to exist with N as a normal subgroup and $G/N \cong H$. Let us take symbols $\bar{u}a$, $u \in H$, $a \in N$ and define a system G with a binary operation of product given by the rule

$$\bar{u}a \cdot \bar{v}b = \overline{uv}(u, v)a^vb. \tag{15.1.9}$$

This product is associative, since

$$\begin{aligned}(\bar{u}a\cdot\bar{v}b)\cdot\bar{w}c &= \overline{uv}(u, v)a^v b\cdot\bar{w}c = \overline{uvw}(uv, w)(u, v)^w(a^v)^w b^w c\\ &= \overline{uvw}(uv, w)(u, v)^w(v, w)^{-1}a^{vw}(v, w)b^w c \quad \text{[by (15.1.5)]}\\ &= \overline{uvw}(u, vw)a^{vw}(v, w)b^w c \quad \text{[by (15.1.6)]}\\ &= \bar{u}a\cdot\overline{vw}(v, w)b^w c\\ &= \bar{u}a\cdot(\bar{v}b\cdot\bar{w}c).\end{aligned}$$

It is convenient (but the reader may verify not necessary) to assume for the converse, besides (15.1.5) and (15.1.6), also

$$(1, 1) = 1, \tag{15.1.10}$$

a particular case of (15.1.8). If in (15.1.5) we put $u = v = 1$ and use (15.1.10), we get $(a^1)^1 = a^1$, and since $a^1 = c$ may be an arbitrary element of N, we have $c^1 = c$ for all $c \in N$. In (15.1.6) put $u = v = 1$. Then $1 = (1, 1)^w = (1, w)$. Similarly, from $v = w = 1$, we find $(u, 1) = 1$. Now $\bar{1}1\cdot\bar{w}c = \bar{w}(1, w)c = \bar{w}c$, and $\bar{u}a\cdot\bar{1}1 = \bar{u}(u, 1)a = \bar{u}a$, and so $\bar{1}1$ is the identity for the system G. Since $a \leftrightarrows a^w$ is an automorphism of N, there is an element d of N such that $d^w = (w^{-1}, w)^{-1}c^{-1}$ for given $c \in N$, $w \in H$. Hence, for an arbitrary $\bar{w}c$ of G, we have $\overline{w^{-1}}d\cdot\bar{w}c = \bar{1}(w^{-1}, w)d^w c = \bar{1}1$, the identity. Since every element of G has a left inverse, this is sufficient to prove that G is a group. The product rule (15.1.9) is such that the mapping

$$\bar{u}a \to u \tag{15.1.11}$$

is a homomorphism of G onto H, where the kernel consists of the elements $\bar{1}a$. But we verify

$$\bar{1}a\cdot\bar{1}b = \bar{1}(1, 1)ab = \bar{1}ab$$

whence $\bar{1}a \leftrightarrows a$ is an isomorphism identifying this kernel with N. Since $\bar{u}1\cdot\bar{1}a = \bar{u}(u, 1)a = \bar{u}a$, we may take the $\bar{u} = \bar{u}1$ as coset representatives of N, and we may regard $\bar{u}a$ as the product of $\bar{u}$ and a.

We summarize these results in a theorem.

THEOREM 15.1.1 (SCHREIER). *Given a group G with a normal subgroup N and factor group $H = G/N$. If we choose coset representatives $\bar{u}$ where $\bar{u}N \to u \in H$, taking $\bar{1} = 1$, then automorphisms and a factor set are determined, satisfying*

$$\begin{aligned}&(a^u)^v = (u, v)^{-1}(a^{uv})(u, v), \quad a, (u, v) \in N;\ u, v \in H;\\ &(uv, w)(u, v)^w = (u, vw)(v, w);\ (1, 1) = 1.\end{aligned}$$

Conversely, if for every $u \in H$ there is given an automorphism $a \leftrightarrows a^u$

of N, and if for these automorphisms and the factor set $[(u, v) \in N]$, $(u, v \in H)$, the above conditions hold, then elements $\bar{u}a$, $u \in H$, $a \in N$, with the product rule

$$\bar{u}a \cdot \bar{v}b = \overline{uv}(u, v)a^v b,$$

define a group G with normal subgroup N and $G/N \cong H$.

If the requirement $(1, 1) = 1$ is omitted, then the theorem still holds with $\bar{1}(1, 1)^{-1}$ the unit for G.

The unique extension G determined by N, H, $a \leftrightarrows a^u$ and factor set (u, v) will be designated $E[N, H, a^u, (u, v)]$.

If we change the coset representatives of N in G, taking

$$\bar{\bar{u}} = \bar{u}\alpha(u), \quad u \in H, \quad \alpha(u) \in N, \tag{15.1.12}$$

where by convention $\bar{\bar{1}} = \bar{1} = 1$, and so, $\alpha(1) = 1$. Here the automorphisms are changed and

$$a \leftrightarrows a^{u^1} = \bar{\bar{u}}^{-1}a\bar{\bar{u}} = \alpha(u)^{-1}a^u\alpha(u). \tag{15.1.13}$$

Also the factor set (u, v) is replaced by the factor set $(u, v)^1$ by the rule

$$\begin{aligned} \bar{\bar{u}} \cdot \bar{\bar{v}} &= \bar{u}\alpha(u)\bar{v}\alpha(v) = \overline{uv}(u, v)\alpha(u)^v\alpha(v) \\ &= \overline{\overline{uv}}(u, v)^1 = \overline{uv}\,\alpha(uv)(u, v)^1. \end{aligned} \tag{15.1.14}$$

DEFINITION: *Two extensions $E_1 = E[N, H, a^u, (u, v)]$ and $E_2 = E[N, H, a^{u^1}, (u, v)^1]$ are equivalent if the automorphisms and factor sets are related by*

$$\begin{aligned} a^{u^1} &= \alpha(u)^{-1}a^u\alpha(u), \\ (u, v)^1 &= \alpha(uv)^{-1}(u, v)\alpha(u)^v\alpha(v), \end{aligned}$$

where $\alpha(u)$ is a function of elements $u \in H$ with values in N and $\alpha(1) = 1$. We write

$$E[N, H, a^u, (u, v)] \sim E[N, H, a^{u^1}, (u, v)^1].$$

The equivalence of E_2 to E_1 amounts to a change of coset representatives for N in the same group G, and so, clearly this is a true equivalence and is symmetric, reflexive, and transitive.

If coset representatives $\bar{\bar{u}}$ of N in G may be chosen so that

$$\overline{\overline{uv}} = \bar{\bar{u}}\,\bar{\bar{v}}, \tag{15.1.15}$$

i.e., $(u, v)^1 = 1$, then the coset representatives form a group isomorphic to H, which we may identify with H. If this happens, we shall say that G splits over N or that G is the semi-direct product of N and H.

THEOREM 15.1.2. *The extension* $G = E[N, H, a^u, (u, v)]$ *splits over* N *if, and only if, we can find a function* $\alpha(u) \in N$, $u \in H$ *such that*

$$(u, v)\alpha(u)^v\alpha(v) = \alpha(uv)$$

for all $u, v \in H$.

Proof: If u are coset representatives such that $G = E[N, H, a^u, (u, v)]$ splits over N, then $(u, v)^1 = 1$, and with $\bar{\bar{u}} = \bar{u}\,\alpha(u)$ we find the relation

$$(u, v)\alpha(u)^v\alpha(v) = \alpha(uv). \tag{15.1.16}$$

Conversely, if a function $\alpha(u)$ exists such that (15.1.16) holds, then define a^{u^1} by $a^{u^1} = \alpha(u)^{-1}a^u\alpha(u)$, $\bar{\bar{u}} = \bar{u}\,\alpha(u)$, and the extension $E[N, H, a^{u^1}, (u, v)^1] = G$ will exist and be an equivalent extension with $(u, v)^1 = 1$ for all $u, v \in H$, whence the extension G splits over N.

15.2. Central Extensions.

Let us suppose that all factors (u, v) in an extension of a group A by a group H lie in the center B of A. Then we shall say that $E[A, H, a^u, (u, v)]$ is a *central extension* of A by H. Thus, if A is an Abelian group, $B = A$ and all extensions of A are central extensions.

For a central extension (15.1.5) reduces to

$$(a^u)^v = a^{uv}, \tag{15.2.1}$$

which says that the automorphisms $a \leftrightarrows a^u$ of A form a group homomorphic to H. Let us denote by χ a particular way of assigning to each element of H an automorphism of A, where the automorphisms that are assigned form a group homomorphic to H. Furthermore, if coset representatives $\bar{u}$ are changed only by factors $\alpha(u)$ lying in B, the automorphisms are unchanged. Hence, for such extensions, which we call with Baer [1] H-χ extensions, the automorphisms are fixed and form a group homomorphic to H. This settles condition (15.1.5) for central extensions, and only (15.1.6) need be considered

$$(uv, w)(u, v)^w = (u, vw)(v, w). \tag{15.2.2}$$

Here, for an equivalent extension,

$$(u, v)^1 = \alpha(uv)^{-1}(u, v)\alpha(u)^v\alpha(v) \tag{15.2.3}$$

with $\alpha(u) \in B$.

If factor sets $(u, v)_1$ and $(u, v)_2$ both satisfy (15.2.2) and we define

$$(u, v)_3 = (u, v)_1(u, v)_2 \quad \text{all } u, v \in H, \tag{15.2.4}$$

then the elements $(u, v)_3$ also satisfy (15.2.2) and are a factor set determining an H-χ extension of A. In this definition of product for factor sets there is an identity, the factor set with all $(u, v) = 1$ and an inverse, the set in which (u, v) is replaced by $(u, v)^{-1}$. Moreover, for equivalent factor sets if $(u, v)_1{}^* \sim (u, v)_1$, and $(u, v)^*{}_2 \sim (u, v)_2$, then $(u, v)_1{}^*(u, v)_2{}^* \sim (u, v)_1(u, v)_2$. Hence the totality of all H-χ factor sets forms an Abelian group even if we identify equivalent sets. The group in which equivalent sets are identified will be called the *group of extensions.*

If H is finite, we define

$$f(v) = \prod_u (u, v). \tag{15.2.5}$$

Multiplying (15.2.2) over all $u \in H$, we have

$$f(w)f(v)^w = f(vw)(v, w)^n, \tag{15.2.6}$$

where n is the order of H. On comparison with (15.2.3),

$$(v, w)^n \sim 1. \tag{15.2.7}$$

Again, if m is a multiple of the order of every element of B, since $(u, v) \in B$,

$$(v, w)^m = 1. \tag{15.2.8}$$

Hence the following theorem holds.

THEOREM 15.2.1. *The order of any element of the group of extensions divides the order of H and the least common multiple of orders of elements of B.*

COROLLARY 15.2.1. *If m and n are relatively prime, then all H-χ extensions of A are equivalent to the semi-direct product of A by H.*

As an application of this theorem we may prove Theorem 15.2.2 on extensions which need not be assumed to be central extensions.

THEOREM 15.2.2. *Let G be a group of finite order mn containing a normal subgroup K of order m and having a factor group $H = G/K$ of order n where m and n are relatively prime. Then G splits over K.*

Proof: It is sufficient to show that G possesses a subgroup of order n. We shall proceed by induction on m, the theorem being trivial if $m = 1$. Let $m > 1$ and p be a prime dividing m. All Sylow subgroups S_p belonging to p in G are subgroups of K, since K contains at least one Sylow subgroup S_p, and K being normal, the conjugates of S_p also belong to K. Thus the number of Sylow subgroups S_p in G is the same as the number in K. Hence by Theorem 1.6.1, $[G:N_G(S_p)] = [K:N_K(S_p)]$, whence $[N_G(S_p):N_K(S_p)] = [G:K] = n$, $N_G(S_p)$, and $N_K(S_p)$ being the normalizers of an S_p respectively in G and K. Here, of course, $N_K(S_p) = N_G(S_p) \cap K$, and by Theorem 2.4.1, $N_K(S_p)$ is normal in $N_G(S_p)$. If $N_G(S_p)$ is a proper subgroup of G, by induction it contains a subgroup of order n.

Hence we may assume $G = N_G(S_p)$, and so, $K = N_K(S_p)$. If S_p is a proper subgroup of K, then by induction G contains a subgroup of order $[G:S_p]$ isomorphic to G/S_p, and thus a subgroup isomorphic to G/K of order n, proving the theorem. Hence our proof is reduced to the case in which $K = S_p$. Here, if S_p is Abelian, G is a central extension of S_p, and by the corollary to Theorem 15.2.1, G splits over S_p, proving our theorem. If S_p is not Abelian, then the center Z of S_p is a proper subgroup of S_p and as a characteristic subgroup of S_p necessarily a normal subgroup of G. Hence, by our induction, G/Z contains a subgroup U/Z of order n. But Z is normal and of index n in the corresponding subgroup U of G, and by induction, U contains a subgroup of order n, proving the theorem for this final case.

15.3. Cyclic Extensions.

Let us suppose that H is a cyclic group of finite order m, generated by an element x; the elements of H will be

$$1, x, x^2, \cdots, x^{m-1}. \tag{15.3.1}$$

With $G/N = H$, choosing a representative $\bar{x}$ of the coset of N mapped onto x, we may also choose $\bar{x}^2, \cdots, \bar{x}^{m-1}$ as representatives of the cosets mapped respectively onto $x^2, \cdots, x^{m-1}$, and so,

$$G = N + N\bar{x} + \cdots + N\bar{x}^{m-1}. \tag{15.3.2}$$

Here

$$\bar{x}^m = \alpha, \tag{15.3.3}$$

where α is an element of N.

Thus for the automorphism $a \leftrightarrows a^x$ of N, we must have for its mth power

$$a^{x^m} = \alpha^{-1} a \, \alpha, \quad \alpha \in N. \tag{15.3.4}$$

Moreover, from the identity

$$\bar{x}^{-1}\bar{x}^m\bar{x} = \bar{x}^m, \tag{15.3.5}$$

we have

$$\alpha^x = \alpha. \tag{15.3.6}$$

We shall show that (15.3.4) and (15.3.6) are the only conditions required to define an extension of N by H.

THEOREM 15.3.1. *Let H be a cyclic group of finite order m. Then an extension G of a group N by H exists if, and only if, we have an automorphism $a \leftrightarrows a^x$ of N and an element $\alpha \in N$ such that* (1) *the mth power of the automorphism is the inner automorphism of N given by transformation by α, and* (2) *α is fixed by the automorphism.*

Proof: We have already shown that if an extension exists, then the automorphism $a \leftrightarrows a^x$ and the element α satisfy (15.3.4) and (15.3.6). Conversely, we must show that (15.3.4) and (15.3.6) suffice to determine an extension. The elements of H are $1, x, \cdots, x^{m-1}$, or x^i, $0 \le i \le m$-1. Let us define the automorphisms by

$$a^{x^0} = a, \quad a^{x^i} = (a^{x^{i-1}})^x, \quad i = 1, \cdots, m-2, \tag{15.3.7}$$

and a factor set put

$$(x^i, x^j) = 1 \quad \text{if } i + j \le m - 1, \tag{15.3.8.1}$$

$$(x^i, x^j) = \alpha \quad \text{if } m \le i + j. \tag{15.3.8.2}$$

With these definitions we easily verify that (15.1.5) and (15.1.6) are satisfied, and so from Theorem 15.1.1, an extension is defined.

If H is a cyclic group of infinite order, we may put $(x^i, x^j) = 1$ for all i, j, and we find that there is no restriction on the automorphism $a \leftrightarrows a^x$. This amounts to taking $x^i = \bar{x}^i$ for all i.

15.4. Defining Relations and Extensions.

In the preceding section we have seen that when H is a cyclic group, the conditions for extending a group N are particularly simple, corresponding to the particularly simple defining relation for H. In this section we shall see how the extension conditions depend on the defining relations for the most general group H.

Let the group H be given in terms of generators, $x, y, z, \cdots$ and relations

$$\phi_i(x, y, z, \cdots) = 1 \quad i = 1, 2, \cdots, r. \tag{15.4.1}$$

We can suppose each element h of H to be represented by a definite word $h = h(x, y, z, \cdots)$ in the generators and their inverses. Then if G is an extension of N by H, we can choose representatives of the cosets of N as the corresponding words in $\bar{x}, \bar{y}, \bar{z}, \cdots$, so that in the homomorphism $G \to H$ we have

$$\begin{gathered} G \to H \\ \bar{x} \to x \\ h(\bar{x}, \bar{y}, \cdots) \to h(x, y, \cdots). \end{gathered} \tag{15.4.2}$$

Now let F_1 be the free group with generators $\mathbf{x}, \mathbf{y}, \mathbf{z} \cdots$ corresponding to $x, y, z \cdots$. Then we have homomorphisms defined by

$$\begin{gathered} \mathbf{x} \to \bar{x} \to x, \\ \mathbf{y} \to \bar{y} \to y, \\ \cdot \quad \cdot \quad \cdot \quad \cdot \quad \cdot, \end{gathered} \tag{15.4.3}$$

which map

$$F_1 \to \bar{H} \to H, \tag{15.4.4}$$

where $\bar{H}$ is the subgroup of G generated by $\bar{x}, \bar{y}, \bar{z}, \cdots$, and so must contain at least one element from each coset of N. Hence $G = \bar{H} \cup N$. Hence, if F_2 is a free group which has N as a homomorphic image, we may take a free group $F = F_1 \cup F_2$ and define homomorphisms

$$\begin{gathered} F \to G \to H, \qquad F = F_1 \cup F_2 \\ F_1 \to \bar{H} \to H, \\ F_2 \to N \to 1. \end{gathered} \tag{15.4.5}$$

Every element $\bar{h}$ of $\bar{H}$ induces an automorphism in N by transformation

$$\bar{h}^{-1} a \bar{h} = a^h \quad \bar{h} \in \bar{H}, \quad a \in N. \tag{15.4.6}$$

In the mapping $F_1 \to H$, we have $H = F_1/W$, where W is the least normal subgroup containing the $\phi_i(\mathbf{x}, \mathbf{y}, \cdots)$. Hence in $\overline{H} \to H$ we have $\phi_i(\bar{x}, \bar{y}, \cdots) \to 1$. Thus

$$\phi_i(\bar{x}, \bar{y}, \cdots) = \alpha_i \in N. \tag{15.4.7}$$

We shall have an identity in the free group F

$$u_1 u_2 \cdots u_r = z_1 z_2 \cdots z_s,$$

if the u's and z's are words such that the reduced forms of these two expressions are the same. In the mapping of F onto G, any identity will remain valid. In particular W and F_2 will be mapped onto elements of N. Thus any identity involving u's and z's from the normal subgroup generated by W and F_2 will by means of the replacement rules (15.4.6) and (15.4.7) lead to conditions on the α_i and the automorphisms $a \leftrightarrows a^h$, which can be interpreted as conditions for the existence of an extension G of N by H. Since $\overline{uv}^{-1}\bar{u}\bar{v}$ is in $N \cap \overline{H}$, it is the image in G of an element of W, and hence a product of conjugates of the $\phi_i(\bar{x}, \bar{y}, \cdots) = \alpha_i$. Hence each factor $(u, v) = \overline{uv}^{-1}\bar{u}\bar{v}$ is the image in $\overline{H}$ of an element of W, and if $\bar{u} = h_1(\bar{x}, \bar{y}, \cdots)$, $\bar{v} = h_2(\bar{x}, \bar{y}, \cdots)$, $\overline{uv} = h_3(\bar{x}, \bar{y}, \cdots)$, then (u, v) is the image of

$$h_3(\mathbf{x}, \mathbf{y}, \cdots)^{-1} h_1(\mathbf{x}, \mathbf{y}, \cdots) h_2(\mathbf{x}, \mathbf{y}, \cdots), \tag{15.4.8}$$

an element of W.

The conditions of Theorem 15.1.1 are identities in F paraphrased into conditions on the factor set and automorphisms by the rules (15.4.6) and (15.4.7). Thus the rule (15.1.5)

$$(a^u)^v = (u, v)^{-1}(a^{uv})(u, v)$$

is a paraphrase of the identity

$$\bar{v}^{-1}(\bar{u}^{-1}a\bar{u})\bar{v} = (\overline{uv}^{-1}\bar{u}\bar{v})^{-1}(\overline{uv}^{-1}a\overline{uv})(\overline{uv}^{-1}\bar{u}\bar{v}), \tag{15.4.9}$$

using (15.4.6) for automorphisms and replacing elements of W by elements in N. Similarly, the rule (15.1.6), $(uv, w)\ (u, v)^w = (u, vw)\ (v, w)$, is a paraphrase of the identity

$$(\overline{uvw}^{-1}\overline{uv}\bar{w})\bar{w}^{-1}(\overline{uv}^{-1}\bar{u}\bar{v})\bar{w} = (\overline{uvw}^{-1}\bar{u}\overline{vw})(\overline{vw}^{-1}\bar{v}\bar{w}). \tag{15.4.10}$$

Thus conditions for the existence of an extension of N by H are paraphrases of identities in F. Note that the defining relations for N do not enter into these conditions. The conditions may be regarded as finding elements α_i in N and automorphisms in N consistent with the defining relations of H. Both these conditions become vacuous when H is a free group, for then in every case we can choose $\overline{uv} = \bar{u}\bar{v}$ and take our factors as the identity, and moreover, require merely that the automorphisms form a group.

In practice it may be difficult to determine the identities in F leading to conditions for an extension. In the next section we shall make such a determination for central extensions of N by a group H.

15.5. Group Rings and Central Extensions.†

We shall consider a central extension of a group N with center C by the finite group H. We suppose, as in §15.2, that the automorphisms satisfy

$$(a^u)^v = a^{uv}. \tag{15.5.1}$$

We have assumed that the factors $(u, v) = \overline{uv}^{-1}\bar{u}\bar{v}$ lie in C. But applying Lemma 7.2.2 (for right cosets rather than left cosets), these elements generate the subgroup T of $\overline{H}$ such that $\overline{H}/T = H$. But if

$$\phi_i(x, y, \cdots) = 1. \tag{15.5.2}$$

are the defining relations for H, then

$$\phi_i(\bar{x}, \bar{y}, \cdots) = \alpha_i \in C, \tag{15.5.3}$$

since the α_i surely belong to T and T is generated by elements of C.

If r and s are endomorphisms of C, then we may define an endomorphism $r + s$ by the rule

$$a^r a^s = a^{r+s}. \tag{15.5.4}$$

Thus, by (15.5.1) and (15.5.4), the group ring H^* of H is a ring of operators on C. Here the group ring H^* consists of elements

$$c_1h_1 + \cdots + c_nh_n, \tag{15.5.5}$$

where $h_1, \cdots, h_n$ are elements of H and $c_1, \cdots, c_n$ are integers. Elements of H^* are added by adding coefficients. Multiplication in H^* is given by the multiplication $h_ih_g = h_k$ in H together with the two distributive laws. It is easily verified that H^* is an associative ring and that the identity of H is the identity of H^*.

† See Marshall Hall, Jr. [1].

We shall say that an Abelian group A which admits H^* as an operator ring is *operator free* if A has a basis of elements $a_1, a_2, \cdots, a_r$ such that every element of A is of the form

$$a = a_1^{z_1} a_2^{z_2} \cdots a_r^{z_r} \qquad z_i \in H^* \tag{15.5.6}$$

and has a unique expression of this form. Thus $a = 1$ implies $z_1 = z_2 = \cdots = z_r = 0$.

THEOREM 15.5.1. *The only extension G of an operator-free group A by a finite group H is the semi-direct product of A by H.*

Proof: In A every element b has a unique expression

$$b = a_1^{z_1} \cdots a_r^{z_r} \qquad z_i \in H^*.$$

Now if $z_i = c_{i1}h_1 + \cdots + c_{in}h_n$, $i = 1, \cdots r$, put

$$(b; h_j) = a_1^{c_{1j}} a_2^{c_{2j}} \cdots a_r^{c_{rj}}.$$

Thus, with $t = h_1, h_2, \cdots, h_n$, b has a unique expression

$$b = \prod_t (b; t)^t \qquad t = h_1, \cdots, h_n. \tag{15.5.7}$$

Hence, for a factor set,

$$(u, v) = \prod_t (u, v; t)^t, \tag{15.5.8}$$

and the rule (15.1.6) because of the uniqueness of (15.5.7) becomes

$$(uv, w; t)(u, v; tw^{-1}) = (u, vw; t)(v, w; t). \tag{15.5.9}$$

If we now put $\overline{\overline{u}} = \bar{u} \prod_t (u, t^{-1}; 1)^{-t}$ for all u of H, we may verify from direct calculation and substitution from (15.5.9) that

$$\overline{\overline{u}}\, \overline{\overline{v}} = \overline{\overline{uv}}. \tag{15.5.10}$$

Hence the new representatives form a group and G is the semi-direct product of A and H.

By the results of §15.4 the conditions for a central extension of a group N by a group H are (15.5.1) and conditions of the form

$$\prod_i \alpha_i^{u_i} = 1 \qquad u_x \in H^* \tag{15.5.11}$$

with $\phi_i(\bar{x}, \bar{y}, \cdots) = \alpha_i$ as in (15.5.3). Now suppose that N is an operator-free group. We know that $\alpha_i = 1$, $i = 1, \cdots, r$ yields a solution and that all others are obtainable by changing representatives. If we put $\bar{x} = \xi\bar{\bar{x}}$, $\bar{y} = \eta\,\bar{\bar{y}}, \cdots$, then $\phi_i(\xi\bar{\bar{x}}, \eta\,\bar{\bar{y}}, \cdots) = 1$. Here, using the rule

$$(15.5.12) \qquad \alpha\bar{\bar{z}} = \bar{\bar{z}}\alpha^z, \quad \alpha \in N,$$

we may write

$$(15.5.13) \qquad 1 = \phi_i(\xi\bar{\bar{x}}, \eta\bar{\bar{y}}, \cdots) = \phi_i(\bar{\bar{x}}, \bar{\bar{y}}, \cdots)\xi^{x_i}\eta^{y_i}\cdots.$$

Hence $\alpha_i^{-1} = \xi^{x_i}\eta^{y_i}\cdots$ must also satisfy conditions (15.5.11), since these values are given by merely changing the representatives in the semi-direct product. Taking $\xi, \eta \cdots$ as independent basis elements of N, we have the following equations holding in H^*:

$$(15.5.14) \qquad \sum_i x_i u_i = 0, \quad \sum_i y_i u_i = 0 \cdots, \quad i = 1, \cdots, r.$$

The elements $x_i, y_i, \cdots$ of H^* are easily computable by the rule (15.5.12) from the relations $\phi_i(x, y, \cdots) = 1$ defining H. Hence the u_i of (15.5.11) are restricted to quantities of H^* satisfying (15.5.14). If we can show conversely that u_i satisfying (15.5.14) yield conditions (15.5.11), we shall have reduced the determination of the conditions (15.5.11) to the solution of (15.5.14). The proof of this, given here, will depend on methods due to W. Magnus [2].

THEOREM 15.5.2. *Given groups H and N. Conditions for the existence of a central extension of N by H are that there be automorphisms $a \leftrightarrows a^h$ associated with elements of H satisfying* (15.5.1); *that elements α_i of C, the center of N, exist with $\phi_i(\bar{\bar{x}}, \bar{\bar{y}}, \cdots) = \alpha_i$, $i = 1, \cdots, r$, where $\phi_i(x, y, \cdots) = 1$ are the defining relations of H; and that* (15.5.11) *hold for the α_i, where the u_i are any elements satisfying* (15.5.14) *in H^*.*

Proof: The preceding discussion has shown all parts of the theorem except that every set of u_i satisfying (15.5.14) determines a condition (15.5.11).

Consider the free group F_1 generated by $x, y, \cdots$, as discussed in §15.4, and let $H = F_1/W$, where W is the least normal subgroup containing $\phi_i(x, y, \cdots)$. Let W' be the derived group of W. Then W' as a characteristic subgroup of W will be a normal subgroup of F_1. Here $T = F_1/W'$ will be the group with the properties that (1) T is generated by $x, y, \cdots$; (2) T has a normal subgroup $V = W/W'$ such that $T/V = H$; and (3) V is Abelian. Finally, it is clear that any group with these properties is a homomorphic image of T, since any such group must be a homomorphic image of F_1 in which the elements of W' are mapped onto the identity. We shall use a lemma for our proof, postponing the proof of the lemma until the end of the main proof.

LEMMA 15.5.1. *Given matrices of the form* $\begin{pmatrix} h, & 0 \\ L, & 1 \end{pmatrix}$ *with* $h \in H$ *and* L *a linear form in indeterminates with coefficients from* H^* *subject to the product rule,*

$$\begin{pmatrix} h_1, & 0 \\ L_1, & 1 \end{pmatrix}\begin{pmatrix} h_2, & 0 \\ L_2, & 1 \end{pmatrix} = \begin{pmatrix} h_1h_2 & , 0 \\ L_1h_2 + L_2, & 1 \end{pmatrix}.$$

Then, corresponding to $x, y, \cdots$, *we have matrices* $\bar{x} = \begin{pmatrix} x, & 0 \\ t_x, & 1 \end{pmatrix} \cdots$, *and these matrices generate a group isomorphic to* $T = F_1/W'$.

Note that since matrices $\begin{pmatrix} 1, & 0 \\ L, & 1 \end{pmatrix}$ generate an additive, and hence Abelian, group, this group is in any event a homomorphic image of T.

In a central extension $\overline{H}$ is an image of T. Hence, if a relation

$$\phi_i(\bar{x}, \bar{y}, \cdots)^{u_i} = 1, \quad u_i \in H^* \tag{15.5.15}$$

holds in T, then the corresponding relation (15.5.11) must hold in $\overline{H}$.

Assuming the lemma we have in T as elements of V,

$$\phi_i(\bar{x}, \bar{y}, \cdots) = \begin{pmatrix} 1 & , 0 \\ L_i, & 1 \end{pmatrix} \qquad i = 1, \cdots, r, \tag{15.5.16}$$

with L_i a linear form in $t_x, t_y, \cdots$ with coefficients from H^*. Let us adjoin to V further elements:

$$\xi = \begin{pmatrix} 1, & 0 \\ t_\xi, & 1 \end{pmatrix}, \quad \eta = \begin{pmatrix} 1, & 0 \\ t_\eta, & 1 \end{pmatrix}, \cdots, \tag{15.5.17}$$

with t_ξ, t_η being new indeterminates. With $\bar{u}$ as before ($u \in H$),

$$\bar{u}^{-1}\xi\bar{u} = \begin{pmatrix} 1 & , 0 \\ t_\xi u, & 1 \end{pmatrix}. \tag{15.5.18}$$

Thus, adjoining the elements of (15.5.17) to V, we have adjoined an operator-free group. Now

(15.5.19)
$$\xi\bar{x} = \begin{pmatrix} x & , 0 \\ t_\xi x + t_x, & 1 \end{pmatrix}.$$

Hence we obtain $\phi_i(\xi\bar{x}, \eta\bar{y}, \cdots)$ by substituting in the L_i of (15.5.16), replacing t_x by $t_\xi x + t_x$, and so on. Hence, from

(15.5.20)
$$\phi_i(\xi\bar{x}, \eta\bar{y}, \cdots) = \phi_i(\bar{x}, \bar{y}, \cdots)\xi^{x_i}\eta^{y_i} \cdots$$

and the linearity of the L_i, we have

(15.5.21)
$$\xi^{x_i}\eta^{y_i} \cdots = \begin{pmatrix} 1 & , 0 \\ L_i(t_\xi x), & 1 \end{pmatrix}.$$

Hence if the equations $\sum_i x_i u_i = 0$, etc., of (15.5.14) hold, then

(15.5.22)
$$\sum_i L_i(t_\xi x) u_i = 0.$$

Here, since the t_ξ were indeterminates satisfying no relations, it must follow that

(15.5.23)
$$\sum_i L_i u_i = 0$$

for any arguments for the L_i. Applying this to (15.5.16), it follows that

(15.5.24)
$$\prod_i \phi_i(\bar{x}, \bar{y}, \cdots)^{u_i} = 1.$$

Since this relation holds in T, it must also hold in $\bar{H}$, and so we have shown that $\prod_i \alpha_i^{u_1} = 1$ in $\bar{H}$ whenever (15.5.14) holds. Hence (15.5.11) is a consequence of (15.5.14), and the proof of the theorem is complete except for the establishment of the lemma.

Proof of the Lemma: With $H = F_1/W$, suppose coset representatives of W chosen so that they are the earliest possible with respect to an alphabetical ordering of the elements of F_1. Then the same alphabetical ordering may be carried over to H, and if $h = h(\mathbf{x}, \mathbf{y}, \cdots)$ is the earliest element the coset Wh, then $h = h(x, y, \cdots) \in H$ is a canonical form for h as its earliest alphabetical expression. Hence the same form may be used for an element of H and the corresponding coset representatives of W, and we may speak of the length of an element of H, this being the length of its canonical form. Now consider the correspondences with matrices

$$\bar{x} \to \begin{pmatrix} x, & 0 \\ t_x, & 1 \end{pmatrix}, \quad y \to \begin{pmatrix} y, & 0 \\ t_y, & 1 \end{pmatrix}, \cdots,$$

with $x, y, \cdots$ generators of H and the rule of composition of the matrices

$$\begin{pmatrix} h_1, & 0 \\ L_1, & 1 \end{pmatrix}\begin{pmatrix} h_2, & 0 \\ L_2, & 1 \end{pmatrix} = \begin{pmatrix} h_1h_2, & 0 \\ L_1h_2 + L_2, & 1 \end{pmatrix}$$

with $h_1, h_2 \in H$ and L_1, L_2 linear forms $t_x, t_y, \cdots$ with coefficients from H^*. As remarked before, the group K generated by these matrices is at least a homomorphic image of T, since $\begin{pmatrix} h, & 0 \\ L, & 1 \end{pmatrix} \to h$ is clearly a homomorphism of K onto H, and the kernel of this homomorphism consists of elements $\begin{pmatrix} 1, & 0 \\ L, & 0 \end{pmatrix}$, which form an additive Abelian group.

By Theorem 7.2.3, W is a free subgroup of F_1, with free generators those elements c_{ij},

$$\text{(15.5.25)} \qquad c_{ij} = h_i x h_j^{-1} \neq 1, \quad h_j = \phi(h_i x)$$

in F_1, x a generator of F_1.

Moreover, by Lemma 7.2.3, h_i does not end in $\mathbf{x}^{-1}$ nor $\boldsymbol{h}_j$ in $\mathbf{x}$. The group W' will be the group given by all commutators of the c_{ij}, and W/W' will be the free Abelian group with the c_{ij} as a basis modulo W'. It will follow that K is a faithful representation of $T = F_1/W'$ if we can show that the elements c_{ij} corresponding to $\bar{h}_i\bar{x}\bar{h}_j^{-1}$ are independent in K. In the mapping $F_1 \to H$ we have $\boldsymbol{c}_{ij} \to 1$, $\boldsymbol{h}_i \to h_i$, $\boldsymbol{h}_j \to h_j$ and $\mathbf{x} \to x$, and so, $h_i x = h_j$ in H. Now let

$$\bar{h}_i \to \begin{pmatrix} h_i & , 0 \\ L(h_i), & 1 \end{pmatrix} \qquad \bar{x} \to \begin{pmatrix} x, & 0 \\ t_x, & 1 \end{pmatrix}$$

$$\bar{h}_j \to \begin{pmatrix} h_j & , 0 \\ L(h_j), & 1 \end{pmatrix}.$$

Then

$$\bar{c}_{ij} \to \begin{pmatrix} h_i x h_j^{-1} & , 0 \\ L(h_i)xh_j^{-1} + t_x h_j^{-1} - L(h_j)h_j^{-1}, & 1 \end{pmatrix} \tag{15.5.26}$$

$$= \begin{pmatrix} 1 & , 0 \\ L(h_i)h_i^{-1} + t_x h_j^{-1} - L(h_j)h_j^{-1}, & 1 \end{pmatrix},$$

using $h_i x = h_j$ in H.

We must examine more closely the linear form $L(h_i)$ occurring in the matrix for an element $\bar{h}_i$. Here

$$\bar{x} \to \begin{pmatrix} x, & 0 \\ t_x, & 1 \end{pmatrix} \qquad \bar{x}^{-1} \to \begin{pmatrix} x^{-1} & , 0 \\ -t_x x^{-1}, & 1 \end{pmatrix}.$$

Let us write $q(a)\ t_a$ if $a = x$ is a generator and $(qa) = -t_a a^{-1}$ if $a^{-1} = x$ is a generator. Then if $\bar{h} = a_1 a_2, \cdots, a_r$ is any word where each a_i is one of $\bar{x}, \bar{y}, \cdots$ or $\bar{x}^{-1}, y^{-1}, \cdots$, we shall have

$$\bar{h} \to \begin{pmatrix} h & , 0 \\ L(h), & 1 \end{pmatrix},$$

with

(15.5.27)

$$L(h) = q(a_1)a_2 \cdots a_r + q(a_2)a_3 \cdots a_r + \cdots q(a_{r-1})a_r + q(a_r),$$

where $h = a_1 a_2 \cdots a_r$.

This formula is easily established by induction on r and the product rule for the matrices. We now note from (15.5.27) the further rule:

$$L(h)h^{-1} = q(a_1)a_1^{-1} + q(a_2)a_2^{-1}a_1^{-1} + \cdots + q(a_r)a_r^{-1}a_{r-1}^{-1} \cdots a_1^{-1}. \tag{15.5.28}$$

If h is in canonical form we note that $q(a_i)$ is multiplied by the inverse of $a_1 a_2, \cdots, a_i$, which is, by the Schreier property of the represent-tives of W, again in canonical form. Thus as a basis for the group ring H^*, it is convenient to use the inverses of canonical forms of elements of H.

With each c_{ij} of W there is a unique h_j and x. Hence we may associate c_{ij} with the term $t_x h_j^{-1}$. This term may be characterized by noting that h_j is in canonical form, but that $h_j x^{-1}$ although in reduced form is not in canonical form, being equal to the canonical form h_i. But in (15.5.26) the linear form $L(h_i)h_i^{-1} + t_x h_j^{-1} - L(h_j)h_j^{-1}$

contains no other term of this type. For, by (15.5.28), the other terms arising from $L(h_i)h_i^{-1}$ or $L(h_j)h_j^{-1}$ are of the type $q(a_k)a_k^{-1} \cdots a_1^{-1}$, where $a_1 \cdots a_k$ is an initial section of h_i or h_j, and so, by the Schreier condition on the h's, will itself be an h in canonical form. Here if $a_k = y$, a generator, $q(a_k)a_k^{-1} \cdots a_1^{-1} = t_y y^{-1} \cdots a_1^{-1} = t_y h^{-1}$, where h ends in y so that hy^{-1} is not in reduced form. But if $a_k = y^{-1}$, where y is a generator, $q(a_k)a_k^{-1} \cdots a_1^{-1} = -ty \cdot a_{k-1}^{-1} \cdots a_1^{-1} = -ty\, h^{-1}$, where $hy^{-1} = a_1 \cdots a_k$ is in canonical form. Thus the term $t_x h_j^{-1}$ is the only term of its type in the linear form associated with $\overline{c_{ij}}$. Also, different, $\overline{c_{ij}}$'s yield different associated terms. Hence the linear forms $L(c_{ij})$ are linearly independent and the c_{ij} generate a free Abelian group, which is of course isomorphic to W/W'. Hence the group K of matrices generated by the matrices corresponding to $\bar{x}, \bar{y}, \cdots$, etc., is a faithful representation of $T = F_1/W'$ and the lemma is proved.

15.6. Double Modules.

Let Ω be any multiplicative group and let A be a *double Ω-module*, i.e., an Abelian group written additively which satisfies the following conditions:

1) A admits Ω as group of operators both on the right and on the left, so that ξa and $a\xi$ are uniquely determined elements of A for given $a \in A$ and $\xi \in \Omega$.

2) Distributivity,

$$\xi(a_1 + a_2) = \xi a_1 + \xi a_2$$
$$(a_1 + a_2)\xi = a_1\xi + a_2\xi,$$

whence

$$-\xi a = \xi(-a), \quad -a\xi = (-a)\xi$$
$$\xi 0 = 0\xi = 0.$$

3) $1a = a1 = a$ where 1 is the unit element of Ω.

4) Associativity, $\xi(\eta a) = (\xi\eta)a$, $(\xi a)\eta = \xi(a\eta)$, and $(a\xi)\eta = a(\xi\eta)$.

These rules are to hold for all $a_1, a_2, a \in A$, and all $\xi, \eta \in \Omega$.

Effectively, then, a double Ω-module is the same as an additive Abelian group admitting the elements (ξ, η) of $\Omega \times \Omega$ as distributive operators.

In the applications, it often happens that Ω acts trivially on one side, e.g., on the left. This means that $\xi a = a$ for all $\xi \in \Omega$ and $a \in A$. In this case, we shall simply omit the left-hand operators. Call this the one-sided case.

For example, let A be a normal Abelian subgroup of some group G, and write $\Omega = G/A$. If $\xi = Au_\xi$, then $u_\xi^{-1}au_\xi$ depends only on a and ξ, but not on the choice of u_ξ in its coset. Hence we may write $u_\xi^{-1}au_\xi = a^\xi$ without ambiguity. We then have an example of the one-sided case, but with A written multiplicatively. In developing the general theorems of cohomology, however, it is more convenient to write A in additive notation.

15.7. Cochains, Coboundaries, and Cohomology Groups.*

Given a double Ω-modulo A, we define $C^n = C^n(A, \Omega)$ to be the additive group of all functions f of n variables which range independently over Ω, and taking values in A, subject to the condition

$$f(\xi_1, \cdots, \xi_n) = 0, \tag{15.7.1}$$

whenever at least one of the $\xi_i = 1$. The elements of C^n are called *n-dimensional cochains.* $C^0 = A$ by definition and a zero dimensional cochain is simply any element of A.

The *coboundary operator* δ maps each C^n into the next, C^{n+1}, in accordance with the rule

$$\begin{aligned}(\delta f)(\xi_0, \xi_1, \cdots, \xi_n) &= \xi_0 f(\xi_1, \cdots, \xi_n)\\ &+ \sum_{t=1}^{n} (-1)^t f(\xi_0, \xi_1, \cdots, \xi_{t-2}, \xi_{t-1}\xi_t, \xi_{t+1}, \cdots, \xi_n)\\ &+ (-1)^{n-1} f(\xi_0, \xi_1, \cdots, \xi_{n-1})\xi_n.\end{aligned} \tag{15.7.2}$$

Here $f \in C^n$ and it is immediately verified that $\delta f \in C^{n+1}$. The map $f \to \delta f$ is homomorphic with respect to addition. The only cases genuinely useful in group theory appear to be the cases $n = 0, 1, 2$. Here the coboundary formulae are

$$\begin{aligned}(\delta f)(\xi) &= \xi f - f\xi, \quad f = a \in A\\ &= \xi a - a\xi\\ (\delta f)(\xi, \eta) &= \xi f(\eta) - f(\xi\eta) + f(\xi)\eta,\\ (\delta f)(\xi, \eta, \zeta) &= \xi f(\eta, \zeta) - f(\xi\eta, \zeta) + f(\xi, \eta\zeta) - f(\xi, \eta)\zeta.\end{aligned} \tag{15.7.3}$$

THEOREM 15.7.1. *If f is any cochain, then $\delta^2 f = 0$.*

Proof: Choose n so that $f \in C^{n-2}$. Then $\delta f \in C^{n-1}$. Therefore when

* See Eilenberg and MacLane [1, 2] and MacLane [2].

we express $(\delta^2 f)(\xi_1, \xi_2, \cdots, \xi_n)$ in terms of the values of δf, using the definition, we obtain $n + 1$ terms with alternating signs, say,

$$u_0 - u_1 + u_2 - \cdots + (-1)^n u_n.$$

Each u_i, when expressed in terms of the values of f, is an alternating sum of n terms, which we may write as

$$u_i = u_{i0} - u_{i1} + \cdots + (-1)^{i-1} u_{i,i-1}$$
$$+ (-1)^i u_{i,i+1} + \cdots + (-1)^{i+j-1} u_{i,i+j} + \cdots.$$

Hence

$$\delta^2 f(\xi_1, \cdots, \xi_n) = \sum_{i<j} (-1)^{i+j-1} u_{ij} + \sum_{i>j} (-1)^{i+j} u_{ij},$$

with i and j running from 1 to n. But it is easy to verify that $u_{ij} = u_{ji}$ for all i, j. Thus the above sum vanishes.

If $f \in C^n$ and is such that $\delta f = 0$, then f is called an n-dimensional *cocycle*. These cocyces form the kernel $Z^n = Z^n(A, \Omega)$ of the homomorphism of C^n into C^{n+1} induced by δ.

If $f \in C^n$ and if there exists an element $g \in C^{n-1}$ such that $\delta g = f$, then f is called an n-dimensional *coboundary*. These coboundaries form the image $B^n = B^n(A, \Omega)$ of C^{n-1} under the mapping δ. We define $B^0 = 0$.

According to the theorem, *every coboundary is a cocycle*, so that $B^n \subseteq Z^n$ for all n. The quotient group Z^n/B^n is called the n-dimensional *cohomology group* of the double Ω-module A. We write it

$$H^n(A, \Omega) = Z^n/B^n.$$

In our definition of cochains $f(\xi_1, \cdots, \xi_n)$ we imposed the restriction (15.7.1) that the cochain vanish if one or more of the arguments is the identity. This is a desirable restriction in many cases, in particular in the application to factor sets as we have defined them. Let us call such cochains *normalized*. If the restriction (15.7.1) is omitted, we speak of *unnormalized cochains*. Theorem 15.7.1, is of course, valid in either case, since it makes no use of the restriction (15.7.1). The distinction is a matter of convenience, since we shall show that the cohomology groups of every dimension for unnormalized cochains are isomorphic to those for normalized cochains.

THEOREM 15.7.2. *The cohomology groups $H^n(A, \Omega)$ of every dimension n for unnormalized cochains are isomorphic to those for normalized cochains.*

Proof: Let us designate the normalized cochains, boundaries, and cocycles of dimension n by C^n, B^n, and Z^n, respectively, and for the unnormalized case use the designations C'^n, B'^n, and Z'^n.

For $n = 0$ and $n = 1$ we readily verify that $B^0 = B'^0 = 0$, $Z^0 = Z'^0$, and $B^1 = B'^1$, $Z^1 = Z'^1$, whence $H^0(A, \Omega)$ and $H^1(A, \Omega)$ are the same in both cases. The principal verification involved here is that if $f(\xi) \in Z'^1$, then $\xi f(\eta) - f(\xi\eta) + f(\xi)\eta = 0$, whence putting $\xi = \eta = 1$, we find that $f(1) = 0$, whence $f(\xi)$ is normalized, and so, $Z'^1 = Z^1$.

Now suppose $n > 1$. Clearly, $B^n \subseteq B'^n$ and $Z^n \subseteq Z'^n$. Hence a cohomology class for C^n, i.e., a coset of B^n in C^n, corresponds to a unique cohomology class for C'^n, namely, the coset of B'^n which contains it. This correspondence is, of course, a homomorphism of $H^n(A, \Omega)$ into $H'^n(A, \Omega)$. To prove isomorphism, we must show that this correspondence is one to one and for this two lemmas will suffice. Let us say that two cochains are *cohomologous* if their difference is a coboundary. Thus two cocycles are cohomologous if they belong to the same cohomology class.

LEMMA 15.7.1. *Every unnormalized cocycle is cohomologous to a normalized cocycle.*

LEMMA 15.7.2. *If the coboundary of some cochain is normalized, then it is the coboundary of a normalized cochain.*

Proof of the lemmas: Let us say that a cochain $f(x_1, \cdots, x_n)$ is i-normalized, $i = 0, \cdots, n$ if it is zero whenever one of the first i arguments is the identity. The 0-normalized cochains are then the unnormalized cochains c'^n and the n-normalized are the normalized cochains C^n. For $f(x_1, \cdots, x_n)$ let us define cochains $f = f_0$, and recursively,

$$f_{i+1} = f_i - \delta g_{i+1} \qquad i = 0, \cdots, n-1 \tag{15.7.4}$$

where

$$g_{i+1}(x_1, \cdots, x_{n-1}) = (-1)^i f_i(x_1, \cdots, x_i, 1, x_{i+1}, \cdots, x_{n-1}). \tag{15.7.5}$$

We note that $f = f_0$ and f_n differ by a coboundary, and also, since $\delta f_i = \delta f_{i+1}$, $f = f_0, f_1, \cdots, f_n$, all have the same coboundary δf.

LEMMA 15.7.3. *If δf is normalized, then f_i is i-normalized.*

This will prove both lemmas, since for Lemma 15.7.1, if f is an unnormalized cocycle, then $\delta f = 0$, which is trivially normalized whence f_n, which is cohomologous to $f_0 = f$, will be a normalized cocycle. For Lemma 15.7.2, if $g = \delta f$ is a coboundary and g is normalized, say, $g \in C^{n+1}$, then $g = \delta f_0, = \cdots = \delta f_n$, where f_n is normalized.

We prove Lemma 15.7.3 by induction on i, the statement being trivially true for $i = 0$. Suppose the lemma to be true for i, and consider it for $i + 1$, it being necessary to prove that

$$f_{i+1}(x_1, \cdots, x_i, 1, x_{i+2}, \cdots, x_n) = 0. \tag{15.7.6}$$

From the definition of f_{i+1} in (15.7.4), we have

$$\begin{aligned} & f_{i+1}(x_1, \cdots, x_i, 1, x_{i+2}, \cdots, x_n) \\ & = f_i(x_1, \cdots, x_i, 1, x_{i+2}, \cdots, x_n) \\ & - x_1 g_{i+1}(x_2, \cdots, x_i, 1, x_{i+2}, \cdots, x_n) \\ & + \sum_{j=1}^{i-1} (-1)^{j-1} g_{i+1}(x_1, \cdots, x_j x_{j+1}, \cdots, x_i, 1, x_{i+2}, \cdots, x_n). \\ & + (-1)^{i-1} g_{i+1}(x_1, \cdots, x_{i-1}, x_i \cdot 1, x_{i+2}, \cdots, x_n) \\ & + (-1)^i g_{i+1}(x_1, \cdots, x_i, 1 \cdot x_{i+2}, \cdots, x_n) \\ & + \sum_{j=i+2}^{n-1} (-1)^{j-1} g_{i+1}(x_1, \cdots, x_i, 1, x_{i+2}, \cdots, x_j x_{j+1}, \cdots, x_n) \\ & + (-1)^n g_{i+1}(x_1, \cdots, x_i, 1, x_{i+2}, \cdots, x_{n-1}) x_n. \end{aligned} \tag{15.7.7}$$

From (15.7.5), since by induction f_i is i-normalized, g_{i+1} is i-normalized. This means that in (15.7.7) the term with the factor x_1 on the left and the sum with $j = 1$ to $i - 1$ are all zero. The next two terms cancel. Now let us take the remaining terms replacing g_{i+1} according to its definition (15.7.5). This gives for (15.7.7)

$$\begin{aligned} & f_{i+1}(x_1, \cdots, x_i, 1, x_{i+2}, \cdots, x_n) \\ & = f_i(x_1, \cdots, x_i, 1, x_{i+2}, \cdots, x_n) \\ & + \sum_{j=i+2}^{n-1} (-1)^{i+j-1} f_i(x_1, \cdots, x_i, 1, 1, x_{i+2}, \cdots, x_j x_{j+1}, \cdots, x_n) \\ & + (-1)^{n+i} f_i(x_1, \cdots, x_i, 1, 1, x_{i+2}, \cdots, x_{n-1}) x_n. \end{aligned} \tag{15.7.8}$$

But, by hypothesis, $\delta f_i = \delta f$ is normalized, whence

(15.7.9) $$(-1)^{i+1}\delta f(x_1, \cdots, x_i, 1, 1, x_{i+2}, \cdots, x_n) = 0,$$

and since by induction f_i is i-normalized, the right-hand side of (15.7.8) consists of all the terms of the expansion of (15.7.9) which remain. Thus

(15.7.10) $$f_{i+1}(x_1, \cdots, x_i, 1, x_{i+2}, \cdots, x_n) = 0,$$

proving Lemma 15.7.3 by induction and thus Lemmas 15.7.1 and 15.7.2, and in turn, the theorem.

15.8. Applications of Cohomology to Extension Theory.

If A is a normal Abelian subgroup of some group G, let $\Omega = G/A$ be the factor group. If the coset $Au_\xi = \xi$ is an element of Ω, then for $a \in A$, $u_\xi^{-1}au_\xi$ depends only on a and ξ but not on the choice of u_ξ in its coset. Hence we may write $u_\xi^{-1}au_\xi = a\xi$ without ambiguity, and in this way Ω is a group of operators on the right for A and we regard Ω as acting trivially on the left. With the operators fixed, and A written additively, if we put $f(u, v) = (u, v)$ for the factors of a factor set, (15.2.2) becomes

(15.8.1) $$f(uv, w) + f(u, v)w = f(u, vw) + f(v, w).$$

Let us rearrange the terms thus

(15.8.2) $$f(v, w) - f(uv, w) + f(u, vw) - f(u, v)w = 0,$$

whence we see that a factor set is a cocycle of dimension two. From (15.2.3), the condition for the equivalence of two factor sets $f(u, v)$ and $f_1(u, v)$ is

(15.8.3) $$f_1(u, v) = f(u, v) + \alpha(v) - \alpha(uv) + \alpha(u)v,$$

or that f_1 and f differ by the coboundary $\alpha(v) - \alpha(uv) + \alpha(u)v$. Here we note that Ω operates trivially on the left. Hence the group of extensions is the second cohomology group $H^2(A, \Omega)$. We state this as a theorem.

THEOREM 15.8.1. *The group of extensions of an Abelian group A by a group Ω is the second cohomology group $H^2(A, \Omega)$, where*

1) *Ω operates trivially on the left in A.*
2) *On the right Ω operates to induce automorphisms in A.*
3) *Factor sets $f(u, v)$ are the cocycles of Z^2.*
4) *Equivalent factor sets differ by coboundaries of B^2.*

The choice of the identity as the representative of A in writing G as a sum of cosets of A leads to the normalization $f(1, 1) = 0$. Putting $u = v = 1$ in (15.8.1), we have

$$f(1, w) + f(1, 1)w = f(1, w) + f(1, w), \tag{15.8.4}$$

whence

$$f(1, w) = 0. \tag{15.8.5}$$

Similarly, putting $v = w = 1$, we have

$$f(1, 1) - f(u, 1) + f(u, 1) - f(u, 1) = 0, \tag{15.8.6}$$

whence also

$$f(u, 1) = 0, \tag{15.8.7}$$

showing that we deal with normalized cocycles.

We shall prove a general theorem in cohomology which includes Theorem 15.2.1 as a special case.

Suppose that Ω is finite of order m. Then for each $n > 0$ we can define an additive homomorphism σ which maps C^n into C^{n-1} by the formula

$$(\sigma f)(x_2, \cdots, x_n) = \sum_{x \in \Omega} x^{-1}f(x, x_2, \cdots, x_n). \tag{15.8.8}$$

Here $f \in C^n$, and it is immediately clear that $\sigma f \in C^{n-1}$.

Write $g = \sigma f$ and let us calculate $(\delta g)(x_1, \cdots, x_n)$:

$$\begin{aligned} \delta g(x_1, &\cdots, x_n) \\ &= x_1 \sum x^{-1}f(x, x_2, \cdots, x_n) \\ &- \sum x^{-1}f(x, x_1x_2, \cdots, x_n) \\ &\cdots\cdots\cdots\cdots\cdots\cdots \\ &+ (-1)^{j-1} \sum x^{-1}f(x, x_1, \cdots, x_{j-1}x_j, \cdots, x_n) \qquad (15.8.9) \\ &\cdots\cdots\cdots\cdots\cdots\cdots \\ &+ (-1)^{n-1} \sum x^{-1}f(x, x_1, \cdots, x_{n-1}x_n) \\ &+ (-1)^n\Big[\sum x^{-1}f(x, x_1, \cdots, x_{n-1})\Big]x_n. \end{aligned}$$

The summation is over all $x \in \Omega$.

Now consider $(\delta f)(x, x_1, \cdots, x_n)$.

$$(15.8.10)\qquad \begin{aligned}(\delta f)(x, x_1, &\cdots, x_n)\\ &= xf(x_1, \cdots, x_n)\\ &- f(x\cdot x_1, x_2, \cdots, x_n)\\ &+ f(x, x_1\cdot x_2, \cdots, x_n)\\ &\cdot\quad\cdot\quad\cdot\quad\cdot\quad\cdot\quad\cdot\quad\cdot\\ &+ (-1)^n f(x, x_1, \cdots, x_{n-1}x_n)\\ &+ (-1)^{n+1} f(x, x_1, \cdots, x_{n-1})x_n.\end{aligned}$$

Let us now calculate, using (15.8.10),

$$(15.8.11)\qquad \begin{aligned}\sigma(\delta f)(x_1, \cdots, x_n) &= \sum_{x\in\Omega} x^{-1}(\delta f)(x, x_1, \cdots, x_n)\\ &= mf(x_1, \cdots, x_n)\\ &- \sum x^{-1}f(xx_1, x_2, \cdots, x_n)\\ &+ \sum x^{-1}f(x, x_1\cdot x_2, \cdots, x_n)\\ &\cdot\quad\cdot\quad\cdot\quad\cdot\quad\cdot\quad\cdot\quad\cdot\\ &+ (-1)^n \sum x^{-1}f(x, x_1, \cdots, x_{n-1}x_n)\\ &+ (-1)^{n+1} \sum x^{-1}f(x, x_1, \cdots, x_{n-1})x_n.\end{aligned}$$

In the sum $S = \sum_{x\in\Omega} x^{-1}f(xx_1, x_2, \cdots, x_n)$, put $y = xx_1$, whence

$$S = x_1 \sum y^{-1}f(y, x_2, \cdots, x_n) = x_1\, \sigma f(x_2, \cdots, x_n),$$

since as x_1 is a fixed element of Ω, y ranges over Ω when x does. Hence (15.8.11) becomes

$$(15.8.12)\qquad (\sigma(\delta f)) = mf - \delta(\sigma f).$$

This gives:

THEOREM 15.8.2. *If $f \in C^n$, then $\sigma(\delta f) + \delta(\sigma f) = mf$.*

COROLLARY 15.8.1. *If $f \in Z^n$, then $mf \in B^n$.*

For $f \in Z^n$ means that $\delta f = 0$ so that mf is the coboundary of σf. We conclude that *if Ω has order m then every element of the cohomology group $H^n = Z^n/B^n$ has an order dividing m.* Theorem 15.2.1 is the special case $n = 2$ of this result.

We have a further theorem, due to Gaschütz [1] for factor sets and in more general form given by Eckmann [1]. This relates the cohomology of a group Ω and a subgroup B. We suppose that B is of finite index m in Ω.

$$\Omega = B\cdot 1 + Bs_2 + \cdots + Bs_m, \quad s_1 = 1. \tag{15.8.13}$$

Here if $a_1, a_2, \cdots, a_n$ are elements of Ω, we write $\overline{s_i a_1} = s_{i1}$, where as usual the bar designates the coset representative. Then write also

$$\overline{s_{i1}a_2} = s_{i2}, \cdots, \overline{s_{in-1}a_n} = s_{in}.$$

We define the *transfer*, $T(f(a_1, \cdots, a_n))$ of $f(a_1, \cdots, a_n) \in C^n$ by the formula

$$T(f(a_1, \cdots, a_n)) = \sum_{i=1}^{m} s_i^{-1} f(s_i a_1 s_{i1}^{-1}, s_{i1}a_2 s_{i2}^{-1}, \cdots, s_{in-1}a_n s_{in}^{-1})s_{in}. \tag{15.8.14}$$

Note that $s_{ij-1}a_j s_{ij}^{-1} \in B$ in every case, whence for an $f \in C^n(A, \Omega)$, Tf is in the subgroup $\Omega C^n(A, B)\Omega$.

THEOREM 15.8.3 (THEOREM OF GASCHÜTZ). *If $f(a_1, \cdots, a_n) \in Z^n$ and B is a subgroup of index m in Ω, then*

$$Tf(a_1, \cdots, a_n) \equiv mf(a_1, \cdots, a_n) \bmod B^n.$$

COROLLARY 15.8.2. *The cohomology class of the transfer is independent of the choice of the coset representatives s_i in* (15.8.13).

Proof of the Theorem: Consider

$$\sum_{i=1}^{m} (\delta f)(s_i^{-1}, s_i a_1 s_{i1}^{-1}, \cdots, s_{in-1}a_n s_{in}^{-1})s_{in}$$
$$- \sum_{i=1}^{m} (\delta f)(a_1, s_{i1}^{-1}, s_{i1}a_2 s_{i2}^{-1}, \cdots, s_{in-1}a_n s_{in}^{-1})s_{in}$$
$$\cdots\cdots\cdots\cdots\cdots\cdots$$
$$+ (-1)^{j-1} \sum_{i=1}^{m} (\delta f)(a_1, \cdots, a_{j-1}, s_{ij-1}^{-1}, s_{ij-1}a_j s_{ij}^{-1}, \cdots,)s_{in} \tag{15.8.15}$$
$$+ (-1)^{j} \sum_{i=1}^{m} (\delta f)(a_1, \cdots, a_{j-1}, a_j, s_{ij}^{-1}, s_{ij}a_{j+1}s_{ij+1}^{-1}, \cdots)\, s_{in}$$
$$\cdots\cdots\cdots\cdots\cdots\cdots$$
$$+ (-1)^n \sum_{i=1}^{m} (\delta f)(a_1, \cdots, a_n, s_{in}^{-1})s_{in} = 0.$$

This is a sum of terms each of which is zero, since $f \in Z^n$, whence $\delta f = 0$. Consider the effect of expanding the coboundary in each of the $n + 1$ lines above and taking the sum from $i = 1$ to m for corresponding terms of the δf's. The first terms of the first line yield

(15.8.16) $$\sum s_i^{-1} f(s_i a_1 s_{i1}^{-1}, \cdots, s_{in-1} a_n s_{in}^{-1}) s_{in} = Tf(a_1, \cdots, a_n).$$

The last terms of the last line yield

(15.8.17)
$$(-1)^n(-1)^{n+1} \sum f(a_1, \cdots, a_n) s_{in}^{-1} \cdot s_{in} = -mf(a_1, \cdots, a_n).$$

Since $s_{ij-1}^{-1} \cdot s_{ij-1} a_j s_{ij}^{-1} = a_j \cdot s_{ij}^{-1}$, the $(j + 1)$st terms of the jth and $(j + 1)$st lines cancel each other, for $j = 1, \cdots, n$. Let us now take the first j terms of the $(j + 1)$st line and terms $j + 2, \cdots, n + 2$ of the jth line together. These are given by

$$(-1)^j a_1 \quad \sum_{i=1}^{m} f(a_2, \cdots, a_j, s_{ij}^{-1}, s_{ij} a_{j+1} s_{ij+1}^{-1}, \cdots) s_{in}$$

$$(-1)^{j+1} \quad \sum_{i=1}^{m} f(a_1 a_2, \cdots, a_j, s_{ij}^{-1}, s_{ij} a_{j+1} s_{ij+1}^{-1}, \cdots) s_{in}$$

$$\cdots\cdots\cdots\cdots\cdots\cdots$$

(15.8.18) $$(-1)^{2j} \quad \sum_{i=1}^{m} f(a_1, \cdots, a_{j-1} a_j, s_{ij}^{-1}, s_{ij} a_{j+1} s_{ij+1}^{-1}, \cdots) s_{in}$$

$$(-1)^{2j+1} \quad \sum_{i=1}^{m} f(a_1, \cdots, a_{j-1}, s_{ij-1}^{-1}, a_j a_{j+1} s_{ij}^{-1}, \cdots) s_{in}$$

$$\cdots\cdots\cdots\cdots\cdots\cdots$$

$$(-1)^{n+j+1} \sum_{i=1}^{m} f(a_1, \cdots, a_{j-1}, s_{ij-1}^{-1}, s_{ij-1} a_j s_{ij}^{-1}, \cdots) s_{in-1} a_n.$$

But if for arguments $u_1, \cdots, u_{n-1}$ we define a function $F_j(u_1, \cdots, u_{n-1}) \in C^{n-1}$ by the formula

(15.8.19) $$F_j(u_1, \cdots, u_{n-1}) = \sum_{i=1}^{m} f(u_1, \cdots, u_{j-1}, \sigma_i^{-1}, \sigma_i u_j \sigma_{ij}^{-1}, \cdots, \sigma_{in-1} u_{n-1} \sigma_{in}^{-1}) \sigma_{in}$$

where $\sigma_i = s_i$ and recursively $\sigma_{it} = \overline{\sigma_{it-1}u_t}$, then we see that the terms of (15.8.18) are the coboundary $(-1)^j(\delta F_j)(a_1, \cdots, a_n)$, since, summing over i, s_{ij} or s_{ij-1} will serve equally well as σ_i. Letting j run from 1 to n, the coboundaries $(-1)^j(\delta F_j)(a^1, \cdots, a_n)$ account for all terms of (15.8.15) except those which cancel and those of (15.8.16) and (15.8.17), giving our theorem.

The group theoretical form of the theorem of Gaschütz is the following:

THEOREM 15.8.4 (THEOREM OF GASCHÜTZ). *Let $F = [(u, v)]$, $u, v \in H$, $(u, v) \in A$ be the factor set of an H-χ extension of an Abelian group A by a finite group H. Let B be a subgroup of index m in H, and*

$$H = B \cdot 1 + Bs_2 + \cdots + Bs_m, \quad s_1 = 1.$$

Then

$$(u, v)^m \sim \prod_{i=1}^{m} (s_i u\, \overline{s_i u}^{-1}, \quad \overline{s_i u} v \overline{s_i uv}^{-1})^{\overline{s_i uv}}$$

The arguments $s_i u \overline{s_i u}^{-1}$ and $\overline{s_i u} v \overline{s_i uv}^{-1}$ are, of course, elements of B.

COROLLARY 15.8.3. *If $(x, y) = 1$ whenever $x, y \in B$, then $(u, v)^m \sim 1$.*

There are many consequences of this theorem, but a very useful consequence is the way in which this relates the $H - \chi$ extensions of A to the $S(p) - \chi$ extensions of A, where $S(p)$ is a Sylow p-subgroup of H. Let H be of order $n = p^e m$, where $S(p)$ is of order p^e. Let $E = E(H)$ be the group of $H - \chi$ extensions of A, as defined in §15.2. Each element of E is a class of equivalent factor sets $F_i = [(u, v)_i]$. By Theorem 15.2.1 every element of E has order dividing n. Thus E is a periodic Abelian group and is the direct product of its Sylow subgroups $E(p)$.

THEOREM 15.8.5. *A Sylow p-subgroup $E(p)$ of $E = E(H)$, the group of $H - \chi$ extensions of an Abelian group A by a finite group H, is isomorphic to E_p, the group of $S(p) - \chi$ extensions given by restricting factor sets $F = [(u, v)]$ of $H - \chi$ extensions of A to arguments (x, y), $x, y \in S(p)$, where $S(p)$ is a Sylow p-subgroup of H.*

Proof: On the factor sets $F = [(u, v)]$ for $H - \chi$ extensions of A, let us define a p-equivalence

$$(u, v) \underset{p}{\sim} (u, v)_1$$

if, when restricted to arguments $x, y \in S(p)$ to yield an $S(p) - \chi$ extension we have

$$(x, y) \sim (x, y)_1.$$

This is readily seen to be a true equivalence. Furthermore, let E_1 be the subgroup of E of those factor sets $F = [(u, v)]$ for which $(x, y) \sim 1$, x, y restricted to $S(p)$. Then elements of E correspond to p-equivalent factor sets if, and only if, they are in the same coset of E_1. Hence E/E_1 is isomorphic to the group E_p of $S(p)$-χ extensions obtained by restricting the factor sets $F = [(u, v)]$ to arguments $x, y \in S(p)$. By the corollary to the Theorem of Gaschütz, with $S(p)$ as the subgroup B, every element of E_1 is of order dividing m, the index of $S(p)$, and by Theorem 15.2.1, every element of E_p is of order dividing p^e. Since $(p^e, m) = 1$, it follows that both E_p and $E(p)$, a Sylow p-subgroup of E are isomorphic to E/E_1, and hence to each other, thus proving our theorem.

THEOREM 15.8.6. *An $H - \chi$ extension of A splits over A if, and only if, for each prime p dividing the order of H, the extension splits when restricted to some Sylow p-subgroup $S(p)$ of H.*

Proof: Trivially, the splitting of H over A implies the splitting of every $S(p)$ over A. We must prove the converse. Let $F = [(u, v)]$ be the factor set determining the $H - \chi$ extension. By hypothesis $(u, v) \sim (u, v)_1$, where $(x, y)_1 = 1$ for $x, y \in S(p)$. By the corollary we have $(u, v)^m \sim (u, v)_1{}^m \sim 1$, where $n = p^e m$. But this must happen for every p dividing n. The different m's for which we have $(u, v)^m \sim 1$ have greatest common divisor 1, and so $(u, v) \sim 1$ and H splits over A, the conclusion of our theorem.

16. GROUP REPRESENTATION

16.1. General Remarks.

We shall call a *representation* of a group G any homomorphism α of G into some group W. Of particular value are representations of G by groups W which lend themselves readily to calculation. Thus the permutation representations of a group G discussed in Chap. 5 are homomorphisms of G into a symmetric group S_n.

Instead of a symmetric group as a representing group, we may turn to the endomorphisms of a vector space V over a field F. Those endomorphisms which are one-to-one form a group, which, if V is of finite dimension n over F, is called the *full linear group* $L_n(F)$ and may be expressed by the nonsingular $n \times n$ matrices over F. Here we consider representations of G by linear transformations. In such a representation we may regard the elements of G as operators on V. In this context the subspaces of V, taken into themselves by the linear transformations corresponding to G, are the invariant subspaces of the representation, and regarding V as an additive group with both F and G as operators, these are the admissible subgroups N.

The full set of endomorphisms of a vector space V forms a ring. Thus a linear representation of G over V leads through addition and scalar multiplication to a linear representation of R_G, the group ring of G over F, and similarly, any admissible subgroup N of V yields a representation of R_G along with that of G. Hence it is not surprising that there is a close relationship between the decomposition of the group ring R_G and the decomposition of linear representations. Historically, the theory of group representations and the structure theory of rings were developed separately, and only in comparatively recent times has the close relationship between these two theories been recognized.

16.2. Matrix Representation. Characters.*

Definition: *A matrix representation of degree n of a group G is a function ρ defined on G with values in the full linear group $L_n(F)$, for some field F, such that $\rho(xy) = \rho(x)\rho(y)$ for all $x, y \in G$.*

* The properties of matrices, determinants, and the full linear group that are assumed here can be found in Birkhoff and MacLane [1], Chapters VI through IX.

Note that by this definition $\rho(x)$ is a nonsingular matrix and that $x \to \rho(x)$ is a homomorphism of G into $L_n(F)$. Here we must have $\rho(1) = I_n$, the unit $n \times n$ matrix, and thus $\rho(x^{-1}) = [\rho(x)]^{-1}$, the matrix inverse. The kernel K of the homomorphism $x \to \rho(x)$ will be a normal subgroup of G and the matrices $\rho(x)$ will represent G/K faithfully. The representation will be faithful only if the kernel K is 1.

DEFINITION: *The character* χ *of a representation* ρ *is the function defined on* G *by* $\chi(x) = \textit{trace}\ \rho(x)$.

Thus the characters are numbers of the field K. If the representation is of degree 1, then $\chi = \rho$.

We shall say that two representations ρ and ρ^* are *equivalent* if there is a nonsingular matrix $S \in L_n(F)$ such that $\rho^*(x) = S^{-1}\rho(x)S$ for every $x \in G$. We note that if S is any nonsingular matrix of $L_n(F)$ and $\rho(x)$ is a representation in $L_n(F)$, then $S^{-1}\rho(x)S$ is also a representation $\rho^*(x)$. Indeed, if we regard $\rho(x)$, $x \in G$ as a group of linear transformations of the vector space V over F into itself with basis $u_1, u_2, \cdots, u_n$, and if

$$\begin{pmatrix} u_1 \\ u_2 \\ \cdot \\ \cdot \\ \cdot \\ u_n \end{pmatrix} = S \begin{pmatrix} v_1 \\ v_2 \\ \cdot \\ \cdot \\ \cdot \\ v_n \end{pmatrix}$$

then $S^{-1}\rho(x)S = \rho^*(x)$, $x \in G$ is the group of the same linear transformations of V in terms of the basis $v_1, \cdots, v_n$.

LEMMA 16.2.1. *Characters are class functions, i.e., conjugate elements have the same character.*

LEMMA 16.2.2. *Equivalent representations have the same characters.*

For if A is a matrix of degree n, its characteristic polynomial is by definition $f(\lambda) = |A - \lambda I| = (-1)^n[\lambda^n - a_1\lambda^{n-1} \cdots + (-1)^n a_n]$. Here the coefficient a_1 is the trace of A, $a_1 = Tr(A)$, and $a_n = |A|$ the determinant of A. Now if T is a nonsingular matrix $|T^{-1}AT - \lambda I| = |T^{-1}(A - \lambda I)T| = |T^{-1}| \cdot |A - \lambda I| \cdot |T| = |A - \lambda I|$. Thus A and $T^{-1}AT$ have the same characteristic polynomial and *a fortiori* the same trace. Hence $\rho(y^{-1}xy) = \rho(y)^{-1}\rho(x)\rho(y)$ and $\rho(x)$

have the same trace, and so, $\chi(y^{-1}xy) = \chi(x)$ and the character is a class function. In the same way $\rho^*(x) = S^{-1}\rho(x)S$ and $\rho(x)$ have the same trace and so equivalent representations have the same characters.

We recall that a vector space V (or "linear space") over a field F is given by the following laws:

V has a *binary addition:*

$$\text{For } \alpha, \beta \in V, \quad \alpha + \beta \in V.$$

V has a *scalar product* $c\alpha$ for $c \in F$, $\alpha \in V$.

These satisfy

V1) V is an Abelian group under addition.

V2) $c(\alpha + \beta) = c\alpha + c\beta, \quad (c + c')\alpha = c\alpha + c'\alpha.$

V3) $(cc')\alpha = c(c'\alpha); \quad 1\alpha = \alpha.$

Here $\alpha, \beta \in V$, $c, c' \in F$, 1 unit of F.

Vectors $u_1, \cdots, u_r$ of V are said to be *linearly independent* if

$$a_1u_1 + \cdots + a_ru_r = 0, \quad a_i \in F$$

implies $a_1 = \cdots = a_r = 0$. Moreover, $u_1, \cdots, u_n$ are a *basis* for V if they are linearly independent and if every $u \in V$ can be expressed as

$$u = b_1u_1 + \cdots + b_nu_n, \quad b_i \in F.$$

If V has a basis, then every basis has the same number of elements, and this number is called the *dimension* of the vector space.

We shall define an *F-G module M* as a vector space V over F which admits the elements of G as operators on V, the rule being

1) $(u + v)g = ug + vg$, $u, v \in V$, $g \in G$.
2) $u(g_1g_2) = (ug_1)g_2$, $u \in V$, $g_1, g_2 \in G$.
3) $u \cdot 1 = u$, $u \in V$, 1 the unit of G.
4) $(au)g = a(ug)$, $a \in F$, $u \in V$, $g \in G$.

We shall also call M a *representation module* for G.

By an *operator homomorphism* of one F-G module M_1 into another M_2 we mean a mapping $M_1 \to M_2$ such that

1) If $u_1 \to v_1$, $u_2 \to v_2$, then $u_1 + u_2 \to v_1 + v_2$.
2) If $u \to v$, $b \in F$, then $bu \to bv$.
3) If $u \to v$, $g \in G$, then $ug \to vg$.

An *operator isomorphism* of M_1 and M_2 is an operator homomorphism of M_1 onto M_2 which is one to one.

Now if M is an F-G module and has a basis $u_1, \cdots, u_n$ over F, then if for $x \in G$, we take the mapping $v \to vx$, $v \in V$, where $u_i \to u_i x = \sum_{j=1}^{n} a_{ij} u_j$, $i = 1, \cdots, n$, $\rho(x) = (a_{ij})$, $i, j = 1, \cdots, n$ will be a representation for G in M. Conversely, if ρ is a representation for G over a vector space V with basis $u_1, \cdots, u_n$, and if we have $\rho(x) = (a_{ij}) = [a_{ij}(x)]$, let us put

$$u_i x = \sum_{j=1}^{n} a_{ij}(x) u_j, \quad i = 1, \cdots, n$$

for every $x \in G$. Then, since $\rho(1) = I_n$ and $\rho(xy) = \rho(x)\rho(y)$, we see that this rule makes V into an F-G module M. Thus every F-G module of dimension n determines a representation of degree n of G, and conversely.

THEOREM 16.2.1. *Two F-G modules M_1 and M_2 give equivalent representations of G if, and only if, they are operator isomorphic.*

Proof: Suppose we have given two equivalent representations of G;

$$\rho(x) \; x \in G \quad \text{and} \quad S^{-1}\rho(x)S.$$

Then if $\rho(x) = [a_{ij}(x)]$, $x \in G$, this corresponds to a basis $u_1, \cdots, u_n$ of a vector space V with G as operators and the mapping $u_i \to u_i x = \sum a_{ij}(x) u_j$. We have observed already that this mapping

$$w \to wx \qquad w \in V$$

corresponds to the representation $\rho^*(x) = S^{-1}\rho(x)S$, $x \in G$ in terms of a basis $v_1, \cdots, v_n$, where

$$\begin{pmatrix} u_1 \\ \cdot \\ \cdot \\ \cdot \\ u_n \end{pmatrix} = S \begin{pmatrix} v_1 \\ \cdot \\ \cdot \\ \cdot \\ v_n \end{pmatrix}$$

If $S = (s_{ij})$, the equivalent representations $\rho(x)$ and $\rho^*(x)$ are operator isomorphic under the mapping determined by $u_i \to \sum_i s_{ij} v_j$, $i = 1, \cdots, n$. Conversely, suppose that two F-G modules M_1 and M_2 are operator isomorphic. To be isomorphic as vector spaces, M_1 and M_2 must have the same number of basis elements, say, $u_1, \cdots, u_n$ for M_1, and thus in the operator isomorphism $u_1, \cdots, u_n$ are mapped onto a basis $v_1, \cdots, v_n$ of M_2. With $u_i \to v_i$ and $u_ix \to v_ix$ then on these bases, M_1 and M_2 yield identical representations, since if

$$u_ix = \sum_{j=1}^{n} a_{ij}(x)u_j, \quad i = 1, \cdots, n,$$

we must also have

$$v_ix = \sum_{j=1}^{n} a_{ij}(x)v_j, \quad i = 1, \cdots, n.$$

Thus equivalent representations $\rho(x)$ *and* $\rho^*(x)$ correspond precisely to operator isomorphisms of representation modules.

16.3. The Theorem of Complete Reducibility.

Suppose that a representation module M has a submodule M_1 which is also a representation module. Then let us take a basis $u_1, \cdots, u_r$ for M_1 and complete this to a basis for M by taking further elements $u_{r+1}, \cdots, u_n$. The corresponding representation ρ is said to be *reducible*, and in terms of the basis $u_1, \cdots, u_n$, takes the form

$$\rho(x) = \left(\begin{array}{c|c} \sigma(x) & 0 \\ \hline \theta(x) & \tau(x) \end{array}\right),$$

and here σ and τ are representations of G of degrees r and $n - r$, respectively. The representation σ is associated with the F-G module M_1 with basis $u_1, \cdots, u_r$. What about τ? It is the representation defined by the basis $M_1 + u_{r+1}, \cdots, M_1 + u_n$ of the quotient module M/M_1. More generally, if

$$0 = M_0 \subset M_1 \subset M_2 \cdots \subset M_k = M$$

is a chain of submodules and we choose a basis of M adapted to this chain, then the corresponding representation ρ will take the form

$$\rho(x) = \begin{pmatrix} \rho_1(x) & & & \mathbf{0} \\ & \rho_2(x) & & \\ & & \ddots & \\ * & & & \rho_k(x) \end{pmatrix},$$

and $\rho_i(x)$ is the representation corresponding to a suitable basis of

M_i/M_{i-1}, namely, $M_{i-1} + u_j$, where u_j runs through the basis elements which belong to M_i but not to M_{i-1}. For the characters we clearly have

$$\chi(x) = \chi_1(x) + \chi_2(x) + \cdots + \chi_k(x).$$

If we choose our chain to be *maximal*, i.e., so that it cannot be further refined, then the M_i/M_{i-1} have no proper representation submodules and they give rise to *irreducible* representations ρ_i. As an immediate consequence we have:

LEMMA 16.3.1. *Every character is the sum of irreducible characters.*

LEMMA 16.3.2. *The irreducible constituents ρ_i are unique apart from order and operator isomorphism.*

Lemma 16.3.2 follows from the Jordan-Hölder theorem.

If a representation module M has a submodule M_1 which is a representation module, it may happen that there is a complementary representation submodule M_2, so that M is their direct sum, $M = M_1 \oplus M_2$. In this case M_2 is clearly operator isomorphic to M/M_1, and the representation $\rho(x)$ takes the form

$$\rho(x) = \left(\begin{array}{c|c} \rho_1(x) & 0 \\ \hline 0 & \rho_2(x) \end{array}\right).$$

Conversely, if the representation $\rho(x)$ can be put in this form of square blocks down the main diagonal, M is the direct sum of representation submodules M_1 and M_2. We say that the representation is *completely reduced* in this case. Not every representation which is reducible can be completely reduced. Thus the representation of the infinite cyclic group generated by an element b given by

$$\rho(b^i) = \begin{pmatrix} 1, & 0 \\ i, & 1 \end{pmatrix}$$

is reducible, but if it could be completely reduced it would represent every element by the identity, since here both $\rho_1(b^i)$ and $\rho_2(b^i)$ are the identity. But $\begin{pmatrix} 1, & 0 \\ 1, & 1 \end{pmatrix}$ is not conjugate to the identity and this is clearly impossible. However, we have an important class of representations for which reducible representations can be completely reduced.

THEOREM 16.3.1. THEOREM OF COMPLETE REDUCIBILITY. *A reducible representation of a finite group G over a field F whose characteristic does not divide the order of G can be completely reduced.*

Proof: Let M be a representation module for G over F and M_1 a representation submodule. With $w_1, \cdots, w_r$ a basis for M_1, complete this to a basis for M with elements $w_{r+1}, \cdots, w_n$, which will be a basis for a subspace N, which, however, will not in general be a representation module. We have for $x \in G$

$$\rho(x) = \left(\begin{array}{c|c} \sigma(x) & 0 \\ \hline \theta(x) & \tau(x) \end{array}\right).$$

We also have $M = M_1 + N$, and for $u \in M$ uniquely,

$$u = u_1 + v, \quad u_1 \in M_1, v \in N.$$

The map $\eta: u \to v$ is idempotent and linear. If g is the order of G, put

$$u' = \frac{1}{g} \sum_{x \in G} ux\eta x^{-1} = u\zeta.$$

Here $u \to u' = u\zeta$ is a linear mapping. This mapping requires that we be able to divide by g, the order of G, and this is possible because by our hypothesis G is of finite order g, not divisible by the characteristic of F.

If $y \in G$, put $z = y^{-1}x$ for $x \in G$, and then

$$\begin{aligned}(u\zeta)y &= \frac{1}{g} \sum_x ux\eta x^{-1}y = \frac{1}{g} \sum_z (uy)z\eta z^{-1} \\ &= (uy)\zeta,\end{aligned}$$

since z runs over G as x does. This shows that $M_2 = M\zeta$ is a representation module. We wish to show that $M = M_1 \oplus M_2$, for which we must show that every $u \in M$ can be written in the form $u = u_1 + u_2$, $u_1 \in M_1$, $u_2 \in M_2$ and that this expression is unique, i.e., $0 = u_1 + u_2$ implies $u_1 = 0 = u_2$. For any $u \in M$, write

$$u = (u - u\zeta) + u\zeta.$$

Here $u\zeta = u_2 \epsilon M_2$. Now $u - u\zeta = \frac{1}{g} \sum_x (ux - ux\eta)x^{-1}$, since $uxx^{-1} = u$. But $ux - ux\eta = (ux)_1 \epsilon M_1$, whence $u - u\zeta = u_1 \epsilon M_1$. Hence $u = u_1 + u_2$ with $u_1 \epsilon M_1$, $u_2 \epsilon M_2$. Now if $w \epsilon M_1$, $wx \epsilon M_1$, $wx\eta = 0$, whence $w\zeta = 0$. Thus for any $u \epsilon M$, $(u - u\zeta)\zeta = 0$, and $u\zeta^2 = u\zeta$. Hence if $u_1 + u_2 = 0$, $u_1\zeta + u_2\zeta = 0$, and thus $0 + u_2\zeta = u_2 = 0$, and so also $u_1 = 0$. Hence the representation can be completely reduced.

Second Proof by Matrices: With

$$\rho(x) = \left(\begin{array}{c|c} \sigma(x) & 0 \\ \hline \theta(x) & \tau(x) \end{array}\right)$$

and $\sigma(x)$, $\tau(x)$ representations of degrees r and $n - r$, respectively, we wish to find a matrix

$$S = \left(\begin{array}{c|c} I_r & 0 \\ \hline \mu & I_{n-r} \end{array}\right),$$

μ independent of x, such that

$$\left(\begin{array}{c|c} \sigma(x) & 0 \\ \hline \theta(x) & \tau(x) \end{array}\right)\left(\begin{array}{c|c} I_r & 0 \\ \hline \mu & I_{n-r} \end{array}\right) = \left(\begin{array}{c|c} I_r & 0 \\ \hline \mu & I_{n-r} \end{array}\right)\left(\begin{array}{c|c} \sigma(x) & 0 \\ \hline 0 & \tau(x) \end{array}\right)$$

for all $x \epsilon G$. Clearly, if it can be found, S is nonsingular and will yield an equivalent representation

$$\rho^*(x) = S^{-1}\rho(x)S = \left(\begin{array}{c|c} \sigma(x) & 0 \\ \hline 0 & \tau(x) \end{array}\right),$$

all $x \epsilon G$, which is completely reduced. This requires finding an $(n - r) \times r$ matrix μ independent of x such that

$$\mu\sigma(x) - \tau(x)\mu = \theta(x),$$

all $x \epsilon G$.

From $\rho(yx) = \rho(y)\rho(x)$ we have

$$\theta(yx) = \theta(y)\sigma(x) + \tau(y)\theta(x),$$

whence

$$\theta(x) = \tau(y^{-1})\theta(yx) - \tau(y^{-1})\theta(y)\sigma(x),$$

and

$$\theta(x) = \frac{1}{g}\sum_y \left(\tau(x)\tau(x^{-1}y^{-1})\theta(yx) - \tau(y^{-1})\theta(y)\sigma(x)\right).$$

Hence, if we put

$$-\mu = \frac{1}{g}\sum_y \tau(y^{-1})\theta(y) = \frac{1}{g}\sum_y \tau(x^{-1}y^{-1})\theta(yx),$$

we have $\theta(x) = -\tau(x)\mu + \mu\sigma(x)$. Thus we have found a suitable $(n - r) \times r$ matrix μ, and so a transforming matrix S exists and the equivalent representation $\rho^*(x)$ is completely reduced.

By repeated application of this theorem we find the following major result:

THEOREM 16.3.2. *Every representation of a finite group G over a field F whose characteristic does not divide the order of G can be completely reduced to the sum of irreducible representations.*

COROLLARY 16.3.1. *Representations of a finite group G over a field F whose characteristic does not divide the order of G are equivalent if, and only if, they reduce to the sum of the same irreducible representations, each with the same multiplicity.*

When a representation ρ is completely reduced we may write

$$\rho = \rho_1 \oplus \rho_2 \oplus \cdots \oplus \rho_k.$$

The order of the ρ_i is immaterial, since we may permute the elements of the corresponding basis of the representation module M to permute the ρ_i. The ρ_i are, of course, the composition factors of M, taken as an additive group with F and G as operators, and as such unique by the Jordan-Hölder theorem up to order and operator isomorphism. By Theorem 16.2.1, operator isomorphism of irreducible representations means equivalence. Thus by "same irreducible representations" in the corollary we do not distinguish equivalent representations.

16.4. Semi-simple Group Rings and Ordinary Representations.

Given any group G and a field F, we may construct the group ring R_G in the following way:

1) R_G is a vector space over F with the elements $g_i \in G$ as a basis.

2) Products are defined by putting

$$\sum_i a_i g_i \sum_j b_j g_j = \sum_{i,j} a_i b_j g_{ij}$$

where $g_{ij} = g_i g_j$ in G.

It is not difficult to show that this definition makes R_G into an associative ring with a unit $1 \cdot 1 = 1$, the product of the unit of F and the identity of G. Clearly, R_G is a representation module for G elements of G operating on R_G by multiplication on the right. If G is of finite order n, taking the elements of G, $g_1, \cdots, g_n$ as a basis for R, the corresponding representation is

$$\rho(x) = (x_{ij}), \quad i, j = 1, \cdots, n, \quad x \in G,$$

where $x_{ij} = 1$ if $g_i x = g_j$ and $x_{ij} = 0$ otherwise. This we recognize as the right regular representation of G which as a permutation group is given by

$$\pi(x) = \begin{pmatrix} g_1 & \cdots & g_n \\ g_1 x & \cdots & g_n x \end{pmatrix} \qquad x \in G,$$

written in matrix form.

LEMMA 16.4.1. *In the right regular representation $\rho(x)$ of a group G of order n we have $\chi(1) = n$, $\chi(g) = 0$, $g \neq 1$.*

For here $\rho(x) = (x_{ij})$, where $x_{ij} = 1$ if $g_i x = g_j$ and $x_{ij} = 0$ otherwise, and so, $\chi(x) = \sum_i x_{ii}$. If $x = 1$, $g_i 1 = g_j = g_i$ and $x_{ii} = 1$, $i = 1, \cdots, n$, and so, $\chi(1) = n$. But for $x = g \neq 1$, $x_{ii} = 0$, since $g_i x = g_i$ cannot hold for any g_i unless $x = 1$.

Nearly all the results we shall obtain will be for the representations of a finite group G over a field F whose characteristic does not divide the order of G. Such a representation we shall call an *ordinary representation.* Representations of a finite group G over a field F whose characteristic does divide the order of G are called *modular representations.* Properties of modular representations are different from those of ordinary representations. And, of course, representations of infinite groups can be expected to differ from representations of finite groups in many ways.

We say that a ring R is *regular* if for every $u \in R$ there is an element $x \in R$ such that $uxu = u$. A regular ring finite dimensional over a field F is said to be *semi-simple.* An element $e \neq 0$ such that $e^2 = e$ is called an *idempotent.*

THEOREM 16.4.1. *The group ring R_G of a finite group G over a field F is semi-simple if, and only if, the characteristic of F does not divide the order of G.*

Proof: Let G be of finite order g. If the characteristic of F divides g and $x_1, \cdots, x_g$ are the elements of G, then in R_G consider the element $u = x_1 + \cdots + x_g$. Here $x_iu = ux_i = u$. Hence, with $x = a_1x_1 + \cdots + a_gx_g$, we have $ux = (a_1 + \cdots + a_g)u$ and $uxu = (a_1 + \cdots + a_g)gu = 0 \neq u$. Hence R_G is not semi-simple.

Now suppose that the order g of G is not divisible by the characteristic of F. We shall prove that R_G is semi-simple and indeed shall prove further properties of R_G. Let $\mathfrak{a}_1$ be any right ideal of R_G. Then $\mathfrak{a}_1$ is a representation submodule of R_G, and conversely, the representation submodules of R_G are the right ideals. By the theorem of complete reducibility

$$R_G = \mathfrak{a}_1 \oplus \mathfrak{a}_2,$$

where $\mathfrak{a}_2$ is another right ideal. Then $1 = a_1 + a_2$, a_1, $a_1 \in \mathfrak{a}_1$, $a_2 \in \mathfrak{a}_2$, and this representation is unique. But then $a_1 = a_1^2 + a_2a_1$ also holds, and comparing this with the unique representation $a_1 = a_1 + 0$, we have $a_1^2 = a_1$, $a_2a_1 = 0$. Thus $a_1 = e$ is an idempotent, and $a_2 = 1 - e$ is also an idempotent. Thus for $x \in R_G$ we have $x = ex + (1 - e)x$ with $ex \in \mathfrak{a}_1$. Conversely, if $y \in \mathfrak{a}_1$, we have $y = ey + (1 - e)y = y + 0$ by the uniqueness of the representation. Hence for $y \in \mathfrak{a}_1$ we have $ey = y$. Thus $\mathfrak{a}_1$ is the principal right ideal eR_G of the idempotent e, and so every right ideal of R_G is the principal right ideal of an idempotent. In particular, for every element u there is an idempotent e such that $uR_G = eR_G$. Hence for some x, $ux = e$, and for some y we have $ey = u$, $eu = e^2y = ey = u$. Here $u = eu = uxu$, and so, R_G is regular.

THEOREM 16.4.2. *A regular ring R of finite dimension over a field F has a unit and every right (left) ideal is the principal ideal of an idempotent. Every two-sided ideal is the principal ideal of an idempotent in the center.*

Proof: Let R be a regular ring of finite dimension over a field F. If u is any element, then by regularity there is an x such that $uxu = u$. Here, with $e = ux$, we have $e^2 = uxux = ux = e$, and with $f = xu$, we have $f^2 = xuxu = xu = f$. Moreover, $u = uxu = eu$

$= uf$, whence $eR = uR$ and $Ru = Rf$. Thus principal right or left ideals are principal right or left ideals of idempotents. Consider a left ideal $\mathfrak{a}$. If $\mathfrak{a} \neq 0$, then it contains some idempotent $e_1 \neq 0$, and thus $Re_1 \subseteq \mathfrak{a}$. Suppose $\mathfrak{a} \neq Re_1$. Then there is some $x \in \mathfrak{a}$, $x \notin Re_1$.

$$x = xe_1 + (x - xe_1),$$

with $x_1 = xe_1 \in Re_1$, and

$$x_2 = x - xe_1,$$

where $x_2e_1 = 0$.

Let f be an idempotent such that $Rx_2 = Rf$. Then $f = wx_2$, $fe_1 = wx_2e_1 = 0$. Now put $e_2 = e_1 + f - e_1f$. Here $e_1e_2 = e_1$, $fe_2 = f$. Thus

$$e_2{}^2 = (e_1 + f - e_1f)e_2 = e_1 + f - e_1f = e_2,$$

and so, e_2 is an idempotent belonging to $\mathfrak{a}$ and Re_2 includes both e_1 and f, whence it includes Re_1 and the element $x \notin Re_1$. Hence Re_2 has a greater dimension than Re_1. Continuing, we may construct further idempotents $e_3, e_4, \cdots$ in $\mathfrak{a}$ with each ideal Re_i of greater dimension than the last until we reach an idempotent e such that $\mathfrak{a} = Re$. This proves that every left ideal is the principal left ideal of an idempotent. A similar argument shows that every right ideal is the principal right ideal of an idempotent. If $x \in eR$, then for some w, $x = ew$ and $ex = e^2w = ew = x$, and e is a left unit for elements of eR. Similarly, for $x \in Rf$, we have $xf = x$. Now considering the entire ring R as both a left ideal and a right ideal, there are idempotents e and f such that $R = eR = Rf$. Hence $ef = f = e$ and $ex = xe = x$, whence $e = 1$ is a unit for R.

The multiples of an idempotent e in the center of R will surely form a two-sided ideal. We wish to show that conversely an arbitrary two-sided ideal $\mathfrak{a}$ is the principal ideal of an idempotent in the center. Now for appropriate idempotents we have $\mathfrak{a} = eR = Rf$. Hence $ef = f = e$, and $\mathfrak{a} = eR = Re$. Now for an arbitrary x of R we have $ex \in \mathfrak{a}$ whence $ex = exe$. Also $xe \in \mathfrak{a}$, whence $xe = exe$. Thus $ex = xe$, and so, e is in the center of R.

Let us call a ring R *simple* if it is semi-simple and contains no two-sided ideals except 0 and R. The direct sum of right ideals we indicate with $\oplus$; of two-sided ideals, we indicate with $\boxplus$.

THEOREM 16.4.3. *A semi-simple ring R is the direct sum $R = R_1 \boxplus R_2 \boxplus \cdots \boxplus R_s$ of simple rings. The simple rings R_i are unique apart from order.*

Proof: Let R_1 be a minimal two-sided ideal contained in R. Then R_1 is the principal ideal of an idempotent e_1 in the center of R. Then for $x \in R$, $x = xe_1 + x(1 - e_1)$. Hence $R = R_1 \boxplus \bar{R}_1$, where $\bar{R}_1$ consists of all elements of the form $x(1 - e_1)$. Now e_1 is the unit for R_1, $\bar{e}_1 = 1 - e_1$, the unit for $\bar{R}_1$, and for $x, y \in R_1$, $z, w \in \bar{R}_1$, we have $(x + z) + (y + w) = (x + y) + (z + w)$ and $(x + z)(y + w) = xy + zw$, since $zy = ze_1(1 - e_1)y = 0$, and similarly, $xw = 0$. Thus in this direct sum $R_1 \boxplus \bar{R}_1$ both sums and product may be computed by combining the components separately. Hence, in particular, regularity of R implies regularity for R_1 and $\bar{R}_1$ separately. Continuing, take R_2 as a minimal two-sided ideal of $\bar{R}_1$ and find $\bar{R}_1 = R_2 \boxplus \bar{R}_2$. Proceeding in this way, we ultimately obtain

$$R = R_1 \boxplus R_2 \boxplus \cdots \boxplus R_s,$$

where $1 = e_1 + e_2 + \cdots + e_s$ and e_i, $i = 1, \cdots, s$ are idempotents in the center of R and $e_ie_j = 0$, $i \neq j$.
For

$$x = x_1 + x_2 + \cdots + x_s,$$
$$y = y_1 + y_2 + \cdots + y_s,$$

with $x_i, y_i \in R_i$, we have

$$x + y = (x_1 + y_1) + (x_2 + y_2) + \cdots + (x_s + y_s),$$
$$xy = x_1y_1 + x_2y_2 + \cdots + x_sy_s.$$

Thus, conversely, the direct sum of simple rings R over the same field F will be regular and hence semi-simple. Now if $\mathfrak{a}$ is any two-sided ideal in R, it is the principal ideal of an idempotent e in the center. Here

$$e = e_1e + e_2e + \cdots + e_se.$$

Thus $e_ie \neq 0$ for some i. But if $\mathfrak{a}$ is minimal and $e_ie \neq e_i$, then the principal ideal of e_ie would be properly contained in R_i, and if $e_ie \neq e$, it would be properly contained in $\mathfrak{a}$. Hence $e_ie = e_i = e$, and so, $\mathfrak{a} = R_i$. This proves the uniqueness of the direct sum.

THEOREM 16.4.4. *Any ordinary irreducible representation of a finite* *group G is equivalent to the representation on some minimal right ideal of R_G. Two minimal right ideals of R_G give equivalent representations if, and only if, they belong to the same simple component of R_G.*

Proof: We note that any representation ρ of G yields a representation of R_G, since if $h = \sum_{x \in G} a_x x$, $a_x \in F$, $x \in G$ is any element of R_G, we may take $\rho(h) = \sum a_x \rho(x)$ and this is a representation of R_G. Equivalent representations of G give equivalent representations of R_G, and conversely.

The regular representation of G is the representation of G with R_G as an F-G module. In its completely reduced form

$$R_G = e_1 R_G \oplus e_2 R_G \oplus \cdots \oplus e_t R_G,$$

where $1 = e_1 + e_2 + \cdots + e_t$ and the e_i are idempotents which are *orthogonal*, i.e., $e_i e_j = 0$ if $i \neq j$. Here each $e_i R_G$ is a minimal right ideal. Now let $\rho(x)$ be an ordinary irreducible representation of G and thus of R_G. Let M be an irreducible F-G module giving the representation of ρ. Then $M = M \cdot 1 = M(e_1 + \cdots + e_t)$. Hence for some e_i, $Me_i \neq 0$. Let m be some vector in M such that $me_i \neq 0$. Then $me_i R_G \neq 0$ is a representation module for G different from zero and contained in M. As M is irreducible, we must have $M = me_i R_G$. The correspondence

$$me_i\Big(\sum_{x \in G} a_x x\Big) \leftrightarrows e_i \sum_{x \in G} a_x x,$$

is one to one, since the elements $e_i h$ for which $me_i h = 0$ form a right ideal properly contained in $e_i R_G$ and hence are zero. We have an operator isomorphism between the representation module M and the representation module $e_i R_G$, and so, by Theorem 16.1.1, $\rho(x)$ is equivalent to the representation of G on the minimal right ideal $e_i R_G$.

When do two minimal right ideals yield equivalent representations? A minimal right ideal must be contained in a unique minimal two-sided ideal. Suppose that

$$R_G = R_1 \boxplus R_2 \boxplus \cdots \boxplus R_s$$

is the decomposition of R_G as a sum of simple ideals, i.e., minimal two-sided ideals. Here $1 = e_1 + e_2 + \cdots e_s$, where the e_i are an orthogonal set of idempotents in the center. Suppose $e_{i1} R_G$ and $e_{i2} R_G$ are two minimal right ideals in the same simple ideal R_i. Then all finite sums $u_1e_{i1}v_1 + \cdots + u_me_{i1}v_m$, $u_k, v_k \in R_G$ form a two-sided ideal, and hence, since $e_i(e_{i1})e_i = e_{i1} \neq 0$ is in this set, the set is R_i. Hence, for appropriate u's and v's,

$$u_1e_{i1}v_1 + \cdots + u_me_{i1}v_m = e_{i2}.$$

Since $e_{i2}^2 = e_{i2}$, we have for some j,

$$e_{i2}u_je_{i1}v_j \neq 0.$$

But then $e_{i1}v_jR_G \neq 0$, and since it is a right ideal contained in the minimal ideal $e_{i1}R_G$, we have $e_{i1}v_jR_G = e_{i1}R_G$. Similarly $e_{i2}u_je_{i1}v_jR_G = e_{i2}R_G$. Thus with $w = e_{i2}u_j$ we have $we_{i1}R_G = e_{i2}R_G$. Hence for $h \in R_G$ we have $we_{i1}h \leftrightarrows e_{i1}h$, an operator isomorphism between the right ideals $e_{i1}R_G$ and $e_{i2}R_G$, and so their representations are equivalent. This shows that minimal right ideals in the same simple ideal give the same representation.

Now suppose that $e_{i1}R_G$ and $e_{j1}R_G$ are minimal right ideals from the simple ideals R_i and R_j, $i \neq j$. Representing on $e_{i1}R_G$, we have the mappings for e_i, e_j, respectively,

$$\begin{aligned} e_i\colon e_{i1}h &\to e_{i1}he_i = e_{i1}e_ih = e_{i1}h \\ e_j\colon e_{i1}h &\to e_{i1}he_j = e_{i1}e_jh = 0, \end{aligned}$$

whence $\rho(e_i) = 1$ and $\rho(e_j) = 0$. Similarly, on $e_{j1}R_G$, e_i is represented by 0 and e_j by 1. Hence the representations are inequivalent.

We have related the decomposition of R_G into a direct sum of simple ideals to finding orthogonal idempotents in the center Z of R_G. What is the center of R_G? This is easily answered.

THEOREM 16.4.5. *The elements $C_i = x_{i1} + \cdots + x_{ih}$, where x_{i1}, $x_{i2}, \cdots, x_{ih}$ are a class of conjugates in a group G are a basis for the center of R_G.*

Proof: If $C_i = x_{i1} + \cdots + x_{ih}$, where $x_{i1}, x_{i2}, \cdots, x_{ih}$ are a class of conjugates in G, then for $y \in G$, $y^{-1}C_iy = C_i$, since transformation by an element merely permutes the elements of a class among themselves. Since C_i permutes with every $y \in G$, it permutes with every element of R_G and is in the center of R_G. Conversely if u is in the center of R_G and $u = \sum_{x \in G} a_xx$, $y \in G$, we have $y^{-1}uy = u = \sum_{x \in G} a_xy^{-1}xy$, and so in u, conjugate elements have equal coefficients and thus u is a linear combination of the C_i.

16.5. Absolutely Irreducible Representations. Structure of Simple Rings.

We have seen that all irreducible ordinary representations occur as components of the regular representation $R(G)$ of a group G. Thus their determination is a matter of finding the complete reduction of $R(G)$, or what comes to the same thing, finding the irreducible right ideals of the group ring R_G.

Irreducibility of a representation is a relative matter, depending on the field. Thus if G is the cyclic group of order 3, with elements 1, x, x^2 over the rational field, R_G has a decomposition $R_G = R_1 \boxplus R_2$ with $1 = e_1 + e_2$, where

$$e_1 = \frac{1 + x + x^2}{3}, \quad e_2 = \frac{2 - x - x^2}{3}$$

are idempotents, R_1 has e_1 as a basis, and R_2 has a basis e_2, e_2x. This gives the representation

$$\rho(x) = \left(\begin{array}{c|cc} 1 & 0, & 0 \\ \hline 0 & 0, & 1 \\ 0 & -1, & -1 \end{array}\right)$$

and R_1 and R_2 give irreducible representations of degrees 1 and 2. But if we extend the rational field by adjoining the cube root of unity $\epsilon = (-1 + \sqrt{-3})/2$, the irreducible representation on R_2 now becomes reducible, and with a basis

$$e_1 = \frac{1 + x + x^2}{3},$$

$$\bar{e}_2 = \frac{1 + \epsilon x + \epsilon^2 x^2}{3}, \quad \bar{e}_3 = \frac{1 + \epsilon^2 x + \epsilon x^2}{3},$$

we have $\bar{e}_2 + \bar{e}_3 = e_2$; on this new basis we have

$$\rho(x) = \begin{pmatrix} 1, & 0, & 0 \\ 0, & \epsilon^2, & 0 \\ 0, & 0, & \epsilon \end{pmatrix} \qquad \epsilon^3 = 1.$$

Clearly, further extension of the field will not reduce $\rho(x)$.

A representation ρ of degree n is *absolutely irreducible* if it cannot be reduced by extending the field F. Now, clearly, if ρ can be reduced over $K \supset F$,

$$\rho(x) = \left(\begin{array}{c|c} \sigma(x) & 0 \\ \hline 0 & \tau(x) \end{array}\right),$$

all $x \in G$, with $\sigma(x)$ an $s \times s$ matrix and $\tau(x)$ an $(n - s) \times (n - s)$ matrix, then $\rho(h)$, $h \in R_G$ as an algebra over K has dimension at most $s^2 + (n - s)^2 < n^2$. Hence, if $\rho(h)$, $h \in R_G$ as an algebra over F has dimension n^2, then ρ is absolutely irreducible over F. We shall show that by appropriate algebraic extension of the field, any ordinary representation is the direct sum of irreducible representations in which an irreducible representation of degree n has dimension n^2 over the field.

THEOREM 16.5.1. *A division ring D of finite dimension over a field F will not be a division ring over algebraic extensions of F unless D is of dimension one over F and then $D = F$.*

Proof: Let D have a basis $u_1, \cdots, u_n$ over F, where we may take $u_1 = 1$ as the unit of D. If $n > 1$, consider $1 = u_1, u_2, u_2^2, \cdots, u_2^n$. These must be linearly dependent over F, and we have a relation

$$u_2^r + a_1 u_2^{r-1} \cdots + a_r = 0, \quad a_i \in F,$$

and so, if we adjoin the roots $\alpha_1, \cdots, \alpha_r$ of $f(x) = x^r + a_1 x^{r-1} \cdots + a_r$ to F, we have $(u_2 - \alpha_1 u_1) \cdots (u_2 - \alpha_r u_1) = 0$. Hence, over this algebraic extension of F, the $u_2 - \alpha_i u_1$ are divisors of zero. Hence D remains a division ring over algebraic extension of F only if $n = 1$, and here $D = F$.

THEOREM 16.5.2. *A simple ring R is a complete matrix ring over a division ring D, contained in R.*

Proof: Let $e_{11}R$ be a minimal right ideal of R, e_{11} being an idempotent. Then $1 - e_{11}$ is an idempotent and $R = e_{11}R \oplus (1 - e_{11})R$. If e_2R is a minimal right ideal in $(1 - e_{11})R$, e_2 being an idempotent, we have $(1 - e_{11})e_2 = e_2$, whence $e_{11}e_2 = 0$. Also $e_{22} = e_2 - e_2e_{11}$ is an idempotent, and $e_{22}R = e_2R$ and $e_{11}e_{22} = 0$, $e_{22}e_{11} = 0$. Here $R = e_{11}R \oplus e_{22}R \oplus (1 - e_{11} - e_{22})R$. Suppose we have found orthogonal idempotents $e_{11}, \cdots, e_{ii}$ such that $e_{ii}R$ are minimal and $R = e_{11}R + e_{22}R \oplus \cdots \oplus e_{ii}R \oplus (1 - e_{11} \cdots - e_{ii})R$. Let e_{i+1} be an idempotent such that $e_{i+1}R$ is a minimal right ideal in $(1 - e_{11} \cdots - e_{ii})R$. Then $e_{i+1} = (1 - e_{11} \cdots - e_{ii})e_{i+1}$, whence $e_{jj}e_{i+1} =$

$e_{jj}(1 - e_{11} \cdots - e_{ii})e_{i+1} = 0$, $j = 1, \cdots, i$. If we put $e_{i+1,i+1} = e_{i+1}(1 - e_{11} \cdots - e_{ii})$, we have $e_{i+1,i+1}$, an idempotent $e_{i+1,i+1}R = e_{i+1}R$, and also $e_{i+1,i+1}$ orthogonal to $e_{11}, \cdots, e_{ii}$. Continuing, we find

$$R = e_{11}R \oplus e_{22}R \oplus \cdots \oplus e_{nn}R,$$

where the e_{ii} are orthogonal idempotents, the $e_{ii}R$ are minimal right ideals, and $1 = e_{11} + e_{22} \cdots + e_{nn}$.

LEMMA 16.5.1. *$e_{ii}Re_{jj} \neq 0$ for $i, j = 1, \cdots, n$.*

Proof: All finite sums $\sum_k u_k e_{ii} v_k$ form a two-sided ideal including $e_{ii} \neq 0$, whence these sums are the entire ring R. Thus for appropriate elements u_k, v_k, we have $\sum u_k e_{ii} v_k = e_{jj}$, and so, $\sum u_k e_{ii} v_k e_{jj} = e_{jj}$. Hence for some v, $e_{ii}ve_{jj} \neq 0$.

LEMMA 16.5.2. *$e_{ii}Re_{ii}$ is a division ring D_i.*

Proof: $e_{ii}Re_{ii}$ is certainly closed under addition and multiplication, and so is a subring of R. It has e_{ii} as a unit. It suffices to find inverses for elements different from 0. If $e_{ii}xe_{ii} \neq 0$, then $e_{ii}xe_{ii}R$ is a right ideal $\neq 0$ contained in, and so equal to, the minimal right ideal $e_{ii}R$. Hence for some y, $e_{ii}xe_{ii}y = e_{ii}$, and so, $e_{ii}xe_{ii} \cdot e_{ii}ye_{ii} = e_{ii}$. Thus $e_{ii}ye_{ii}$ is an inverse for $e_{ii}xe_{ii}$ in $e_{ii}Re_{ii}$, which is therefore a division ring D_i.

Choose for each $i = 2, \cdots, n$ an element $e_{11}be_{ii} \neq 0$, and write $e_{11}be_{ii} = e_{1i}$. Then

$$e_{11}e_{1i} = e_{1i}e_{ii} = e_{1i}.$$

Now $e_{1i}R \subseteq e_{11}R$, whence $e_{1i}R = e_{11}R$. Hence for some y, $e_{1i}y = e_{11}$, $e_{1i}(e_{ii}ye_{11}) = e_{11}$. Write $e_{i1} = e_{ii}ye_{11}$. Then for our elements $i = 2, \cdots, n$.

$$e_{ii}e_{i1} = e_{i1}e_{11} = e_{i1}, \quad e_{1i}e_{i1} = e_{11}.$$

Hence $e_{1i}e_{i1}e_{1i}e_{i1} = e_{11}^2 = e_{11}$, and so, $e_{i1}e_{1i} \neq 0$. But $(e_{i1}e_{1i})^2 = e_{i1}e_{1i}$ is an idempotent in $e_{ii}Re_{ii}$, whence $e_{i1}e_{1i} = e_{ii}$, the unit being the only nonzero idempotent in a division ring. Now put $e_{i1}e_{1j} = e_{ij}$ if $i \neq j$. Then we have $e_{ij}e_{jk} = e_{i1}e_{1j}e_{j1}e_{1k} = e_{i1}e_{1j}e_{1k} = e_{i1}e_{1k} = e_{ik}$.

Also if $j \neq k$, $e_{ij}e_{kt} = e_{ij}e_{jj}e_{kk}e_{kt} = 0$. Thus for our n^2 units e_{ij}, we have shown in all cases

$$e_{ij}e_{kt} = \delta_{jk}e_{it} \qquad \delta_{jj} = 1, \quad \delta_{jk} = 0, \quad j \neq k,$$

and so, the e_{ij} have the multiplication properties of the $n \times n$ matrix elements

$$i, j = 1, \cdots, n, \quad E_{ij} = (a_{rs}), \quad a_{ij} = 1, \quad a_{rs} = 0,$$

if $(r, s) \neq (i, j)$.

Now from the division ring $D_1 = e_{11}Re_{11}$, define a ring D by putting

$$d = d_1 + e_{21}d_1e_{12} + \cdots + e_{n1}d_1e_{1n},$$

for each $d_1 \in D_1$. We verify without trouble that D is isomorphic to D_1 and hence is a division ring. Corresponding to e_{11} the unit of D_1, we have $e_{11} + e_{22} + \cdots + e_{nn} = 1$, the unit of D and also of R. Also for $d \in D$, we find $e_{ij}d = e_{i1}d_1e_{1j} = de_{ij}$.

Finally, for an arbitrary $x \in R$, we have $x = 1\,x\,1 = (e_{11} + \cdots + e_{nn})x\,(e_{11} + \cdots + e_{nn}) = \sum_{i,j} e_{ii}xe_{jj}$. But here

$$x_{ij} = e_{ii}xe_{jj} = e_{i1}e_{1i}xe_{j1}e_{1j} = e_{i1}u_1e_{1j}$$

for some $u_1 \in D_1$, whence for $u \in D$ we have $x_{ij} = ue_{ij} = e_{ij}u$. This completes our theorem. We have shown that a simple ring R can be exhibited in the explicit form of an $n \times n$ matrix ring over a division ring D whose unit coincides with the unit of R.

Theorem 16.5.3. *If R_G is a semi-simple group ring over a field F, there is an algebraic extension F^* of F in which R_G is the direct sum of complete matrix rings over F^*. We can take F^* to be a finite algebraic extension of F.*

Proof: R_G is semi-simple over a field F if, and only if, the characteristic of the field does not divide the order of the group G. This property is unaltered if F is replaced by an algebraic extension F^* of F. If in the decomposition of R_G over F as the direct sum of simple rings $R_1 \boxplus \cdots \boxplus R_s$, there is some simple R_k whose corresponding division ring D is not the field F, then by an algebraic extension of F to some F^*, the ring D loses the property of being a division ring. This alters the decomposition of R_G in one of two ways: (1) we may increase (but surely not decrease) the number of idempotents in the center of R and thus break up a simple ring as the direct sum of several simple rings; or (2) in a simple ring R we find a division ring D^* of smaller dimension and express R as a larger matrix ring over D^*. Both these situations can arise. We have already seen the first case in representing the group of order 3. The second case arises in the ring R_Q of the quaternion group Q over the rational field F. R_Q over F is the direct sum of four simple rings of dimension 1 and one of dimension 4 which is a division ring (the quaternion algebra). If we adjoin i to F, the division ring becomes the ring of 2×2 matrices over the complex rational field.

In any event the algebraic closure $\bar{F}$ of F is a field over which every simple ring R_k arising in R_G is a matrix ring over $\bar{F}$. The matrix units $e_{ij}{}^k$ of the simple rings R_k can be expressed in terms of the elements x of G, and any field F^* containing all the coefficients appearing in these expressions will be such that the R_k are complete matrix rings over F^*. F^* is clearly finite over F.

THEOREM 16.5.4. *The center of a complete matrix ring R_k over a field F consists of the scalar multiples of the unit of R_k, which is the identity matrix. The center of the direct sum of matrix rings $R = R_1 + \cdots + R_r$ over a field F has as a basis the r units of $R_1, \cdots, R_r$.*

Proof: Let R_k be the complete $n \times n$ matrix ring over F. Then suppose

$$x = \sum_{i,j} a_{ij}e_{ij}, \quad a_{ij} \in F$$

is in the center of R_k. From $e_{rs}x = xe_{rs}$ we find

$$\sum_{j} a_{sj}e_{rj} = \sum_{i} a_{ir}e_{is}.$$

Hence $a_{sj} = 0$ for $j \neq s$ and $a_{ss} = a_{rr}$. Thus $x = a_{11}(e_{11} + \cdots + e_{nn}) = a_{11} \cdot 1$, and all such elements are in the center of R_k. *If*

$$R = R_1 + \cdots + R_r,$$

then the center of R is the direct sum of the centers of the R_k and as such has as a basis the r units of the R_k.

We now have a number of theorems which relate the ordinary representations of G to the semi-simple group ring R_G. We combine these results in a theorem.

THEOREM 16.5.5. *Every irreducible ordinary representation of a finite group G occurs as a component of the right regular representation $R(G)$. The number of inequivalent absolutely irreducible representations is the number of classes in G. If $\rho_1, \cdots, \rho_r$ are the distinct absolutely irreducible representations and ρ_i is of degree n_i, $i = 1, \cdots, r$, then ρ_i is of dimension n_i^2 over F and ρ_i occurs n_i times in $R(G)$. The only matrices permuting with $\rho_i(x)$, all $x \in G$ are scalar multiples of the identity. If g is the order of G, then $g = n_1^2 + n_2^2 + \cdots + n_r^2$.*

Proof: By Theorem 16.4.4, every ordinary irreducible representation is equivalent to a representation on some minimal right ideal of R_G and as such occurs as a component of $R(G)$. Also, there are as many inequivalent irreducible representations as there are simple ideals in R_G. Extending the field F if necessary to F^*, the center of R_G has a basis of r idempotents where by Theorem 16.5.4, R_G is the direct sum of r matrix rings. But by Theorem 16.4.5, the center of R_G has the class sums C_i as a basis, and so r is the number of classes in G. Over F^* a minimal right ideal occurring in R_i will be $e_{11}R$, and if R_i is an $n_i \times n_i$ matrix ring, this will have a basis $e_{11}, e_{12}, \cdots, e_{1n_i}$. The corresponding representation ρ_i will be of degree n_i, and ρ_i extended to a representation of R_G will represent faithfully the simple ring R_i and will represent all other R_j's by 0, since, as shown in the proof of Theorem 16.4.4, we shall have $\rho_i(e_j) = 0$ if e_j is the unit of $R_j, j \neq i$. Thus $\rho_i(R_G)$ is the full matrix ring of dimension n_i^2 over F^*, and so, is surely absolutely irreducible since a further reduction would be possible only if it were of lower dimension over F^*. Also, being of dimension n_i^2, the only matrices permuting with every $\rho_i(x)$, $x \in G$ will be scalar multiples of the identity. Finally, as each R_i has a basis of n_i^2 elements, their direct sum has a basis of $n_1^2 + \cdots + n_r^2$ elements. But R_G has a basis of the g elements of G. Hence

$$g = n_1^2 + \cdots + n_r^2.$$

R_i is the direct sum of the n_i right ideals $e_{11}R, \cdots e_{n_in_i}R$, and so, ρ_i occurs n_i times in $R(G)$.

16.6. Relations on Ordinary Characters.

The preceding section gave information on representations of G which depended on the nature of R_G and the fact that a representation of G gives a representation of R_G. In this section we find relations on the characters $\chi(x)$, $x \in G$. These are more intimately related to G itself than to R_G. We assume throughout this section that we are dealing with ordinary representations.

THEOREM 16.6.1. *Let A and B be two F-G modules. If A is of dimension m and yields the representation $\rho(x)$, $x \in G$, and B is of dimension n and yields the representation $\sigma(x)$, then the additive group of operator homomorphisms of A into B is isomorphic to the additive group of all $m \times n$ matrices α such that $\rho(x)\alpha = \alpha\sigma(x)$, all $x \in G$.*

COROLLARY 16.6.1. *The ring of operator endomorphisms of A into itself is isomorphic to the ring of $m \times m$ matrices α such that $\rho(x)\alpha = \alpha\rho(x)$.*

Proof: Let A have a basis $u_1, \cdots, u_m$ and B have a basis $v_1, \cdots, v_n$. Then any linear mapping of A into B is determined by the images of the basis, say,

$$\begin{array}{l} u_1 \to a_{11}v_1 + \cdots + a_{1n}v_n \\ \cdot \quad \cdot \quad \cdot \quad \cdot \quad \cdot \quad \cdot \quad \cdot \quad \cdot \quad \cdot \\ u_i \to a_{i1}v_1 + \cdots + a_{in}v_n \\ \cdot \quad \cdot \quad \cdot \quad \cdot \quad \cdot \quad \cdot \quad \cdot \quad \cdot \quad \cdot \\ u_m \to a_{m1}v_1 + \cdots + a_{mn}v_n, \end{array}$$

and let us write $\alpha = (a_{ij})$, $i = 1, \cdots, m; j = 1, \cdots, n$. These linear mappings form an additive group isomorphic to the additive group of the matrices α. If in addition to being a linear mapping, the mapping is to be an operator homomorphism, whenever $u \to v$ we must have also $ux \to vx$ for $x \in G$. This means that the mappings $u \to ux \xrightarrow{\alpha} (vx)$ and $u \xrightarrow{\alpha} v \to vx$ are identical, but this is the relation

$$\rho(x)\alpha = \alpha\sigma(x),$$

for all $x \in G$.

If we map A into itself, the mappings are called *endomorphisms* and here if α, β are two operator endomorphisms, we have

$$\rho(x)(\alpha\beta) = [\rho(x)\alpha]\beta = [\alpha\rho(x)]\beta = (\alpha\beta)\rho\,(x),$$

and thus the matrices α with $\rho(x)\alpha = \alpha\rho(x)$ are isomorphic to the ring of operator endomorphisms of A.

Theorem 16.6.2 holds for any Ω module, where Ω is any ring of operators, but, of course, we are interested mainly in F-G modules.

THEOREM 16.6.2 (SCHUR'S LEMMA). *If A, B are two irreducible Ω modules, then unless they are operator isomorphic, the only operator homomorphism of A into B maps A onto 0. If A is irreducible, every operator endomorphism of A not identically zero is an operator isomorphism.*

Proof: Let $u \in A$, $v \in B$, $\omega \in \Omega$. Then if for some $u \in A$, an operator homomorphism maps $u \to v \neq 0$, then $u\omega \to v\omega$ for all $\omega \in \Omega$. Here $u\Omega$ is a submodule of A, and hence, as A is irreducible, all of A. Thus $A = u\Omega \to v\Omega \neq 0$, whence $A \to v\Omega = B$. The mapping must be one to one, since otherwise, nonzero elements of A are mapped onto zero and these form an Ω submodule of A, contrary to the assumption that A was irreducible. Hence the mapping is an isomorphism, and so, in particular, every operator endomorphism of A into itself is an operator isomorphism.

THEOREM 16.6.3. *If ρ and σ are irreducible and inequivalent representations of the finite group G of degree m and n, respectively, and ξ is any $m \times n$ matrix, then*

$$\sum_{y \in G} \rho(y)\xi\sigma(y^{-1}) = 0.$$

Proof: Write

$$\alpha = \sum_{y \in G} \rho(y)\xi\sigma(y^{-1}).$$

Then for $x \in G$, $xy = z$, $y^{-1} = z^{-1}x$,

$$\begin{aligned} \rho(x)\alpha &= \sum_{y} \rho(x)\rho(y)\xi\sigma(y^{-1}) \\ &= \sum_{y} \rho(xy)\xi\sigma(y^{-1}) \\ &= \sum_{z} \rho(z)\xi\sigma(z^{-1})\sigma(x) \\ &= \alpha\sigma(x), \end{aligned}$$

all $x \in G$.

Hence, by Theorems 16.6.1 and 16.6.2, if ρ and σ are irreducible and inequivalent, we must have $\alpha = 0$. Note that these theorems hold for representations of G over any field.

If $f_1(y)$ and $f_2(y)$ are any two functions defined for $y \in G$ with values in F (where we now assume that the characteristic of F does not divide the order of G), then we define the *symmetric bilinear scalar product:*

$$(f_1, f_2) = \frac{1}{g} \sum_{y \in G} f_1(y) f_2(y^{-1}).$$

We verify, noting that y^{-1} runs over G as y does, that

1) $(f_1, f_2) = (f_2, f_1)$.
2) $(f_1 + f_2, f_3) = (f_1, f_3) + (f_2, f_3)$.
3) $(af_1, f_2) = a(f_1, f_2)$, $a \in F$.

Now suppose that $\rho(x)$ and $\sigma(x)$ are irreducible inequivalent representations. If in Theorem 16.6.3 we take $\xi = e_{rs}$, the $m \times n$ matrix with 1 in position (r, s) and 0 elsewhere, we find $\alpha = (\alpha_{ij})$, where $\alpha_{ij} = \sum_{y \in G} \rho_{ir}(y)\sigma_{sj}(y^{-1})$ where

$$\begin{aligned} \rho(x) &= (\rho_{ij}(x)) \quad i, j = 1, \cdots, m, \\ \sigma(x) &= (\sigma_{ij}(x)) \quad i, j = 1, \cdots, n. \end{aligned}$$

Since $\alpha = 0$ by Theorems 16.6.1 and 16.6.2, we have $(\rho_{ir}, \sigma_{sj}) = 0$. We may show even more.

THEOREM 16.6.4. *If ρ and σ are inequivalent ordinary irreducible representations of a finite group G, then the symmetric bilinear scalar product $(\rho_{ir}, \sigma_{sj}) = 0$ for all i, r, s, j. If ρ is an absolutely irreducible ordinary representation of degree n then $(\rho_{ir}, \rho_{sj}) = 0$ unless $i = j$, $r = s$, and then $(\rho_{ij}, \rho_{ji}) = 1/n$ for all i, j.*

Proof: We have already shown the first part. Now consider an absolutely irreducible ordinary representation ρ of G. Let n be the degree of ρ. If ξ is an arbitrary $n \times n$ matrix, then we verify as before that

$$\alpha = \frac{1}{g} \sum_{y \in G} \rho(y)\xi\rho(y^{-1}),$$

where ξ is an arbitrary $n \times n$ matrix, satisfies $\rho(x)\alpha = \alpha\rho(x)$ for all $x \in G$. Hence by Theorem 16.5.5, α is a scalar multiple of the identity $\alpha = \lambda I_n$, where the scalar λ depends on ξ. If $\xi = e_{rs}$, write $\lambda = \lambda_{rs}$. Then we find $\lambda_{rs}\delta_{ij} = (\rho_{ir}, \rho_{sj})$. But $(\rho_{ir}, \rho_{sj}) = (\rho_{sj}, \rho_{ir})$, so $\lambda_{rs}\delta_{ij} = \lambda_{ji}\delta_{sr} = 0$ unless both $i = j$ and $r = s$, while $(\rho_{ij}, \rho_{ji}) = \lambda_{ii} = (\rho_{ji}, \rho_{ij}) = \lambda_{jj}$. Hence $\lambda_{11} = \lambda_{22} \cdots = \lambda_{nn} = \lambda$ has the same value for all subscripts. Thus

$$\begin{aligned} n\lambda &= \sum_j \lambda_{jj} \\ &= \frac{1}{g} \sum_{y,j} \rho_{ij}(y)\rho_{ji}(y^{-1}) \\ &= \frac{1}{g} \sum_y \rho_{ii}(1) = 1. \end{aligned}$$

Hence $\lambda = 1/n$. This proves the rest of the theorem. Note that $n\lambda = 1$ shows that the degree n is not divisible by the characteristic of F.

These results carry over to the characters.

THEOREM 16.6.5. *If χ, ψ are distinct irreducible characters, then $(\chi, \psi) = 0$. If χ is an absolutely irreducible character, then $(\chi, \chi) = 1$.*

Proof: If χ and ψ are irreducible characters of the representations ρ and σ, then for $y \in G$, $\chi(y) = \sum_i \rho_{ii}(y)$, $\psi(y) = \sum_i \sigma_{jj}(y)$. Since the scalar product is bilinear,

$$(\chi, \psi) = \sum_{i,j} (\rho_{ii}, \sigma_{jj}) = 0,$$

since each individual summand is zero. Now let χ be an absolutely irreducible character of the representation ρ of degree n. Here

$$(\chi, \chi) = \sum_{i,j} (\rho_{ii}, \rho_{jj}) = \sum_i (\rho_{ii}, \rho_{ii}) = \sum_i \frac{1}{n} = 1.$$

This completes the proof.

COROLLARY 16.6.2. $\sum_{x \in G} \chi(x) = g$ *for the identical representation.* $\sum_{x \in G} \chi(x) = 0$ *for any other irreducible representation.*

For $\chi(x) = 1$ for any x in the identical representation, whence here $\sum_{x \in G} \chi(x) = g$. But if χ is the character of any other irreducible representation, take ψ as the identical character. Then $(\chi, \psi) = 0$ gives $1/g \sum \chi(x) = 0$.

THEOREM 16.6.6. *If* χ, ψ *are characters and* $\chi = \sum a_i\chi_i$, $\psi = \sum b_i\chi_i$, *where the* χ_i, $i = 1, \cdots, r$ *are the absolutely irreducible characters, then* $(\chi, \psi) = \sum a_ib_i$. *Thus for a character* ϕ, $(\phi, \phi) = 1$ *is necessary and sufficient for* ϕ *to be an absolutely irreducible character if the field* F *is of characteristic zero.*

Proof: This is essentially a corollary to Theorem 16.6.5, using the bilinearity of the scalar product. If $\phi = \sum c_i\chi_i$, then the c's are non-negative integers, and if $(\phi, \phi) = \sum c_i^2 = 1$, we may conclude in a field of characteristic zero that one c_i is 1 and the rest are zero.

THEOREM 16.6.7. *The absolutely irreducible representations of an Abelian group* G *are all of degree one.*

Proof: Since G is Abelian, every element is a class and so if g is its order, we have g absolutely irreducible representations of degrees $n_1, \cdots, n_g$, where $g = n_1^2 + n_2^2, \cdots + n_g^2$. Hence $n_1 = n_2 \cdots = n_g = 1$. Here for every representation ρ of degree one, we have $\chi(x) = \rho(x)$. Thus the representations coincide with the characters and are indeed the same as the characters of an Abelian group as treated in Chap. 13.

THEOREM 16.6.8. *Let* x *be an element of order* m *in a group* G *and let* ρ *be a representation of* G *of degree* n. *Then, adjoining the* m^{th} *roots of unity to* F, *if necessary,* $\rho(x)$ *is similar to a diagonal matrix whose elements are* m^{th} *roots of unity. If* F *is the complex field,* $\chi(x^{-1}) = \overline{\chi(x)}$ *the complex conjugate of* $\chi(x)$.

Proof: The matrices $1, \rho(x), \cdots, \rho(x^{m-1})$ are a representation of the cyclic group C of order m. But the absolutely irreducible representations of C are of degree 1, with $\sigma(x) = (b)$, where since $1 = x^m$, we must have $b^m = 1$, and so b is an mth root of unity. In R_c we easily verify that $1/m(1 + \omega x + \omega^2x^2 \cdots + \omega^{m-1}x^{m-1})$, since ω ranges over all mth roots of unity, are idempotents yielding the irreducible representations. Hence, adjoining the mth roots of unity to the field F (whose characteristic is, of course, not a divisor of m), the representation $\rho(x)$ of C reduces completely and we have a matrix similar to $\rho(x)$ which is a diagonal matrix, diag. $(b_1, \cdots, b_n)$, where each b_i is an mth root of unity. Hence $\chi(x) = b_1 + \cdots + b_n$. Here $\rho(x^{-1})$ must be similar to diag. $(b_1^{-1}, \cdots, b_n^{-1})$ and $\chi(x^{-1}) = b_1^{-1} + \cdots + b_n^{-1}$. But if F is the complex field, then the inverse of any root of unity is its complex conjugate, $b_i^{-1} = \overline{b_i}$, and so, $\chi(x^{-1}) = \overline{\chi(x)}$.

Let ρ be any representation of a group G and for each $x \in G$ let us define $\hat{\rho}(x) = \rho(x^{-1})^T$, where we designate the transpose of a matrix by the superscript T. Then

$$\begin{aligned}\hat{\rho}(xy) &= \rho(y^{-1}x^{-1})^T = [\rho(y^{-1})\rho(x^{-1})]^T \\ &= \rho(x^{-1})^T\rho(y^{-1})^T = \hat{\rho}(x)\hat{\rho}(y).\end{aligned}$$

Thus $\hat{\rho}$ is also a representation of G, called the *contragredient* representation.

Suppose that L is the representation module for ρ with a basis $u_1, \cdots, u_n$ over a field F. Take another space $\hat{L}$ over F with a basis $v_1, \cdots, v_n$, and define a scalar product $u \cdot v$ for $u = a_1u_1 + \cdots + a_nu_n \in L$ and $v = b_1v_1 + \cdots + b_nv_n \in \hat{L}$ by the rule

$$u \cdot v = a_1b_1 + a_2b_2 + \cdots + a_nb_n \in F.$$

This scalar product is the bilinear function on $(L, \hat{L})$ defined by $u_i \cdot v_j = \delta_{ij}$.

We make $v_1, \cdots, v_n$ a representation basis for $\hat{\rho}$ by the rule

$$v_ix = \sum_j \hat{\rho}_{ij}(x)v_j = \sum_j \rho_{ji}(x^{-1})v_j.$$

Then

$$\begin{aligned}u_ix \cdot v_jx &= \sum_{k,s} \rho_{ik}(x)\rho_{sj}(x^{-1})(u_k \cdot v_s) \\ &= \sum_k \rho_{ik}(x)\rho_{kj}(x^{-1}) = \delta_{ij}\end{aligned}$$

since $\rho(x)\rho(x^{-1}) = I_n$. Thus $ux \cdot vx = u \cdot v$ for all $u \in L$, $v \in \hat{L}$, and $x \in G$, and the scalar product is preserved by operation on both factors by the same element of G. To any subspace M' of $\hat{L}$ let us make correspond the subspace M of L of all $u \in L$, such that $u \cdot v = 0$ for all $v \in M'$. Hence dim. M' + dim. $M = n$, and this is a dual correspondence between the subspaces of L and $\hat{L}$. If M' is a representation submodule of $\hat{L}$, and $v \in M'$, then for $x \in G$, $vx^{-1} \in M'$, and so, $u \cdot vx^{-1} = 0$ and $ux \cdot v = 0$ for all $v \in M'$, $u \in M$, whence $ux \in M$ and M

is a representation submodule of L. Hence, in particular, $\hat{L}$ is irreducible if, and only if, L is irreducible. If ρ is an absolutely irreducible $n \times n$ representation, then since ρ is of dimension n^2 over F, it would follow that $\hat{\rho}$ is also of dimension n^2 over F, and so it is clearly absolutely irreducible.

From the definition $\hat{\hat{\rho}} = \rho$. Also if ρ and σ are equivalent, then for some S,

$$S^{-1}\rho(x^{-1})S = \sigma(x^{-1}),$$

all $x \in G$.

Then taking transposes,

$$S^T\rho(x^{-1})^T S^{T-1} = \sigma(x^{-1})^T,$$

all $x \in G$, and so $\hat{\rho}$ and $\hat{\sigma}$ are equivalent.

Let r be the number of classes in G. Let $\rho_1, \cdots, \rho_r$ be the absolutely irreducible representations of G over F, where by convention we take ρ_1, the identical representation, $\rho_1(x) = 1$, all $x \in G$. (This corresponds to the idempotent $\frac{1}{g} \sum_{x \in G} x$.) Then $\hat{\rho}_1 = \rho_1, \cdots, \hat{\rho}_r$ will be the same representations in some order. Similarly, let $C_1, \cdots, C_r$ be the classes of G, where by convention we take $C_1 = 1$, the class consisting of the identity alone. The inverses of the elements in a class C_i will themselves be a class C_i'. Hence $C_1' = C_1, \cdots, C_r'$ will again be the classes of C.

If $\chi(x)$ is the character of $\rho(x)$, let us designate the character of $\hat{\rho}(x)$ by $\bar{\chi}(x)$. Here $\chi(x) = \text{trace } \rho(x)$:

$$\bar{\chi}(x) = \text{trace } \rho(x^{-1})^T = \text{trace } \rho(x^{-1}) = \chi(x^{-1}),$$

and we have noted in Theorem 16.6.8 that for the complex field, $\chi(x^{-1}) = \overline{\chi(x)}$, the complex conjugate. Thus this notation agrees with that for complex conjugates over the complex field. We note that over the complex field, $\rho = \hat{\rho}$ only if all characters $\chi(x)$ for ρ are real.

Let us designate by χ_i^a the absolutely irreducible character of an element of the class C_i in the representation ρ_a. We also write h_i for the number of elements in C_i. The number h_i is the index of the normalizer of an element $x \in C_i$, and if its order is g_i, we have $g_i h_i = g$.

THEOREM 16.6.9. *The following orthogonality relations hold for absolutely irreducible characters of a group G:*

$$\sum_{i=1}^{r} \frac{\chi_i{}^a\overline{\chi_i{}^b}}{g_i} = \delta_{ab}$$

$$\sum_{a=1}^{r} \overline{\chi_i{}^a}\chi_j{}^a = \delta_{ij}g_j.$$

Proof: By Theorem 16.6.5 we have

$$\frac{1}{g}\sum_{x \in G} \chi^a(x)\chi^b(x^{-1}) = \delta_{ab}.$$

But $\chi(x) = \chi(y)$ if x and y are in the same class C_i, and then x^{-1} and y^{-1} are in the same class C_i'. Here $\chi^b(x^{-1}) = \overline{\chi^b}(x)$. Hence for x in C_i, the above sum will contain h_i terms equal to $\chi_i{}^a\overline{\chi_i{}^b}$. Hence

$$\sum_{i=1}^{r} \frac{h_i}{g} \chi_i{}^a\overline{\chi_i{}^b} = \delta_{ab},$$

or

$$\sum_{i=1}^{r} \frac{\chi_i{}^a\overline{\chi_i{}^b}}{g_i} = \delta_{ab}.$$

But this says that if M is the matrix, $M = (m_{ai})$ $a, i = 1, \cdots, r$, where $m_{ai} = \chi_i{}^a$, then the matrix

$$M' = (r_{ij}) \quad i, j = 1, \cdots, r,$$

with

$$r_{ib} = \frac{1}{g_i}\overline{\chi_i{}^b},$$

is such that

$$MM' = I_r,$$

and so M' is the inverse of M. But then it is also true that $M'M = I_r$, and then it follows

$$\sum_{a=1}^{r} \frac{1}{g_i} \chi_i{}^a\overline{\chi_j{}^a} = \delta_{ij}$$

from which the second relation follows.

The structure of the group ring yields further relations on characters. In the decomposition of R_G as a direct sum of simple rings

$$R_G = R_1 \boxplus \cdots \boxplus R_a \boxplus \cdots \boxplus R_r,$$

let $e_{ij}{}^a$, $i, j = 1, \cdots, n$ be the matrix units for R_a whose unit is

$e_a = e_{11}{}^a + e_{22}{}^a + \cdots + e_{nn}{}^a$. The irreducible representation $\rho_a = \rho^a$ associated with R_a is equivalent to that on a minimal right ideal of R_a. Let this be associated in a specific way, using the minimal ideal $e_{11}{}^a R$ with the basis

$$e_{11}{}^a, e_{12}{}^a, \cdots, e_{1n}{}^a.$$

Then

$$e_{1i}{}^a x = \sum_j \rho_{ij}{}^a(x) e_{1j}{}^a, \quad i = 1, \cdots, n.$$

Now $x = x_1 + x_2 + \cdots + x_a + \cdots + x_r$ with $x_a \in R_a$. Here

$$x_a = e_a x e_a = e_a x = x e_a.$$

Then if

$$x_a = \sum_{i,j} x_{ij}{}^a e_{ij}{}^a,$$

we have $e_{1i}{}^a x = e_{1i}{}^a e_a x = e_{1i}{}^a x_a$, and

$$e_{1i}{}^a x e_{jj}{}^a = x_{ij}{}^a e_{1j}{}^a.$$

But the definition of the representation gives

$$e_{1i}{}^a x e_{jj}{}^a = \rho_{ij}{}^a(x) e_{1j}{}^a.$$

Hence $x_{ij}{}^a = \rho_{ij}{}^a(x)$ in all cases, and so,

$$x_a = \sum_{i,j} \rho_{ij}{}^a(x) e_{ij}{}^a.$$

We write $C_k = \sum_{C_k} x$, since no ambiguity will arise in using the same letter for the class and the sum of the elements regarded as an element of R_G. Then $C_1, C_2, \cdots, C_r$ are a basis for the center Z_G of R_G. Let

$$C_k = C_k{}^1 + \cdots + C_k{}^a + \cdots + C_k{}^r,$$

with $C_k{}^a \in R_a$. Then, since $C_k{}^a$ is in the center of R_a, it is a scalar multiple of e_a. Here

$$C_k{}^a = u_k{}^a e_a.$$

But

$$\text{Trace } \rho^a(C_k{}^a) = \sum_{x \in C_k} \text{trace } \rho^a(x),$$

whence $n_a u_k{}^a = h_k \chi_k{}^a$, where n_a is the degree of ρ_a. Hence

$$u_k{}^a = \frac{h_k \chi_k{}^a}{n_a},$$

and so,

$$C_k^a = \frac{h_k \chi_k^a}{n_a} e_a.$$

The elements $C_1, \cdots, C_r$ of R_G as a basis of $Z(R_G)$ will have a multiplication table

$$C_j C_i = C_i C_j = \sum_k c_{ijk} C_k,$$

where over a field F of characteristic zero the c_{ijk} are non-negative integers, since $C_i C_j = C_j C_i$ contains no negative terms.

As R_G is the direct sum of the simple rings R_a, the components C_i^a will satisfy the same relations as the C_i. Thus:

Theorem 16.6.10.

$$C_i^a C_j^a = C_j^a C_i^a = \sum_k c_{ijk} C_k^a, \quad a = 1, \cdots, r,$$

and

$$\frac{h_i \chi_i^a}{n_a} \frac{h_j \chi_j^a}{n_a} = \sum_k c_{ijk} \frac{h_k \chi_k^a}{n_a}, \quad a = 1, \cdots, r.$$

In the proof of Theorem 16.6.4 we found the relation $n\lambda = 1$, where $n = n_a$ was the degree of an absolutely irreducible representation. Hence the division by n_a in Theorem 16.6.10 is permissible.

Given any two linear spaces L and M over a field F, we define their *tensor product* $L \times M$ in the following way: If $u_1, \cdots, u_m$ are a basis for L, and $v_1, \cdots, v_n$ are a basis for M, then $L \times M$ is the linear space over F with a basis $u_i v_j$, $i = 1, \cdots, m$, $j = 1, \cdots, n$. If

$$u = a_1 u_1 + \cdots + a_m u_m \in L,$$

and

$$v = b_1 v_1 + \cdots + b_n v_n \in M,$$

the *product* $uv = \sum_{i,j} a_i b_j u_i v_j$ is defined as an element of $L \times M$. We verify that a change of basis for L or M corresponds to a change of basis for $L \times M$.

If L is an F-G module for the representation ρ of the group G and M for the representation σ of G, we define the *Kronecker product* $\rho \times \sigma$ of the representations as the representation of G on $L \times M$ given by

$$(uv)x = (ux)(vx), \quad \text{all } u \in L, v \in M, x \in G.$$

Thus if ρ_1 is equivalent to ρ and σ_1 is equivalent to σ, we have $\rho_1 \times \sigma_1$ equivalent to $\rho \times \sigma$, since this corresponds to a change of basis for L and M.

THEOREM 16.6.11. *If ρ and σ are representations of G with characters χ and ψ, respectively, and if ϕ is the character of $\rho \times \sigma$, then for every $x \in G$ we have $\phi(x) = \chi(x)\psi(x)$.*

Proof: If $u_i x = \sum_j \rho_{ij}(x)u_j$, $i = 1, \cdots, m$, $v_i x = \sum_j \sigma_{ij}(x)v_j$, $i = 1, \cdots, n$,

we have

$$\chi(x) = \sum_i \rho_{ii}(x), \quad \psi(x) = \sum_i \sigma_{ii}(x).$$

But with

$$(u_i v_j)x = \sum_{k,t} [\rho_{ik}(x)u_k][\sigma_{jt}(x)v_t],$$

then

$$\begin{aligned} \phi(x) &= \sum_{i,j} \rho_{ii}(x)\sigma_{jj}(x) \\ &= [\sum_i \rho_{ii}(x)][\sum_j \sigma_{jj}(x)] \\ &= \chi(x)\psi(x). \end{aligned}$$

From their definitions we see that the tensor product and the Kronecker product are commutative and associative. Hence if ρ_a and ρ_b are absolutely irreducible representations of G, then

$$\rho_b \times \rho_a = \rho_a \times \rho_b = \sum_{c=1} g_{abc}\rho_c,$$

where the g_{abc} are non-negative integers giving the decomposition of $\rho_a \times \rho_b$ as the direct sum of irreducible representations ρ_c with multiplicity g_{abc}. The same relation will hold for the characters. Thus we state as a theorem:

THEOREM 16.6.12. *The absolutely irreducible characters of a group G satisfy*

$$\chi_i^a \chi_i^b = \sum_c g_{abc}\chi_i^c,$$

where the g_{abc} are non-negative integers, being the multiplication constants of a commutative and associative algebra.

We summarize the character relations we have found. Let $C_1 = 1$, $C_2, \cdots, C_r$ be the classes of G, $\rho_1 =$ the identity representation, and $\rho_2, \cdots, \rho_r$ be the absolutely irreducible representations, where χ_i^a is the character of an element of the ith class in the ath representation

$$\begin{array}{c|ccccc} & C_1 & \cdots & C_i & \cdots & C_r \\ \hline \rho_1 & \chi_1^1 & \cdots & \chi_i^1 & \cdots & \chi_r^1 \\ & \cdot & \cdot & \cdot & \cdot & \cdot \\ \rho_a & \chi_1^a & \cdots & \chi_i^a & \cdots & \chi_r^a \\ & \cdot & \cdot & \cdot & \cdot & \cdot \\ \rho_r & \chi_1^r & \cdots & \chi_i^r & \cdots & \chi_r^r. \end{array}$$

Here we have, $g_i h_i = g$, where there are h_i elements in the class C_i.

1) On the rows: $$\sum_{i=1}^{r} \frac{\chi_i^a \overline{\chi_i^b}}{g_i} = \delta_{ab}.$$

2) On the columns: $$\sum_{a=1}^{r} \chi_i^a \overline{\chi_j^a} = \delta_{ij} g_i.$$

3) Within each row:
$$\frac{h_i \chi_i^a}{n_a} \frac{h_j \chi_j^a}{n_a} = \sum_k c_{ijk} \frac{h_k \chi_k^a}{n_a}.$$

4) Within each column:
$$\chi_i^a \chi_i^b = \sum_c g_{abc} \chi_i^c.$$

Here c_{ijk} and g_{abc} are non-negative integers which are the multiplication constants of commutative and associative algebras.

Every representation of a group G as a permutation group $\pi(G)$ can also be regarded as a matrix representation, since if

$$\pi(x) = \begin{pmatrix} u_1 & \cdots & u_n \\ u_{i_1} & \cdots & u_{i_n} \end{pmatrix}$$

for $x \in G$, we may regard this as the representation ρ on a basis $u_1, \cdots, u_n$, where

$$u_j x = u_{i_j}.$$

Here $\chi(x)$ is the number of letters fixed by $\pi(x)$.

THEOREM 16.6.13. *In a permutation representation $\pi(G)$ of a group G of order g, $\sum_{x \in G} \chi(x) = kg$, where k is the number of transitive constituents. Here the representation as a matrix representation contains the identical representation exactly k times.*

Proof: Let $n_1, n_2, \cdots, n_k$ be the number of letters in the k transitive constituents. Then a subgroup H_j fixing a letter a_j of the jth transitive constituent will be of index n_j and of order g/n_j. Hence the letter a_j is fixed g/n_j times in all the elements of G. Thus the number of times letters of the jth constituent are fixed is $n_j \cdot g/n_j = g$. Hence the number of times letters of any of the k constituents are fixed is kg or $\sum_{x \in G} \chi(x) = kg$. If $\chi = \sum_a m_a \chi^a$ gives χ as a sum of absolutely irreducible characters, then $\sum_{x \in G} \chi(x) = m_1 g$ by the corollary to Theorem 16.6.5. Hence the representation contains the identical representation $m_1 = k$ times.

THEOREM 16.6.14. *If χ is the character of a transitive permutation group G, then $\sum_{x \in G} \chi^2(x) = tg$, where t is the number of transitive constituents of a subgroup H fixing a letter. t is also the number of double cosets $H\ x\ H$ in G.*

Proof: Let G be a transitive permutation group on letters $1, 2, \cdots, n$. Let H_i be the subgroup fixing i, $i = 1, \cdots, n$. We may take $H = H_1$, since all the H_i are conjugate. Let h be the order of H. Then

$$\sum_{x \in H_i} \chi(x) = th$$

by the previous theorem. Hence

$$\sum_i \sum_{x \in H_i} \chi(x) = tnh = tg.$$

But on the left we have counted $\chi(x)$ once for every H_i containing x. But x fixes $\chi(x)$ letters and so is contained in $\chi(x)$ different H_i's. (This number is zero if x displaces all letters.) Hence

$$tg = \sum_i \sum_{x \in H_i} \chi(x) = \sum_{x \in G} \chi^2(x).$$

But we easily see that t is the number of double cosets $H\ x\ H$ in G. For let $G = H + Hx_2 \cdots + Hx_n$, where H is the subgroup fixing 1 and $x_{i,} = (1, i, \cdots)\ i = 2, \cdots, n$. Then if $Hx_iH = Hx_jH$, we have $x_i = h_1x_jh_2$ with $h_1, h_2 \in H$. Here the element h_2 must take j into i, whence i and j are in a transitive constituent of H. Conversely, suppose that i and j are in a transitive constituent of H. Then for some $h_2 \in H$, h_2 takes j into i, x_jh_2 takes 1 into i, whence $x_jh_2 \in Hx_i$ and $x_i = h_1x_jh_2$ and $Hx_iH = Hx_jH$. Now every double coset of H is one of Hx_iH. Hence there are exactly as many double cosets HxH as there are transitive constituents in H.

THEOREM 16.6.15. *A doubly transitive permutation representation of a group G over the complex field is the sum of the identical representation and an absolutely irreducible representation.*

Proof: For a doubly transitive representation,

$$\sum \chi^2(x) = 2g,$$

since a subgroup H fixing a letter 1, has exactly two transitive constituents, 1, and the remaining letters. Since $\chi(x)$ is real, we may write this

$$\sum_{x \in G} \chi(x)\overline{\chi(x)} = 2g.$$

But if $\chi = \sum_a c_a\chi^a$ expresses χ as a sum of absolutely irreducible characters, we have

$$\begin{aligned}\sum_{x \in G} \chi(x)\overline{\chi(x)} &= \sum_x \Big[\sum_a c_a\chi^a(x)\Big]\Big[\sum c_a\overline{\chi^a(x)}\Big] \\ &= g\sum_a c_a^2\end{aligned}$$

by the orthogonality relations. Hence $\sum_a c_a^2 = 2$, whence $c_1 = 1$, as we already know, and for exactly one further c_a we have $c_a = 1$.

16.7. Imprimitive Representations.

Suppose that we have a representation module M for a group G

which is the direct sum of subspaces $M_1, M_2, \cdots, M_n$, on which the representation is *transitive* but *imprimitive*. By this we mean:

1) For any M_i and M_j there is an $x \in G$ such that
$$M_ix = M_j.$$
2) For every M_i and every $x \in G$ there is an M_j such that
$$M_ix = M_j.$$

The first of these is the transitive property; the second, the imprimitivity.

Choose a particular subspace M_1. The set of all x such that $M_1x = M_1$ surely includes the identity $x = 1$ and so is not vacuous and is readily seen to be a subgroup H of G. Thus for $h \in H$,
$$M_1h = M_1,$$
and so M_1 is a representation module for H. If $b_i \in G$ is an element such that
$$M_1b_i = M_i,$$
then the elements x such that $M_1x = M_i$ are the elements of the coset Hb_i. Thus we have
$$G = H + Hb_2 + Hb_3 \cdots + Hb_n,$$
where
$$M_1(hb_i) = M_i \qquad i = 1, \cdots, n,$$
and we have associated the subspaces with the left cosets of H. If x is such that
$$M_ix = M_i,$$
then
$$M_1b_ix = M_1b_i,$$
and
$$M_1b_ixb_i^{-1} = M_1,$$
whence $b_ixb_i^{-1} \in H$ or $x \in b_i^{-1}Hb_i$, a subgroup conjugate to H. Finally, if
$$M_ix = M_j,$$
then
$$x \in b_i^{-1}Hb_j.$$

Let ρ_1 be the representation of H associated with a basis $v_1, \cdots, v_m$ of M_1. We may then take $v_1b_i, \cdots, v_mb_i$ as a basis of M_i. Then for an arbitrary $x \in G$ we have
$$M_1x = M_{j_1} \cdots, \quad M_ix = M_{j_i} \cdots, \quad M_nx = M_{j_n}.$$
Here $M_{j_1}, \cdots, M_{j_n}$ must be a permutation of $M_1, \cdots, M_n$, since operating on them with x^{-1} we must get back to $M_1, \cdots, M_n$. Here

if $M_i x = M_j$ ($j = j_i$), we have $x \in b_i^{-1}Hb_j$, or $b_i x b_j^{-1} = h_{ij} \in H$. Hence, with $v_k b_i$, $k = 1, \cdots, m$ as a basis of M_i, we have

$$v_k b_i \cdot x = (v_k \cdot h_{ij})b_j \qquad k = 1, \cdots, m.$$

In other words this part of the representation is completely determined by the representation of h_{ij} on M_1:

$$v_k(b_i x b_j^{-1}) = v_k h_{ij}.$$

Hence

$$\rho(x) = (\rho_1(b_i x b_j^{-1})) \quad i,j = 1, \cdots, n,$$

with the convention that $\rho_1(y) = 0$ if $y \notin H$. Here $\rho\ (x)$ is of degree mn, made up of n^2 matrices of degree m. Thus every representation transitive on subspaces $M_1, \cdots, M_n$ and imprimitive on these is determined by a representation ρ_1 of a subgroup H of index n in G. The converse is also true. Let ρ_1 be any representation of H where

$$G = H + Hb_2 + \cdots + Hb_n.$$

Then define

$$\rho(x) = (\rho_1(b_i x b_j^{-1}))\ i,j = 1, \cdots, n,$$

with the convention that $\rho(y) = 0$ if $y \notin H$. Then, using block multi plication of matrices,

$$\begin{aligned} \rho(x)\rho(y) &= (\rho_1(b_i x b_j^{-1}))(\rho_1(b_k y b_t^{-1})) \\ &= (\rho_1(b_i x b_j^{-1}))(\rho_1(b_j y b_t^{-1})) \\ &= (\rho_1(b_i x y b_t^{-1})) = \rho(xy), \end{aligned}$$

and trivially,

$$\rho(1) = (\rho_1(b_i 1 b_i^{-1})) = (\rho_1(1)) = I.$$

Thus $\rho(x)$ is a representation for G.

THEOREM 16.7.1. *Given a representation ρ_1 of degree m of a subgroup H of a group G, if $G = H + Hb_2 + \cdots + Hb_n$, then*

$$\rho(x) = (\rho_1(b_i x b_j^{-1}))\ i,j = 1, \cdots, n,$$

with the convention $\rho_1(y) = 0$ if $y \notin H$ is a representation of G of degree mn on a module M with subspaces $M_1, M_2, \cdots, M_n$ corresponding to H, $Hb_2, \cdots, Hb_n$, respectively. ρ is transitive and imprimitive on $M_1, \cdots, M_n$. Conversely, any representation transitive and imprimitive on subspaces of a module is of this type.

Proof: If $u_1, \cdots, u_m, \cdots, u_{mn}$ are a basis for the representation module of ρ of the theorem, then $u_1, \cdots, u_m$ are a basis for ρ_1 on H and $u_{m(i-1)+j} = u_j b_i$, $i = 1, \cdots, n$. It follows that the module M with basis $u_1, \cdots, u_{mn}$ has subspaces $M_1, \cdots, M_n$ on which ρ is transitive and imprimitive. We say that the representation ρ of G is *induced* by the representation ρ_1 of H.

COROLLARY 16.7.1. *The representation ρ of G induced by the representation ρ_1 of H does not depend on the choice of coset representatives of H in G.*

This follows since a change of representatives does not alter the subspaces $M_1, \cdots, M_n$ but merely changes the bases for them.

THEOREM 16.7.2. *Let χ be the character of the representation ρ of G induced by the representation ρ_1 of H, whose character is χ_1. Let x be in the class C_j of conjugates in G with h_j elements, and let $g = g_j h_j$, where g is the order of G. Let h be the order of H. Then*

$$\chi(x) = \frac{g_j}{h} \sum_{z \in C_j \cap H} \chi_1(z).$$

Proof: $\chi(x) = \sum_{i=1}^{n} \chi_1(b_i x b_i^{-1})$ with the convention $\chi_1(w) = 0$ if $w \notin H$. Then

$$\chi(x) = \frac{1}{h} \sum_{y \in G} \chi_1(yxy^{-1}),$$

since every element y of Hb_i contributes the same amount to the sum on the right, viz., $\chi_1(b_i x b_i^{-1})$. Here yxy^{-1}, as y ranges over G, ranges over C_j and gives each $z \in C_j$ exactly g_j times. Thus

$$\sum_{y \in G} \chi_1(yxy^{-1}) = g_j \sum_{z \in C_j \cap H} \chi_1(z),$$

proving the theorem.

THEOREM 16.7.3 (RECIPROCITY THEOREM). *Let ρ, ρ_1 be absolutely irreducible representations of a group G and a subgroup H, respectively over a field of characteristic zero. Then the multiplicity of ρ_1 occurring in ρ restricted to H is the same as the multiplicity of ρ in the representation ρ^* of G induced by ρ_1.*

Proof: Let $\chi = \chi^a$ be the character of ρ, and $\chi_1 = \chi_1{}^c$ be the character of ρ_1. Let χ^* be the character of ρ^*, $\chi^* = \sum_b m_b\chi^b$, χ^b the irreducible characters of G. When restricted to H, let $\chi = \chi^a = \sum_d n_d\chi_1{}^d$, $\chi_1{}^d$ the irreducible characters of H. Here the multiplicity of ρ in ρ^* is m_a, and the multiplicity of ρ_1 in ρ restricted to H is n_c. By the previous theorem,

$$\frac{1}{g_j}\chi_j{}^* = \frac{1}{g_j}\sum_b m_b\chi_j{}^b = \frac{1}{h}\sum_{z\in C_j\cap H}\chi_1{}^c(z).$$

Here the convention is that a void sum is zero. Multiply this by $\overline{\chi_j{}^a}$ and sum over j. We have

$$\sum_{j,b} m_b\frac{\overline{\chi}_j{}^a\chi_j{}^b}{g_j} = \frac{1}{h}\sum_j\overline{\chi_j{}^a}\sum_{z\in C_j\cap H}\chi_1{}^c(z),$$

whence, using the orthogonality relations in G and also in H,

$$\begin{aligned} m_a &= \frac{1}{h}\sum_{d,j} n_d\overline{\chi}_{1,j}{}^d\sum_{z\in C_j\cap H}\chi_1{}^c(z) \\ &= \frac{1}{h}\sum_{d,z\in H} n_d\overline{\chi}_1{}^d(z)\chi_1{}^c(z) \\ &= \frac{1}{h}hn_c = n_c, \end{aligned}$$

the statement of the theorem.

16.8. Some Applications of the Theory of Characters.

We shall assume throughout this section that we are dealing with the field F of complex numbers, though it will be clear to the reader that a number of the results carry over to all fields whose characteristic is not divisible by the order of the group G being represented.

First some facts will be needed about algebraic numbers.† A number θ is said to be an *algebraic number* if it is the root $x = \theta$ of a monic polynomial:

$$x^n + a_1x^n + \cdots + a_n = 0,$$

where $a_1, \cdots, a_n$ are rational numbers. θ is said to be an *algebraic integer* if it is the root of such a polynomial where $a_1, \cdots, a_n$ are rational integers.

† These facts are covered in Birkoff and MacLane [1], pp. 410–422.

THEOREM 16.8.1. *A rational number which is an algebraic integer is a rational integer.*

Proof: Suppose that $\theta = r/s$ is a rational number expressed in its lowest terms and that it satisfies

$$x^n + a_1x^{n-1} + \cdots + a_n = 0,$$

where $a_1, \cdots, a_n$ are integers. Then

$$r^n = -s(a_1r^{n-1} + a_2sr^{n-2} \cdots + a_ns^{n-1}).$$

Hence any prime dividing s must divide r^n and hence r. This cannot happen if r/s is in its lowest terms and $s \neq 1$. Hence $s = 1$, and so $\theta = r$ is a rational integer.

THEOREM 16.8.2. *Algebraic numbers form a field. The sum or product of two algebraic integers is an algebraic integer.*

Proof: Let θ be an algebraic number satisfying $x^n + a_1x^{n-1} + \cdots + a_n = 0$, and ϕ be an algebraic number satisfying $x^m + b_1x^{m-1} + \cdots + b_m = 0$. Let

$$v_{i,j} = \theta^i\phi^j \qquad i = 0, \cdots n-1, j = 0, \cdots m-1.$$

Then

$$\theta v_{i,j} = v_{i+1,j} \quad \text{for } i = 0, \cdots n-2,$$

and

$$\theta v_{n-1,j} = -a_1v_{n-1,j} \cdots -a_nv_{0j}.$$

Similarly,

$$\phi v_{i,j} = v_{i,j+1} \quad \text{for } j = 0, \cdots, m-2,$$

and

$$\phi v_{i,m-1} = -b_1v_{i,m-1} \cdots -b_mv_{i0}.$$

LEMMA 16.8.1. *If $y_1, \cdots, y_N$ are numbers not all zero and if z is a number such that*

$$zy_i = \sum_j a_{ij}y_j \qquad i = 1, \cdots, N,$$

with all a_{ij} rational, then z is an algebraic number. If the a_{ij} are integers, then z is an algebraic integer.

Proof: The hypothesis gives us a system of equations:

$$\begin{aligned}(a_{11} - z)y_1 + a_{12}y_2 + \cdots + a_{1N}y_N &= 0,\\ a_{21}y_1 + (a_{22} - z)y_2 + \cdots + a_{2N}y_N &= 0,\\ \cdots\cdots\cdots\cdots\cdots\cdots\cdots\cdots&\\ a_{N1}y_1 + a_{N2}y_2 + \cdots + (a_{NN} - z)y_N &= 0,\end{aligned}$$

which, when regarded as linear equations for the y's, has the solution $y_1, \cdots, y_N$, where not all y's are zero. Hence the determinant of the coefficients must be zero.

Thus
$$\begin{vmatrix} a_{11} - z, & a_{12}, & \cdots, & a_{1N} \\ a_{21}, & a_{22} - z, & \cdots, & a_{2N} \\ \cdots & \cdots & \cdots & \cdots \\ a_{N1}, & \cdots, & \cdots, & a_{NN} - z \end{vmatrix} = 0.$$

But this, on expansion, is

$$(-1)^N z^N + p_1 z^{N-1} \cdots + p_N = 0,$$

where the p's are integral polynomials in the a's. Hence if the a's are rational, z is an algebraic number, and if the a's are integers, z is an algebraic integer.

We may use this lemma to prove our theorem. We exclude the trivial cases when θ or ϕ is 0. Then we take $y_1, \cdots, y_N$ to be the v_{ij}, and since $v_{00} = 1$, the v_{ij} are not all zero. Here take z as $\theta + \phi$ or as $\theta\phi$. The a_{ij} of the lemma in these cases will be integral polynomials in the $a_1, \cdots, a_n$ and $b_1, \cdots, b_m$. Hence $z = \theta + \phi$ and $z = \theta\phi$ will be algebraic numbers, and if $a_1, \cdots, a_n$ and $b_1, \cdots, b_m$ are integers, then $\theta + \phi$ and $\theta\phi$ will be algebraic integers. Thus the sum and product of algebraic numbers are algebraic numbers, and the sum and product of algebraic integers are algebraic integers. Finally, if θ is an algebraic number $\neq 0$ satisfying $z^n + a_1 z^{n-1} \cdots + a_n = 0$, we may, if necessary, divide by a power of z to get a constant term $a_n \neq 0$. Here

$$w^n + \frac{a_{n-1}}{a_n} w^{n-1} + \cdots + \frac{1}{a_n} = 0$$

is an equation which $1/\theta$ satisfies. Trivially, $-\theta$ satisfies $z^n - a_1 z^{n-1} \cdots + (-1)^n a_n = 0$. Hence algebraic integers form an integral domain and algebraic numbers form a field.

THEOREM 16.8.3. *Every character $\chi(x)$ is an algebraic integer. The numbers $h_i \chi_i^a / n_a$ of Theorem* 16.6.10 *are algebraic integers.*

Proof: An mth root of unity satisfies $x^m - 1 = 0$ and so is an algebraic integer. Thus, by Theorem 16.6.8, every character $\chi(x)$ is a sum of roots of unity and so is an algebraic integer. Since the c_{ijk} are integers in Theorem 16.6.10, we may apply Lemma 16.8.1 with the

$$\frac{h_i\chi_i^a}{n_a} = \eta_i^a, \quad i = 1, \cdots, r$$

as the y's of the lemma and also any one of them as z, and we conclude that η_i^a are algebraic integers.

THEOREM 16.8.4. *The degrees n_a of the absolutely irreducible representations of the finite group G are divisors of its order g.*

Proof: From our orthogonality relations

$$\sum_{i=1}^{r} \frac{\chi_i^a\overline{\chi_i^a}}{g_i} = 1.$$

This becomes, since $g_ih_i = g$,

$$\sum_{i=1}^{r} \frac{\chi_i^a h_i\overline{\chi_i^a}}{g} = 1,$$

or

$$\sum_{i=1}^{r} \frac{h_i\chi_i^a}{n_a}\,\overline{\chi_i^a} = \frac{g}{n_a}.$$

But the left side is a sum of products of algebraic integers. Hence g/n_a is an algebraic integer and, being rational, is a rational integer. Thus n_a divides g.

For our use of algebraic numbers we need a little of the theory of symmetric functions. If we expand

$$\begin{aligned}(z - x_1)(z - x_2) \cdots (z - x_n)\\ = z^n - E_1z^{n-1} + E_2z^{n-2} \cdots + (-1)^nE_n,\end{aligned}$$

we have

$$\begin{aligned}E_1 &= \sum x_i\\ E_2 &= \sum x_ix_j\\ &\cdots\cdots\\ E_r &= \sum x_{i_1}x_{i_2}\cdots x_{i_r}\\ &\cdots\cdots\\ E_n &= x_1x_2\cdots x_n.\end{aligned}$$

Here $E_1, \cdots, E_n$ are clearly unchanged by any permutation of $x_1, \cdots, x_n$ and are called the elementary symmetric functions of $x_1, \cdots, x_n$. A polynomial $P(x_1, \cdots, x_n)$ over a field F is called a *symmetric function* if it is unchanged by the entire symmetric group of permutations of $x_1, \cdots, x_n$.

THEOREM 16.8.5. *Every symmetric function $P(x_1, \cdots, x_n)$ is a polynomial $Q(E_1, \cdots, E_n)$ in the elementary symmetric functions $E_1, \cdots, E_n$, and the coefficients in Q are integral polynomials in the coefficients of P.*

Proof: If P is symmetric, then its terms of each degree are separately symmetric functions. The theorem is trivially true for degree 1, the only symmetric functions being cE_1, $c \in F$. Moreover, P is a sum of symmetric polynomials, each determined by a single term in it, say,

$$c(x_1 \cdots x_r)^a(x_{r+1} \cdots x_{r+s})^b \cdots (x_{u+1} \cdots x_{u+v})^t,$$

where the exponents $a > b \cdots > t$ are strictly decreasing. It is enough to prove the theorem for symmetric sums:

$$K = \sum(x_1 \cdots x_r)^a(x_{r+1} \cdots x_{r+s})^b \cdots (x_{u+1} \cdots x_{u+v})^t,$$

with $a > b \cdots > t$. We proceed by induction on (1) the degree of K; (2) the value of a; and (3) the value of r. If $x_1, \cdots, x_n$ appear in every term, we factor out E_n and have the remaining factor symmetric of lower degree. Hence we may assume $u + v < n$. If $a = 1$, then $K = E_r$. Otherwise consider

$$\sum x_1 \cdots x_r \cdot \sum(x_1 \cdots x_r)^{a-1}(x_{r+1} \cdots x_{r+s})^b \cdots (x_{u+1} \cdots x_{u+v})^t$$
$$= E_r \cdot K^*.$$

Here $E_r \cdot K^* = K +$ other terms. Both K^* and the other terms appearing precede K in our induction, and so our theorem is proved.

The rational polynomial of lowest degree which has an algebraic number θ as a root is called the *minimal polynomial* for θ. If

$$f(x) = x^n + a_1x^{n+1} + \cdots + a_n$$

is the minimal polynomial, it is a divisor of any rational polynomial $h(x)$ which has θ as a root. Now if

$$f(x) = (x - \theta_1)(x - \theta_2) \cdots (x - \theta_n),$$

where $\theta = \theta_1$, we say that $\theta_1, \cdots, \theta_n$ are the *conjugates* of θ. Hence the conjugates of θ also satisfy any rational equation $h(x) = 0$ which θ satisfies. Hence if θ is an algebraic integer, its conjugates are also algebraic integers, and so the coefficients of the minimal polynomial for θ, being the symmetric functions of the conjugates of θ, are algebraic integers and hence rational integers.

In the study of representations we are mostly concerned with roots of unity. The primitive mth roots of unity are $\omega = \exp(2\pi i/m)$ and powers ω^j where $(j, m) = 1$. ω and the other primitive mth roots of unity satisfy $x^m - 1 = 0$, and no equation $x^r - 1 = 0$ with $0 < r < m$. The remaining mth roots of unity satisfy equations $x^d - 1 = 0$, where d runs over the divisors of m. Removing all factors from $x^m - 1$ which it has in common with $x^d - 1$, we are left with a rational $f(x)$ which has its roots precisely the primitive mth roots of unity. Hence

$$f(x) = \prod_j (x - \omega^j) \qquad (j, m) = 1,$$

and $f(x)$ is rational and integral of degree $\phi(m)$, this being the Euler ϕ function. $f(x)$ is in fact irreducible, but this is difficult to prove without using more theory of algebraic numbers than we can prove here. We need only know that the elementary symmetric functions of the primitive mth roots are rational integers.

THEOREM 16.8.6. *Let ρ_a be an absolutely irreducible representation of G of degree n, and let there be a class C_i where $(h_i, n) = 1$. Then either* (1) $\chi_i^a = 0$, *or* (2) $\chi_i^a = n\omega$, *where ω is a root of unity and C_i is in the center of ρ_a.*

Proof: For a particular $x \in C_i$ we may transform ρ_a so that $\rho_a(x)$ is in diagonal form. If all the n characteristic roots of x are equal, say, to some mth root of unity ω, then

$$\rho_a(x) = \omega I_n, \quad \chi(x) = n\omega,$$

and x is in the center of ρ_a. This is the second alternative of the theorem. Thus we must show that if the characteristic roots of x are not all equal, then, under the hypotheses of the theorem, $\chi(x) = 0$. Now in this case, χ being of order m, $\chi_i^a = \chi(x) = \omega^{e_1} + \cdots + \omega^{e_n}$ and $|\chi_i^a| < n$, since the ω^{e_i} do not all have the same argument. Here

$$\frac{h_i \chi_i^a}{n}$$

is an algebraic integer, and since $(h_i, n) = 1$, there are integers r and s so that $rh_i + sn = 1$. Hence

$$r\left(\frac{h_i\chi_i{}^a}{n}\right) + s\cdot\chi_i{}^a = \frac{\chi_i{}^a}{n}$$

is an algebraic integer. Here

$$\left|\frac{\chi_i{}^a}{n}\right| < 1,$$

and also

$$\xi = \frac{\chi_i{}^a}{n} = \frac{\omega^{e_1} + \cdots + \omega^{e_n}}{n}.$$

Replacing ω by its conjugates ω^j, we have

$$\prod_{(j,m)=1} [z - q(\omega^j)],$$

a polynomial whose coefficients are symmetric functions of the conjugates of ω and hence rational. Thus the conjugates of ξ lie among the numbers.

$$\frac{\omega^{je_1} + \cdots + \omega^{je_n}}{n},$$

and so, for every conjugate $\xi^{(i)}$ of ξ we have $|\xi^{(i)}| \leq 1$, and every conjugate is an algebraic integer. Now $|\xi| = |\xi^{(1)}| < 1$, and so the product $|\xi^{(1)} \cdots \xi^{(s)}| < 1$, this being the product of all conjugates of ξ. This must be a rational integer and hence must be 0. Thus $\xi^{(1)} \cdots \xi^{(s)} = 0$. Thus at least one of the conjugates is 0. But 0 is its own only conjugate and so $\xi = \xi^{(1)} = 0$, and so,

$$\xi = \frac{\chi_i{}^a}{n} = 0,$$

whence $\chi_i{}^a = 0$, as we were to prove.

THEOREM 16.8.7*. (1) *If the number h_i of elements in a class C_i of a group G is a prime power, then G is not a simple group. More explicitly, there is a homomorphic image of G in which the elements of C_i are in the center.* (2) *Groups of order p^aq^b, p, q primes are solvable.*

Proof: (1) Let $n_1 = 1, n_2, \cdots, n_r$ be the degrees of the absolutely

* W. Burnside [2], pp. 322–323.

irreducible representations of G. Let $h_i = p^s$ be the number of elements in C_i. For the regular representation of G we have $\chi(x) = 0$ for $x \in C_i$ (since $x \neq 1$); also

$$\chi(x) = \sum_{a=1}^{r} n_a \chi_i^a$$

by the decomposition of the regular representation.

Here $n_1\chi_i^1 = 1$. For the remaining terms if $p \nmid n_a$, then by Theorem 16.8.6, either $\chi_i^a = 0$ or C_i is in the center of the homomorphic image $\rho_a(G)$. But if $\chi_i^a = 0$ in every instance where $p \nmid n_a$, we would have

$$0 = 1 + \sum_{a=2}^{r} n_a \chi_i^a = 1 + p\alpha,$$

where α is an algebraic integer. This would make $-(1/p)$ an algebraic integer, which is a conflict. Hence for some ρ_a, C_i is in the center of $\rho_a(G)$.

(2) Let G be a group of order $p^a q^b$. We proceed by induction on the order of such groups, p-groups being solvable. An element in the center of a Sylow q-group is either in the center of G or has a number of conjugates which is a power of p. In either event G has a proper normal subgroup H and both H and G/H are solvable by our induction, and so G itself is solvable.

THEOREM 16.8.8 (FROBENIUS). *If G is a transitive permutation group of degree n whose permutations other than the identity leave at most one of the symbols invariant, then those permutations of G which displace all the symbols form together with the identity a normal subgroup of order n.*

Proof: Let G permute $1, 2, \cdots, n$ and let H_i be the subgroup fixing i. Then by hypothesis $H_i \cap H_j = 1$ for $i \neq j$. If $H = H_1$ has order h, then all H_i have order h and the elements $x \neq 1$ belonging to the H_i will number $(h - 1)n$. $[G:H] = n$ so that G is of order hn. This leaves exactly n other elements, the identity and $n - 1$ elements displacing all letters.

Let ψ be an absolutely irreducible character of H and ψ' the induced character of G. G is given as a representation of itself and as such has a character $\theta_1 = \psi_1'$, where ψ_1 is the unit character of H. By Theorem 16.6.13, θ_1 is the sum of the unit character of G and another character, say, θ. Denote by r_G the character of the regular representation of G. Let us put $\omega = r_G - h\theta$. Our theorem will depend on proving that ω is a character of G. In the following character table let x be a typical element $\neq 1$ of the H's, and y a typical element displacing all letters.

	1	x	y
ψ'	mn,	$\psi(x)$,	0
θ	$n-1$,	0,	-1
r_G	nh,	0,	0
ω	h,	0,	h

Here ψ', θ, r_G are known to be characters. If ω is a character, then since $\omega(1) = h$, it is the character of a representation of degree h. Since $\omega(y) = h$, $\omega(x) = 0$, every y but no x is represented by the identity. Hence ω is a representation of G homomorphic to G with a kernel consisting of precisely 1 and the elements y displacing all letters. As the kernel of a homomorphism, the identity and the $n - 1$ elements displacing all letters form a normal subgroup of G.

We now prove that ω is a character. Let s be the number of classes in H and ψ^a, $a = 1 \cdots s$ the absolutely irreducible characters of H of degree m_a. Then

$$r_H = \sum_{a=1}^{s} m_a\psi^a.$$

But $r_G = (r_H)'$ and $h = \sum_a m_a^2$. Hence

$$\omega = \sum_a m_a[(\psi^a)' - m_a\theta].$$

Hence it is enough to prove $\psi' - m\theta$ is a character of G for any $\psi = \psi^a$, $m = m_a$.

We calculate the scalar product

$$(\psi' - m\theta, \psi' - m\theta) = (\psi', \psi') - 2m(\psi', \theta) + m^2(\theta, \theta).$$

Now $g = nh$, so from our preceding character table,

$$\begin{aligned}(\psi', \psi') &= \frac{m^2n}{h} + \frac{1}{nh}\sum_{x\neq 1} \psi(x)\overline{\psi(x)} \\ &= \frac{m^2n}{h} + \frac{1}{h}\sum_{\substack{x \in H \\ x \neq 1}} \psi(x)\overline{\psi(x)}.\end{aligned}$$

But

$$\sum_{x \in H} \psi(x)\overline{\psi(x)} = h,$$

whence

$$(\psi', \psi') = [m^2 n + (h - m^2)]/h.$$

Similarly, from our table, $(\psi', \theta) = m(n - 1)/h$, and

$$(\theta, \theta) = [(n - 1)^2 + (n - 1)]/nh = (n - 1)/h.$$

Hence

$$(\psi' - m\theta, \psi' - m\theta) = 1.$$

But $\psi' - m\theta$ is in any event a linear combination of characters with integral coefficients, say, $\psi' - m\theta = \sum c_a \chi^a$, which gives

$$(\psi' - m\theta, \psi' - m\theta) = \sum c_a^2,$$

whence $\sum c_a^2 = 1$, and there is exactly one $c_a = \pm 1$ and the remainder zero. Thus $\psi' - m\theta = \pm\psi^a$. But $(\psi' - m\theta)(1) = m > 0$, whence $\psi' - m\theta = \psi^a$ is a character of G. This proves that $\psi' - m\theta$ is a character, and so, ω is a character, proving our theorem.

16.9. Unitary and Orthogonal Representations.

With an arbitrary $n \times n$ matrix A,

$$A = (a_{ij}) \quad i, j = 1, \cdots, n, \tag{16.9.1}$$

we may associate a bilinear form $B(y, x)$,

$$B(y, x) = \sum_{i,j} a_{ij} y_i x_j, \quad i, j = 1, \cdots, n, \tag{16.9.2}$$

and, of course, we may also associate the matrix A with the bilinear form. We are interested in the way in which linear transformation of the x's and y's in a bilinear form affects the corresponding matrix. Let us put

$$x_j = \sum_k c_{jk} x_k' \qquad j, k = 1, \cdots, n, \tag{16.9.3}$$

$$y_i = \sum_s d_{is} y_s' \qquad i, s = 1, \cdots, n.$$

Then

$$B(y, x) = B'(y', x') = \sum_{i,j,k,s} d_{is} a_{ij} c_{jk} y_s' x_k'. \tag{16.9.4}$$

Thus we see that $B'(y', x')$ corresponds to

(16.9.5) $$A' = D^T AC, \quad D = (d_{is}), \quad C = (c_{jk}).$$

We are not interested here in the most general bilinear forms, but in certain special kinds. We shall throughout this section take our coefficients from the field of complex numbers. If

(16.9.6) $$a_{ji} = \overline{a_{ij}}, \quad i, j = 1, \cdots, n,$$

where a_{ij} denotes the complex conjugate of a_{ij}, we say that the matrix A is *Hermitian*. Thus a Hermitian matrix is associated with a *Hermitian form* $H(\bar{x}, x)$:

(16.9.7) $$H(\bar{x}, x) = \sum_{i,j} a_{ij}\bar{x}_i x_j, \quad a_{ji} = \overline{a_{ij}}, \quad i, j = 1, \cdots, n.$$

Note that in a Hermitian form or matrix, the coefficients $a_{ii} = \bar{a}_{ii}$ are real. A real Hermitian matrix is a real symmetric matrix and corresponds to a real quadratic form $Q(x)$:

(16.9.8) $$Q(x) = \sum_{i,j} a_{ij} x_i x_j, \quad a_{ji} = a_{ij}, \quad i, j = 1, \cdots, n.$$

The non-singular linear transformations which leave a Hermitian form (or a quadratic form) invariant clearly form a group. Here it is understood that the conjugates $\bar{x}_i$ undergo the transformation conjugate to that applied to the x_i.

DEFINITION: *A matrix U satisfying*

(16.9.9) $$\overline{U}^T I U = I$$

is called a unitary matrix.

DEFINITION: *A matrix V satisfying*

(16.9.10) $$V^T I V = I$$

is called an orthogonal matrix.

It is immediately evident that unitary and orthogonal matrices are nonsingular, and that they form groups. The unitary matrices are those giving the group of linear transformations leaving $\bar{x}_1 x_1 + \cdots + \bar{x}_n x_n$ invariant, and the orthogonal matrices are those giving the group of linear transformations leaving $x_1^2 + \cdots + x_n^2$ invariant. A real unitary matrix is orthogonal, but in a strict sense there are also orthogonal matrices which are not real; for example,

(16.9.11) $$V = \begin{pmatrix} i, & \sqrt{2} \\ -\sqrt{2}, & i \end{pmatrix}.$$

But here when we speak of an orthogonal matrix, it will be understood that we mean a real matrix.

We say that a representation $\rho(g)$ of a group G is *unitary* if all the matrices $\rho(g)$, $g \in G$ are unitary, and *orthogonal* if all the matrices $\rho(g)$, $g \in G$ are orthogonal.

THEOREM 16.9.1. *Every representation of a finite group G over the complex (real) field is equivalent to a unitary (orthogonal) representation.*

Proof: If in a Hermitian form

(16.9.12) $$H(\bar{x}, x) = \sum a_{ij}\bar{x}_i x_j, \quad a_{ji} = \bar{a}_{ij},$$

the variables x_i are given complex values, then H represents a real number, since we may pair the terms:

(16.9.13) $$a_{ji}\bar{x}_j x_i + a_{ij}\bar{x}_i x_j = \bar{a}_{ij}\bar{x}_j x_i + a_{ij}\bar{x}_i x_j,$$

each pair being the sum of a complex number and its complex conjugate. The diagonal terms $a_{ii}\bar{x}_i x_i$ are, of course, real. We say that $H(\bar{x}, x)$ is *positive definite* if it represents zero only when all variables are zero and when it otherwise represents positive numbers. Clearly, the property of being positive definite is unchanged if the variables are transformed by a nonsingular transformation. A particular positive definite form with n variables is

(16.9.14) $$I(\bar{x}, x) = \bar{x}_1 x_1 + \cdots + \bar{x}_n x_n,$$

the form associated with the identity matrix. $I(\bar{x}, x)$ is positive definite, since each term $\bar{x}_j x_j$ is positive unless $x_j = 0$, for $j = 1, \cdots, n$.

LEMMA 16.9.1. *A positive definite Hermitian form $H(\bar{x}, x)$ in n variables can be transformed into $\bar{x}_1 x_1 + \cdots + \bar{x}_n x_n$.*

We note that if $H(\bar{x}, x)$ is positive definite, where

$$H(\bar{x}, x) = \sum_{i,j} a_{ij}\bar{x}_i x_j, \; a_{ji} = \overline{a_{ij}},$$

then every diagonal coefficient, a_{rr} is positive. For, otherwise, with $a_{rr} \leq 0$, putting $x_r = 1$, $x_j = 0$, for $j \neq r$, $H(\bar{x}, x) = a_{rr} \leq 0$, contrary to the property of being positive definite. Now in

(16.9.15) $$H = \sum_{i,j} a_{ij}\bar{x}_i x_j, \quad a_{ji} = \overline{a_{ij}}, \quad i, j = 1, \cdots, n$$

put

(16.9.16) $$x'_1 = \sqrt{a_{11}}\left(x_1 + \frac{a_{12}x_2}{a_{11}} + \cdots + \frac{a_{1n}x_n}{a_{11}}\right),$$
$$x'_j = x_j, \quad j = 2, \cdots, n,$$

this being a permissible transformation since a_{11} is real and positive. We readily calculate

(16.9.17) $$H = \bar{x}_1'x_1' + \sum b_{ij}\bar{x}_i'x_j', \quad i, j = 2, \cdots, n,$$

where the terms in $x_2', \cdots, x_n'$ are a positive definite form in these variables. Continuing, we finally transform H into the form

(16.9.18) $$H = \bar{x}_1x_1 + \cdots + \bar{x}_nx_n.$$

This proves the lemma. We note that if H were a real quadratic form Q to start with, the same procedure would transform Q into $x_1^2 + \cdots + x_n^2$.

Now let $\rho(g)$, $g \in G$ be a complex representation of degree n of the finite group G, and let the elements of G be $g_1 = 1, g_2, \cdots, g_t$. Then

(16.9.19) $$M = I + \overline{\rho(g_2)}^T I\rho(g_2) + \cdots + \overline{\rho(g_t)}^T I\rho(g_t)$$

is a matrix corresponding to a positive definite Hermitian form, since each of the summands separately corresponds to a positive definite form. Furthermore, for any $g \in G$,

(16.9.20) $$\overline{\rho(g)}^T M\rho(g) = \sum_{i=1}^{t} \overline{\rho(g)}^T\overline{\rho(g_i)}^T\rho(g_i)\rho(g)$$
$$= \sum_{i=1}^{t} \overline{\rho(g_ig)}^T\rho(g_ig) = \sum_{i=1}^{t} \overline{\rho(g_i)}^T\rho(g_i)$$
$$= M.$$

Hence M corresponds to a positive definite Hermitian form H left invariant by $\rho(g)$ for every $g \in G$. By a change of variables,

(16.9.21) $$x_j = \sum_{k=1}^{n} c_{jk}x_k', \quad j = 1, \cdots, n.$$

H becomes $\bar{x}_1'x_1' + \cdots + \bar{x}_n'x_n' = I(\bar{x}', x')$, and correspondingly,

$$\rho'(g) = C^{-1}\rho(g)C, \quad C = (c_{jk}) \tag{16.9.22}$$

is a representation equivalent to $\rho(g)$ which is unitary, as we wished to prove.

We may see this in matrix form. We have

$$C^TMC = I, \quad M = \overline{C^{-1T}}C^{-1}, \tag{16.9.23}$$

and for every $g \in G$,

$$\overline{\rho(g)}^TM\rho(g) = M, \tag{16.9.24}$$

or

$$\overline{\rho(g)}^T\overline{C^{-1T}}C^{-1}\rho(g) = \overline{C^{-1T}}C^{-1}, \tag{16.9.25}$$

whence

$$\overline{(C^{-1}\rho(g)C)}^TC^{-1}\rho(g)C = I, \tag{16.9.26}$$

and so, $\rho'(g) = C^{-1}\rho(g)C$ is unitary.

16.10. Some Examples of Group Representation.

We begin with an example, of interest to physicists, of the representation of an infinite matrix group by another group of matrices. The two-dimensional unitary and unimodular matrices U_2 are of the form

$$\begin{pmatrix} \alpha, & \beta \\ -\bar{\beta}, & \bar{\alpha} \end{pmatrix} \qquad \alpha\bar{\alpha} + \beta\bar{\beta} = 1, \tag{16.10.1}$$

where α and β are complex numbers arbitrary except for the condition $\alpha\bar{\alpha} + \beta\bar{\beta} = 1$. Thus U_2 is the group of the linear transformations

$$\begin{aligned} u &= \alpha u' + \beta v' \\ v &= -\bar{\beta}u' + \bar{\alpha}v' \end{aligned} \qquad \alpha\bar{\alpha} + \beta\bar{\beta} = 1, \tag{16.10.2}$$

and these leave $\bar{u}u + \bar{v}v$ invariant. In terms of the complex variables u and v we may define three real variables

$$\begin{aligned} x_1 &= \bar{u}v + \bar{v}u, \\ x_2 &= \frac{1}{i}(\bar{u}v - \bar{v}u), \\ x_3 &= \bar{u}u - \bar{v}v. \end{aligned} \tag{16.10.3}$$

We note that

(16.10.4) $$x_1^2 + x_2^2 + x_3^2 = (\bar{u}u + \bar{v}v)^2.$$

A linear transformation (16.10.2) when substituted into (16.10.3) induces a real linear transformation of the x's, which by (16.10.4) belongs to the real orthogonal group O_3. The linear transformation of the x's induced by (16.10.2) is

(16.10.5)

$$\begin{aligned} x_1 &= \frac{1}{2}(\alpha^2+\bar{\alpha}^2-\beta^2-\bar{\beta}^2)x_1' + \frac{i}{2}(-\alpha^2+\bar{\alpha}^2-\beta^2+\bar{\beta}^2)x_2' + (-\alpha\beta-\overline{\alpha\beta})x_3'. \\ x_2 &= \frac{i}{2}(\alpha^2-\bar{\alpha}^2-\beta^2+\bar{\beta}^2)x_1' + \frac{1}{2}(\alpha^2+\bar{\alpha}^2+\beta^2+\bar{\beta}^2)x_2' + i(\overline{\alpha\beta}-\alpha\beta)x_3'. \\ x_3 &= (\alpha\bar{\beta}+\beta\bar{\alpha})x_1' + i(\alpha\bar{\beta}-\beta\bar{\alpha})x_2' + (\alpha\bar{\alpha}-\beta\bar{\beta})x_3'. \end{aligned}$$

Thus the group U_2 has a representation ρ given by (16.10.5) in the real three dimensional orthogonal group O_3.

This representation is not faithful, but is two to one, both $\begin{pmatrix}1, & 0\\ 0, & 1\end{pmatrix}$ and $\begin{pmatrix}-1, & 0\\ 0, & -1\end{pmatrix}$ of U_2 being represented by the identity of O_3. U_2 is represented by the entire group of proper rotations (those of determinant $+1$). The inverse image in U_2 of a group G of proper rotations is known as the *double group* $2G$. Group $2G$ is an extension of a subgroup of order 2 in its center by a factor group isomorphic to G.

It was found by Pauli [1] that when a physical system S has a certain subgroup K of O_3 associated with it, then the wave functions for the electron spin of S are associated with the doubled group $2K$.

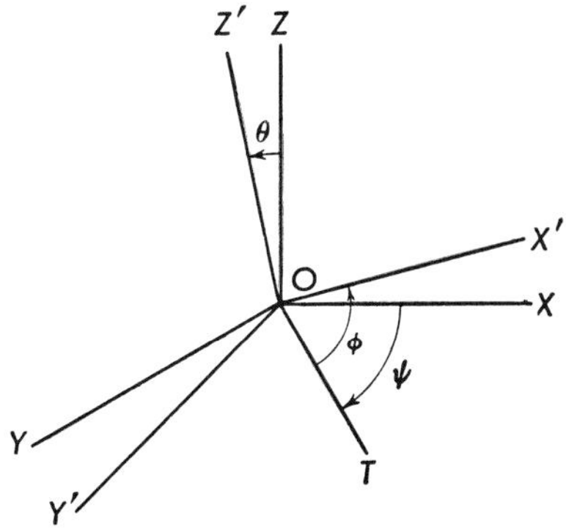

Fig. 7. Three dimensional rotation.

Besides (16.10.5) there is another formula giving an explicit connection between U_2 and this representation of it in O_3. Given a proper rotation of Euclidean three-space (i.e., an element of O_3), let OT be the line in which the coordinate plane XOY intersects its image $X'OY'$. Then (see figure) if ϕ is the angle $X'OT$, ψ the angle XOT, and θ is the angle ZOZ', we may find corresponding elements of U_2 in (16.10.1) by putting

$$(16.10.6) \qquad \alpha = \cos\frac{\theta}{2}\, exp\, i\left(\frac{\phi+\psi}{2}\right), \quad \beta = i\sin\frac{\theta}{2}\, exp\, i\left(\frac{\phi-\psi}{2}\right).$$

This is also valid when XOY and $X'OY'$ coincide (i.e., for a rotation about the Z-axis) if we put $\theta = 0$, $\phi = 0$ and take ψ as the angle of rotation about the Z-axis.

The group of proper rotations of the regular tetrahedron can be faithfully represented as a permutation group on its four vertices. For each vertex there is a subgroup fixing this vertex and rotating the three vertices of the opposite face, this being a group of order 3. The symmetry fixing two vertices and interchanging the other two is a reflection about a plane and not a proper rotation, since it reverses orientation. Thus the group of proper rotations, the tetrahedral group, is of order 12 and is isomorphic to A_4, the alternating group on four letters. Let us list the elements by classes.

$$(16.10.7) \qquad \begin{aligned} C_1 &= (1), \\ C_2 &= (12)(34),\ (13)(24),\ (14)(23), \\ C_3 &= (123),\ (142),\ (134),\ (243), \\ C_4 &= (132),\ (124),\ (143),\ (234). \end{aligned}$$

The multiplication table of the classes is

$$(16.10.8) \qquad \begin{aligned} C_1C_i &= C_iC_1 = C_i, \quad i = 1, 2, 3, 4. \\ C_2^2 &= 3C_1 + 2C_2. \\ C_2C_3 &= C_3C_2 = 3C_3. \\ C_2C_4 &= C_4C_2 = 3C_4. \\ C_3^2 &= 4C_4. \\ C_3C_4 &= C_4C_3 = 4C_1 + 4C_2. \\ C_4^2 &= 4C_3. \end{aligned}$$

We now look for the idempotents in the center Z of the group ring which form an orthogonal basis for Z. An idempotent e such that eZ is a minimal ideal in Z will be one of these, and conversely. This process may be regarded as a decomposition of Z into the direct sum of (two-sided) ideals. In particular, any proper divisor of zero will yield a proper ideal in Z. Thus by finding proper divisors of zero in two-sided ideals, we find smaller two-sided ideals, and ultimately the minimal ideals, and from these the orthogonal idempotents. In general let f be an idempotent in Z. If $fC_i = a_i f$ for every class, then fZ is a minimal ideal in Z and f is one of the orthogonal idempotents. If for some class C we have fC linearly independent of f, then let s be the smallest integer such that $fC^j\ j = 0, \cdots, s - 1$ are independent, but fC^s is dependent on these. Then we have a relation $f(C^s + a_1C^{s-1} + \cdots + a_s) = 0$. Adjoin to the field of coefficients, if necessary, a root of $x^s + a_1x^{s-1} + \cdots + a_s = 0$. If u is such a root, then $f(C - u)$ is a proper divisor of zero in the ideal fZ and leads to a smaller two-sided ideal. This general procedure is illustrated in our study of the tetrahedral group.

In any group G of order g, the sum of the elements divided by g is an idempotent e whose ideal is minimal in Z. This idempotent corresponds to the identical representation of G. Here this is $e_1 = (C_1 + C_2 + C_3 + C_4)/12$. We also note the relation, from (16.10.8),

$$C_2^2 - 2C_2 - 3C_1 = (C_2 - 3C_1)(C_2 + C_1) = 0. \tag{16.10.9}$$

Thus both $C_2 - 3C_1$ and $C_2 + C_1$ are proper divisors of zero. We find in fact that $(C_2 - 3C_1)Z$ is a minimal ideal and that $e_2 = (3C_1 - C_2)/4$ is the idempotent generating this minimal ideal. Also $e_1e_2 = e_2e_1 = 0$. We now construct the idempotent $f = 1 - e_1 - e_2$ which must be the unit for the part of Z which remains. Indeed fZ must be of dimension 2 and $f = e_3 + e_4$, where e_3 and e_4 are the remaining orthogonal idempotents of a basis for Z. Here we find

$$\begin{aligned} f &= (2C_1 + 2C_2 - C_3 - C_4)/12, \\ fC_1 &= f, \\ fC_2 &= 3f, \\ fC_3 &= (-4C_1 - 4C_2 + 8C_3 - 4C_4)/12, \\ fC_4 &= (-4C_1 - 4C_2 - 4C_3 + 8C_4)/12. \end{aligned} \tag{16.10.10}$$

The ideal fZ is of dimension 2, as it should be, and we note

$$fC_4 + fC_3 + 4f = 0, \tag{16.10.11}$$

a linear dependency, showing the dimension of fZ to be 2. We have the following relation:

$$f(C_3^2 + 4C_3 + 16) = 0. \tag{16.10.12}$$

If we adjoin to the rational field of coefficients, the complex cube root of unity, $\omega = (-1 + \sqrt{3}i)/2$, then (16.10.12) takes the form

$$f(C_3 - 4\omega)(C_3 - 4\omega^2) = 0. \tag{16.10.13}$$

Thus the principal ideal of each of $f(C_3 - 4\omega)$ and $f(C_3 - 4\omega^2)$ is smaller than the principal ideal fZ. The first of these differs by a scalar factor from the idempotent e_3 in (16.10.14) and the second from e_4.

$$\begin{aligned} e_1 &= (C_1 + C_2 + C_3 + C_4)/12, \\ e_2 &= (3C_1 - C_2)/4, \\ e_3 &= (C_1 + C_2 + \omega C_3 + \omega^2 C_4)/12, \\ e_4 &= (C_1 + C_2 + \omega^2 C_3 + \omega C_4)/12, \\ e_1 + e_2 + e_3 + e_4 &= C_1 = 1. \end{aligned} \tag{16.10.14}$$

From the expressions for the minimal orthogonal idempotents in terms of the classes, we may immediately write down the character table, and conversely. Immediately preceding Theorem 16.6.10 we established a relation which may be written

$$C_k = h_k \sum_a \frac{\chi^a{}_k}{n_a} e_a. \tag{16.10.15}$$

In this multiply by $\overline{\chi}_k{}^b$ and sum over k. This gives

$$\sum_k \overline{\chi}_k{}^b C_k = \sum_{a,k} \frac{h_k \overline{\chi}_k{}^b \chi_k{}^a e_a}{n_a}. \tag{16.10.16}$$

If we sum the right-hand side first with respect to k and use the orthogonality relation

$$\sum_k h_k \overline{\chi}_k{}^b \chi_k^a = \delta_{ab} g, \tag{16.10.17}$$

then when we sum with respect to a, all terms are zero except that for $a = b$, and our relation (16.10.16) becomes

$$\sum_k \overline{\chi}_k{}^b C_k = g e_b / n_b. \tag{16.10.18}$$

This we write in the form

$$e_b = \frac{n_b}{g} \sum_k \overline{\chi}_k{}^b C_k. \tag{16.10.19}$$

Since we know that the character of the unit class is the degree, $\bar{\chi}_1^b = n_b$, the coefficient of C_1 in the expression (16.10.19) for e_b must be n_b^2/g. This determines n_b, and we may then use (16.10.19) to read off the remaining characters.

Using the general rule (16.10.19) and the table (16.10.14) for the tetrahedral group, we may write down the character table for it:

(16.10.20)

	C_1	C_2	C_3	C_4
ρ_1	1	1	1	1
ρ_2	3	-1	0	0
ρ_3	1	1	ω^2	ω
ρ_4	1	1	ω	ω^2

Conversely, using the character table (16.10.20), we could use (16.10.19) to write down the minimal orthogonal idempotents in (16.10.14).

The three irreducible representations of degree one, ρ_1, ρ_3, and ρ_4, may of course be written down directly from the character table.

By Theorem 16.6.15 the representation of the tetrahedral group as A_4 is the sum of the identical representation and an irreducible representation which, being of degree 3, must be ρ_2. The permutations of A_4 on variables x_1, x_2, x_3, and x_4 fix the linear form $x_1 + x_2 + x_3 + x_4$ and take the complementary space spanned by $y_1 = x_1 - x_4$, $y_2 = x_2 - x_4$, $y_3 = x_3 - x_4$ into itself. Here the permutation (12)(34) is the linear transformation

$$(16.10.21) \qquad \begin{aligned} x_1 &= x_2', \\ x_2 &= x_1', \\ x_3 &= x_4', \\ x_4 &= x_3', \end{aligned}$$

and on the y's this is

$$(16.10.22) \qquad \begin{aligned} y_1 &= y_2' - y_3' \\ y_2 &= y_1' - y_3' \\ y_3 &= - y_3' \end{aligned}$$

Thus

$$(16.10.23) \qquad \rho_2[(12)(34)] = \begin{pmatrix} 0, & 1, & -1 \\ 1, & 0, & -1 \\ 0, & 0, & -1 \end{pmatrix}$$

Similarly,

$$(16.10.24) \qquad \rho_2[(123)] \qquad \begin{pmatrix} 0, 1, 0 \\ 0, 0, 1 \\ 1, 0, 0 \end{pmatrix}$$

Since the two elements (12)(34) and (123) generate the entire group, the entire representation is determined. This is not the orthogonal form of ρ_2, but this may be obtained from the next example, since the tetrahedral group is a subgroup of the octahedral group.

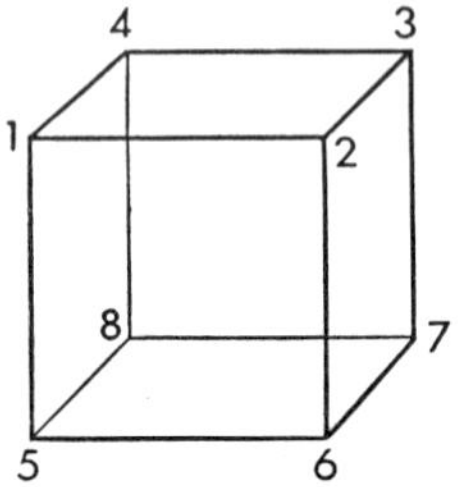

Fig. 8. Symmetries of the cube.

The group of symmetries of the cube was discussed in Example 2 of Chap. 1. The proper rotations form a group G_{24} of order 24 generated by

$$a = \begin{pmatrix} 1, 2, 3, 4, 5, 6, 7, 8 \\ 2, 3, 4, 1, 6, 7, 8, 5 \end{pmatrix} = (1234)(5678),$$

and

$$b = \begin{pmatrix} 1, 2, 3, 4, 5, 6, 7, 8 \\ 1, 4, 8, 5, 2, 3, 7, 6 \end{pmatrix} = (1)(245)(7)(386).$$

The group G_{24} is also the group of symmetries of the regular octahedron, which may be regarded as inscribed in the cube with one vertex in the middle of each face of the cube. Thus G_{24} may also be represented as a permutation group on the six faces of the cube, corresponding to the six vertices of the inscribed octahedron. We shall use six variables, corresponding to the six vertices of the inscribed octahedron, whose three dimensional coordinates we give. We write:

Variable	Face of Cube	Vertex of Octahedron
x_1	Face 1234	(0, 0, 1)
x_2	Face 1256	(0, 1, 0)
x_3	Face 1458	(−1, 0, 0)
x_4	Face 5678	(0, 0, −1)
x_5	Face 3478	(0, −1, 0)
x_6	Face 2367	(1, 0, 0)

Then $a = (x_1)(x_4)(x_2, x_6, x_5, x_3)$, $b = (x_1, x_3, x_2)(x_4, x_6, x_5)$. The 24 elements of G_{24} are the following: We give each element both as a permutation of the x's and also as a monomial linear transformation of the x, y, z coordinates which yields the appropriate permutation of the vertices of the octahedron.

Class	g_i	Permutation	$x,\ y,\ z$
C_1	g_1	(x_1)	$x,\ y,\ z$
C_2	g_2	$(x_1)(x_4)(x_2, x_5)(x_3, x_6)$	$-x,\ -y,\ z$
	g_3	$(x_1, x_4)(x_2)(x_5)(x_3, x_6)$	$-x,\ y,\ -z$
	g_4	$(x_1, x_4)(x_2, x_5)(x_3)(x_6)$	$x,\ -y,\ -z$
C_3	g_5	$(x_1)(x_4)(x_2, x_6, x_5, x_3)$	$-y,\ x,\ z$
	g_6	$(x_1)(x_4)(x_2, x_3, x_5, x_6)$	$y,\ -x,\ z$
	g_7	$(x_2)(x_5)(x_1, x_3, x_4, x_6)$	$z,\ y,\ -x$
	g_8	$(x_2)(x_5)(x_1, x_6, x_4, x_3)$	$-z,\ y,\ x$
	g_9	$(x_3)(x_6)(x_1, x_2, x_4, x_5)$	$x,\ -z,\ y$
	g_{10}	$(x_3)(x_6)(x_1, x_5, x_4, x_2)$	$x,\ z,\ -y$
C_4	g_{11}	$(x_1, x_2)(x_3, x_6)(x_4, x_5)$	$-x,\ z,\ y$
	g_{12}	$(x_1, x_3)(x_2, x_5)(x_4, x_6)$	$-z,\ -y,\ -x$
	g_{13}	$(x_1, x_4)(x_2, x_6)(x_3, x_5)$	$y,\ x,\ -z$
	g_{14}	$(x_1, x_4)(x_2, x_3)(x_5, x_6)$	$-y,\ -x,\ -z$
	g_{15}	$(x_1, x_5)(x_2, x_4)(x_3, x_6)$	$-x,\ -z,\ -y$
	g_{16}	$(x_1, x_6)(x_2, x_5)(x_3, x_4)$	$z,\ -y,\ x$
C_5	g_{17}	$(x_1, x_2, x_3)(x_4, x_5, x_6)$	$-z,\ -x,\ y$
	g_{18}	$(x_1, x_2, x_6)(x_3, x_4, x_5)$	$z,\ x,\ y$
	g_{19}	$(x_1, x_3, x_2)(x_4, x_6, x_5)$	$-y,\ z,\ -x$
	g_{20}	$(x_1, x_3, x_5)(x_2, x_4, x_6)$	$y,\ -z,\ -x$
	g_{21}	$(x_1, x_5, x_3)(x_2, x_6, x_4)$	$-z,\ x,\ -y$
	g_{22}	$(x_1, x_5, x_6)(x_2, x_3, x_4)$	$z,\ -x,\ -y$
	g_{23}	$(x_1, x_6, x_2)(x_3, x_5, x_4)$	$y,\ z,\ x$
	g_{24}	$(x_1, x_6, x_5)(x_2, x_4, x_3)$	$-y,\ -z,\ x$

We may map the even permutations of this representation of G_{24} (those of classes C_1, C_2, C_5) onto $+1$ and the odd permutations (those of classes C_3, C_4) onto -1. This gives us, besides the identical representation, still another representation of degree one. The degrees of the irreducible representations, n_i, satisfy the relation

$$n_1^2 + n_2^2 + n_3^2 + n_4^2 + n_5^2 = 24. \tag{16.10.25}$$

With $n_1 = 1$, $n_2 = 1$, we find that n_3, n_4, n_5 must be 2, 3, 3. The three-dimensional representation we have given above has the following characters:

$$\chi_1 = 3, \quad \chi_2 = \chi(g_2) = -1, \quad \chi_3 = \chi(g_5) = 1,$$
$$\chi_4 = \chi(g_{11}) = -1, \quad \chi_5 = \chi(g_{17}) = 0. \tag{16.10.26}$$

In general, if

$$\rho = \sum_a c_a \rho_a \tag{16.10.27}$$

is the expression of a representation ρ, as a sum of irreducible representations ρ_a, then for the ith class we have

$$\chi_i = \sum_a c_a \chi_i^a, \tag{16.10.28}$$

and by the orthogonality relations,

$$\sum_i h_i \chi_i \bar{\chi}_i = g \sum_a c_a^2. \tag{16.10.29}$$

Since for our representation of degree 3, we have, from (16.10.26)

$$\sum_i h_i \chi_i \bar{\chi}_i = 24, \tag{16.10.30}$$

it follows that $\sum_a c_a^2 = 1$, and so our representation (which we shall designate ρ_4) is irreducible. Let us note that ρ_4 is orthogonal, and since the even permutations (elements of classes C_1, C_2, C_5) form a subgroup isomorphic to the tetrahedral group, this gives the orthogonal representation of degree 3, promised when we discussed the tetrahedral group.

We now have a partial table of the characters:

(16.10.31)

	$h_1 = 1$	$h_2 = 3$	$h_3 = 6$	$h_4 = 6$	$h_5 = 8$
	C_1	C_2	C_3	C_4	C_5
ρ_1	1	1	1	1	1
ρ_2	1	1	-1	-1	1
ρ_3	2	y_2	y_3	y_4	y_5
ρ_4	3	-1	1	-1	0
ρ_5	3	z_2	z_3	z_4	z_5

From the orthogonality relations involving the second column with the first and with itself, we have

$$\begin{aligned} 1 + 1 + 2y_2 - 3 + 3z_2 &= 0, \\ 1 + 1 + y_2^2 + 1 + z_2^2 &= 8. \end{aligned} \tag{16.10.32}$$

These equations give either $y_2 = 2$ and $z_2 = -1$, or $y_2 = -22/13$ and $z_2 = 19/13$. As y_2 is the sum of two square roots of unity,

$y_2 = \pm 1 \pm 1$, the first values, $y_2 = 2$ and $z_2 = -1$ are the true values.

The orthogonality between the third column and the first two columns gives

(16.10.33)
$$\begin{aligned} 1 - 1 + 2y_3 + 3 + 3z_3 &= 0, \\ 1 - 1 + 2y_3 - 1 - z_3 &= 0. \end{aligned}$$

These equations give $y_3 = 0$, $z_3 = -1$. In the same way we may find the missing values in the remaining columns. The complete table of characters is:

(16.10.34)

	$h_1 = 1$ C_1	$h_2 = 3$ C_2	$h_3 = 6$ C_3	$h_4 = 6$ C_4	$h_5 = 8$ C_5
ρ_1	1	1	1	1	1
ρ_2	1	1	-1	-1	1
ρ_3	2	2	0	0	-1
ρ_4	3	-1	1	-1	0
ρ_5	3	-1	-1	1	0

The representation ρ of degree 8 of G_{24} as a permutation group on the vertices of a cube has the following characters:

(16.10.35)

	C_1	C_2	C_3	C_4	C_5
χ	8	0	0	0	2

From these characters we may find the decomposition of ρ, since if

(16.10.36)
$$\rho = \sum_a c_a \rho_a,$$
$$\chi_i = \sum_a c_a \chi_i^a,$$

then from orthogonality,

(16.10.37)
$$\sum_i h_i \chi_i \bar{\chi}_i^b = \sum_{i,\,a} c_a h_i \chi_i^a \bar{\chi}_i^b = g \sum_a c_a \delta_{ab} = g c_b.$$

This gives the multiplicity c_b of each irreducible representation ρ_b occurring in ρ. In this way we find from (16.10.35) that

(16.10.38)
$$\rho = \rho_1 + \rho_2 + \rho_4 + \rho_5.$$

We may use this to find ρ_5 explicitly from ρ. Using (16.10.19) and our table of characters (16.10.34), we find

(16.10.39) $$e_5 = (3C_1 - C_2 - C_3 + C_4)/4.$$

The representation ρ of G_{24} also represents the group ring R_G, and we find the representation of e_5 to be

(16.10.40) $$\rho(e_5) = \frac{1}{4}\begin{pmatrix} 3, & -1, & -1, & -1, & -1, & -1, & 3, & -1 \\ -1, & 3, & -1, & -1, & -1, & -1, & -1, & 3 \\ -1, & -1, & 3, & -1, & 3, & -1, & -1, & -1 \\ -1, & -1, & -1, & 3, & -1, & 3, & -1, & -1 \\ -1, & -1, & 3, & -1, & 3, & -1, & -1, & -1 \\ -1, & -1, & -1, & 3, & -1, & 3, & -1, & -1 \\ 3, & -1, & -1, & -1, & -1, & -1, & 3, & -1 \\ -1, & 3, & -1, & -1, & -1, & -1, & -1, & 3 \end{pmatrix}$$

(i)

Taking $x_i = (0, \cdots, 1, \cdots, 0)$, $i = 1, \cdots, 8$ as the basis of the representation module M for ρ, then the rows of the matrix $\rho(e_5)$ span the submodule Me_5, which is, of course, of dimension 3. We may take the first three rows, r_1, r_2, r_3 as a basis. It is, however, more convenient to take as a basis vectors proportional to $r_1 + r_3$, $r_1 + r_2$, and $r_2 + r_3$. This basis is

(16.10.41) $$\begin{aligned} y_1 &= x_1 - x_2 + x_3 - x_4 + x_5 - x_6 + x_7 - x_8, \\ y_2 &= x_1 + x_2 - x_3 - x_4 - x_5 - x_6 + x_7 + x_8, \\ y_3 &= x_1 - x_2 - x_3 + x_4 - x_5 + x_6 + x_7 - x_8. \end{aligned}$$

With this basis we find

(16.10.42) $$\begin{aligned} \rho_5(a) &= \begin{pmatrix} -1, & 0, & 0 \\ 0, & 0, & -1 \\ 0, & 1, & 0 \end{pmatrix} = \begin{pmatrix} y_1, & y_2, y_3 \\ -y_1, & -y_3, y_2 \end{pmatrix} \\ \rho_5(b) &= \begin{pmatrix} 0, & 1, & 0 \\ 0, & 0, & 1 \\ 1, & 0, & 0 \end{pmatrix} = \begin{pmatrix} y_1, & y_2, y_3 \\ y_2, & y_3, y_1 \end{pmatrix} \end{aligned}$$

These generate the entire representation ρ_5. Thus ρ_5 is a monomial and orthogonal representation of G_{24} as given by (16.10.42.) But this is not a group of proper rotations, as we note that the determinant of $\rho_5(a)$ is -1. A representation equivalent to ρ_5 may be obtained by taking ρ_4 and multiplying the matrices for elements of C_3 and C_4 by -1. Thus ρ_5 is the Kronecker product of ρ_4 and ρ_2.

The representation ρ_3 is not faithful but has a kernel of order 4 consisting of the identity and the elements of C_2. We may take as generators

$$\rho_3(a) = \begin{pmatrix} 0, & 1 \\ 1, & 0 \end{pmatrix}, \quad \rho_3(b) = \begin{pmatrix} 0, & 1 \\ -1, & -1 \end{pmatrix}. \tag{16.10.43}$$

In our representation ρ_4 of G_{24} as a group of proper rotations we had

$$\rho_4(a) = \begin{pmatrix} 0, & -1, & 0 \\ 1, & 0, & 0 \\ 0, & 0, & 1 \end{pmatrix}, \quad \rho_4(b) = \begin{pmatrix} 0, & -1, & 0 \\ 0, & 0, & 1 \\ -1, & 0, & 0 \end{pmatrix}. \tag{16.10.44}$$

Thus ρ_4 is a subgroup of O_3, and using (16.10.5), which gives the mapping of U_2 on O_3, we find that

(16.10.45)

$$\frac{\pm 1}{\sqrt{2}} \begin{pmatrix} 1-i, & 0 \\ 0\ , & 1+i \end{pmatrix} \to \rho_4(a), \qquad \frac{\pm 1}{\sqrt{2}} \begin{pmatrix} 1-i, & -1+i \\ 1+i, & 1+i \end{pmatrix} \to \rho_4(b).$$

The entire double group $D = 2G_{24}$ has eight classes. In general, in a double group $2G$, the inverse image of a class C_i in G is either a class in $2G$ with twice as many elements as C_i in G, or two classes C_i' and $C_i'' = -C_i'$, each with the same number of elements as C_i in G.

Following Bethe [1] we may list the elements of D, expressing them in terms of four matrices

(16.10.46)

$$1 = \begin{pmatrix} 1, & 0 \\ 0, & 1 \end{pmatrix}, \quad \sigma_x = \begin{pmatrix} 0, & 1 \\ 1, & 0 \end{pmatrix}, \quad \sigma_y = \begin{pmatrix} 0, & -i \\ i, & 0 \end{pmatrix}, \quad \sigma_z = \begin{pmatrix} 1, & 0 \\ 0, & -1 \end{pmatrix}.$$

Now the list of elements of D by classes is

(16.10.47)

$$\begin{aligned} E &= [1], \\ R &= [-1], \\ C_2 &= \pm[-i\sigma_x,\ -i\sigma_y,\ -i\sigma_z], \\ C_3' &= \left[\frac{1 \pm i\sigma_x}{\sqrt{2}}\ \ \frac{1 \pm i\sigma_y}{\sqrt{2}}\ \ \frac{1 \pm i\sigma_z}{\sqrt{2}}\right], \\ C_3'' &= -C'_3 \end{aligned}$$

$$C_4 = \pm\left[\frac{-i(\sigma_y \pm \sigma_z)}{\sqrt{2}}, \quad \frac{-i(\sigma_z \pm \sigma_x)}{\sqrt{2}}, \quad \frac{-i(\sigma_x \pm \sigma_y)}{\sqrt{2}}\right],$$

$$\begin{aligned} C_5' &= [\tfrac{1}{2}\{1 \pm i(\sigma_x + \sigma_y + \sigma_z)\}, \ \tfrac{1}{2}\{1 \pm i(-\sigma_x + \sigma_y - \sigma_z)\} \\ &\qquad \tfrac{1}{2}\{1 \pm i(-\sigma_x - \sigma_y + \sigma_z)\}, \ \tfrac{1}{2}\{1 \pm i(\sigma_x - \sigma_y - \sigma_z)\}], \\ C_5'' &= -C_5'. \end{aligned}$$

As G_{24} is a homomorphic image of D, every irreducible representation of G_{24} is also an irreducible representation of D. But D has three further irreducible representations which are in fact faithful representations of D. The further characters are:

(16.10.48)

	E	R	C_2	C_3'	C_3''	C_4	C_5'	C_5''
ρ_6	2	-2	0	$\sqrt{2}$	$-\sqrt{2}$	0	1	-1
ρ_7	2	-2	0	$-\sqrt{2}$	$\sqrt{2}$	0	1	-1
ρ_8	4	-4	0	0	0	0	-1	1

17. FREE AND AMALGAMATED PRODUCTS

17.1. Definition of Free Product.

Let G_i be a set of groups indexed by letters $i \in I$, where we shall assume I well ordered. We shall define the *free product* $\pi_i{}^*G_i$ of the groups G_i in a manner similar to that for defining the free group with given generators.

Consider words (or strings)

$$(17.1.1) \qquad a_1a_2 \cdots a_t$$

which are either void (written 1) or in which each a_i, $i = 1 \cdots t$ is an element of some G_j. For these strings we define operations of *elementary equivalence.*

$$(E1) \qquad a_1a_2 \cdots a_{i-1}a_ia_{i+1} \cdots a_t$$

is equivalent to $a_1 \cdots a_{i-1}a_{i+1} \cdots a_t$ if a_i is the identity of some group G_j.

$$(E2) \qquad a_1a_2 \cdots a_{i-1}a_ia_{i+1} \cdots a_t$$

is equivalent to $a_1a_2 \cdots a_{i-1}a_i{}^*a_{i+2} \cdots a_t$ if a_i and a_{i+1} belong to the same group G_j and if $a_ia_{i+1} = a_i{}^*$ in G_j.

It is to be understood that the elementary equivalences are symmetric. We say that two words, x and y, are *equivalent* if there is a finite sequence $x_1 = x, x_2, x_3, \cdots x_n = y$, with x_i and x_{i+1} elementary equivalent for $i = 1, 2, \cdots n - 1$. All equivalent words form a class.

A word $a_1a_2 \cdots a_t$ is *reduced* if it is the void word or if (1) no a_i is the identity of its group G_j and (2) a_i and a_{i+1} belong to different groups G for $i = 1, \cdots t - 1$. As in §7.1 we may define a W-process for a word $f = a_1a_2 \cdots a_t$, taking:

$W_0 = 1$.

$W_1 = 1$, if a_1 is the identity of its group.

$W_1 = a_1$, otherwise.

If W_i is of the reduced form $b_1b_2 \cdots b_s$, we take

1) $W_{i+1} = b_1 \cdots b_sa_{i+1}$, if a_{i+1} is not the identity of its group and is not in the same group as b_s.

2) $W_{i+1} = b_1 \cdots b_s$, if $a_{i+1} = 1$ in its group.

3) If b_s and a_{i+1} are in the same group and $b_s a_{i+1} = 1$, take $W_{i+1} = b_1 \cdots b_{s-1}$.

4) If b_s and a_{i+1} are in the same group and $b_s a_{i+1} = b_s^* \neq 1$, take $W_{i+1} = b_1 \cdots b_{s-1} b_s^*$.

Here $W(f) = W_t$, as in §7.1, can be shown to be reduced, and it can be shown that the W process gives the same result for elementary equivalent words and thus for an entire class of equivalent words. It thus follows that there is a unique reduced word in each class. If $f = a_1 \cdots a_t$ is reduced, we call t the *length* of f.

We may now define a product for classes of words, putting

$$[f_1]\ [f_2] = [f_1 f_2], \tag{17.1.2}$$

and we can show that this product is independent of the representatives chosen, following the proof of Theorem 7.1.1. This product is associative and forms a group with the void words as the identity. This group is the free product $\overset{*}{\prod_i} G_i$ of the groups G_i. We note that by $(E1)$ the identity of every group G_i is equivalent to the void word 1. We shall not further distinguish these identities. Elements $\neq 1$ of different groups are, however, distinct reduced words, and so are not identified.

THEOREM 17.1.1. *Let G be a group which is the union of subgroups H_i, $i \in I$, where H_i are isomorphic to groups G_i. Then G is a homomorphic image of the free product $Q = \overset{*}{\prod_i} G_i$.*

Proof: As in the proof of Theorem 7.1.2, consider an element $a_1 a_2 \cdots a_t$ of Q. If a_i is an element of G_j, then let b_i be the corresponding element of H_j and map $a_1 a_2 \cdots a_x$ onto $b_1 b_2 \cdots b_t$. We see that equivalent words of Q are mapped onto the same element of G. This mapping of elements of Q onto elements of G also preserves products and hence is a homomorphism of Q onto the union of the H_j, which was given to be G.

17.2. Amalgamated Products.

Let G_i, $i \in I$ be a set of groups indexed by the set I. Let us suppose that each G_i contains a subgroup U_i and that all U_i are given as isomorphic to a group U. It is to be emphasized that there is a specific isomorphism given between each U_i and U. We wish to consider the most general group generated by the G_i in which all U_i are identified with each other so that all U_i form the same group U^1 isomorphic to U. This is clearly the image of the free product of the G_i obtained by identifying in every case $u_i \in U_i$ and $u_j \in U_j$ if, in the given isomorphisms between U_i, U_j, and U, u_i and u_j correspond to the same element u. There must, of course, be some such group, but it is not at all obvious how much identification results from these basic identifications. In particular, it is conceivable that this might result in identifying all elements with the identity. This is not the case, and in fact, this identification has essentially no effect on elements not in the U_i.

We shall construct the group generated by the G_i with all U_i's identified with each other, and this we shall call the amalgamated product of the G_i. Consider words $a_1a_2 \cdots a_t$, with each a_i from some G_j. We define elementary equivalences:

($E1$) *If* $a_i = 1$, *then*

$$a_1a_2 \cdots a_{i-1}a_ia_{i+1} \cdots a_t$$

is equivalent to $a_1a_2 \cdots a_{i-1}a_{i+1} \cdots a_t$.

($E2$) *If* a_i *and* a_{i+1} *belong to the same* G_j *and* $a_ia_{i+1} = a_i^*$ *in* G_j, *then*

$$a_1a_2 \cdots a_ia_{i+1} \cdots a_t$$

is equivalent to $a_1a_2 \cdots a_i^* \cdots a_t$.

($E3$) *If* $a_i = u_i$ *is an element of* $U_j \subseteq G_j$ *and if* $b_i = u_k \in U_k$ *is such that in the isomorphisms between* U_j, U_k, *and* U, u_i *and* u_k *correspond to the same element* u, *then*

$$a_1 \cdots a_{i-1}a_ia_{i+1} \cdots a_t$$

is equivalent to $a_1 \cdots a_{i+1}b_ia_{i+1} \cdots a_t$.

We now define words x and y as equivalent if there is a finite sequence $x = x_1, x_2 \cdots y_n = y$ such that x_i and x_{i+1} are elementary equivalent for $i = 1, \cdots n - 1$. Classes of equivalent words form the elements of a group if the product of $[f]$ and $[g]$ is defined as $[fg]$. As in Theorem 7.1.1, this product is well defined for the classes, and with respect to this product, the classes form a group T, the free product of the G_i with the amalgamated subgroup U. More briefly,

T is the amalgamated product of the G_i. But as yet we have no knowledge of the nature of T. For this we need a canonical form for the elements of T.

We shall define a canonical form associated with words $f = a_1a_2 \cdots a_t$, $a_i \in G_j$. It will then be necessary to show that this canonical form is the same for equivalent words, and thus be shown to be a canonical form for elements of T.

For each G_i, $i \in I$, let us choose representatives x_{ik} for left cosets of U_i, choosing the identity of G_i as the representative for U_i, but leaving the choice otherwise arbitrary.

$$(17.2.1) \qquad G_i = U_i + U_i x_{i2} + \cdots + U_i x_{in_i}, \quad i \in I.$$

From the elementary equivalence $(E1)$ the void word is the identity of T and the identity of all G_i. Also from $(E3)$ we may regard all U_i $i \in I$ as identified with U. Thus we may write instead of (17.2.1),

$$(17.2.2) \qquad G_i = U + Ux_{i2} + \cdots + Ux_{in_i}, \quad i \in I.$$

Hence an element $g_i \in G_i$ can be written

$$(17.2.3) \qquad g_i = u \quad \text{or} \quad g_i = uz,\ u \in U,\ z = x_{ik} \neq 1.$$

In an amalgamated product it is convenient to modify the usual definition of length of words. We define $l(a_0a_1 \cdots a_t) = t$ if $a_0 \in U$, and $l(a_1 \cdots a_t) = t$ if $a_1 \neq U$. Thus we do not count the first letter if it is an element of U.

We say an element of T is in canonical form if it is of the form

$$(17.2.4) \qquad f = uz_1z_2 \cdots z_t,$$

where $u \in U$, z_j, $j = 1 \cdots t$ are coset representatives, $x_{ik} \neq 1$ as given in (17.2.2), and if z_j, z_{j+1} for $j = 1, \cdots t - 1$ are in different G's.

THEOREM 17.2.1. *In the amalgamated product of groups G_i with the amalgamated subgroup U, there is in each class of equivalent words one, and only one, element in canonical form $f = uz_1z_2 \cdots z_t$. Here $u \in U$ and the z_i, $i = 1 \cdots t$ are coset representatives $x_{jk} \neq 1$ of U in the G_i, taken from some arbitrary but fixed selection of coset representatives. z_1 and z_{i+1}, $i = 1 \cdots t - 1$ belong to different G's.*

The proof of this theorem is very similar to the proof of the lemma in §7.1. Lack of space prevents giving the details. For more advanced work on this subject see H. Neumann [1,2].

17.3. The Theorem of Kurosch.

It was shown by Kurosch† that every subgroup of a free product is itself a free product. This result will be proved here. Subgroups of a product with an amalgamated subgroup need not themselves be of this type. If U is the amalgamated subgroup, then if there are more than two groups G_i in the product, we can take subgroups H_i of the G_i which have various different intersections with U. Here the different H_i will amalgamate in various ways and we are dealing with what is called a *generalized amalgamated product.* A number of complications arise. At present the theory is still incomplete.

THEOREM 17.3.1 (THEOREM OF KUROSCH). *A subgroup $H \neq 1$ of a free product*

$$G = \prod_{v}^{*} A_v$$

is itself a free product.

$$H = F^* \prod_{j}^{*} x_j^{-1} U_j x_j$$

where F is a free group and each $x_j^{-1} U_j x_j$ is the conjugate of a subgroup U_j of one of the free factors A_v of G.

Proof: The elements of the free factors of G may be well ordered by beginning with the identity, then taking an ordering of the free factors, and within a free factor taking an ordering of the elements $\neq 1$. Based on this ordering we define an alphabetical ordering for the elements of G. Write

$$g = a_1 a_2 \cdots a_t$$

as the reduced form of an element g of G. The void product is the identity; and for $g \neq 1$ each a_i is an element $\neq 1$ of one of the free factors A_v, and no two consecutive terms a_i, a_{i+1} $(i = 1, \cdots, t - 1)$ belong to the same free factor A_v. The length $l(g)$ of an element g is defined as zero for $g = 1$, and for $g \neq 1$ as the number t of terms in its reduced form (1). We define the alphabetical ordering of elements by ordering successively on:

1) The length of g;

2.1) The order of the first term a_1 if $g = a_1 a_2 \cdots a_t$ is its reduced form;

† A. Kurosch [1]. The proof given here is that of the author [6].

2.2) The order of a_2;

. . . .

2.t) The order of a_t;

This is clearly a well ordering of the elements of G.

We now define a second ordering for the elements of G, the semi-alphabetical ordering. For this we write an element g of even length $t = 2r$ in the form $g = \alpha\beta^{-1}$, where $l(\alpha) = l(\beta) = r$ and an element g of odd length $t = 2s + 1$ in the form $g = \alpha a_{s+1}\beta^{-1}$, where $l(\alpha) = l(\beta) = s$. The semi-alphabetical ordering for elements g is determined successively by:

1) The length of g;

2) For $g = \alpha\beta^{-1}$ of even length by (2.1), the alphabetical order of α, and by (2.2), the alphabetical order of β;

3) For $g = \alpha a_{s+1}\beta^{-1}$ of odd length by (3.1), the alphabetical order of α, and by (3.2), the alphabetical order of β, and by (3.3), the order of a_{s+1}.

The proof that the subgroup H of G is a free product will be carried out by selecting, in terms of the semi-alphabetical ordering, a subset K of the elements of H and showing (1) that the elements of K generate H and then (2) that the elements K generate a free product,

$$F * \prod_j^{*} x_j^{-1} U_j x_j \tag{17.3.1}$$

where F is a free group and each U_j is a subgroup of some free factor A.

The set K of elements shall consist of all elements $k \neq 1$ such that (1) $k \in H$, and (2) k does not belong to the group generated by the elements of H which precede k in the semi-alphabetical ordering.

Since $H \neq 1$, the first $h \neq 1$ of H belongs to the set K, and so K is not vacuous. Consider the group $|K|$ generated by the set K. Clearly $|K| \subseteq H$. If $|K| \neq H$, there must be a first $h \in H$ such that $h \in |K|$. Such an h does not belong to K, and so is a product of elements h_i preceding h and belonging to H. But these h_i belong to $|K|$, and so h as a product of these h_i's also belongs to $|K|$. Hence $|K| = H$, and this covers the first part of the proof.

We shall use the sign $<$ for numerical inequalities and for both the alphabetical and semi-alphabetical orderings. It will be clear from the context which meaning is appropriate, the semi-alphabetical ordering applying to entire words, the alphabetical to beginnings or endings of words. Writing $u \neq 1$ in the form $u = \alpha\beta^{-1}$ or $u = \alpha a\beta^{-1}$, we cannot have $\beta = \alpha$ for words of even length since $\alpha\alpha^{-1} = 1$. For elements of odd length, $\beta = \alpha$ is possible; and those elements of H of the form $\alpha a\alpha^{-1}$ for fixed α, and a's belonging to some fixed A_v, together with the identity, form a subgroup $\alpha B\alpha^{-1}$ conjugate to $B \subseteq A_v$. Let us call elements $\alpha a\alpha^{-1}$ *transforms*. Let us extend the set K to a larger set T which consists of K, and for each α and A_v, those transforms $\alpha a^1\alpha^{-1}$, $a^1 \in A_v$, generated by transforms $\alpha a\alpha^{-1}$, $a \in A_v$, belonging to K. Hence T consists of elements of H not generated by their predecessors and transforms $\alpha a^1\alpha^{-1}$ generated by earlier transforms of the same kind.

An element $h \in H$ can be written in the form

$$h = u_1u_2 \cdots u_t, \tag{17.3.2}$$

where $u_i \in T$ or T^{-1} (the set of inverses of elements in T). Moreover, we can take (17.3.2) so that (*a*) $u_iu_{i+1} \neq 1$ ($i = 1, \cdots, t - 1$) and (*b*) no two consecutive u_i, u_{i+1} belong to the same conjugate group $\alpha B\alpha^{-1}$, $B \subseteq A_v$. If these conditions are satisfied, then we shall say that $u_1 \cdots u_t$ is in half-reduced form.

The theorem will follow immediately if it can be shown that any nonvacuous half-reduced form cannot be the identity. For then it will follow that the elements of K that are not transforms generate a free group F, and that H is the free product of F and the conjugates $\alpha\beta\alpha^{-1}$, $\beta \subseteq A_v$.

If u is an element of K, and $u^{-1} \neq u$, then $u < u^{-1}$, since $u = (u^{-1})^{-1}$ and u^{-1} cannot be a predecessor of u. Also, if $u \neq v$ are elements of K, then $w = u^\epsilon v^\eta (\epsilon, \eta = \pm 1)$ will follow both u and v, since any two of u, v, w generate the third, and by the choice of K neither u nor v is generated by predecessors. These two principles are the main tools in studying the way in which the elements of T and T^{-1} combine. In reducing a product $a_1a_2 \cdots a_m$ in G, where each a_i belongs to one of the free factors, we say that a_i and a_{i+1} amalgamate into a_i^1 if a_i and a_{i+1} belong to the same free factor A and $a_ia_{i+1} = a_i^1 \neq 1$, and that they cancel if $a_ia_{i+1} = 1$.

Lemma 17.3.1 *If* $u = \alpha\beta^{-1}$ *or* $\alpha a\beta^{-1} \in T$, *and* $\beta \neq \alpha$, *then* $\alpha < \beta$.

Proof: Since $\beta \neq \alpha$, we have $u \in K$ and if $\beta < \alpha$, we would have $u^{-1} < u$. Thus the elements of T are of three kinds:

1) $l(u)$ even, $u = \alpha\beta^{-1}$, $\alpha < \beta$, $u \in K$.

2) $l(u)$ odd, $u = \alpha a\beta^{-1}$, $\alpha < \beta$, $u \in K$.

3) $l(u)$ odd, $u = \alpha a\alpha^{-1}$, generated by transforms of the same kind in K.

LEMMA 17.3.2. *If $u \neq v$ belong to T and are not both in the same conjugate $\alpha B\alpha^{-1}$, and w is any one of $u^\epsilon v^\eta$ or $v^\eta u^\epsilon$ ($\epsilon, \eta = \pm 1$), then w follows both u and v in the semi-alphabetical ordering. This leads to the following restrictions on cancelation and amalgamation in the product w:*

1) If $u = \alpha\beta^{-1}$, β^{-1} does not cancel, and if α cancels, then the adjacent term of β^{-1} does not amalgamate.

2) If $u = \alpha a\beta^{-1}$, $\alpha < \beta$, α and a do not cancel, and if β^{-1} cancels, then a does not amalgamate.

3) If $u = \alpha a\alpha^{-1} \in \alpha B\alpha^{-1}$, α, and a do not cancel, and if $v^\eta = \alpha a^{-1}\sigma$, with a, $a^1 \in A_v$, then a^{-1} is the earliest element in the coset Ba^1.

Proof: Of the two different elements u and v belonging to T, let the letter u represent the earlier, so that $u < v$. If w does not follow both u and v, then $w < v$. Here the possibility $w = v$ may be eliminated at once, since it would imply $u = 1$, which cannot hold, or $u = v^2$ or v^{-2}. Now the square of a transform v is 1, or is a similar transform, while if v is not a transform then $l(v^2) > l(v)$. In either case $u = v^2$ or v^{-2} is impossible.

Since any two of u, v, w generate the third, w must be the third if v belongs to K. Thus we need only consider cases with $v = \alpha a\alpha^{-1} \in \alpha B\alpha^{-1}$ a transform. Now with $u < v$, since $u \notin \alpha B\alpha^{-1}$, we also have $u < v^*$, where v^* is any transform in $\alpha B\alpha^{-1}$. Hence, if the canceling between u and $v = \alpha a\alpha^{-1}$ involves only α (or α^{-1}), the same will hold for u and some $v^* = \alpha a^*\alpha^{-1} \in K$, yielding a product $w^* = u^\epsilon v^*$ or $v^* u^\epsilon$, with $w^* < v^*$ contrary to $v^* \in K$. Hence, the canceling between u and $v = \alpha a\alpha^{-1}$ involves all of α and cancellation or amalgamation with the center term a. Thus $u^\epsilon = \sigma a''^{-1}\alpha^{-1}$, where a'' amalgamates or cancels with a. Since $u < v = \alpha a\alpha^{-1}$, either $l(\sigma) < l(\alpha)$ or $l(\sigma) = l(\alpha)$; and $u = \sigma a''^{-1}\alpha^{-1}$, with $\sigma < \alpha$. In either event, u and $\sigma(a''^{-1}a^*a'')\sigma^{-1}$ precede and generate a $v^* = \alpha a^*\alpha^{-1} \in K$, a contradiction. Thus in all cases we reach a contradiction if $w < v$, and so w follows both u and v.

In consequence of the fact that all eight products $u^\epsilon v^\eta$ and $v^\eta u^\epsilon$ follow both u and v, we have the restrictions on canceling and amalgamating listed in the theorem. These say explicitly that not more than half of either u or v cancels, and that in cases where canceling and amalgamating with one replaces an initial (or final) segment of the other with another segment of the same length, the result is an element later in the ordering.

LEMMA 17.3.3. *In a product* $u_1u_2 \cdots u_t$ *with* $u_i \in T \cup T^{-1}(i = 1, \cdots, t)$, $u_iu_{i+1} \neq 1(i = 1, \cdots, t - 1)$, *and* u_i, u_{i+1} *not both in the same group* $\alpha B\alpha^{-1}$ $(B \subseteq A_v)$, *the reduced form will end as follows:*

1) β^{-1} *if* $u_t = \alpha\beta^{-1}$.
2) $b^*\alpha^{-1}$ *if* $u_t = (\alpha\beta^{-1})^{-1}$.
3) $a^*\beta^{-1}$ *if* $u_t = \alpha a\beta^{-1}$, $\alpha < \beta$.
4) $a^{-1}\alpha^{-1}$ *if* $u_t = (\alpha a\beta^{-1})^{-1}$, $\alpha < \beta$.
5) $a^*\alpha^{-1}$ *if* $u_t = \alpha a\alpha^{-1}$.

Here b^* *in* (2) *and* a^* *in* (5) *are either the term immediately preceding in* u_t *or are amalgamations with a similar term in* u_{t-1}. *In* (3), a^* *can involve amalgamation with* u_{t-1} *and* u_{t-2}.

Proof: This lemma will be proved by induction on t, being trivial for $t = 1$. For $t = 2$, the results come directly from Lemma 17.3.2 with the added observation that for $u = \alpha\beta^{-1}$ or $\alpha a\beta^{-1}$, the cancellation in u^2 does not go through α or β. In proving the induction from t to $t + 1$, we need only apply Lemma 17.3.2 to each of the five cases listed, as well as to each of the five possibilities for u_{t+1}, using only one additional property not an immediate consequence of Lemma 17.3.2. This is as follows: It may happen that $u_t = \alpha a\alpha^{-1}$, that α cancels, and that a amalgamates with $u_{t-1} = \sigma a'^{-1}\alpha^{-1}$, and similarly, with $u_{t+1} = \alpha a''\lambda$. Now by Lemma 17.3.2 each of a' and a'' is the earliest element in its own coset Ba', Ba''. If $a'^{-1}aa'' = 1$, this would mean that a' and a'' were in the same coset, and hence $a' = a''$, $a = 1$, $u_t = 1$, would be a contradiction. Hence $a'^{-1}aa'' \neq 1$, and the reduced form of $u_{t-1}u_tu_{t+1}$ is $\sigma(a'^{-1}aa'')\lambda$. This is the only way in which amalgamation can involve as many as three consecutive terms in any product $u_1u_2, \cdots, u_m$ which is half-reduced.

In establishing the ending of the reduced form for the half-reduced expression $h = u_1u_2, \cdots, u_t$, we have shown *a fortiori* that $h \neq 1$, and hence that H is the free product of the infinite cyclic groups generated by the elements $\alpha\beta^{-1}$ and $\alpha a^{-1}\beta^{-1}(\alpha < \beta)$ and the conjugates $\alpha B\alpha^{-1}$ of subgroups B of free factors A.

18. THE BURNSIDE PROBLEM

18.1. Statement of the Problem.

In 1902 Burnside [1] wrote "A still undecided point in the theory of discontinuous groups is whether the order of a group may be not finite while the order of every operation it contains is finite." He is, of course, discussing finitely generated groups. The question is still undecided. In this generality the problem has not really been attacked. He considers a more specialized form of the problem in which it is assumed that the given group is finitely generated and that the orders of the elements are bounded.

If G is generated by r elements and n is the least common multiple of the orders of elements of G, then the problem is: Is G a finite group? This problem is known as the Burnside problem. If $x_1, \cdots, x_r$ generate a group $B(n, r)$ with relations $g^n = 1$ for every $g \in B(n, r)$, then this group is called the *Burnside group of order n with r generators.* Clearly, every group with r generators and elements of orders dividing n will be a homomorphic image of this particular group. Thus the Burnside problem reduces to the question: Which of the groups $B(n, r)$ are finite?

If F_r is the free group generated by $x_1, \cdots, x_r$ and N is the fully invariant subgroup generated by all z^n, with $z \in F_r$, then $B(n, r) = F_r/N$.

18.2. The Burnside Problem for $n = 2$ and $n = 3$.

If every element of a group G besides the identity is of order 2, then from $x^2 = 1$, $y^2 = 1$, $(xy)^2 = 1$ we have $xyxy = 1$, $xy = y^{-1}x^{-1} = yx$, whence G is Abelian. Thus the Burnside group $B(2, r)$ generated by $x_1, \cdots, x_r$, in which the square of every element is the identity, is Abelian of order 2^r, having $x_1, \cdots, x_r$ as a basis. This settles $n = 2$.

When $n = 3$ it is easy to show that $B(3, r)$ is finite. We proceed by induction on r. $B(3, 1)$ is the cyclic group of order 3. Suppose that $B_h = B(3, h)$ is of order $3^{m(h)}$ We use the relation

$$(18.2.1) \qquad yxy = x^{-1}y^{-1}x^{-1},$$

which is a consequence of $(xy)^3 = 1$. B_{h+1} is obtained by adjoining a new generator z to B_h. Hence the elements of B_{h+1} are of the form

$$g = u_1 z^{\pm 1} u_2 z^{\pm 1} \cdots z^{\pm 1} u_n, \tag{18.2.2}$$

with $u_i \in B_h$. We show that g can be expressed, using at most two z's. If in (18.2.2) we have two consecutive terms with the same exponent, we use (18.2.1) to put $zu_iz = u_i^{-1}z^{-1}u_i^{-1}$ or $z^{-1}u_jz^{-1} = u_j^{-1}zu_j^{-1}$, thus reducing the number of z's by one. Thus g can be expressed in the form (18.2.2) with the exponents of z alternating in sign. Here if g has as many as three z terms, then in $g = u_1zu_2z^{-1}u_3z \cdots$ we write $g = u_1zu_2z \cdot zu_3z \cdots = u_1u_2^{-1}z^{-1}u_2^{-1}u_3^{-1}z^{-1}u_3^{-1} \cdots$, reducing the number of z's by one. A similar argument holds if $g = u_1z^{-1}u_2zu_3z^{-1} \cdots$. Hence g can be expressed, using at most two z's. We may also write $u_1z^{-1}u_2zu_3 = u_1z^{-1}u_2z^{-1}z^{-1}u_3 = u_1u_2^{-1}zu_2^{-1}z^{-1}u_3$. Hence every element of B_{h+1} can be put in one of the forms:

$$\begin{aligned} &u_1, \\ &u_1zu_2, \\ &u_1z^{-1}u_2, \\ &u_1zu_2z^{-1}u_3. \end{aligned} \tag{18.2.3}$$

Hence B_{h+1} has at most $3^m + 2\cdot 3^{2m} + 3^{3m} < 3^{3m+1}$ elements, and so, $m(h+1) \leq 3m(h)$, whence generally $m(r) \leq 3^{r-1}$ and $B(3, r)$ is of order at most $3^{3^{r-1}}$. In his original paper, using a more complicated method, Burnside obtains the better restriction $m(r) \leq 2^r - 1$. We shall, however, proceed to obtain the exact result $m(r) = \binom{r}{3} + \binom{r}{2} + r$ as obtained by Levi and van der Waerden [1].

We begin with a formula, applying (18.2.1) three times,

$$\begin{aligned} x^{-1}yxzx^{-1} &= (x^{-1}yx^{-1})(x^{-1}zx^{-1}) \\ &= y^{-1}(xy^{-1}z^{-1}x)z^{-1} \\ &= y^{-1}zyx^{-1}zyz^{-1}. \end{aligned} \tag{18.2.4}$$

As a special case of (18.2.4) with $z = y$, we find $x^{-1}yxyx^{-1} = yx^{-1}y$, whence

$$(x^{-1}yx)y = y(x^{-1}yx). \tag{18.2.5}$$

Thus any element y permutes with any of its conjugates $x^{-1}yx$. Hence y also permutes with $y^{-1}x^{-1}yx$, and so, in the notation of commutators,

$$(y, x, y) = 1. \tag{18.2.6}$$

This also leads to

$$(18.2.7) \qquad \begin{aligned}(x, y)^{-1} &= (x^{-1}y^{-1}xy)^{-1} = y^{-1}x^{-1}yx \\ &= x^{-1}yxy^{-1} = (x, y^{-1}).\end{aligned}$$

From these we also have

$$(18.2.8) \qquad \begin{aligned}(x^{-1}, y) &= (y, x^{-1})^{-1} = (y, x) = (x, y)^{-1}, \\ (y, x, x) &= ((x, y)^{-1}, x) = (x, y, x)^{-1} = 1.\end{aligned}$$

We now consider $(a, c, b)^{-1} = b^{-1}(c^{-1}a^{-1}caba^{-1}c^{-1})ac$ and apply (18.2.4) to the part in parentheses with $x = c$, $y = a^{-1}$, $z = aba^{-1}$. This yields

$$\begin{aligned}(a, c, b)^{-1} &= b^{-1}(a \cdot aba^{-1}a^{-1}c^{-1}aba^{-1}a^{-1}ab^{-1}a^{-1})ac \\ &= b^{-1}a^{-1}bac^{-1}aba^{-1}b^{-1}c \\ &= (a, b, c).\end{aligned}$$

We also have $(a, c, b)^{-1} = ((a, c)^{-1}, b) = (c, a, b)$. These results combine to give

$$(18.2.9) \qquad \begin{aligned}(a, b, c) &= (c, a, b) = (b, c, a), \\ (a, c, b)^{-1} &= (a, b, c).\end{aligned}$$

We are now in a position to show that every commutator of weight four is the identity. Consider first the complex commutator $(a, b; c, d)$. Using (18.2.9), we have

$$\begin{aligned}(a, b; (c, d)) &= ((c, d), a, b) = ((c, d, a), b) = (a, c, d, b) \\ &= (a, c, b, d)^{-1} = ((a, b, c)^{-1}, d)^{-1} \\ &= (a, b, c, d).\end{aligned}$$

But also $\quad (c, d; a, b) = ((a, b), c, d) = (a, b, c, d)$.

Hence $(a, b; c, d) = (c, d; a, b) = (a, b; c, d)^{-1}$, whence

$$(18.2.10) \qquad (a, b; c, d) = (a, b, c, d) = 1.$$

From the preceding results a commutator of weight 3 is the identity if it involves a repeated element. With three distinct generators, a commutator of weight 3, by (18.2.9), may be put in the form (x_i, x_j, x_k) $i < j < k$ or the inverse of this form. By (18.2.10) commutators of weight 3 are in the center, and the commutator subgroup is Abelian. Hence every element of $B(3, r)$ may be put in the form:

(18.2.11)

$$g = x_1^{a_1} \cdots x_r^{a_r}(x_1, x_2)^{b_{12}} \cdots (x_i, x_j)^{b_{ij}} \cdots (x_i, x_j, x_k)^{c_{ijk}},$$

where for (x_i, x_j) $i < j$, for (x_i, x_j, x_k) $i < j < k$, and the exponents are 0, 1, or 2. The total number of such expressions is $3^{m(r)}$, where $m(r) = r + \binom{r}{2} + \binom{r}{3}$ is the number of ways of choosing combinations of the r generators $x_1, \cdots, x_r$ one, two, or three at a time. Hence the Burnside group $B(3, r)$ is of order at most $3^{m(r)}$, and this will be the exact order unless some of the elements of (18.2.11) whose exponents are not all zero reduce to the identity. For if two different expressions g_1 and g_2 represent the same element of $B(3, r)$, they do so in every homomorphic image of $B(3, r)$, in particular in the elementary Abelian group of order 3^r whence their exponents a_i, $i = 1, \cdots, r$ agree. The commutator subgroup being Abelian $g = g_1g_2^{-1} = 1$ would be an element with some b_{ij} or c_{ijk} different from zero modulo 3 representing the identity. Regarding $g = 1$ as a relation of $B(3, r)$, this remains valid if we add additional relations $x_s = 1$, $s \neq i, j, k$. Hence, to show that the exact order of $B(3, r)$ is $3^{m(r)}$, it is sufficient to show that the exact order of $B(3, 3)$ is 3^7.

We shall use the normal product as given in Theorems 6.5.1 and 6.5.2 to construct $B(3, 3)$ as a group of order 3^7. Let us write

(18.2.12) $$C_1 = x, \quad C_2 = y, \quad C_3 = z, \quad C_4 = (x, y),$$
$$C_5 = (x, z), \quad C_6 = (y, z), \quad C_7 = (x, y, z).$$

First construct $A = \{C_4, C_5, C_6, C_7\}$ as an elementary Abelian group of order 3^4. We then extend A by adjoining C_3 with relations

(18.2.13) $$C_3^3 = 1, \quad C_3^{-1}C_4C_3 = C_4C_7, \quad C_3^{-1}C_5C_3 = C_5,$$
$$C_3^{-1}C_6C_3 = C_6, \quad C_3^{-1}C_7C_3 = C_7.$$

From Theorems 6.5.1 and 6.5.2 the group $B = \{A, C_3\}$ will be of order 3^5 and an extension of A by the cyclic group C_3, if we verify that from the relations (18.2.13) transformation by C_3 induces an automorphism of order 3 in A. In the same way we may extend B by C_2 to obtain $H = \{B, C_2\}$ of order 3^6, using relations

(18.2.14) $$C_2^3 = 1, \quad C_2^{-1}C_3C_2 = C_3C_6^{-1}, \quad C_2^{-1}C_4C_2 = C_4,$$
$$C_2^{-1}C_5C_2 = C_5C_7^{-1}, \quad C_2^{-1}C_6C_2 = C_6, \quad C_2^{-1}C_7C_2 = C_7,$$

and finally extend H by C_1 to $G = \{C_1, H\}$ of order 3^7, using

(18.2.15)

$$C_1^3 = 1, \quad C_1^{-1}C_2C_1 = C_2C_4^{-1}, \quad C_1^{-1}C_3C_1 = C_3C_5^{-1},$$
$$C_1^{-1}C_4C_1 = C_4, \quad C_1^{-1}C_5C_1 = C_5, \quad C_1^{-1}C_6C_1 = C_6C_7, \quad C_1^{-1}C_7C_1 = C_7.$$

From these relations G is of class 3 and the collection formula,

(18.2.16) $$(PQ)^3 = P^3Q^3(Q,P)^3(Q, P, P)\,(Q, P, Q)^5,$$

holds. Taking $P = z$ and Q an arbitrary element of A, it follows that B is of exponent 3. Similarly, H and G may in turn be shown to be of exponent 3. Thus $G = B(3, 3)$ is of order 3^7. We have observed above that this leads to a general theorem.

THEOREM 18.2.1. *The Burnside group* $B(3, r)$ *is of order* $3^{m(r)}$, *where* $m(r) = r + \binom{r}{2} + \binom{r}{3}$. *An element of* $B(3, r)$ *has a unique expression of the form given by* (18.2.11).

18.3. Finiteness of $B(4, r)$.

It was shown by Burnside in his original paper that $B(4, 2)$ is of order at most 2^{12}. It was proved by Sanov [1] that $B(4, r)$ is finite for every r. The order of $B(4, r)$ is not known exactly, but $B(4, 2)$ is indeed of order 2^{12}.

THEOREM 18.3.1. *The groups* $B(4, r)$ *are finite.*

Proof: Let H be any finite group whose elements are all of orders dividing 4. We wish to show that adjoining an element b of order 4 to H and putting the fourth power of every element equal to the identity in the extended group $G = H \cup (b)$ requires that G be finite. We can accomplish this adjunction in two steps, first adjoining b^2 to H to yield a group $H_1 = H \cup (b^2)$, and then adjoin b to H_1 to yield $G = H_1 \cup (b) = H \cup (b)$. Each of these extensions is such that we are adjoining an element whose square is in the preceding group. Hence it is enough to show that adjoining to a finite group H an element x with $x^2 \in H$ and putting $z^4 = 1$ for $z \in H \cup (x)$ implies that $H \cup (x)$ is finite.

Every element g of $H \cup (x)$, with $x^2 \in H$ is of the form

(18.3.1) $$g = h_1xh_2xh_3x \cdots h_{n-1}xh_n, \quad h_i \in H.$$

From the relation $(xh)^4 = 1$ we get

(18.3.2) $$xhx = h^{-1}x^{-1}h^{-1}x^{-1}h^{-1} = h^{-1}x(x^2\,h^{-1}x^2)xh^{-1}$$
$$= h^{-1}xh^*xh^{-1},$$

where h^* also belongs to H. Thus, without increasing the length n of the word in (18.3.1), we may use (18.3.2) to alter its form to

$$(18.3.3) \qquad h_1xh_2 \cdots xh_{i-1}h_i^{-1}xh_i^*xh_i^{-1}h_{i+1}x \cdots xh_n.$$

If any h_j is 1 in (18.3.1), $2 \leq j \leq n - 1$, we may reduce the length by putting $x^2 = h \in H$. We may also be able to use (18.3.2) a number of times to change some h_j to 1.

Sanov observes that, using (18.3.2) repeatedly, we may replace h_{i-1} by $h_{i-1}h_i^{-1}$, then h_{i-2} by $h_{i-2}(h_{i-1}h_i^{-1})^{-1} = h_{i-2}h_ih_{i-1}^{-1}$, and so on. In this way we may replace h_2 by any one of h_2, $h_2h_3^{-1}$, $h_2h_4h_3^{-1}$, $h_2h_4h_5^{-1}h_3^{-1}, \cdots, h_2h_4 \cdots h_{2s}h_{2s-1}^{-1} \cdots h_3^{-1}, h_2h_4 \cdots h_{2s}h_{2s+1}^{-1} \cdots h_3^{-1}$. If any one of these is 1, we may reduce the length of g. But if H is of order M and $n \geq M + 2$, then either one of these expressions is 1 or there will be a repeated value, say, $h_2 \cdots h_{2r}h_{2r+1}^{-1} \cdots h_3^{-1} = h_2 \cdots h_{2r} \cdots h_{2s}h_{2s+1}^{-1} \cdots h_{2r+1}^{-1} \cdots h_3^{-1}$, whence $h_{2r+2} \cdots h_{2s}h_{2s+1}^{-1} \cdots h_{2r+3}^{-1} = 1$. But this is one of the values with which we could replace h_{2r+2}. Similarly, if the repetition involves $h_2h_4 \cdots h_{2r}h_{2r-1}^{-1} \cdots h_3^{-1}$, h_{2r+1} can be replaced by a combination whose value is 1. In any event if $n \geq M + 2$, we may reduce the length. Hence any g may be represented by a word of length $n \leq M + 1$. Thus $H \cup (x)$ is of order at most M^{M+1}.

18.4. The Restricted Burnside Problem. Theorems of P. Hall and G. Higman. Finiteness of $B(6, r)$.

A weaker form of the Burnside conjecture is the following proposition; its proof is known as the restricted Burnside problem:

> R_n: *For each positive integer r there is an integer $b_{n,r}$ such that every finite group of exponent n that can be generated by r elements has order at most $b_{n,r}$.*

If for some value of n, R_n is true, it is conceivable that there may be an infinite group of exponent n with r generators. But R_n being true, there will be a largest finite group $R(n, r)$ of exponent n generated by r elements. For each finite group of exponent n generated by r elements is a factor group F_r/N_i, where F_r is the free group with r generators and N_i is some normal subgroup containing all nth powers of elements of F_r. If R_n is true, there can be only a finite number of such normal subgroups N_i, and their intersection is a normal subgroup N of finite index and $F_i/N = R(n, r)$ will be a finite group of exponent

n generated by r elements such that all others are homomorphic images of it.

Let G be a group with lower central series:

$$G = G_1 \supseteq G_2 \supseteq G_3 \supseteq \cdots. \tag{18.4.1}$$

Suppose that a relation $G_s = G_{s+1}$ holds. Then from the properties of the lower central series we have $G_s = G_{s+1} = \cdots = G_{s+i} = \cdots$. If G is nilpotent, then some $G_{s+i} = 1$, whence $G_s = 1$. Since a finite group G of prime power exponent $n = p^t$ is nilpotent, the relation $G_s = G_{s+1}$ in such a group implies $G_s = 1$. Now suppose that G is of exponent p^t and is generated by r elements. Then each G_i/G_{i+1} is a finite Abelian group. If we can show that for every such group G there is an integer $s = s(p^t, r)$ such that $G_s = G_{s+1}$, then we shall have solved the restricted Burnside problem for exponent $n = p^t$.

In the collecting process (Theorem 12.3.1) as applied to $(xy)^n$ we have found

$$(xy)^n = x^n y^n c_1^{a_1(n)} \cdots c_t^{a_t(n)} \cdots, \tag{18.4.2}$$

where if c_i is of weight m, then its exponent, $a_i(n)$, is of the form

$$u_i \cdot n + u_{i2}\binom{n}{2} + \cdots + u_{im}\binom{n}{m}. \tag{18.4.3}$$

And if c_i is of the form

$$c_i = (y, \overbrace{x, \cdots, x}^{s}), \tag{18.4.4}$$

the exponent $a_i(n)$ is the number of ways of choosing indices $j_1, j_2, \cdots, j_{s+1}$ such that in

$$(y_{j_1}, x_{j_2}, x_{j_3}, \cdots, x_{j_{s+1}}) \tag{18.4.5}$$

we have

$$j_1 < j_2 < j_3 \cdots < j_{s+1}, \tag{18.4.6}$$

and

$$1 \le j_k \le n.$$

But this is merely the number of ways of choosing $s + 1$ distinct numbers from $1, 2, \cdots, n$ and is $\binom{n}{s+1}$.

If $n = p$ is a prime, the exponents for commutators of weights at most $p - 1$ are all multiples of p, since the binomial coefficients

$\binom{p}{i}$ with $1 \leq i \leq p - 1$ are all multiples of p. But for the commutator

(18.4.7) $$(y, \overbrace{x, \cdots, x}^{p-1}),$$

the exponent is $\binom{p}{p} = 1$. Hence in a group G of exponent p we have

(18.4.8) $$1 = (xy)^p = (y, \overbrace{x, x, \cdots x}^{p-1})\, v_1 \cdots v_t,$$

where $v_1, v_2, \cdots, v_t$ are commutators of weight at least p, and for those of weight p the weight in y is at least 2.

This gives the relation in G_p modulo G_{p+1}.

(18.4.9) $$(y, \overbrace{x, \cdots, x}^{p-1})\, v_1 \cdots v_s \equiv 1 \pmod{G_{p+1}}$$

where $v_1, \cdots, v_s$ are commutators of weight p in x and y, and of weight at least 1 and at most $p - 2$ in x. From our rules (10.2.1) we have generally in any group that if (u, v) is of weight m, then

(18.4.10) $$(u^i, v^j) \equiv (u, v)^{ij} \pmod{G_{m+1}}.$$

Using this we find that if a v in (18.4.9) is of weight r in x, the replacement of x by x^i in (18.4.9) replaces v by v^{i^r}. Putting $i = 1, 2, \cdots, p - 1$ in turn and multiplying, we have for the exponent of v in the product

(18.4.11) $$1^r + 2^r + 3^r + \cdots + (p - 1)^r \equiv 0 \pmod{p}$$

for $1 \leq r \leq p - 2$, but for the leading term $(y, \overbrace{x, \cdots x}^{p-1})$, $r = p - 1$ and $i^r \equiv 1 \pmod{p}$. Hence the product takes the form

(18.4.12) $$(y, \overbrace{x, \cdots, x}^{p-1})^{p-1} \equiv 1 \pmod{G_{p+1}},$$

and so,

(18.4.13) $$(y, \overbrace{x, \cdots, x}^{p-1}) \equiv 1 \pmod{G_{p+1}}.$$

This relation has been the key to investigation of the restricted Burnside problem for groups of prime exponent p. Starting with this, Kostrikin [1] has solved the restricted Burnside problem for the group G of exponent 5 with two generators. He has shown that $G_{13} = G_{14}$ and that G, if finite, has order at most 5^{34}.

A number of authors have studied the restricted Burnside problem, and for this it has often been convenient to carry out calculations in the associated Lie ring of a group. We shall describe this now.

In an associative ring R let us define a *Lie product* $[x, y]$ by the rule

$$[x, y] = xy - yx. \tag{18.4.14}$$

Then with respect to the addition in R and the Lie product, the elements of R form a Lie ring L. A Lie ring L satisfies the following laws:

$L0$. *Addition $x + y$, and Lie product $[x, y]$ are well-defined operations.*
$L1$. *Addition is an Abelian group with zero element* 0.
$L2$. $[x + y, z] = [x, z] + [y, z]$.
$[x, y + z] = [x, y] + [x, z]$.
$L3$. $[x, x] = 0$.
$L4$. $[[x, y], z] + [[y, z], x] + [[z, x], y] = 0$.

It is easy to check that $[x, y]$ as defined by (18.4.14) satisfies these laws.

From $L2$ and $L3$ we find

$$\begin{aligned} 0 = [x + y, x + y] &= [x, x] + [x, y] + [y, x] + [y, y] \\ &= [x, y] + [y, x] \end{aligned} \tag{18.4.15}$$

whence

$$[y, x] = -[x, y]. \tag{18.4.16}$$

If R is generated by elements $x_1, \cdots, x_r$, then the elements generated from $x_1, \cdots, x_r$ by addition and the Lie product $[x, y]$ will not in general include all the elements generated in R by addition and the associative product. The elements generated by the Lie product are called *Lie elements*. Thus x_1^2 is not a Lie element, but $x_1^2x_2 - 2x_1x_2x_1 + x_2x_1^2 = x_1(x_1x_2 - x_2x_1) - (x_1x_2 - x_2x_1)x_1$ is a Lie element. It may, of course, happen because of relations in R that x_1^2 is equal to a Lie element.

We may take the laws $L0$, $L1$, $L2$, $L3$, $L4$ as the definition of a Lie ring L. It has been shown by Garrett Birkhoff [1] and E. Witt [1] that every Lie ring L can be represented as a ring of Lie elements of an appropriate associative ring R. This important result is not, however, needed here.

If G is a group with lower central series,

$$G = G_1 \supseteq G_2 \supseteq G_3 \supseteq \cdots \supseteq G_n \supseteq \cdots. \tag{18.4.17}$$

The *associated Lie ring* L of G is formed in the following way:

1) L is the Cartesian sum of the additively written factor groups G_i/G_{i+1}, and this Cartesian sum gives the addition in L.

2) The elements of G_i/G_{i+1} are regarded as homogeneous of degree i.

3) The Lie product of a homogeneous element A of degree i with a homogeneous element B of degree j is the group commutator (A, B) modulo G_{i+j+1}.

4) The Lie product of general elements of L is given by (3) and the distributive laws.

We shall not prove here that these rules do define a Lie ring. We note only that $L2$ corresponds to the commutator identities (10.2.1.2) and (10.2.1.3) and that $L4$ corresponds to (10.2.1.5). The results of §11.2 can be restated to show that the Lie ring corresponding to a free group with r generators is the free Lie ring with r generators except that some infinite sums are allowed. A modification of the methods of §11.2 may be used to prove that these rules define a Lie ring.

In a Lie ring L let us write monomials in left normed form, i.e., write x_1x_2 for $[x_1, x_2]$, and recursively, $x_1x_2, \cdots, x_n$ for $[x_1 \cdots x_{n-1}, x_n]$. The following theorem is due to Graham Higman [1].

THEOREM 18.4.1. *The associated Lie ring of a group of prime exponent p satisfies the identical relation $yx^{p-1} = 0$.*

Proof: The relation (18.4.13) holds in a group G of exponent p, and this we write in the form

$$(y, \overbrace{x, \cdots, x}^{p-1}) = c_1c_2 \cdots c_t, \tag{18.4.18}$$

where $c_1, c_2, \cdots, c_t$ are commutators in x and y of total weight $p + 1$ or higher, and naturally of weight at least one in y.

In this put $x = x_1x_2 \cdots x_{p-1}$. Using the rules (10.2.1) and collecting so as to leave on the left-hand side only the commutators with distinct x's, we have

$$X = \prod_{\sigma} (y, x_{1\sigma}, \cdots, x_{(p-1)\sigma}) = d_1 \cdot d_2 \cdots d_s, \tag{18.4.19}$$

where σ runs over the $(p-1)!$ permutations of $1, 2, \cdots, p-1$ in some order, and $d_1, d_2, \cdots d_s$ are commutators which are of positive weight in y and either (1) of total weight at least $p + 1$ in $y, x_1, \cdots, x_{p-1}$, or (2) of total weight p in $y, x_1, \cdots, x_{p-1}$ and having some x_j missing.

We may suppose that each d_i is of positive weight in each of y, $x_1, \cdots, x_{p-1}$. This is proved inductively. Suppose, in fact, that we already have such a relation with each d_i of positive weight in y, $x_1, \cdots, x_{j-1}$. At the cost of introducing further commutators, we may evidently suppose that the d_i which are of zero weight in x_j form the initial segment $d_1 \cdots d_t$. Putting $x_j = 1$, we see that $d_1 d_2 \cdots d_t = 1$, and so they can be omitted. Hence we may suppose that the d's are of positive weight in each of $y, x_1, \cdots, x_{p-1}$ and of total weight $p + 1$. The commutators of weight p originally present had some x_j missing, and these have been omitted at some stage. But this now means in terms of the associated Lie ring L that if $y, x_1, x_2, \cdots, x_{p-1}$ are any homogeneous elements of whatever weight, we have

(18.4.20) $$\sum_\sigma y x_{1\sigma} x_{2\sigma} \cdots x_{(p-1)\sigma} = 0.$$

But (18.4.20) is an identity in L valid for homogeneous elements y, $x_1, \cdots, x_p$. Since it is linear in each argument, the identity therefore will be valid for any arguments. Thus with $x_1 = x_2 = \cdots = x_p = x$, and y arbitrary, (18.4.20) becomes

(18.4.21) $$(p - 1)!\, y x^{p-1} = 0,$$

and as L is easily seen to be of characteristic p, we have

(18.4.22) $$y x^{p-1} = 0,$$

proving our theorem.

Using the relation $yx^4 = 0$ in a Lie ring L of characteristic 5 (more precisely of characteristic prime to 2 or 3), Graham Higman [1] has shown that if L is generated by r elements, then in L, products of degree Nr or higher are zero, where N is some integer not depending on r. A little work shows that he has in fact proved this with $N = 25$, but he states that with further calculations he believes it possible to prove the result with $N = 9$, though even this is probably not the best possible result.

In a very important paper, Philip Hall and Graham Higman [1] have, among other things, related the restricted Burnside problem for general exponents to that for prime power exponents. But for this it is necessary to restrict ourselves to finite solvable groups. Here the conjecture, weaker than R_n, takes the following form:

S_n: *For each positive integer* r *there is an integer* $b_{n,r}$ *such that every finite solvable group of exponent n that can be generated by r elements has order at most* $b_{n,r}$.

Their result takes the precise form:

THEOREM 18.4.2. *If* $n = p_1^{e_1}p_2^{e_2}, \cdots, p_s^{e_s}$ *and if* $S_{p_i^{e_i}}$ *is true for* $i = 1, \cdots, s$, *then* S_n *is true.*

We shall not give the proof of this theorem here, since it depends on some long and complicated preliminary results. Since a finite group of order $p^a q^b$ is solvable (Theorem 16.8.7), R_n and S_n are the same statement when n is divisible by at most two distinct primes. But since the Burnside groups $B(2, r)$, $B(3, r)$, and $B(4, r)$ are known to be finite, and Graham Higman has shown R_5 to be true, Theorem 18.4.2 proves the truth of R_6, R_{12}, R_{10}, R_{15}, R_{20}, and also S_{30} and S_{60}. Encouraged by these results, the author has shown that the groups $B(6, r)$ are finite, and a sketch of this proof will be given below.

We shall give here a small part of the results of Philip Hall and Graham Higman, and some indication of the lines along which the rest proceeds.

Let a group be called a p'-group, where p is a prime if its order is prime to p, and as usual, a p-group if its order is a power of p.

DEFINITION: *A finite group G is called p-solvable if it has a normal series.*

$$1 = V_0 \subset V_1 \subset \cdots \subset V_n = G, \tag{18.4.23}$$

in which each factor group V_{i+1}/V_i *is either a p-group or a p'-group.* We note from Theorem 9.2.4 that a finite solvable group G is p-solvable for every prime p. For a p-solvable group G we define the *upper p-series*

$$1 = P_0 \subseteq N_0 \subset P_1 \subset N_1 \subset P_2 \cdots \subset P_l \subseteq N_l = G \tag{18.4.24}$$

recursively by the rule that N_k/P_k is the greatest normal p'-subgroup of G/P_k, and P_{k+1}/N_k is the greatest normal p-subgroup of G/N_k. The number l, which is the least integer such that $N_l = G$, we call the p-length of G, and we write this l_p or $l_p(G)$. It is easy to see that l_p is the smallest number of p factor groups that can occur in any normal series for G, such as (18.4.23), in which the factor groups V_{i+1}/V_i are either p-groups or p'-groups.

The purpose of the Hall-Higman paper is to relate the p-length of a p-solvable group G to properties of a Sylow p-subgroup $S(p)$ of G. In particular let p^{e_p} be the *exponent* of $S(p)$, i.e., the highest order of an element of $S(p)$. Then the *exponent* of G, i.e., the least common multiple n of the orders of elements of G is $n = \prod_p p^{e_p}$. Their main theorems apply to odd primes p, and the results are slightly different for the Fermat primes p, which are of the form $p = 2^n + 1$, and for primes which are not Fermat primes. The theorem, which is relevant to the Burnside problem, is the following:

THEOREM 18.4.3. *If G is a p-solvable group where p is an odd prime, then*

1) $e_p \geq l_p$, *if p is not a Fermat prime, and*
2) $e_p \geq [\frac{1}{2}(l_p + 1)]$, *if p is a Fermat prime.*

We may readily deduce Theorem 18.4.2 from Theorem 18.4.3. Let $n = p_1^{e_1}p_2^{e_2} \cdots p_s^{e_s}$. We can take $p_1 = 2$ if n is even and proceed by induction on s, assuming S_m to be true for $m = p_1^{e_1}p_2^{e_2} \cdots p_{s-1}^{e_{s-1}}$. Then by Theorem 18.4.3 a finite solvable group G of exponent n has a bound of at most $2e_s$ on its p_s-length; $l = l_{p_s} \leq 2e_s$. Then if G is generated by r elements, from S_m, the order of G/P_l is bounded by $b_{m,r}$, and so also P_l has a bound on the number of its generators (corollary to Lemma 7.2.2), say, r_1. Then from $S_{p_s^{e_s}}$, P_l/N_{l-1} has a bound on its order and on the number of the generators of N_{l-1}. Continuing, each of N_i/P_i and P_i/N_{i-1} is of an order bounded by some $b_{m,k}$ or $b_{p^{e_s},k}$ and hence, since $l \leq 2e_s$, we find a bound on the order of G.

THEOREM 18.4.4. *In the upper p-series for a finite p-solvable group G,*

$$1 = P_0 \subseteq N_0 \subset P_1 \subset N_1 \subset P_2 \cdots \subset P_l \subseteq N_l = G,$$

P_1/N_0 contains its centralizer in G/N_0.

COROLLARY 18.4.1. *P_1 contains its centralizer in G.*

Proof of corollary: If x centralizes P_1 in G, then in G/N_0, xN_0/N_0 centralizes P_1/N_0. Hence, by the theorem xN_0/N_0 lies in P_1/N_0, whence in G the coset xN_0 lies in P_1, and so x lies in P_1.

Proof of theorem: In the group $G_1 = G/N_0$ there is no normal p'-subgroup, since N_0 was the greatest normal p'-subgroup in G. The subgroup $\overline{P}_1 = P_1/N_0$ of G_1 is the greatest normal p-subgroup of G_1, by its construction. Let Z be the centralizer of $\overline{P}_1$ in G_1 and suppose, contrary to the theorem, that $Z \not\subseteq \overline{P}_1$. Now Z is a normal subgroup of G_1, and so, $Z \cup \overline{P}_1 = Z\overline{P}_1$ is a normal subgroup of G_1. Since we are supposing $Z\overline{P}_1 \supset \overline{P}_1$, let M be a minimal normal subgroup of G such that $\overline{P}_1 \subset M \subseteq Z\overline{P}_1$. Then $M/\overline{P}_1$ cannot be a p-group, since $\overline{P}_1$ was a maximal normal p-subgroup of G_1. Hence $M/\overline{P}_1$ is a p'-subgroup because G is p-solvable. But then $\overline{P}_1$ and $M/\overline{P}_1$ are of relatively prime orders, and by Theorem 15.2.2, M splits over $\overline{P}_1$, i.e., $M = K\overline{P}_1$, where $K \cap \overline{P}_1 = 1$ for a subgroup K of M isomorphic to $M/\overline{P}_1$. Since $K \subseteq Z\overline{P}_1$, transformation of $\overline{P}_1$ by an element y of K induces an inner automorphism in $\overline{P}_1$, and since the orders of K and $\overline{P}_1$ are relatively prime, this inner automorphism can only be the identity. Hence M is the direct product of K and $\overline{P}_1$, $M = K \times \overline{P}_1$. But then K, as a characteristic subgroup of M, is a normal subgroup of G_1, contrary to the fact that G_1 contains no normal p'-subgroup. Thus the assumption that $Z \not\subseteq \overline{P}_1$ leads to a contradiction and our theorem is proved.

Our next step is essentially a refinement of the preceding theorem.

Theorem 18.4.5. *If G is a p-solvable group with upper p-series,*

$$1 = P_0 \subseteq N_0 \subset P_1 \subset N_1 \subset P_2 \cdots \subset P_l \subseteq N_l = G,$$

and if F/N_0 is the Frattini subgroup of P_1/N_0, the automorphisms of P_1/F induced by transformation with elements of G represent G/P_1 faithfully.

Proof: F/N_0 is the intersection of the maximal subgroups of the p-group P_1/N_0, and P_1/F is an elementary Abelian p-group (Theorem 12.2.1). Since F/N_0 contains the derived group of P_1/N_0, every element of P_1 induces by transformation the identical automorphism in P_1/F. Hence the set of elements of G that induce the identical automorphism in P_1/F is a subgroup K of G (necessarily normal in G) and $K \supseteq P_1$. We show that $K \supset P_1$ leads to a contradiction, and hence $K = P_1$, and so, G/P_1 is faithfully represented by transformation on P_1/F. If $K \supset P_1$, then K/P_1 is not a p-group, since by construction, P_1/N_0 is the maximal normal p-subgroup of G/N_0. Then K contains an element x not in P_1 of order prime to p which induces the identical automorphism in P_1/F by transformation. But, by Theorem 12.2.2, an automorphism of a p-group P_1/N_0 which is the identity on P_1/F is of order a power of p. But then, since the order of x is prime to p, x induces the identical automorphism in P_1/N_0, and by Theorem 18.4.4, this means that $x \in P_1$, which conflicts with our choice of x. Thus $K \supset P_1$ leads to a contradiction, and so, $K = P_1$, giving our theorem.

From Theorem 18.4.5 we have G/P_1 faithfully represented by transformation of the elementary Abelian p-group P_1/F. Here G/P_1 is a p-solvable group and $l_p(G/P_1) = l_p(G) - 1$. Further, by definition, G/P_1 contains no normal p-subgroup. The rest of the Hall-Higman paper consists in studying properties of the representation of G/P_1 on P_1/F, i.e., in effect, by linear transformations in a vector space over the field with p elements. G/P_1 is a p-solvable group which contains no normal p-subgroup, and the theory depends upon what can be said about such groups which can be faithfully represented by a representation over a field of characteristic p. These results also depend on the induction on p-length, using $l_p(G) = l_p(G/P_1) + 1$.

Apart from the results already quoted, we shall restrict our attention to finite groups G of exponent 6 in order to determine the nature of the Burnside groups $B(6, r)$ if they can be shown to be finite.

THEOREM 18.4.6. *In a finite group G of exponent* 6, $l_2(G) \leq 1$ *and* $l_3(G) \leq 1$.

Proof: A finite group G of exponent 6 is necessarily of order 2^a3^b, and so is solvable. In the upper 2-series for G, P_1/N_0 is a 2-group containing its own centralizer in G/N_0. But since G is of exponent 6, a Sylow 2-subgroup of G is of exponent 2, and so is elementary Abelian. Hence P_1/N_0 is centralized by a Sylow 2-subgroup of G/N_0 containing it. Thus P_1/N_0 is a Sylow 2-subgroup of G/N_0, and so $l_2(G) = 1$, and the upper 2-series for G is of the form

$$1 = P_0 \subseteq N_0 \subset P_1 \subseteq N_1 = G, \tag{18.4.25}$$

where N_0/P_0 is a 3-group; P_1/N_0, a 2-group; and N_1/P_1, a 3-group.

Since (18.4.25) is a normal series for G, with at most two factor groups which are 3-groups, we have $l_3(G) \leq 2$. We show that $l_3(G) = 2$ implies that G contains an element of order 9, conflicting with the hypothesis that G is of exponent 6, and we conclude that $l_3(G) \leq 1$. The upper 3-series is of the form

$$1 = A_0 \subseteq B_0 \subset A_1 \subset B_1 \subset A_2 \subseteq B_2 = G, \tag{18.4.26}$$

where B_0/A_0, B_1/A_1, and B_2/A_2 are 2-groups, and A_1/B_0 and A_2/B_1 are 3-groups. We note that

$$(18.4.27) \qquad 1 = B_0/B_0 \subset A_1/B_0 \subset B_1/B_0 \subset A_2/B_0 \subseteq G/B_0$$

is the upper 2-series for G/B_0, and since its 2-length is one, we have $A_2 = B_2 = G$. By Theorem 18.4.4 the 2-group B_1/A_1 is its own centralizer in A_2/A_1. Hence in A_2/A_1, given an element x of order 3, there is an element u of order 2 in B_1/A_1 such that x does not permute with u. If we now write

$$(18.4.28) \qquad u = u_1, \quad x^{-1}u_1x = u_2, \quad x^{-1}u_2x = u_3,$$

then $x^{-1}u_3x = u_1$ since $x^3 = 1$. Let us put $y = y_1 = u_1u_2$ and $y_2 = u_2u_3$. Since u_1, u_2, u_3 belong to an elementary Abelian 2-group, $u_3u_1 = (u_1u_2)(u_2u_3) = y_1y_2$. Hence the group $C = \{x, y_1, y_2\}$ satisfies the relations

$$(18.4.29) \qquad \begin{aligned} x^3 = 1, \quad y_1{}^2 = y_2{}^2 = 1, \quad y_2y_1 = y_1y_2, \\ x^{-1}y_1x = y_2, \quad x^{-1}y_2x = y_1y_2. \end{aligned}$$

Since x does not permute with u_1, $u_2 \neq u_1$, and so, $y_1 = u_1u_2 \neq 1$. Also, $y_2 = y_1$ would imply $x^{-1}y_2x = 1$ and $1 = y_2 = y_1$. Hence $y_2 \neq y_1$ and the group C is seen, from the relations (18.4.29), to be of order 12 and in fact isomorphic to the alternating group on four letters. By Theorem 18.4.5, if F/B_0 is the Frattini subgroup of A_1/B_0, G/A_1 is faithfully represented by transformation of the elementary Abelian 3-group A_1/F. In particular, C is faithfully represented by transformation of $A_1/F = W$. If we write W additively, then transformation of W by an element z of G/A_1 may be represented by taking z as an operator on the right. Here not only C but also the group ring C^* operates on W. Those operators of C^*, which map every element of W onto zero, are easily seen to form a two-sided ideal in C^*. We consider C as generated by x and $y = y_1$ subject to the relations

$$(18.4.30) \qquad x^3 = 1, \quad y^2 = 1, \quad (xy)^3 = 1.$$

Then a two-sided ideal of C^* containing $1 + x + x^2$ also contains

$$(18.4.31) \qquad \begin{aligned} &x^2y(1 + x + x^2)yx - (1 + x + x^2)y \\ &-y(1 + x + x^2) + xy(1 + x + x^2)yx^2 \\ &= 2 - 2y. \end{aligned}$$

If for every $w \in W$ we had $w(1 + x + x^2) = 0$, then from (18.4.31) we would also have $w(2 - 2y) = 0$, and since elements of W are of order 3, this means $wy = w$ for every $w \in W$. But then y is not faithfully represented by transformation of $W = A_1/F$, contrary to Theorem 18.4.5. Hence, for some $w \in W$, we have $w(1 + x + x^2) \neq 0$. In multiplicative form this means that for $\bar{x}$ a representative of the coset $\bar{x}A_1$, this being the element x in A_2/F, we have

$$w(\bar{x}^{-1}w\bar{x})(\bar{x}^{-2}w\bar{x}^2) \neq 1. \tag{18.4.32}$$

But this is the element

$$(w\bar{x}^{-1})^3\bar{x}^3 \neq 1. \tag{18.4.33}$$

Both $\bar{x}^3$ and $(w\bar{x}^{-1})^3$ are elements of W. Not both of these can be the identity by (18.4.33). Hence either $\bar{x}$ or $w\bar{x}^{-1}$ is an element of order 9. Thus $l_3(G) = 2$ implies the existence of an element of order 9 in conflict with the hypothesis that G is of exponent 6. Hence $l_3(G) \leq 1$, and our theorem is proved.

With the help of Theorem 18.4.6 we can find the exact order of the largest finite group of exponent 6 generated by r elements.

Theorem 18.4.7. *The order of $R(r, 6)$ is*

$$2^a 3^{b+\binom{b}{2}+\binom{b}{3}} \tag{18.4.34}$$

where

$$a = 1 + (r - 1)3^{r+\binom{r}{2}+\binom{r}{3}}, \quad b = 1 + (r - 1)2^r.$$

Proof: Let F_r be the free group with r generators. The subgroup S generated by the squares of elements of F_r is such that F_r/S is the elementary Abelian group of order 2^r. Hence, by Theorem 7.2.8, S is a free group with $b = 1 + (r - 1)2^r$ generators. In S the fully invariant subgroup T generated by the cubes of elements of S is such that S/T is $B(3, b)$, and so, T is of index $3^{b+\binom{b}{2}+\binom{b}{3}}$ in S. Here F_r/T is a finite group of exponent 6, since for $g \in F_r$, $g^2 \in S$ and $(g^2)^3 \in T$. Similarly, F_r has a subgroup C of index $3^{r+\binom{r}{2}+\binom{r}{3}}$ generated by the cubes of elements of F_r, and by Theorem 7.2.8, C has $a = 1 + (r - 1)3^{r+\binom{r}{2}+\binom{r}{3}}$ free generators. The fully invariant subgroup D generated by the squares of elements of C is of index 2^a in C. Now let $X = D \cap T$. F_r/X is a finite group of exponent 6, since both D and T contain g^6 for every $g \in F_r$. We easily see that the upper 2-series for F_r/X is

(18.4.35) $$1 = X/X \subset D/X \subset C/X \subset F_r/X,$$

and the upper 3-series for F_r/X is

(18.4.36) $$1 = X/X \subset T/X \subset S/X \subset F_r/X.$$

Here C/D is isomorphic to a Sylow 2-subgroup of F_r/X and S/T to a Sylow 3-subgroup, whence the order of F_r/X is given by (18.4.34).

Let G be any finite group of exponent 6 generated by r elements. Then if

(18.4.37) $$1 = P_0 \subseteq N_0 \subseteq P_1 \subseteq N_1 = G$$

is the upper 3-series for G, we see that N_1/P_1 is of order at most 2^r, and so, P_1 is generated by at most b elements, whence P_1/N_0, isomorphic to a Sylow 3-subgroup of G, is of order at most $3^{b+\binom{b}{2}+\binom{b}{3}}$. Similarly, the order of G is divisible by at most 2^a.

There is not space here to give the complete proof of the Burnside conjecture for exponent 6, since it involves long calculations. The proof does not depend on the preceding work and, as it stands, the proof gives the right power of 2 dividing the order of $B(6, r)$, but to get the right power of 3 it is necessary to apply Theorem 18.4.7. The proof consists in showing that a finitely generated group G of exponent 6 has 2-length one, and so by the finiteness of $B(2, r)$ and $B(3, r)$ is finite. This takes the form of showing that there is a normal chain

(18.4.38) $$G \supset M \supset M' \supset 1,$$

in which G/M is a finite 3-group, M/M' is a finite 2-group, and M' is a finitely generated group of exponent 3 and so finite. Almost all the difficulty consists in showing that M' is of exponent 3.

THEOREM 18.4.8. *A group G of exponent* 6 *generated by r elements is finite.*

COROLLARY 18.4.2. *The order of $B(6, r)$ is given by* (18.4.34).

LEMMA 18.4.1. *The cubes of the elements of G generate a normal subgroup M of index at most* $3^{r+\binom{r}{2}+\binom{r}{3}}$.

LEMMA 18.4.2. *M is generated by a finite number of elements of order* 2. *The derived group M' of M is of index a power of* 2 *in M and is generated by a finite number of elements of the form abab, where* $a^2 = b^2 = 1$.

LEMMA 18.4.3. *If a group H is generated by $x_1, \cdots, x_n$ and if every subgroup of H generated by four of the x's is of exponent* 3, *then H is of exponent* 3.

LEMMA 18.4.4. *If $H = \{a, b, c, d\}$ is of exponent* 6 *and $\{a, b, c\}$, $\{a, b, d\}$, $\{a, c, d\}$, and $\{b, c, d\}$ are of exponent* 3, *then H is of exponent* 3.

LEMMA 18.4.5. *If $H = \{x, a, b\}$ is of exponent* 6 *and if $x^2 = 1$, $a^3 = b^3 = 1$, $xax = a^{-1}$, $xbx = b^{-1}$, then $\{a, b\}$ is of exponent* 3.

This is the most difficult lemma of all and involves some complicated calculations with relations.

LEMMA 18.4.6. If $H = \{x, a, b\}$ is of exponent 6 and if $x^2 = 1$, $a^3 = b^3 = 1$, $xax = a^{-1}$, $xbx = b$, *then $\{a, b\}$ is of exponent* 3.

LEMMA 18.4.7. *If $H = \{x, a, b, c\}$ is of exponent* 6 *and if $x^2 = 1$, $a^3 = b^3 = c^3 = 1$, $xax = a^{-1}$, $xbx = b^{-1}$, $xcx = c^{-1}$, then $\{a, b, c\}$ is of exponent* 3.

LEMMA 18.4.8. *If $H = \{x, a_i\}$ $i = 1, \cdots, n$ is of exponent* 6 *and if* $(x^2 = 1, a_i^3 = 1, xa_ix = a_i^{-1}, i = 1, \cdots n)$, *then $\{a_i\}$ $i = 1, \cdots, n$ is of exponent* 3.

This is easily seen to follow from Lemmas 18.4.3,4, and 7.

LEMMA 18.4.9. *If $H = \{a, b, c\}$ is of exponent* 6 *and $a^2 = b^2 = c^2 = 1$, then H' is of exponent* 3.

LEMMA 18.4.10. *If $H = \{a, b, c, d\}$ is of exponent* 6 *and $a^2 = b^2 = c^2 = d^2 = 1$, then $\{abab, cdcd\}$ is of exponent* 3.

LEMMA 18.4.11. *If $H = \{a, b, c, d, e, f\}$ is of exponent* 6 *and $a^2 = b^2 = c^2 = d^2 = e^2 = f^2 = 1$, then $\{abab, cdcd, efef\}$ is of exponent* 3.

LEMMA 18.4.12. *M' is of exponent* 3 *and so finite. Hence G is finite.*

This lemma is an immediate consequence of Lemmas 18.4.2,3,4, and 11.

19. LATTICES OF SUBGROUPS

19.1. General Properties.

The subgroups of a group G may be taken as the elements of a lattice $L(G)$ under the operations of union and intersection. Every cyclic group of prime order has as subgroups only the entire group and the identity subgroup, whence all these cyclic groups have the same subgroup lattice, consisting merely of a two-element chain. We have already shown (Theorem 1.5.4) that, conversely, a group with no proper subgroups is the identity alone or a finite cyclic group of prime order. We note also that the non-Abelian group of order pq with $p < q$, $p \mid q - 1$ and the elementary Abelian group of order q^2 both have the same subgroup lattice, consisting of the identity, $q + 1$ subgroups of prime order, and the whole group, where any two of the proper subgroups have the identity as their intersection and the whole group as their union.

Thus, although G determines $L(G)$ uniquely, in general $L(G)$ does not determine G uniquely. Moreover, it is easy to find examples of lattices L which are not the lattice $L(G)$ for any group G. But many groups G are indeed determined uniquely by $L(G)$, this being true, for example, for the alternating and symmetric groups on four letters. It may even be true that, except for a few groups of relatively simple types, $L(G)$ does determine G uniquely.

In the terminology of lattice theory, $L(G)$ is complete. This means that infinite unions and intersections always exist, for the set of elements common to all subgroups of a family of subgroups will itself form a group, and this is clearly the intersection of the family. Similarly, the set of all finite products of elements chosen from a family of subgroups will itself form a group, and this is the union of the family.

For a fuller treatment of the study of lattices of subgroups, the reader is referred to the monograph by Suzuki [1].

19.2. Locally Cyclic Groups and Distributive Lattices.

In a lattice the two distributive laws

$D1.\ a \cap (b \cup c) = (a \cap b) \cup (a \cap c),$
$D2.\ a \cup (b \cap c) = (a \cup b) \cap (a \cup c)$

are equivalent to each other. Let us show that $D1$ implies $D2$. Here, using $D1$,

$$\begin{aligned}(a \cup b) \cap (a \cup c) &= [(a \cup b) \cap a] \cup [(a \cup b) \cap c] \\ &= a \cup [(a \cap c) \cup (b \cap c)] \\ &= [a \cup (a \cap c)] \cup (b \cap c) \\ &= a \cup (b \cap c),\end{aligned}$$

which is the law $D2$. Similarly, the law $D2$ implies $D1$. The distributive law is a very strong condition in lattices. We shall show that for groups G the condition that $L(G)$ be distributive is very strong and implies that G is locally cyclic.

DEFINITION: *A group G is a locally cyclic group if, and only if, every finite set of elements in G generates a cyclic group.* (This is to be compared with §13.1.)

Since an element of finite order greater than 1 and an element of infinite order cannot generate a cyclic group, it follows that in a locally cyclic group every element $\neq 1$ is of infinite order or every element is of finite order. The additive group of rationals R_+ is a locally cyclic group which is aperiodic, and the group R_+ modulo 1 is a periodic locally cyclic group. It is not too difficult to show that a locally cyclic group is a subgroup of one of these two groups.

THEOREM 19.2.1. *The lattice $L(G)$ is distributive if, and only if, G is a locally cyclic group.*

Proof: Let us first suppose that G is a locally cyclic group. Let A, B, C be any three subgroups of G. We wish to prove that $D1$ holds. Now generally

$$\begin{aligned} A &\supseteq A \cap B, \\ B \cup C \supseteq B &\supseteq A \cap B,\end{aligned}$$

whence $U = A \cap (B \cup C) \supseteq A \cap B$. Also,

$$\begin{aligned} A &\supseteq A \cap C, \\ B \cup C \supseteq C &\supseteq A \cap C,\end{aligned}$$

whence $U = A \cap (B \cup C) \supseteq A \cap C$. Combining these inclusions,

$$U = A \cap (B \cup C) \supseteq (A \cap B) \cup (A \cap C).$$

Thus it is necessary to show only the inclusion

$$U = A \cap (B \cup C) \subseteq (A \cap B) \cup (A \cap C) = V.$$

Now consider an arbitrary element $g \in U$. Here g is of the form

$$g = a = bc \qquad a \in A, b \in B, c \in C,$$

where we note that, since G is Abelian, $B \cup C = BC$. Since G is locally cyclic, the elements b and c generate a cyclic group $\{u\}$ and $u^r = b$, $u^s = c$, and since for some m and n, $b^m c^n = u$, it follows that $u^{rm+sn} = u$. Also, $a = bc = u^{r+s}$. Now $x = u^{r(r+s)} = a^r = b^{r+s} \in A \cap B$, and $y = u^{s(r+s)} = a^s = c^{s+s} \in A \cap C$. Hence $a = u^{r+s} = u^{mr(r+s)+ns(r+s)} = x^m y^n$ is an element of $(A \cap B) \cup (A \cap C)$, as we wished to prove. Hence in all cases

$$A \cap (B \cup C) = (A \cap B) \cup (A \cap C)$$

for subgroups of a locally cyclic group.

Now, to prove the converse, we assume that $L(G)$ is distributive satisfying both $D1$ and $D2$. Let b and c be any two elements of G and write $a = bc$ and

$$A = \{a\}, \quad B = \{b\}, \quad C = \{c\}.$$

Then, since $a \in B \cup C$, we have

$$A = A \cap (B \cup C) = (A \cap B) \cup (A \cap C).$$

As subgroups of cyclic groups both $A \cap B$ and $A \cap C$ are cyclic and, say,

$$A \cap B = \{u\}, \quad A \cap C = \{v\},$$

where for appropriate exponents,

$$a^x = b^y = u, \quad a^z = c^w = v.$$

Here u and v as powers of a will permute, and since $A = \{u\} \cup \{v\}$, we must have, since $a \in A$, $a = u^r v^s = v^s u^r$, or remembering that $a = bc$, $bc = u^r v^s = b^{yr} c^{ws} = c^{ws} b^{yr}$. Here $b^{1-yr} = c^{ws-1}$, whence

$c^{-ws+1} = b^{yr-1}$ or $v^{-s}c = u^r b^{-1}$, whence $cb = v^s u^r = u^r v^s = bc$. Hence b and c must permute and G is Abelian.

We now note that G cannot contain an element $a \neq 1$ of finite order and an element b of infinite order. For putting $c = ab$, c is also of infinite order and $\{a\} = \{a\} \cap (\{b\} \cup \{c\})$, since $a = b^{-1}c$, while $(\{a\} \cap \{b\}) \cup (\{a\} \cap \{c\}) = (1) \cup (1) = 1 \neq \{a\}$, since the infinite cyclic groups $\{b\}$ and $\{c\}$ containing no elements $\neq 1$ of finite order must intersect $\{a\}$ in the identity. Thus we may consider only two cases; first, that in which G is aperiodic, and second, that in which G is periodic. In either case if two elements do not generate a cyclic group, then, by the basis theorem for Abelian groups, they generate the direct product of two cyclic groups, say, $\{b\}$ and $\{c\}$. Here with $a = bc$ and $A = \{a\}$, $B = \{b\}$, $C = \{c\}$, we have $A = A \cap (B \cup C)$, and $(A \cap B) \cup (A \cap C) = (1) \cup (1) = (1)$, and hence $D1$ does not hold. In the periodic case if b and c have relatively prime orders, then $\{b\} \cup \{c\} = \{bc\}$, and they do generate a cyclic group. But the direct product of two cyclic groups $\{b\}$ and $\{c\}$ whose orders do have a common factor, say, a prime p, will not have a distributive lattice, since for $b_1 \in \{b\}$ and $c_1 \in \{c\}$ both of order p and $a_1 = b_1c_1$, the law $D1$ fails as above with $A = \{a_1\}$, $B = \{b_1\}$, $C = \{c_1\}$. Here again for the distributive law to hold it is necessary that any two elements generate a cyclic group. But if any two elements generate a cyclic group, then it follows immediately that any finite number of elements generate a cyclic group, and thus G is locally cyclic, proving the converse part of the theorem.

19.3. The Theorem of Iwasawa.

One of the properties of a composition series (or chief series) shown in §8.5 is that all series have the same length. This property is a result of the modularity of the lattice of normal subgroups and of a weak form of modularity in the composition series. But, in general, maximal chains of unrestricted subgroups may differ in length. By Theorem 10.5.5 it follows that in a finite supersolvable group, all maximal chains of subgroups have the same length. The following theorem, due to Iwasawa [1], shows the converse to be true.

THEOREM 19.3.1. *The maximal subgroup chains of a finite group G all have the same length if, and only if, G is supersolvable.*

Proof: As noted above, Theorem 10.5.5 shows that in a supersolvable group G, all maximal subgroup chains have the same length, this being the total number of primes, counting repetitions, dividing the order of G.

Let us call the property of having all maximal chains the same length the *equi-chain condition.* This property is clearly inherited by subgroups and factor groups. Let G be a finite group with the equi-chain property. Since a group whose lattice is the chain of length one is a cyclic group of prime order and thus supersolvable, we may assume by induction on the length of maximal chains that all subgroups and factor groups of G are supersolvable.

We need first a lemma on supersolvable groups.

LEMMA 19.3.1. *Let G be a finite supersolvable group of order $n = p_1p_2 \cdots p_m$ where $p_1 \leq p_2 \leq \cdots \leq p_m$. Then G has a chief series*

$$K_0 = 1 \subset K_1 \subset K_2 \subset \cdots \subset K_m = G.$$

where K_i/K_{i-1} is of order p_{m-i+1}, $i = 1, \cdots, m$.

This is Corollary 10.5.2.

The next and most difficult step in the proof consists in showing that G has a normal subgroup. For this we need a choice of methods. Lemma 19.3.2 guarantees the possibility of this choice for any finite group.

LEMMA 19.3.2. *If G is a finite group of order divisible by the prime p, then either* (1) *G is p-normal or* (2) *G has a subgroup P of order a power of p which is normal in one Sylow subgroup $S_1(p)$ but is a non-normal subgroup of another Sylow subgroup $S_2(p)$.*

Proof: We recall that a group G is by definition p-normal if the center Z of a Sylow subgroup $S_1(p)$ is the center of any other Sylow subgroup $S_2(p)$ which contains it. Hence if G is not p-normal, then the center Z of some $S_1(p)$ is contained in another $S_2(p)$ but is not the center of $S_2(p)$. In this case we show that the second alternative holds, taking P as Z by showing that Z is not normal in $S_2(p)$. Assume to the contrary that Z is normal in $S_2(p)$. Then both $S_1(p)$ and $S_2(p)$ are contained in $N = N_G(Z)$, and hence as Sylow subgroups of N, will be conjugate in N. Thus for some $x \in N$, $x^{-1}S_1(p)x = S_2(p)$. Since Z is the center of $S_1(p)$, the center of $S_2(p) = x^{-1}S_1(p)x$ will be $x^{-1}Zx = Z$, since $x \in N_G(Z)$, and this conflicts with the assumption that Z was not the center of $S_2(p)$. This proves the lemma. We note that in the second alternative the Burnside Theorem 4.2.5 applies.

Let our group G be of the order $n = p_1^{e_1}p_2^{e_2} \cdots p_r^{e_r}$, where $p_1 < p_2 < \cdots < p_r$ are distinct primes. We use the alternatives of Lemma 19.3.2 for p_1, the smallest prime dividing n. We show that the second alternative cannot arise in G. Here, by Theorem 4.2.5, there are $t \not\equiv 0 (\mathrm{mod}\ p_1)$ p_1-groups $h_1, h_2, \cdots, h_t$ normal in their union H and conjugate in the normalizer of H in G, $N = N_G(H)$. If H is normal in G, we have a normal subgroup, as we wished to find. Suppose N is a proper subgroup of G and thus supersolvable by induction. Applying Lemma 19.3.1 to N, we find that N has a normal subgroup Q which has index in N, the highest power of p_1 dividing the order of N. But then both Q and H are normal subgroups in N, and so, since $Q \cap H = 1$ (H is a p_1-group), then $Q \cup H = Q \times H$. Thus Q permutes with every element of H, and so, the normalizer of h_1 contains Q and cannot have an index t prime to p_1. Strictly, we have shown only that the second alternative cannot arise when N is a proper subgroup, but when we have finished and proved that G is supersolvable, the above argument holds also for $N = G$.

Now we consider the first alternative, namely, that G is p_1-normal. Let Z be the center of a Sylow subgroup $S_1(p_1)$. Let $K = N_G(Z)$. If $K = G$, then we have Z as a proper normal subgroup of G, as we wished to show. Hence suppose K is a proper subgroup of G and therefore, by our induction hypothesis, supersolvable. Then by Lemma 19.3.1 applied to K, K must have a normal subgroup W of index p_1, and since K/W is the cyclic group of order p_1, $W \supseteq K'$, K has a nontrivial homomorphic image which is an Abelian p_1,-group, this being denoted by $K/K'(p_1)$. But by Theorem 14.4.5, since G is p_1-normal, $G/G'(p_1)$ is isomorphic to $K/K'(p_1)$, and thus $G'(p_1,)$ is a proper normal subgroup. Having shown in all cases that G has a proper normal subgroup, and since by induction both the normal subgroup and factor group are supersolvable, we conclude that G is solvable.

Now with G of order $n = p_1^{e_1}p_2^{e_2} \cdots p_r^{e_r}$, $p_1 < p_2 < \cdots < p_r$ and $m = e_1 + e_2 + \cdots + e_r$, and having shown G solvable, we now know that all maximal chains are of length m and that every covering $A > B$ is such that $[A:B]$ is a prime. Let $S(p_r)$ be a Sylow subgroup of order $p_r^{e_r}$, and let $1 \subset A_1 \subset A_2 \subset \cdots \subset A_{e_r} = S(p_r) \subset B_1 \subset \cdots \subset B_{m-e_r} = G$ be a maximal chain in which $S(p_r)$ is at the bottom. We wish to show that $S(p_r)$ is normal in G. Now $B_1/S(p_r)$ is of order some prime $p_j < p_r$. Since the number of conjugates of $S(p_r)$ in B_1 must divide p_j and be of the form $1 + kp_r$ by the third Sylow theorem, this number must be 1, and so, $S(p_r) \triangleleft B_1$. Similarly, if we have shown that $S(p_r)$ is normal in some B_i, then the number of conjugates of $S(p_r)$ in B_{i+1} must be of the form $1 + kp_r$ and also a divisor of $[B_{i+1}:B_i] = p_j < p_r$. Hence $S(p_r)$ is normal in B_{i+1} and, by continuing this argument, $S(p_r) \triangleleft G$. As a solvable group G possesses a Sylow complement C to $S(p_r)$ of order $p_1^{e_1} \cdots p_{r-1}^{e_{r-1}}$. This is given by Theorem 9.3.1. Let Z be the center of $S(p_r)$. As a characteristic subgroup of $S(p_r)$, Z is normal in G. Thus $C \cup Z = CZ = U$. In U a maximal chain of C can be extended to a maximal chain of U, and in this, $C \subset V$ for a group V such that $[V:C] = p_r$. By the same argument that showed $S(p_r)$ normal in G, it follows that V has a normal subgroup R of order p_r. Now R must be contained in Z, which is the unique Sylow subgroup for p_r in U. Thus R is normalized by C and, belonging to the center of $S(p_r)$, is also normalized by $S(p_r)$. Hence R is normal in G. Having shown that G possesses a normal subgroup R of prime order p_r, since by induction G/R is supersolvable, it follows at once that G is supersolvable. This proves the theorem.

20. GROUP THEORY AND PROJECTIVE PLANES

20.1. Axioms.

A *projective plane* is a set of points, of which certain distinguished subsets are called lines, satisfying the following axioms:

$P1$. *Any two distinct points are contained in one and only one line.*

$P2$. *Any two distinct lines contain one and only one point in common.*

$P3$. *There exist four points, no three of which are contained in one line.*

The unique line k containing two distinct points A and B will be called *the line joining A and B*. The unique point P contained in two distinct lines k and t will be called *the intersection of k and t*.

Let A_1, A_2, A_3, A_4 be four points, no three on a line, whose existence is given by $P3$. Then there are six distinct lines joining the different pairs:

$$L_1\colon A_1A_2B_1.$$
$$L_2\colon A_1A_3B_2.$$
$$L_3\colon A_1A_4B_3.$$
$$L_4\colon A_2A_3B_3.$$
$$L_5\colon A_2A_4B_2.$$
$$L_6\colon A_3A_4B_1.$$

Here the points B_1, B_2, B_3 are the intersections of these lines, and from the distinctness of the lines we easily find that the B's are different from the A's and from each other.

LEMMA 20.1.1. *Every line contains at least three points.*

Proof: The lines $L_1, \cdots, L_6$ as constructed each contain at least three points. A further line L, if it does not contain A_1, will intersect L_1, L_2, L_3 in three distinct points. If L does not contain A_2, then L intersects L_1, L_4, L_5 in three distinct points. If L contains both A_1 and A_2, then L is L_1, which does contain at least the three points A_1, A_2, B_1.

LEMMA 20.1.2. *There exist four lines, no three of which contain the same point.*

Proof: Here L_1, L_2, L_5, L_6 are four lines, no three of which intersect in a common point.

If we interchange the roles of points and lines and replace "contains" by "is contained in," Axioms $P1$ and $P2$ are interchanged and Axiom $P3$ and Lemma 20.1.2 are interchanged. This leads to the concept of *duality*. More precisely if π is any projective plane, then there is a plane π^* *dual* to π which may be constructed as follows: Let $\{P_i\}$ be the set of points of π and $\{k_j\}$ the set of lines of π. Then in π^* we have lines $\{p_i\}$ in a one-to-one correspondence with the points $\{P_i\}$ of π and points $\{K_j\}$ in a one-to-one correspondence with the lines $\{k_j\}$ of π. Further, if $P_i \in k_j$ in π, we put $K_j \in p_i$ in π^*, where $k_j \rightleftarrows K_j$ and $P_i \rightleftarrows p_i$. Our observations show that if π satisfies the axioms for a projective plane, then π^* also does so. Furthermore, the dual of π^* is π, i.e., $(\pi^*)^* = \pi$. Hence, interchanging the roles of points and lines and reversing inclusions, every statement about a plane π becomes a statement about its dual π^*. This is the *principle of duality*. In particular from the principle of duality, if some statement is true of every projective plane π, then its dual is also true. Thus, applying the principle of duality, Lemma 20.1.1 becomes:

LEMMA 20.1.3. *Every point is on at least three lines.*

The reader will not find it difficult to verify that the axioms given here are equivalent, for planes, to the axioms of projective geometry as given in Veblen and Young, "Projective Geometry," vol. I, pages 16–18.

Suppose that one line L_1 of a projective plane π contains a finite number of points. Call this number $n + 1$, where $n \geq 2$ by Lemma 20.1.1. By Axiom $P3$ there are at least two points, say, P_3 and P_4, not on L_1. Let P_3P_4 intersect L_1 in B_1, and let P_1 and P_2 be two other points of L_1. Then P_1P_3 and P_2P_4 intersect in a further point B_2 not on L_1 and not on P_3P_4. If P is any point not on L_1, joining P to the $n + 1$ points of L_1, we have $n + 1$ lines through P and these are all the lines through P since every line through P must intersect L_1. In particular there are $n + 1$ lines through each of P_3, P_4, and B_2. If there are $n + 1$ lines through a point P, these intersect a line L not through P in $n + 1$ points and these are all the points on L, since every point on L is joined to P by a line. Hence every line L of π contains $n + 1$ points, since at least one of P_3, P_4, or B_2 is not on L. Also there are $n + 1$ lines through every point P of π, these being the lines joining P to the $n + 1$ points of some line L not through P. We have now proved the major part of the following theorem.

THEOREM 20.1.1. *Let $n \geq 2$ be an integer. In a projective plane the following properties are equivalent:*

1) *One line contains exactly $n + 1$ points.*
2) *One point is on exactly $n + 1$ lines.*
3) *Every line contains exactly $n + 1$ points.*
4) *Every point is on exactly $n + 1$ lines.*
5) *There are exactly $n^2 + n + 1$ points in π.*
6) *There are exactly $n^2 + n + 1$ lines in π.*

Proof: We have already shown that (1) implies (2), (3), and (4). To prove (5), let P_0 be one point of π and let $L_1, \cdots, L_{n+1}$ be the $n + 1$ lines through P_0. These lines include all the points of π, and each of them contains P_0 and n other points. P_0 is the only point common to any two of $L_1, \cdots, L_n$. Hence π contains $1 + (n + 1)n = n^2 + n + 1$ points. To prove (6), let L_0 be a line of π, and P_1, $P_2, \cdots, P_{n+1}$ the $n + 1$ points of L_0. Each of $P_1, \cdots, P_{n+1}$ is on L_0 and n other lines. In this way we obtain all the lines of π, and there are $1 + (n + 1)n = n^2 + n + 1$ lines in π. Hence property (1) implies the remaining properties. By duality (2) implies the remaining properties. Trivially, (3) implies (1), and (4) implies (2). If (5) holds, then some line has $m + 1$ points, where m is an integer, and we conclude that π has $m^2 + m + 1 = n^2 + n + 1$ points, whence $m = n$ and (5) implies (1). Similarly, (6) implies (2).

20.2. Collineations and the Theorem of Desargues.*

A plane π_1 is said to be *isomorphic* to a plane π_2 if there is a one-to-one correspondence $P_1 \rightleftarrows P_2 = (P_1)\alpha$ between the points $\{P_1\}$ of π_1 and the points $\{P_2\}$ of π_2 and a one-to-one correspondence $k_1 \rightleftarrows k_2 = (k_1)\beta$ between the lines $\{k_1\}$ of π_1 and the lines $\{k_2\}$ of π_2, such that if $P_1 \in k_1$, then $(P_1)\alpha \in (k_1)\beta$. Clearly, each of the correspondences α or β fully determines the other, and a one-to-one correspondence of points $P_1 \rightleftarrows (P_1)\alpha$ will determine an isomorphism if whenever three points P_1, Q_1, R_1 of π_1 are on a line, then $(P_1)\alpha$, $(Q_1)\alpha$, and $(R_1)\alpha$ are on a line. Similarly, a one-to-one correspondence

* For the properties of three-dimensional spaces used here, see Veblen and Young [1], pp. 20–25.

β of lines will determine an isomorphism if every set of three concurrent lines is mapped onto a set of three concurrent lines. A homomorphism of planes would be a many-to-one correspondence of points and lines preserving incidence, but this does not seem to be as valuable a concept in planes as in other subjects.

An isomorphism α of a plane π with itself is called a *collineation.* The collineations of a plane form a group.

A plane π_1 which can be embedded in a three-dimensional space E_3 always has a large family of collineations. Let π_2 be another plane in E_3, and let L be the line in which π_2 intersects π_1. Take P_1 and P_2 as any two points of E_3 not lying in either π_1 or π_2. We define a *perspectivity* of π_1 onto π_2 with center P_1. This is a mapping of an arbitrary point Q of π_1 onto a point R of π_2, written as

$$Q \xrightarrow{P_1} R, \tag{20.2.1}$$

where R is defined as that point of π_2 in which the line P_1Q pierces π_2. Here P_1QR are on a line, with $Q \in \pi_1$, $R \in \pi_2$. The perspectivity (20.2.1) is an isomorphism of π_1 onto π_2, since if M_1 is a line of π_1, the plane π_3 containing M_1 and P_1 intersects π_2 in a line M_2, and the perspectivity maps the points of M_1 onto the points of M_2. Furthermore, every point of L is mapped onto itself, this being the intersection of π_1 and π_2. If we follow the perspectivity (20.2.1) by a perspectivity with center P_2 mapping π_2 onto π_1,

$$R \xrightarrow{P_2} S, \tag{20.2.2}$$

this is also an isomorphism of π_2 onto π_1, leaving all points of L fixed. The combination of the two perspectivities

$$Q \xrightarrow{P_1} R \xrightarrow{P_2} S, \tag{20.2.3}$$

will be a collineation α of π_1 fixing all points of L. In addition let O be the point in which the line P_1P_2 intersects π_1. We shall have

$$O \xrightarrow{P_1} T \xrightarrow{P_2} O, \tag{20.2.4}$$

since P_1P_2OT are on a line, whence $(O)\alpha = O$. Furthermore, let k be any line through O. If Q is a point of k, then in (20.2.3) R will be a point in the plane π_4 containing the intersecting lines k and P_1P_2OT, whence also S in (20.2.3) will be a point of π_4 and hence a point of k. Thus α takes every line through O into itself. Such a collineation α is called a *perspective* collineation and sometimes a *perspectivity.* The line L, all of whose points are fixed by α, is called the *axis* of the collineation, and the point O through which every line is fixed is called the *center* of the collineation. The center O may or may not lie on the axis L. If we wish to make a distinction in these two cases, then if the center O lies on the axis L, the collineation is called an *elation,* while if O does not lie on L, the collineation is called a *homology.* The nature of perspective collineations can be seen in the accompanying diagram, labeled Theorem of Desargues.

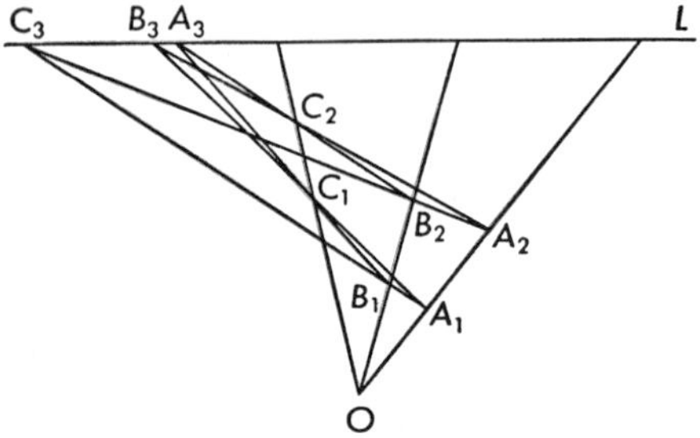

Fig. 9. Theorem of Desargues.

Suppose we have a perspective collineation α with center O and axis L and that A_1 is a point of π not on L and different from O. Then $(A_1)\alpha = A_2$ must be on the line OA_1. Now given O, L, and A_1 and A_2 with OA_1A_2 on a line but neither A_1 nor A_2 on L or equal to O, then we assert that with O as center, L as axis, and $(A_1)\alpha = A_2$, the collineation α is completely determined, for let B_1 be any further point of π not on OA_1A_2 or L. Let A_1B_1 intersect L in C_3. Then $(B_1)\alpha$ must lie on OB_1. But also, since A_1, B_1, C_3 were collinear, $(A_1)\alpha = A_2$, $(B_1)\alpha$, and $(C_3)\alpha = C_3$ must also be collinear. Hence $(B_1)\alpha$ must lie on both OB_1 and C_3A_2, whence $(B_1)\alpha = B_2$ is the intersection of OB_1 and C_3A_2. Thus from $(A_1)\alpha = A_2$ the image of every point B_1 not on OA_1A_2 is uniquely determined. But with $(B_1)\alpha = B_2$, the images of points on OA_1A_2 can be uniquely determined.

Now let C_1 be a point not on OA_1A_2 or on OB_1B_2. Then let B_1C_1 meet L in A_3 and A_1C_1 meet L in B_3. Then $(C_1)\alpha = C_2$ is determined as the intersection of A_2B_3 and OC_1. But since $B_1C_1A_3$ are on a line, then also $(B_1)\alpha = B_2$, $(C_1)\alpha = C_2$, and $(A_3)\alpha = A_3$ are on a line. This gives us the nontrivial configuration of our figure, known as the *configuration of Desargues.* The existence of this configuration is called the Theorem of Desargues. We shall say that triangles A_1,

B_1, C_1 and A_2,B_2,C_2 are perspective with respect to a center O if corresponding vertices are on lines through O, i.e., that OA_1A_2, OB_1B_2, and OC_1C_2 are lines. The triangles are perspective with respect to an axis L if corresponding sides intersect in points of L.

THEOREM 20.2.1 (THEOREM OF DESARGUES). *If two triangles A_1,B_1,C_1 and A_2,B_2,C_2 are perspective with respect to a center O, then corresponding sides A_1B_1 and A_2B_2, A_1C_1 and A_2C_2, B_1C_1 and B_2C_2 meet in points C_3,B_3, and A_3 lying on a line L.*

The validity of the Theorem of Desargues in a plane π is equivalent to the existence of all possible perspective collineations in π. This we see in the following theorems:

THEOREM 20.2.2. *In a plane π given a line L, a point O, and two points A_1, A_2 different from O and not lying on L, and such that OA_1A_2 are on a line. Then there is at most one perspective collineation α of π with center O and axis L such that $(A_1)\alpha = A_2$. If π can be embedded in a three-dimensional space there is one such collineation.*

Proof: We saw above that given the center O, axis L, and $(A_1)\alpha = A_2$, where OA_1A_2 are on a line, then the perspective collineation α is completely determined. Hence there is at most one such collineation in π. Now suppose that π can be embedded in a three-dimensional space E_3. Take a plane π_2 in E_3 intersecting π in L and choose some point P_1 of E_3 not on π or π_2. Join P_1 to O and let P_1O intersect π_2 in T. (See the figure.)

If A_1P_1 intersects π_2 in Q, then Q and A_2 are in the plane π_3 of OP_1 and OA_1A_2. π_3 is the plane of the figure. Hence A_2Q intersects OP_1 in a point P_2. Now P_2 is not in π_2, since then it would be T and A_2 would coincide with X, the intersection of OA_1A_2 and the line L, contrary to our assumption that A_2 was not on L. Similarly, since A_2 is not O, P_2 is not on π_1. Now we see that the perspective collineation α

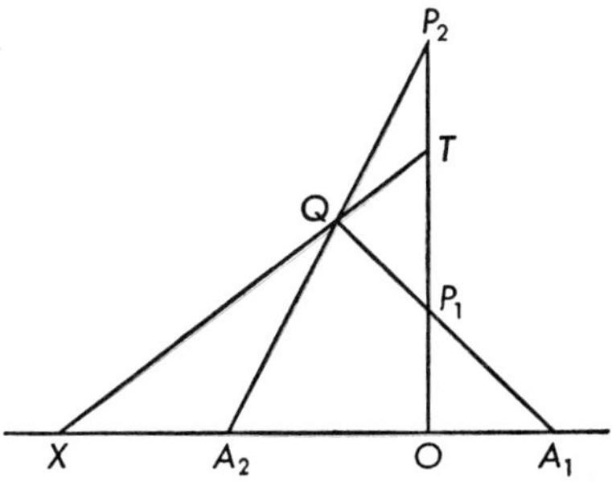

Fig. 10. Perspectivities.

$$\pi \xrightarrow{P_1} \pi_2 \xrightarrow{P_2} \pi$$

has L as its axis and O as its center; also we have $A_1 \xrightarrow{P_1} Q \xrightarrow{P_2} A_2$.

Hence $A_2 = (A_1)\alpha$, as was required, and the collineation of the theorem does exist.

THEOREM 20.2.3. *The Theorem of Desargues is valid in a plane π if, and only if, all possible perspective collineations exist in π.*

COROLLARY 20.2.1. *The Theorem of Desargues is valid in any plane π which can be embedded in a projective three space.*

Proof: Once the theorem is established, the corollary will follow from the previous theorem.

Suppose first that all possible perspective collineations exist in π. We are given the two triangles $A_1B_1C_1$ and $A_2B_2C_2$ such that the three lines A_1A_2, B_1B_2, and C_1C_2 meet in a point O. (See figure for Theorem of Desargues.) Let A_1B_1 and A_2B_2 meet in a point C_3, and A_1C_1 and A_2C_2 meet in a point B_3. Call the line joining B_3 and C_3 the line L. Then, by our hypothesis, there is a perspective collineation α with center O, axis L, and such that $(A_1)\alpha = A_2$. Then, by our construction, $(B_1)\alpha = B_2$ and $(C_1)\alpha = C_2$. Let B_1C_1 meet L in A_3. Then $(B_1)\alpha = B_2$, $(C_1)\alpha = C_2$, and $(A_3)\alpha = A_3$ lie on a line whence A_3, the intersection of B_1C_1 and B_2C_2, lies on the line L with B_3 and C_3. This proves the Theorem of Desargues.

Now suppose conversely that the Theorem of Desargues is valid in π. We are given the line L and the points O, A_1, A_2 which are distinct and on a line, and A_1 and A_2 are not on L (though O may be). We define a mapping α of the points of π and show that it is a collineation. (We refer again to the figure of the Theorem of Desargues.) For any point X on L we put $(X)\alpha = X$. We also put $(O)\alpha = O$ and $(A_1)\alpha = A_2$. If B_1 is not on L or on OA_1A_2, let A_1B_1 meet L in C_3. If A_2C_3 intersects OB_1 in B_2, put $(B_1)\alpha = B_2$. This defines the mapping α for all points of π except those of OA_1A_2. Now if A_1C_1 meets L in B_3, then if A_2B_3 meets OC_1 in C_2, we put $C_2 = (C_1)\alpha$. If OA_1, OB_1, and OC_1 are distinct lines, then the triangles $A_1B_1C_1$ and $A_2B_2C_2$ are perspective with respect to the center O, and by the Theorem of Desargues, corresponding sides meet in points on a line. But C_3 and B_3 are on L, whence B_1C_1 and B_2C_2 intersect in a point A_3 of L. But if we had started with $(B_1)\beta = B_2$, we would have defined $(C_1)\beta$ as the intersection of B_2A_3 with OC_1. But this is C_2, and hence we have $(C_1)\alpha = (C_1)\beta = C_2$ as a consequence of either $(A_1)\alpha = A_2$ or $(B_1)\beta = B_2$. Thus the mappings $\alpha = \beta$ agree on all lines such as OC_1C_2 for which they are both defined. But α is defined on all of

OB_1B_2, and β is defined on all of OA_1A_2. In this way our mapping α is defined for all points of π.

Again from the same figure with $(A_1)\alpha = A_2$ and k an arbitrary line not through O or A_1, let k intersect L in A_3 and let B_1, C_1 be two further points on k. Then, by our definition, $(B_1)\alpha = B_2$, $(C_1)\alpha = C_2$, and $(A_3)\alpha = A_3$, and by the Theorem of Desargues applied to the triangles $A_1B_1C_1$ and $A_2B_2C_2$, we conclude that B_2, C_2, and A_3 are on a line. This tells us that the mapping α takes points C_1 of $k = A_3B_1$ into points of A_3B_2, except possibly if C_1 is the intersection of A_3B_1 with OA_1A_2. But if $C_1 = D_1$ is the intersection of B_1A_3 with OA_1A_2, then from $(B_1)\alpha = (B_1)\beta = B_2$ we define the image $(D_1)\beta = D_2$ as the intersection D_2 of A_3B_2 with OA_1A_2. Hence the mapping takes all the points of k onto the points of A_3B_2. Clearly, the mapping takes L into itself and lines through O into themselves. Hence α is the required collineation.

We have indeed proved a more precise result than Theorem 20.2.3. This we state as a theorem.

THEOREM 20.2.4. *In a plane π there are all possible perspective collineations with a given center O and given axis L if, and only if, the Theorem of Desargues is valid for all triangles perspective with respect to O and having two pairs of corresponding sides intersect on L whence the third pair also intersect on L.*

Not every plane π can be embedded in a three-dimensional space, and there are cases in which Theorem 20.2.4 is applicable for a limited number of axes L and centers O.

20.3. Introduction of Coordinates.

Let π be any projective plane and choose four points X, Y, O, I, no three on a line. Call the line XY *the line of infinity* L_∞. Call the line OI *the line* $y = x$.

On the line OI give coordinates $(0, 0)$ to O, $(1, 1)$ to I and the single coordinate (1) to the point C which is the intersection of OI and XY. For other points of OI assign coordinates (b, b), taking different symbols b for different points. For a point P not on L_∞ let XP intersect OI in (b, b) and YP intersect OI in (a, a). Then assign coordinates (a, b) to P. This rule reassigns the same coordinates to points OI. Let the line joining $(0, 0)$ and $(1, m)$ intersect L_∞ in a point M. Assign to M the single coordinate (m), which we may think

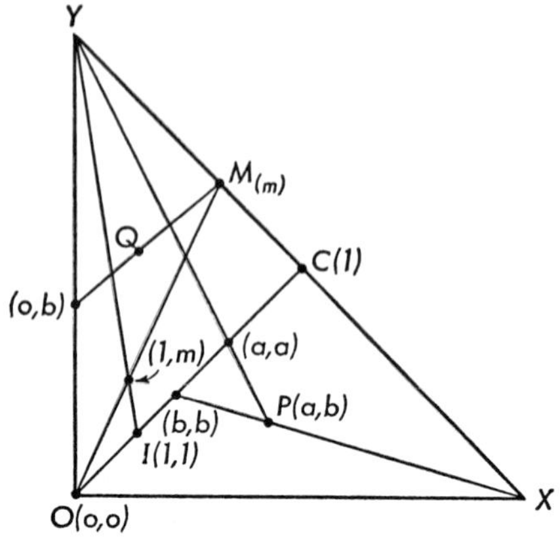

Fig. 11. Introduction of coordinates.

of intuitively as a slope. We have now assigned coordinates to every point except Y, and to this we arbitrarily assign the single coordinate (∞).

We shall use the lines of our plane to define algebraic operations on the system of coordinates. This algebraic system will be a *ternary ring*, and every line of π except L_∞ will have an equation expressible in terms of the operations of the ternary ring. If (x, y) is a finite point of OI, we shall have $y = x$, and so, we take $y = x$ as the equation for OI. A line through Y different from L_∞ will have the property that all its finite points (x, y) have the same x coordinate, say, $x = c$, and this we take as its equation.

If (x, y) is a finite point of the line joining $C = (1)$ and $(0, b)$ we define a binary operation of addition by putting

$$y = x + b, \tag{20.3.1}$$

and taking this as the equation of the line. If (x, y) is a finite point of the line joining $O = (0, 0)$ and (m), we define a binary operation of multiplication, putting

$$y = xm, \tag{20.3.2}$$

and taking this as the equation of the line. In general, any line not through Y will intersect L_∞ in some point (m) and OY in some point $(0, b)$. If $Q = (x, y)$ is a point of this line, we define a ternary operation

$$y = x \cdot m \circ b, \tag{20.3.3}$$

and take this as the equation of the line. Thus both addition and multiplication are special cases of the ternary operation, and we see that

$$\begin{aligned} x + b &= x \cdot 1 \circ b, \\ xm &= x \cdot m \circ 0. \end{aligned} \tag{20.3.4}$$

The elements 0 and 1 have the familiar properties

$$\begin{aligned} 0 + a &= a + 0 = a, \\ 0m &= m0 = 0, \\ 1m &= m1 = m. \end{aligned} \tag{20.3.5}$$

The plane π can be represented by a ternary ring R whose ternary operation satisfies certain properties, and conversely, a ternary ring with these properties uniquely determines a plane. This we state as our main theorem on ternary rings.

THEOREM 20.3.1. *For every choice of four points X, Y, O, I, no three on a line in a plane π, there is determined a ternary ring R. The elements of R include a zero*, 0, *and a unit* $1 \neq 0$. *The ternary operation $x \cdot m \circ b$ satisfies the following laws:*

*T*1. $0 \cdot m \circ c = a \cdot 0 \circ c = c$.
*T*2. $1 \cdot m \circ 0 = m \cdot 1 \circ 0 = m$.
*T*3. *Given a, m, c, there exists exactly one z such that* $a \cdot m \circ z = c$.
*T*4. *If $m_1 \neq m_2$, b_1, b_2 are given, there exists a unique x such that*

$$x \cdot m_1 \circ b_1 = x \cdot m_2 \circ b_2.$$

*T*5. *If $a_1 \neq a_2$, c_1, c_2 are given, then there exists a unique pair m, b such that*

$$a_1 \cdot m \circ b = c_1 \quad \text{and} \quad a_2 \cdot m \circ b = c_2.$$

Proof: Having chosen four points X, Y, O, I, no three on a line in a plane π, we construct a ternary ring as before with an operation $x \cdot m \circ b$. Properties $T1$ and $T2$ are immediate consequence of the definition. $T3$ says that the line joining (m) and (a, c) intersects OY in a unique point $(0, z)$. $T4$ says that two lines $y = x \cdot m_1 \circ b_1$ and $y = x \cdot m_2 \circ b_2$ with different slopes m_1 and m_2 intersect in a unique finite point. $T5$ says that if (a_1, c_1) and (a_2, c_2) are two finite points with $a_1 \neq a_2$, then there is a unique line of the form $y = x \cdot m \circ b$ passing through them.

Conversely, suppose that a ternary ring R is given, satisfying $T1, \cdots, T5$. Let us construct finite points (a, b) and infinite points (m) and (∞), where a, b, m range over the elements of R. A line L_∞ is to contain all the infinite points (m), (∞) and no other points. All points (c, y), with c fixed and also (∞), are to be the points of a line $x = c$. The point (m) and points (x, y) such that $y = x \cdot m \circ b$ for fixed m and b are to be the points of the line $y = x \cdot m \circ b$. There are several cases to be considered, but the net result of the above axioms is easily found to be that two distinct points lie on one and only one line, and that two distinct lines intersect in one and only one point. Also the four points (∞), (0), $(0, 0)$ and $(1, 1)$ are such that no three are on a line. We give only one verification, others being of the same type. Consider two distinct lines $y = x \cdot m \circ b_1$ and $y = x \cdot m \circ b_2$. These both contain the infinite point (m) but no other infinite point. If both also were to contain a finite point (a, c) we would have $a \cdot m \circ b_1 = c = a \cdot m \circ b_2$, which would conflict with $T3$ since $b_1 \neq b_2$.

20.4. Veblen-Wedderburn Systems. Hall Systems.

We shall investigate the properties of planes with certain collineation groups and relate these properties to coordinatizing ternary rings.

LEMMA 20.4.1. *A collineation in a projective plane π which fixes every point on each of two distinct lines is the identical collineation.*

Proof: Let the collineation α fix every point on the lines L_1 and L_2, and let L_1 and L_2 intersect in the point Q. Let P be any point of π not on L_1 or L_2. Take R and S as two points on L_1 distinct from Q. Let PR intersect L_2 in T and PS intersect L_2 in U. Then R, S, T, U are fixed by α, and hence the lines RT and SU are fixed by α and thus also their intersection P. Hence not only the points of L_1 and L_2 but every point P not on L_1 or L_2 is fixed by α, and so, every point of the plane is fixed and α is the identical collineation.

LEMMA 20.4.2. *A collineation in a projective plane which fixes every point on one line and two points not on the line is the identical collineation.*

Proof: Let α be a collineation fixing every point on a line L and two points P_1 and P_2 not on L. Let P be any point of the plane not on L and not on the line P_1P_2. Let P_1P intersect L in Q_1, and P_2P

intersect L in Q_2. Since P was not on P_1P_2 or L, the lines P_1PQ_1 and P_2PQ_2 are distinct. Since P_1, Q_1, P_2, Q_2 are distinct and fixed points, the lines P_1Q_1 and P_2Q_2 are fixed lines of α, and so, their intersection P is fixed. Hence α fixes every point on L and every point not on P_1P_2. Hence α fixes every point on L and on some line, say, P_1Q_1, through P_1 but distinct from P_1P_2. Hence, by Lemma 20.4.1, α is the identical collineation.

Thus we see that a collineation fixing every point on a line can fix at most one point not on the line.

THEOREM 20.4.1. *Given a collineation α of a plane which fixes a line L and every point on L. Then there is a point C such that α fixes C and every line through C. If α is not the identity, then α fixes no further points or lines. Dually, if a collineation α fixes a point C and all the lines through C, then there is a line L such that α fixes L and all the points on L but has no further fixed points or lines if α is not the identity.*

Proof: Let $\alpha \neq 1$ be a collineation of a plane π fixing every point of a line L. Then, from Lemma 20.4.2, α fixes at most one point not on L. Suppose first that α fixes one point C not on L. Then a line through C intersects L in a point Q different from C, and since both C and Q are fixed by α, the line CQ is fixed by α, and so, every line through C is fixed by α. If there were a fixed line L_2 besides L and the lines through C, then every point of L_2 at the intersection of L_2 and a line through C would be fixed, and by Lemma 20.4.1, α would be the identity. Equally, by Lemma 20.4.2, there can be no further points fixed by α.

Now suppose that α fixes no point not on L. Let P be a point not on L. Then $P\alpha$, the image of P under α, is different from P and not on L. Hence the line $M = PP\alpha$ intersects L in a point $C \neq P$, $P\alpha$. Thus $M = PC$ and $M\alpha = P\alpha C\alpha = P\alpha C$. But $PP\alpha C$ are on a line and so $M = M\alpha$. Thus every point P not on L lies on a fixed line M. Moreover, such a P could not lie on two distinct fixed lines, since then it would itself be fixed. Now if $M = PC$ is one fixed line, consider a point Q not on L or on M. Then Q also lies on a unique fixed line N. Now the intersection of M and N is a fixed point, and by hypothesis there are no fixed points not on L. Hence N must intersect M in the point C lying on L. Thus every fixed line passes through C. But an arbitrary line K through C, different from L, contains a point, say, R, not on L. But R lies on a fixed line which passes through C, and so, this fixed line must be $RC = K$. Thus every line through C is a fixed line. Here again if there were any fixed elements besides L and the points on it and C and the lines through, then by Lemmas 20.4.1 and 20.4.2, α would be the identity.

The rest of the theorem follows by duality.

Thus the collineations of the theorem are just the perspective collineations discussed in §20.2. These are sometimes called *central collineations*. If we wish to specify the center C and axis L, we say we have a *C-L collineation*. Clearly all collineations with center C and axis L form a group. As in §20.2 we call the collineation an *elation* if the center C lies on the axis L, a *homology* if C does not lie on L.

LEMMA 20.4.3. *A central collineation α is completely determined by its center C, axis L and the mapping $P \rightarrow P\alpha$ of any point P not on L and different from C. P, $P\alpha$ and C must be collinear.*

Proof: If there were two collineations α_1 and α_2 with center C and axis L and $P\alpha_1 = P\alpha_2$, then $\alpha_1\alpha_2^{-1}$ would fix the points of L, have center C, and also fix P, whence by Theorem 2.4.1, $\alpha_1\alpha_2^{-1} = 1$ and $\alpha_1 = \alpha_2$. This is the assertion of the lemma.

THEOREM 20.4.2. *The product of two elations with the same axis L but different centers C_1 and C_2 is an elation with axis L and center $C_3 \neq C_1, C_2$.*

Proof: Let α_1 be an elation with center C_1 lying on the axis L and α_2 with center $C_2 \neq C_1$ lying on L. Then $\alpha_1\alpha_2$ is a collineation fixing all points on L. Hence, by Theorem 20.4.1, $\alpha_1\alpha_2 = \alpha_3$ is a central collineation with axis L. To show that α_3 is an elation, we must show that α_3 does not fix any point not on L. If $P\alpha_3 = P$, then $P\alpha_1 = P\alpha_2^{-1}$. Here C_1, P, $P\alpha_1$ lie on a line and C_2, P, $P\alpha_2^{-1}$ lie on a line. Hence if $P\alpha_1 = P\alpha_2^{-1}$, these lines would coincide and the intersections with L would coincide, giving $C_1 = C_2$, contrary to hypothesis. Hence $\alpha_3 = \alpha_1\alpha_2$ fixes no point not on L and hence is an elation with center C_3 on L. If $C_3 = C_1$, then $\alpha_2 = \alpha_1^{-1}\alpha_3$ would be an elation with center C_1, contrary to assumption. Hence $C_3 \neq C_1$, and similarly, $C_3 \neq C_2$.

Let us consider the group $G = G(C, L)$ of C-L central collineations. If $P \neq C$ and $P \notin L$, then for any $\alpha \in G$, C, P, $P\alpha$ are on a line. If for every Q on CP with $Q \neq C$, $Q \notin L$ there is an $\alpha \in G$ such that

$P\alpha = Q$, we say that π is *C-L transitive*. This says that every conceivable C-L collineation actually arises. Equivalently, the statement that π is C-L transitive means that for a line M through C, $M \neq L$, the C-L collineations permute transitively the points of M except for C and the intersection of M and L. This will be true for any line $M \neq L$ passing through C.

By Theorem 20.4.2 all elations with axis L form a group $G(L)$. We call this group $G(L)$ the *translation group* with axis L.

THEOREM 20.4.3 (BAER [10]). *If for two different centers C_1 and C_2 on an axis L the elation groups $G(C_1, L)$ and $G(C_2, L)$ are different from the identity, then the entire translation group $G(L)$ is Abelian. Also every element $\neq 1$ of $G(L)$ is either* (1) *of infinite order or* (2) *of the same prime order p.*

Proof: Suppose $\alpha_1 \neq 1 \in G(C_1, L)$ and $\alpha_2 \neq 1 \in G(C_2, L)$. Let P be any point not on L. Then we have the following lines:

L_1: $C_1, P, P\alpha_1$, L_2: $C_2, P, P\alpha_2$.

$L_1\alpha_2$: $C_1, P\alpha_2, P(\alpha_1\alpha_2)$, $L_2\alpha_1$: $C_2, P\alpha_1, P(\alpha_2\alpha_1)$.

But C_2, $P\alpha_1$ and $(P\alpha_1)\alpha_2 = P(\alpha_1\alpha_2)$ are on a line, and C_1, $P\alpha_2$ and $(P\alpha_2)\alpha_1 = P(\alpha_2\alpha_1)$ are on a line. Hence the intersection of the distinct lines $C_2P\alpha_1$ and $C_1P\alpha_2$ is $P(\alpha_1\alpha_2)$ and also $P(\alpha_2\alpha_1)$. Hence $P(\alpha_1\alpha_2) = P(\alpha_2\alpha_1)$ for every $P \notin L$. Hence $\alpha_1\alpha_2 = \alpha_2\alpha_1$. Hence an element $\alpha_1 \in G(C_1, L)$ permutes with every element α_2 of any $G(C_2, L)$ with $C_2 \neq C_1$. Suppose that $\beta_1 \neq 1$ is another element of $G(C_1, L)$. Then $\beta_1\alpha_2$ is an elation with center $C_3 \neq C_1, C_2$. Hence α_1 permutes with $\beta_1\alpha_2$, and since α_1 permutes with α_2, α_1 also permutes with β_1. Hence an $\alpha_1 \neq 1$ of $G(C_1, L)$ permutes with every element of $G(L)$, and so $G(L)$ is Abelian. Examples exist showing that $G(C_1, L)$ need not be Abelian if every other $G(C_i, L) = 1$ with $C_i \in L$.

If every element of $G(L)$ is of infinite order, then (1) holds. If $G(L)$ contains elements of finite order, then there is an element of prime order, say, $\alpha_1 \in G(C_1, L)$, $\alpha_1^p = 1$. Now with $\alpha_2 \neq 1 \in G(C_2, L)$, $C_2 \neq C_1$, we have $\alpha_1\alpha_2 = \alpha_3 \in G(C_3, L)$, $C_3 \neq C_1, C_2$. Here $(\alpha_1\alpha_2)^p = \alpha_2{}^p = \alpha_3^p$ is an element common to $G(C_2, L)$ and $G(C_3, L)$ and hence the identity. Thus $\alpha_2{}^p = 1$. Similarly, from $\alpha_2^p = 1$, follows $\beta_1^p = 1$ for any $\beta_1 \in G(C_1, L)$. Hence every element of $G(L)$ except the identity is of order p.

THEOREM 20.4.4. *If a plane π is C_1-L transitive and C_2-L transitive for two centers $C_1 \neq C_2$ on L, then π is C-L transitive for every $C \in L$.*

Proof: Take a line $M \neq L$ through $C \neq C_1, C_2$ and let P and Q be any two points of M different from C. Let PC_1 and QC_2 intersect in S. Then let $\alpha_1 \in G(C_1, L)$ be such that $P\alpha_1 = S$ and $\alpha_2 \in G(C_2, L)$ be such that $S\alpha_2 = Q$. By C_1-L and C_2-L transitivity α_1 and α_2 exist. Here $\alpha_1\alpha_2 = \alpha_3$ is an elation with axis L and P, $P\alpha_3 = Q$, C are on a line. Hence $\alpha_3 \in G(C, L)$ and π is C-L transitive.

Corollary 20.4.1. *If π is C_1-L and C_2-L transitive with $C_1 \neq C_2$ points of L, then $G(L)$ contains every conceivable elation with center on L.* If π is C-L transitive for every $C \in L$, we say that π is a *translation plane* with respect to the axis L.

What is the meaning of elations in terms of properties of a ternary ring coordinatizing π? Let us consider first the case in which π is C-L transitive for a point C on L. We shall take the axis L as L_∞ and the center as $Y = (\infty)$.

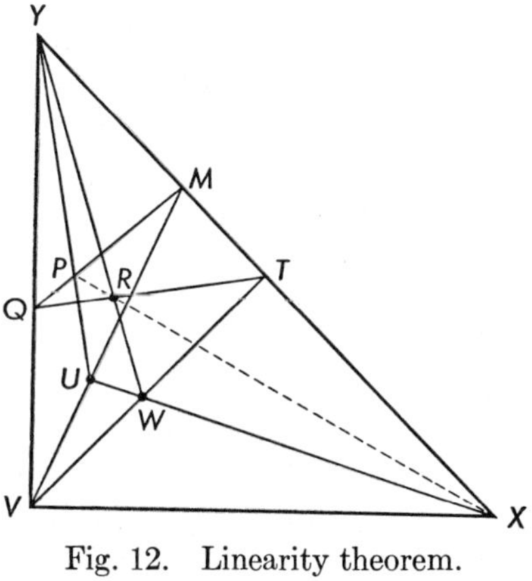

Fig. 12. Linearity theorem.

Theorem 20.4.5. *A plane is Y-L_∞ transitive if, and only if, in the corresponding coordinatizing ternary ring R we have*

1) $a \cdot m \circ b = am + b$, *and*
2) *Addition is a group.*

Proof: Suppose that π is Y-L_∞ transitive. In the figure let us take YQV as $x = 0$, $V = (0, 0)$, $Q = (0, b)$, $X = (0)$, $T = (1)$, $M = (m)$. Here MQ is $y = x \cdot m \circ b$. Take P on MQ as $P = (a, a \cdot m \circ b)$. Draw VM which is $y = xm$, TQ which is $y = x + b$, and YP which is

$x = a$. Then U, the intersection of YP and VM, is $U = (a, am)$. Then draw UX which is $y = am$. UX intersects VT which is $y = x$ in $W = (am, am)$. Then YW, which is $x = am$ intersects QT, which is $y = x + b$, in $R = (am, am + b)$. Now if it is true that PRX lie on a line, since RX is $y = am + b$, we shall have from $P = (a, a \cdot m \circ b)$ $a \cdot m \circ b = am + b$. Hence we must show that PRX lie on a line. By hypothesis π is Y-L_∞ transitive. Let β be the Y-L_∞ collineation fixing all points on L_∞, all lines through Y and on $x = 0$ through Y such that $V\beta = Q$ or $(0, 0)\beta = (0, b)$. Then β fixes the lines YPU, $(x = a)$ YRW, $(x = am)$. Moreover, $(VM)\beta = QM$, $(VT)\beta = QT$. Hence $U\beta = P$, $W\beta = R$, and of course, $X\beta = X$. But UWX were on the line $y = am$. Hence $U\beta$, $W\beta$, $X\beta$, or PRX are on the line $y = am + b$ Hence P is $(a, am + b)$ and $a \cdot m \circ b = am + b$, the first part of our theorem.

What is the effect of the collineation β determined by $(0, 0)\beta = (0, b)$ on a general point (a, c)? This we easily find in a few steps. Thus

$$\begin{aligned} y = x &\to y = x + b, \\ x = c &\to x = c, \\ (c, c) &\to (c, c + b), \\ y = c &\to y = c + b, \\ x = a &\to x = a, \\ (a, c) &\to (a, c + b). \end{aligned}$$

Hence if $(0, 0)\beta = (0, b)$ then

$$(a, c)\beta = (a, c + b).$$

Now if δ is the Y-L_∞ collineation determined by

$$(0, 0)\delta = (0, d),$$

we find in general $(u, v)\delta = (u, v + d)$.

For $\beta\delta$ we find

$$(0, 0)(\beta\delta) = [(0, 0)\beta]\delta = (0, b)\delta = (0, b + d).$$

Hence, generally, $(a, c)(\beta\delta) = [a, c + (b + d)]$. But $[(a, c)\beta]\delta = (a, c + b)\delta = [a, (c + b) + d]$. Hence addition satisfies the associative law

$$c + (b + d) = (c + b) + d.$$

Since addition in a plane always has a zero and is a loop, it follows that addition will be a group. Hence we have proved (2).

Conversely, suppose that a ternary ring R of π satisfies

1) $a \cdot m \circ b = am + b$, *and*
2) *Addition is a group.*

For any $b \in R$ define a mapping $\beta = \beta(b)$ for points:

$$(\infty) \to (\infty),$$
$$(m) \to (m),$$
$$(a, c) \to (a, c + b).$$

For lines:

$$L_\infty \to L_\infty,$$
$$x = a \to x = a,$$
$$y = xm + t \to y = xm + (t + b).$$

This is a collineation, since if (a, c) is on $y = xm + t$, then $c = am + t$, whence

$$c + b = (am + t) + b = am + (t + b)$$

and so $(a, c + b)$ is on $y = xm + (t + b)$. The other verifications needed to show that β is a collineation are immediate. But this is a Y-L_∞ collineation, taking $(0, 0)$ into $(0, b)$. But since b was arbitrary, π is Y-L_∞ transitive.

The following is the corresponding theorem for translation planes:

THEOREM 20.4.6. *A plane π is a translation plane with respect to the axis L_∞ if, and only if, the corresponding ternary ring is a Veblen-Wedderburn system. This means that:*

1) *Addition is an Abelian group.*
2) *Multiplication (excluding 0) is a loop.*
3) $(a + b)m = am + bm$.
4) *If $r \neq s$, $xr = xs + t$ has a unique solution x.*
5) $a \cdot m \circ b = am + b$.

Proof: By Theorem 20.4.4 π will be a translation plane with axis L_∞ if it is Y-L_∞ transitive and also X-L_∞ transitive. From Theorem 20.4.5, we know (5), $a \cdot m \circ b = am + b$ and that addition is a group. In the proof of Theorem 20.4.5 we showed the existence of an elation $\beta(b)$ for every $b \in R$ which maps an arbitrary (a, c) into $(a, c + b)$. By Theorem 20.4.3 the entire translation group is Abelian, whence

$$\beta(b)\beta(d) = \beta(d)\beta(b),$$

and so $(a, c + b + d) = (a, c + d + b)$, whence $b + d = d + b$ and the addition in R is Abelian, proving (1).

Let b be an arbitrary element of R and consider the elation with center X taking $(0, 0)$ into $(b, 0)$. We have, in turn,

$$
\begin{aligned}
(0, 0) &\to (b, 0),\\
y = x &\to y = x - b,\\
y = a &\to y = a,\\
(a, a) &\to (a + b, a),\\
x = a &\to x = a + b,\\
y = am &\to y = am,\\
(a, am) &\to (a + b, am).
\end{aligned}
$$

Also, since $(0, 0) \to (b, 0)$,

$$y = xm \to y = xm - bm.$$

But then, since (a, am) is on $y = xm$, we have $(a + b, am)$ on $y = xm\text{-}bm$, whence

$$am = (a + b)m - bm,$$

and so, $am + bm = (a + b)m$. This proves the distributive law (3).

In a plane multiplication is always a loop, and condition (4) says that if $r \neq s$, the lines $y = xr$ and $y = xs + t$ intersect in a unique finite point.

A system of elements with binary operations of addition and multiplication satisfying conditions (1), (2), (3), (4) is called a Veblen-Wedderburn system, since this was first described in a paper [1] by these authors.

We show that, conversely, any Veblen-Wedderburn system R may be used as the coordinate system of a translation plane with axis L_∞. We take as our points: (1) the finite points (a, b) with a, b arbitrary elements of R; (2) the infinite points (m) with $m \in R$; and (3) the point $Y = (\infty)$. As our lines we take (1) L_∞ with points (∞) and (m); (2) lines $x = c$ containing (∞) and all points (c, d); and (3) lines $y = xm + b$ containing the points (m) and $(a, am + b)$ for every $a \in R$. It is now straightforward to verify that there is a unique line joining any two distinct points, a unique point lying on two distinct lines, and that of the four points $(0, 0)$, $(1, 1)$, (∞), and (0), no three are on a line. This verification involves several cases and we need condition (4) to show that lines $y = xr + b$ and $y = xs + c$ with $r \neq s$ intersect in a unique finite point.

For a Veblen-Wedderburn plane we easily verify that the mapping of finite points $(x, y) \to (x + r, y + s)$ is a collineation for any r and s, fixing all points on L_∞ and for finite lines mapping $x = c \to x = r + c$, $y = xm + b \to y = xm - rm + s + b$. If $s = rt$ this collineation is an elation with axis L_∞ and center (t). Hence a Veblen-Wedderburn plane is a translation plane with axis L_∞.

It is, of course, true that any associative division ring is a Veblen-Wedderburn system, and even any nonassociative division ring. A Veblen-Wedderburn system in which the multiplication is associative is called a *near-field.* This will be considered later. It will be remarked here and illustrated later (in §20.9) that nonisomorphic Veblen-Wedderburn systems may coordinate the same plane. For if we alter our choice of the points X and Y on the line L_∞ for which π is a translation plane, then the coordinatizing ternary ring will, by Theorem 2.4.6, be a Veblen-Wedderburn system, but not all such systems need be isomorphic.

A class of Veblen-Wedderburn systems known as Hall [2] systems is easily constructed. Suppose we have a field F such that there exists a quadratic polynomial $x^2 - rx - s$ irreducible over F. Then we may construct a Veblen-Wedderburn system J over F.

THEOREM 20.4.7. *Given a field F and a quadratic polynomial $f(x) = x^2 - rx - s$ irreducible over F. Then the set of elements $a + ub$, $a, b \in F$ is a Veblen-Wedderburn system J under the following rules:*

1) $(a_1 + ub_1) + (a_2 + ub_2) = (a_1 + a_2) + u(b_1 + b_2)$.

2) *For $c \in F$ put $(a + ub)c = ac + u(bc)$.*

3) *For $z = a + ub$ where $a, b \in F$, $b \neq 0$, and $w = c + zd$, c, $d \in F$, put*

$$\begin{aligned} wz &= ds + z(c + dr) \\ &= ac + adr + ds + u(bc + bdr). \end{aligned}$$

Under these rules J is a Veblen-Wedderburn system satisfying the distributive law $(x + y)z = xz + yz$. Elements $c \in F$ have the property that $cx = xc$, $c(xy) = (cx)y = (xc)y$. Moreover, every $z \notin F$ satisfies the equation $z^2 - rz - s = 0$.

Proof: Addition is well defined and is clearly an Abelian group. Multiplication is well defined, but there are two essentially different rules for a product xy: rule (2) for $y = c \in F$, and rule (3) for $y = z \notin F$.

The distributive law $(x + y)z = xz + yz$ holds, since these products are all calculated by rule (2) if $z \in F$ and all by rule (3) if $z \notin F$, and each of these rules separately satisfies this distributive law.

We note that if $z \notin F$, rule (3) gives the product $z^2 = rz + s$. Also, rule (3) gives $cz = zc$ when $c \in F$. Trivially, $cx = xc$ with $c, x \in F$. The unit 1 of F is a unit for J. To show that multiplication is a loop, we must show that in $xy = v$ any two of x, y, v (all $\neq 0$) uniquely determine the third. Here x and y determine $xy = v$ uniquely, using rule (2) if $y \in F$ and rule (3) if $y \notin F$, and with $x \neq 0$, $y \neq 0$ we readily verify that $v \neq 0$, using in (3) the fact that $b \neq 0$ and $s \neq 0$. Given $y \neq 0$ and $v \neq 0$, the appropriate rule (2) or (3) uniquely determines $x \neq 0$, satisfying $xy = v$.

The situation is a little more complicated if we are given $x \neq 0$ and $v \neq 0$. Write $x = a + ub$, $v = c + ud$ with $a, b, c, d \in F$. If $ad - bc = 0$, then there is a unique $f \neq 0, f \in F$ such that $af = c$ and $bf = d$, whence $xf = v$, and this is the only element of J satisfying this relation. Now suppose $ad - bc \neq 0$. If there were a $y = y_1 + uy_2$ with $y_1, y_2 \in F$, $y_2 \neq 0$, we would have $x = (a - by_1y_2^{-1}) + y(by_2^{-1})$, $xy = sby_2^{-1} + y(a - by_1y_2^{-1} + rby_2^{-1})$, and $v = (c - dy_1y_2^{-1}) + y(dy_2^{-1})$. Here $xy = v$ yields the relations:

$$\begin{aligned} ay_2 - by_1 &= d - rb, \\ cy_2 - dy_1 &= sb. \end{aligned}$$

Since $ad - bc \neq 0$ these equations do have a unique solution for y_1 and y_2, and $y_2 = (d^2 - rbd - sb^2)/(ad - bc)$. Here $y_2 \neq 0$, since because $x^2 - rx - s$ is irreducible over F, we cannot have $d^2 - rbd - sb^2 = 0$ with d, b elements of F not both zero. The values obtained for y_1, y_2 will yield an element $y \notin F$ satisfying $xy = v$. This shows that under multiplication, the nonzero elements of J form a loop.

To show that when $m \neq n$, $xm = xn + v$ has a unique solution, it is enough to find a solution, since if there were two solutions x_1 and x_2, we would have $(x_1 - x_2)m = (x_1 - x_2)n$, contrary to the loop property of multiplication. It is also easy to find a solution if m and n are both in F. Hence suppose $m \notin F$. (If $m \in F$, $n \notin F$, we consider $xn = xm - v$.) We may express n and v in terms of m. Suppose $n = c \in F$, $v = v_1 + mv_2$, $v_1, v_2 \in F$. If $x = x_1 + mx_2$, then we find

$$\begin{aligned} -cx_1 + sx_2 &= v_1, \\ x_1 + (r - c)x_2 &= v_2, \end{aligned}$$

which is solvable, since by the irreducibility of $x^2 - rx - s$ the determinant $c^2 - rc - s$ is not zero. Finally if $n = a + mb$, $v = v_1 + mv_2$ and we put $x = x_1 + mx_2$, then we have $-ax_1 + (a^2b^{-1} - ab^{-1}r - b^{-1}s + s)x_2 = v_1$, $(1 - b)x_1 + ax_2 = v_2$.

Here the determinant is $-b^{-1}[a^2 - ra(1 - b) - s(1 - b)^2]$ and is not zero since $b \neq 0$ and $x^2 - rx - s$ is irreducible. Hence a solution exists and $xm = xn + v$ has a unique solution x. Thus J is a Veblen-Wedderburn system.

20.5. Moufang and Desarguesian Planes.

In the preceding section it was shown that a plane can be coordinatized by a Veblen-Wedderburn system if, and only if, it is a translation plane with respect to one line, taken as L_∞. What can be said of the plane and its coordinates if it is a translation plane with respect to more than one line? This we shall answer in this section.

THEOREM 20.5.1. *If a plane π is a translation plane with respect to two lines intersecting in a point Q, then it is a translation plane for every line of the pencil of lines passing through Q.*

COROLLARY 20.5.1. *If π is a translation plane with respect to three lines not passing through a common point, then it is a translation plane for every line.*

Proof: For the corollary we note that a family of lines, which with any two lines contains the pencil of lines through their intersections, is necessarily the family of all lines in the plane if it contains three lines not through a common point.

Suppose that L_1 and L_2 are two lines intersecting in a point Y and that π is a translation plane with respect to both L_1 and L_2. Let L_3 be a third line passing through Y and let C be any point of L_3 different from Y. Let RCS be any line through C different from L_3 and intersecting L_1 in R and L_2 in S. Then there is an elation α with axis L_1 and center R taking S into C and L_2 into L_3. Since π has all elations with axis L_2 and center S, then the linearity theorem configuration is valid for all cases with S as center and L_2 as axis. The collineation α takes all these configurations into all linearity theorem configurations with center C and axis L_3. Hence in π there are all possible elations with center C and axis L_3. Since this argument is valid for every point of L_3 different from Y, then by the corollary to Theorem 20.4.4, π is a translation plane with axis L_3.

THEOREM 20.5.2. *A plane π is a translation plane for every line passing through the point $Y = (\infty)$ if, and only if, (1) its finite lines are given by linear equations $x = c$ and $y = xm + b$, and (2) the coordinates satisfy the following laws:*

2.1) *Addition is an Abelian group.*
2.2) $(a + b)m = am + bm$.
2.3) $a(s + t) = as + at$.
2.4) *Each $a \neq 0$ has an inverse a^{-1} such that $aa^{-1} = 1 = a^{-1}a$.*
2.5) $a^{-1}(ab) = b$.

Proof: Suppose that π is a translation plane for every line through $Y = (\infty)$. This includes L_∞, and so by Theorem 20.4.6, we know that the linearity conditions (1) are satisfied and that the coordinates are a Veblen-Wedderburn system. This gives conditions (2.1) and (2.2) of the theorem. The remaining three conditions must be proved.

Consider the elation with $Y = (\infty)$ as center, $x = 0$ as axis, which maps the point (0) onto the point (m).

Here all points $(0, b)$ are fixed and the lines L_∞ and $x = c$ are fixed. We find in turn,

$$\begin{aligned} (0) &\to (m), \\ (0, b) &\to (0, b), \\ y = b &\to y = xm + b, \\ x = a &\to x = a, \\ (a, b) &\to (a, am + b). \end{aligned}$$

This gives the mapping for an arbitrary finite point.

In particular $(1, t) \to (1, m + t)$,
and $(0, 0) \to (0, 0)$,
whence $y = xt \to y = x(m + t)$.
But $(a, at) \to (a, am + at)$,

and since (a, at) is on $y = xt$, we have $(a, am + at)$ on $y = x(m + t)$, whence

$$am + at = a(m + t),$$

the distributive law (2.3).

Now consider the elation with center (0, 0) and axis $x = 0$, which takes (0) to $(-1-a, 0)$. Here

$$(0) \to (-1-a, 0),$$
$$(0, 1 + a) \to (0, 1 + \text{a}),$$

whence $y = 1 + a \to y = x + 1 + a$.

$$(0) \to (-1-a, 0),$$
$$(0, b + ab) \to (0, b + ab),$$

whence $y = b + ab \to y = xb + b + ab$.

$$y = 1 + a \to y = x + 1 + a,$$
$$y = x(1 + a) \to y = x(1 + a),$$

whence $\quad (1, 1 + a) \to (d, d + 1 + a)$ if $a \neq 0$,

where $d(1 + a) = d + 1 + a$.

Also

$$(\infty) \to (\infty),$$
$$(1, 1 + a) \to (d, d + 1 + a),$$

whence $x = 1 \to x = d$.

Now

$$y = x(b + ab) \to y = x(b + ab),$$
$$y = b + ab \to y = xb + b + ab,$$

and so $\quad (1, b + ab) \to (d, d[b + ab])$,

where also $\quad d(b + ab) = db + b + ab$.

Here we assume not only $a \neq 0$, but also $(-1-a, 0) \neq (0, 0)$, i.e., $a \neq -1$. For such an a there exists a d such that $d(1 + a) = d + 1 + a$, and for any b, $d(b + ab) = db + b + ab$. If we put $d = u + 1$ and use the distributive laws, we find

$$ua = 1,$$

and

$$u(ab) = b.$$

From the distributive laws we find directly that even for $a = -1$, these relations hold with $u = -1$. Since for $u \neq 0$ there is a v, with

$$vu = 1,$$

and

$$v(ua) = a,$$

we have $v = a$. Hence we write $u = a^{-1}$ and have the laws (2.4) $aa^{-1} = a^{-1}a = 1$, and (2.5) $a^{-1}(ab) = b$.

Conversely, let us suppose that the conditions (1) and (2) hold for the coordinates of a plane π. From Theorem 20.4.6 we know that π is a translation plane with respect to L_∞. Using Theorem 20.5.1, it will be sufficient to show that π has a collineation which maps L_∞ onto some other line through $Y = (\infty)$.

The following mapping is such a collineation:

$$\begin{aligned}
(\infty) &\to (\infty),\\
(m) &\to (1, m),\\
(-1, m) &\to (-m),\\
(0, b) &\to (0, b),\\
(c, d) &\to [(1 + c^{-1})^{-1}, (1 + c)^{-1}d], \quad c \neq 0, -1.\\
L_\infty &\to x = 1,\\
x = -1 &\to L_\infty,\\
x = 0 &\to x = 0,\\
x = c &\to x = (1 + c^{-1})^{-1}, c \neq 0, -1.\\
y = xm + b &\to y = x(m - b) + b.
\end{aligned}$$

To show this, we must prove that incidences are preserved by the mapping; in particular, that if (c, d) is on $y = xm + b$, then the image point is on the image line. This reduces to showing that

$$(1 + c)^{-1}(cm + b) = (1 + c^{-1})^{-1}(m - b) + b$$

is an identity. This follows from the laws of the theorem, both of the following being identities:

$$\begin{aligned}
(1 + c)^{-1}(cm) &= (1 + c^{-1})^{-1}m,\\
(1 + c)^{-1}b &= (1 + c^{-1})^{-1}(-b) + b.
\end{aligned}$$

We shall prove, following R. H. Bruck, a further identity from the laws above. Write:

$$[y^{-1} - (y + z^{-1})^{-1}][y(zy) + y] = t,$$

where we exclude only the values $y = 0$, $y = -z^{-1}$. Then multiplying by $y + z^{-1}$, we have

$$\begin{aligned}
(y + z^{-1})t &= (y + z^{-1})[zy + 1 - (y + z^{-1})^{-1}(y(zy)) - (y + z^{-1})^{-1}y]\\
&= y(zy) + y + y + z^{-1} - y(zy) - y\\
&= y + z^{-1}.
\end{aligned}$$

Hence $t = 1$ and $y^{-1} - (y + z^{-1})^{-1}$, and $y(zy) + y$ are inverses. Then for any x,

$$[y^{-1} - (y + z^{-1})^{-1}][(y(zy))x + yx] = x.$$

Now write

$$[y^{-1} - (y + z^{-1})^{-1}][y(z(yx)) + yx] = w.$$

We now find

$$\begin{aligned}(y + z^{-1})w &= (y + z^{-1})[z(yx) + x] - y[z(yx)] - yx \\ &= yx + z^{-1}x = (y + z^{-1})x.\end{aligned}$$

Hence $w = x$. Now comparing expressions for w and x we see that we must have

$$\text{(M)} \qquad [y(zy)]x = y[z(yx)],$$

the Moufang identity. This identity is clearly valid for the excluded values $y = 0$, $y = -z^{-1}$, and so holds without exception. In particular, the Moufang identity with $z = 1$ reduces to the left alternative law

$$\text{(LA)} \qquad (yy)x = y(yx).$$

If a plane is a translation plane for every line, it is called a *Moufang plane*, after Ruth Moufang [1] who first studied these.

THEOREM 20.5.3. *A plane is a Moufang plane if, and only if, every ternary ring is* (1) *linear and* (2) *is an alternative division ring, i.e., satisfies the following laws:*

2.1) *Addition is an Abelian group.*
2.2) $(a + b)m = am + bm$.
2.3) $a(s + t) = as + at$.
2.4) *Each* $a \neq 0$ *has an inverse* a^{-1} *satisfying* $a^{-1}a = aa^{-1} = 1$.
2.5) $a^{-1}(ab) = b$.
2.6) $(ba)a^{-1} = b$.
In addition the alternative laws are valid:
2.7) $a(ab) = (aa)b$, $(ba)a = b(aa)$.

Proof: From Theorem 20.5.2 we have (1) and also (2.1) $\cdots$ (2.5). We must prove (2.6), since clearly the right alternative law $(ba)a = b(aa)$ follows from (2.6) in just the same way that the left alternative law follows from (2.5).

We consider the elation with axis $y = 0$ and center $(0, 0)$ such that $Y = (\infty)$ is mapped onto $(0, -1)$. We have in turn:

$$(\infty) \to (0, -1),$$
$$(1, 0) \to (1, 0),$$
whence
$$x = 1 \to y = x - 1.$$
$$(\infty) \to (0, -1),$$
$$(a, 0) \to (a, 0),$$
whence
$$x = a \to y = xa^{-1} - 1.$$
$$x = 1 \to y = x - 1,$$
$$y = x(1 - ab) \to y = x(1 - ab).$$
Thus
$$(1, 1 - ab) \to [(ab)^{-1}, (ab)^{-1} - 1].$$
Also
$$(0) \to (0),$$
whence
$$y = 1 - ab \to y = (ab)^{-1} - 1.$$
$$x = a \to y = xa^{-1} - 1,$$
$$y = x(a^{-1} - b) \to y = x(a^{-1} - b),$$
whence
$$(a, 1 - ab) \to (b^{-1}, b^{-1}a^{-1} - 1).$$
Since
$$(0) \to (0)$$
we have
$$y = 1 - ab \to y = b^{-1}a^{-1} - 1.$$

Comparing images of $y = 1 - ab$, we must have

$$(ab)^{-1} = b^{-1}a^{-1}.$$

Using this as a law, we find, since $b^{-1} = a(a^{-1}b^{-1})$,

$$\begin{aligned} b = (b^{-1})^{-1} &= [a(a^{-1}b^{-1})]^{-1} \\ &= (a^{-1}b^{-1})^{-1}a^{-1} \\ &= (ba)a^{-1}. \end{aligned}$$

This proves (2.6).

We have now shown that in a Moufang plane the coordinates satisfy the above laws for an alternative division ring. Conversely, suppose we are given an alternative division ring. Constructing a plane with these coordinates, by Theorem 20.5.2 the plane is a translation plane for every line passing through $Y = (\infty)$. Hence, by Theorem 20.5.1 and its corollary, it will follow that the plane is a translation plane for every line if we can find a collineation moving $Y = (\infty)$. The following reflection is valid:

$$\begin{aligned} (a, b) &\rightleftarrows (b, a), \\ (0) &\rightleftarrows (\infty), \\ (m) &\rightleftarrows (m^{-1}), \quad m \neq 0, \\ x = c &\rightleftarrows y = c, \\ y = xm + b &\rightleftarrows y = xm^{-1} - bm^{-1}, \quad m \neq 0. \end{aligned}$$

This completes the proof of the theorem. A few comments may be in order.

More is true than has been proved here, but the proof requires more of the theory of alternative rings than can be given here. It is true that if a plane is a translation plane for two different lines, then it is a Moufang plane, a translation plane for every line. This means algebraically that the law (2.6), $(ba)a^{-1} = b$, is a consequence of the preceding laws. No simple proof of this fact is known to the author. As we saw, the left alternative law, $x(xy) = (xx)y$, is a consequence of (2.5). Kleinfeld [2] and Skornyakov [2] have shown that for characteristic different from 2 this law (in a division ring) also implies the right alternative law $(yx)x = y(xx)$. This is false for characteristic 2, but with the stronger Moufang law $[y(zy)]x = y[z(yx)]$, which we showed was a consequence of (2.5), San Soucie [1] has shown that even for characteristic 2 the right alternative law is a consequence. Bruck and Kleinfeld [1] studied alternative division rings R and found the surprising result that such a ring R is either associative or is a particular kind of algebra over its center, which is a field F. R is a Cayley-Dickson algebra over F, this being an algebra with eight basis elements, any single element not in F generating a quadratic field over F, any two elements (not in the same quadratic extension of F) generating a quaternion algebra. This detailed knowledge enables them to show that in a Moufang plane every ternary ring not only yields an alternative division ring of coordinates, but also the same alternative division ring. For a complete account of all these results, except for that of San Soucie, the reader is referred to Pickert [1].

THEOREM 20.5.4. *The following conditions in a plane π are equivalent:*

1) *π is X-OY transitive, i.e., π has all homologies with center $X = (0)$ and axis $x = 0$.*

2) *In the ternary ring for π given by the four points X, Y, O, I, we have*

2.1) $x \cdot m \circ b = xm + b$, *and*

2.2) *Multiplication is a group.*

Proof: Assume that (1) holds in π. Consider the homology with axis $x = 0$, center $X = (0)$ and mapping $(m) \to (1)$ on L_∞. We have

$$(0, 0) \to (0, 0),$$
$$(m) \to (1),$$

whence
$$y = xm \to y = x.$$

Now
$$y = am \to y = am,$$

whence
$$(a, am) \to (am, am).$$

As
$$(\infty) \to (\infty),$$

we have
$$x = a \to x = am,$$

and also
$$y = c \to y = c,$$

whence
$$(a, c) \to (am, c).$$

Also
$$(0, b) \to (0, b),$$
$$(m) \to (1),$$

whence
$$y = x \cdot m \circ b \to y = x + b.$$

Hence if (a, c) is on $y = x \cdot m \circ b$, we have (am, c) on $y = x + b$. Thus $c = a \cdot m \circ b$ implies $c = am + b$, whence $a \cdot m \circ b = am + b$, the linearity condition (2.1).

Here the homology determined by $(m) \to (1)$ maps $(a, 1) \to (am, 1)$, and in particular, $(1, 1) \to (m, 1)$. If we follow this by the homology mapping $(n) \to (1)$, we have $(1, 1) \to (m, 1) \to (mn, 1)$ and $(a, 1) \to (am, 1) \to ((am)n, 1)$.

But the product must be the homology $(mn) \to (1)$, whence $(a, 1) \to [a(mn), 1]$, and so $(am)n = a(mn)$, the associative law for multiplication. But since multiplication is a loop, it follows that multiplication is a group.

Now suppose that (2) holds. Then the following is a homology in π for fixed $m \neq 0$.

$$\begin{aligned}(\infty) &\to (\infty),\\ (0) &\to (0),\\ (n) &\to (m^{-1}n),\\ (a, b) &\to (am, b),\\ L_\infty &\to L_\infty\\ x = a &\to x = am,\\ y = xn + b &\to y = x(m^{-1}n) + b.\end{aligned}$$

Letting m range over all values $\neq 0$, we have all homologies with center $X = (0)$ and axis $x = 0$. Hence (2) implies (1).

THEOREM 20.5.5. *In a plane π if the following special cases of the Theorem of Desargues hold:*

1) *The linearity theorem in all cases with three different axes not through a point;*

2) *The general theorem for one axis and one center not on the axis; then π can be coordinatized by an associative division ring. In this case the Theorem of Desargues holds throughout π. The collineation group for π is transitive on quadrilaterals, and every ternary ring for π is the same associative division ring.*

Proof: From the hypothesis we may apply Theorem 20.5.3 and 20.5.4, and find that for one choice of a quadrilateral X, Y, O, I from which a ternary ring is given for π, we have $x \cdot m \circ b = xm + b$, and the coordinates form an associative division ring. Clearly, if there is a collineation of π taking the points of a quadrilateral X_1, Y_1, O_1, I_1 into those of a second X_2, Y_2, O_2, I_2 in this order, the ternary rings will be isomorphic. We prove first that if a plane π is coordinatized by an associative division ring D for one quadrilateral X_1, Y_1, O_1, I, then the collineations of π are transitive on quadrilaterals, whence every quadrilateral yields a coordinate ring isomorphic to D. From Theorem 20.5.3, π is a translation plane for every line in π. Given a triangle ABC and taking AB as an axis, there is an elation fixing A and B and moving C to any point C' not on AB. In this way start from any quadrilateral X_2, Y_2, O_2, I_2. With appropriate use of such elations we find a collineation mapping X_2 onto X_1, Y_2 onto Y_1 and O_2 onto O_1. The new point I_2 does not lie on any side of the triangle X_2, Y_2, O_2, and so, in the coordinates of X_1, Y_1, O_1, I_1 it will be a finite point $I_2 = (a, b)$ with $a \neq 0$, $b \neq 0$. The following collineation fixes X, Y, O and maps I_2 onto I_1:

$$\begin{aligned}
(x, y) &\rightarrow (xa^{-1}, yb^{-1}),\\
(m) &\rightarrow (amb^{-1}),\\
(\infty) &\rightarrow (\infty),\\
y = xm + s &\rightarrow y = x(amb^{-1}) + sb^{-1},\\
x = c &\rightarrow x = ca^{-1},\\
L_\infty &\rightarrow L_\infty.
\end{aligned}$$

Thus all ternary rings yield the same associative division ring D, and π is transitive on quadrilaterals.

It remains only to show that the Theorem of Desargues holds

throughout π. If the center lies on the axis, this is surely true, since π has all possible elations. Now suppose the center is not on the axis. Take the center to be O and the axis L_∞. We show that π has all homologies with center O and axis L_∞. For fixed $a \neq 0$ the following is such a homology:

$$\begin{aligned}
(\infty) &\rightarrow (\infty), \\
(m) &\rightarrow (m), \\
(c, d) &\rightarrow (ac, ad), \\
L_\infty &\rightarrow L_\infty, \\
x = c &\rightarrow x = ac, \\
y = xm + b &\rightarrow y = xm + ab.
\end{aligned}$$

Letting a range over all values different from 0, we get all possible homologies with center O and axis L_∞.

20.6. The Theorem of Wedderburn and the Artin-Zorn Theorem.

We give here some of the properties of finite fields which we shall need. For the proof of these properties see Van der Waerden [1] vol. 1, Chapter V, §37.

1) The number of elements in a finite field is a power of a prime. For each prime power p^r there is a finite field $GF(p^r)$ with p^r elements, and it is unique to within isomorphism.

2) Every element x of $GF(p^r)$ satisfies the relation $x^{p^r} = x$. The multiplicative group $F^*(p^r)$ of the $p^r - 1$ elements of $GF(p^r)$, excluding zero, is cyclic. A generator of this cyclic group is called a *primitive root.*

3) $GF(p^r)$ may be represented as the residue classes of polynomials $P(x)$ with coefficients in the field F_p, with p elements modulo a polynomial $f(x)$ irreducible of degree r over F_p. Equivalently $GF(p^r)$ may be represented as the residue classes of integral polynomials modulo the ideal $[p, f(x)]$.

4) The automorphisms of $GF(p^r)$ are a cyclic group of order r generated by the automorphism $z \rightarrow \alpha(z) = z^p$.

THEOREM 20.6.1 (THE THEOREM OF WEDDERBURN). *A finite division ring R is necessarily commutative and so is a finite field $GF(p^r)$.*

Proof: The following proof is due to Witt [1]. The unit of R generates the characteristic subfield of R, and this must be a finite field F_p for some prime p. Let R have a basis of r elements $x_1 = 1, x_2, \cdots, x_r$

over F_p. Then R has exactly p^r elements. The center Z of R consists of all elements z of R such that $zx = xz$ for every $x \in R$. Z is a commutative subring of R and so is a finite field. Let Z have $q = p^s$ elements. We wish to show that Z is all of R. In any event R is a vector space over Z, and if R has a basis of t elements over Z, then R has $q^t = p^{st} = p^r$ elements in all. Here $t = 1$ if $R = Z$. The normalizer N_x of an element x of R is a subring containing Z. Hence N_x contains q^d elements, and since R is a vector space over N_x, we have necessarily $d|t$. Hence, in the multiplicative group R^* of the $p^r - 1 = q^t - 1$ elements of R excluding 0, an element x not in Z has normalizer of order $q^d - 1$, where d is a divisor of t and $d < t$. Counting the elements of R^* we have

$$(20.6.1) \qquad q^t - 1 = q - 1 + \Sigma(q^t - 1)/(q^d - 1),$$

where $q - 1$ enumerates the elements of the center and each remaining summand counts the elements in a class $(q^t - 1)/(q^d - 1)$, where d/t, $d < t$.

We showed in §16.8 that the polynomial $f(x) = x^t - 1$ has the factor with rational integral coefficients, $k(x) = \prod (x - w^j)$, where w is a primitive tth root of unity and the w^j for $j = 1, \cdots, t$, $(j, t) = 1$ are all the primitive tth roots of unity. Then if $d|t$, $d < t$, we have $x^t - 1 = k(x)\,(x^d - 1)r(x)$, since $k(x)$ does not contain any of the factors of $x^d - 1$. Here $r(x)$ contains all remaining factors of $x^t - 1$, if there are any. Hence $(x^t - 1)/(x^d - 1) = k(x)r(x)$. For $x = q$, $k(q) = \prod_{(j,t)=1} (q - w^j)$, and since $w^j \neq 1$ is a complex root of unity, we have $|q - w^j| > q - 1$. Hence $k(q)$ is a rational integer greater in absolute value than $q - 1$. Hence if $t > 1$ we have $k(q)$ dividing every term in (20.6.1) except $q - 1$. But then $k(q)$ also divides $q - 1$, and this is impossible because $|k(q)| > q - 1$. Hence the only possibility is that $t = 1$, and this means that $Z = R$, and so, R is commutative and is a finite field.

THEOREM 20.6.2 (ARTIN-ZORN).* *A finite alternative division ring is a finite field $GF(p^r)$.*

Proof: We begin by developing a little of the theory of alternative rings. An alternative ring R is a system with a binary addition and multiplication in which (1) the addition is an Abelian group, (2) both

* See Zorn [1].

distributive laws hold, and (3) the multiplication satisfies the two weak associative laws,

$$(xx)y = x(xy), \quad y(xx) = (yx)x. \tag{20.6.1}$$

R is a division ring if the nonzero elements form a loop with respect to multiplication. We noted in the preceding section that a Moufang plane is coordinatized by an alternative division ring. Here, instead of (20.6.1), we had the multiplicative laws,

$$\begin{aligned} aa^{-1} &= a^{-1}a = 1, \\ a^{-1}(ab) &= b = (ba)a^{-1}. \end{aligned} \tag{20.6.2}$$

We showed that laws (20.6.2) implied the laws (20.6.1).

We shall define two quantities for any ring in which the distributive laws hold, the *associator* (x, y, z) and the *commutator* (x, y). These are defined by the rules

$$\begin{aligned} (x, y, z) &= (xy)z - x(yz), \\ (x, y) &= xy - yx. \end{aligned} \tag{20.6.3}$$

Thus the associator vanishes identically in any associative ring and the commutator vanishes identically in any commutative ring. Both the associator and the commutator are linear in each argument. The laws (20.6.1) can be rewritten in the form

$$(x, x, y) = 0, \quad (y, x, x) = 0. \tag{20.6.4}$$

From the linearity of the associator we have

$$\begin{aligned} 0 &= (x, y + z, y + z) \\ &= (x, y, y) + (x, y, z) + (x, z, y) + (x, z, z) \\ &= (x, y, z) + (x, z, y), \end{aligned} \tag{20.6.5}$$

and

$$\begin{aligned} 0 &= (x + y, x + y, z) \\ &= (x, x, z) + (x, y, z) + (y, x, z) + (y, y, z) \\ &= (x, y, z) + (y, x, z). \end{aligned}$$

From this we have

$$\begin{aligned} (x, y, z) &= -(x, z, y) = (z, x, y) = -(z, y, x) \\ &= (y, z, x) = -(y, x, z). \end{aligned} \tag{20.6.6}$$

This says that (x, y, z) under the permutations of the symmetric group on x, y, z is unchanged by the alternating group and is changed in sign by the odd elements. It is this property which has led to the name *alternative* for these rings. The rules (20.6.6) give immediately $(x, y, x) = -(x, x, y) = 0$, whence

$$(xy)x = x(yx). \tag{20.6.7}$$

The law (20.6.7) by itself is called the *reflexive law.*

A function $h(x_1, \cdots, x_n)$ in a ring is said to be *skew symmetric* if it is (1) linear in each of its arguments, and (2) vanishes whenever any two of its arguments are equal. We note that skew symmetry implies the alternating property and that the associator and commutator are symmetric in an alternative ring.

We shall develop here a few formulae valid in alternative rings. More will be given than are really needed to prove our theorem, but all are of sufficient interest to include here. The following identity may be verified as valid in any ring with the distributive laws

(20.6.8)

$$(wx, y, z) - (w, xy, z) + (w, x, yz) = w(x, y, z) + (w, x, y)z.$$

We define the function $f(w, x, y, z)$ by the rule

$$f(w, x, y, z) = (wx, y, z) - x(w, y, z) - (x, y, z)w. \tag{20.6.9}$$

Lemma 20.6.1. *In any alternative ring R the function $f(w, x, y, z)$ of* (20.6.9) *is skew symmetric and satisfies the identities*

$$\begin{aligned} 3f(w, x, y, z) = (w, (x, y, z)) - (x, (y, z, w)) \\ + (y, (z, w, x)) - (z, (w, x, y)). \end{aligned} \tag{20.6.10}$$

$$f(w, x, y, z) = ((w, x), y, z) + ((y, z), w, x). \tag{20.6.11}$$

Proof: From (20.6.6) we may rewrite (20.6.8) in the form

(20.6.12)

$$(wx, y, z) - (xy, z, w) + (yz, w, x) = w(x, y, z) + (w, x, y)z.$$

Substituting for (wx, y, z) its expression in terms of f as given by (20.6.9), and similarly for the other terms on the left of (20.6.12), we get

$$f(w, x, y, z) - f(x, y, z, w) + f(y, z, w, x) = F(x, y, z, w), \tag{20.6.13}$$

where $F(x, y, z, w)$ is the right-hand side of (20.6.10) and thus changes sign when its arguments are permuted cyclically. Hence from (20.6.13), $0 = F(w, x, y, z) + F(x, y, z, w) = f(w, x, y, z) + f(z,$

$w, x, y)$, and so,

$$f(w, x, y, z) = -f(z, w, x, y). \tag{20.6.14}$$

Hence f changes sign when its arguments are permuted cyclically and, from (20.6.9), when its last two arguments are interchanged, and therefore when any two are interchanged. Since $f(w, x, y, y) = 0$, f is skew symmetric. In particular, (20.6.13) reduces to (20.6.10). Subtract from (20.6.8) the result of interchanging w and x and get

$$((w, x), y, z) = -(w, (x, y, z)) + (x, (y, z, w)) + 2f(w, x, y, z).$$

Computed thus, the right-hand side of (20.6.11) in view of (20.6.10) reduces to the left-hand side.

LEMMA 20.6.2. *For all x, y, z of an alternative ring we have:*

$$(x^2, y, z) = x(x, y, z) + (x, y, z)x, \tag{20.6.15}$$
$$(x, xy, z) = (x, y, xz) = (x, y, z)x, \tag{20.6.16}$$
$$(x, yx, z) = (x, y, zx) = x(x, y, z), \tag{20.6.17}$$

and the Moufang identities

$$(xy)(zx) = x((yz)x) = (x(yz))x. \tag{20.6.18}$$
$$x(y(xz)) = ((xy)x)z, \quad ((zx)y)x = z(x(yx)). \tag{20.6.19}$$

Proof: We obtain (20.6.15) from $f(x, x, y, z) = 0$. We get the two parts of (20.6.16) by observing that $f(x, y, z, x) = 0$ and $f(x, z, x, y) = 0$. We obtain (20.6.17) from $f(y, x, x, z) = 0$ and $f(z, x, x, y) = 0$. To prove (20.6.18) we note that:

$$\begin{aligned}(xy)(zx) &= x(y(zx)) + (x, y, zx)\\ &= x(y(zx)) + x(y, z, x)\\ &= x((yz)x)\end{aligned}$$

by (20.6.17). The first equation of (20.6.19) may be derived as follows:

$$\begin{aligned}((xy)x)z &= (xy)(xz) + (xy, x, z)\\ &= x(y(xz)) + (x, y, xz) - (x, xy, z)\\ &= x(y(xz))\end{aligned}$$

by (20.6.16). The second may similarly be derived from (20.6.17).

We can now show that the laws (20.6.2) follow from those of (20.6.1) in a division ring. Given an element $a \neq 0$, then there is a u

such that $au = 1$. Then $a = (au)a = a(ua)$, whence also $ua = 1$, and we may write $u = a^{-1}$, where $a^{-1}a = aa^{-1} = 1$. Given any a, b not zero, determine c by the relation $b = a^{-1}c$. Then

$$\begin{aligned} a^{-1}(ab) &= a^{-1}(a(a^{-1}c)) = ((a^{-1}a)a^{-1})c \\ &= (1a^{-1})c = a^{-1}c = b, \end{aligned}$$

using the first law of (20.6.19). Similarly, $(ba)a^{-1} = b$ follows from the second law of (20.6.19).

We now have more than enough information about alternative rings to prove the Artin-Zorn theorem. Let R be a finite alternative ring. It will be enough to show that a finite division ring R_1 generated by two elements b and c is associative. For then by the Wedderburn theorem, R_1 is a finite field and is generated by a single element d. Hence, if R is generated by $b_1, b_2, \cdots, b_s$, then b_1, b_2 generate a finite field which is generated by the single element c_1. Thus R is generated by $c_1, b_3, \cdots, b_s$. Continuing, we may reduce the number of generators until R itself is generated by a single element and is associative and thus is a finite field $GF(p^r)$.

Consider R_1 generated by b and c, where R_1 is a subsystem of R closed under addition and multiplication. R_1 is finite and has no zero divisors. Let $a_1, \cdots, a_t$ be the elements of R_1 different from zero. Then for any $x \in R_1$, $xa_1, \cdots, xa_t$ are all different, since R_1 contains no zero divisors, and so are $a_1, a_2, \cdots, a_t$ in some order. Hence for some element, say, a_1, we have $xa_1 = x$, whence $a_1 = 1$ is the unit of R. Also for some a_i we have $xa_i = 1$, whence $a_i = x^{-1}$. Hence R_1 is a division ring. The elements of R_1 are sums of monomials $(x_1 \cdots x_r)(x_{r+1} \cdots x_n)$, where each x_i is b or c and the terms are bracketed in any way. Since both distributive laws hold, multiplication will be associative in R_1 if, and only if, the multiplication of the monomials is associative. To show this, we define normal (or left-bracketed) monomials recursively by the rules:

$$\begin{aligned} &[x_1x_2] = x_1x_2 \\ &[x_1 \cdots x_n] = [x_1 \cdots x_{n-1}]x_n. \end{aligned} \tag{20.6.20}$$

If it can be shown that an arbitrary bracketing is equal to the normal monomial, then associativity is immediate, for then

$$\begin{aligned} &([x_1 \cdots x_r][y_1 \cdots y_s])[z_1 \cdots z_t] \\ &= [x_1 \cdots x_ry_1 \cdots y_sz_1 \cdots z_t] \\ &= [x_1 \cdots x_r]([y_1 \cdots y_s][z_1 \cdots z_t]). \end{aligned}$$

That every monomial is equal to the normal monomial with the same factors in order, we prove by induction on the length of the monomial. From (20.6.4) and (20.6.6) this is true for monomials in b and c of length three, since there must be a repeated factor, and so,

$$x_1(x_2x_3) = (x_1x_2)x_3 = [x_1x_2x_3]. \tag{20.6.21}$$

We must show that

$$[u_1 \cdots u_r][v_1 \cdots v_s] = [u_1 \cdots u_rv_1 \cdots v_s], \tag{20.6.22}$$

using induction on $n = r + s$, and for fixed n, using induction on s, this being the definition (20.6.20) for $s = 1$. Suppose $s > 1$ and $v_1 = v_s = b$ or $v_1 = v_s = c$. Then in the second identity of (20.6.19), $z[x(yx)] = [(zx)y]x$, take $z = [u_1, \cdots, u_r]$, $v_1 = v_s = x$ and $[v_2 \cdots v_{s-1}] = y$. This gives

$$\begin{aligned}[u_1 \cdots u_r][xv_2 \cdots v_{s-1}x] &= \{([u_1 \cdots u_r]x)[v_2 \cdots v_{s-1}]\}x \\ &= [u_1 \cdots u_rv_1v_2 \cdots v_{s-1}]v_s \\ &= [u_1 \cdots u_rv_1v_2 \cdots v_{s-1}v_s],\end{aligned} \tag{20.6.23}$$

using our induction and proving (20.6.22) when $v_1 = v_s$. Now suppose $v_1 \neq v_s$, where, say, $v_1 = b$, $v_s = c$. Here u_r is either b or c. If $u_r = b$, write

$$x = [u_1 \cdots u_{r-1}], \quad u_r = b, \quad v_1 = b, \quad [v_2 \cdots y_s] = y.$$

Then $f(x, b, b, y) = 0 = (xb, b, y) - b(x, b, y) - (x, b, b)y$. Here $(x, b, b) = 0$, and by induction on the length, x, b, and y associate, whence $(x, b, y) = 0$. Hence $(xb, b, y) = 0$, whence $(xb)(by) = [(xb)b]y$ or

$$\begin{aligned}[u_1 \cdots u_{r-1}b][bv_2 \cdots v_s] &= ([u_1 \cdots u_{r-1}]bb)[v_2 \cdots v_s] \\ &= [u_1 \cdots u_{r-1}bb][v_2 \cdots v_s] \\ &= [u_1 \cdots u_rv_1 \cdots v_s],\end{aligned} \tag{20.6.24}$$

using our induction on s in the last step.

Similarly, if $u_r = c$, write

$$x = [u_1 \cdots u_{r-1}], \quad u_r = c, \quad [v_1 \cdots v_{s-1}] = z, \quad v_s = c.$$

Now $f(x, c, z, c) = 0 = (xc, z, c) - c(x, z, c) - (c, z, c)x$. Here

$(c, z, c) = 0$, and by induction on length, $(x, z, c) = 0$, whence $(xc, z, c) = 0$ and this gives $(xc)(zc) = [(xc)z]c$ or

$$
\begin{aligned}
(20.6.25) \quad [u_1 \cdots u_{r-1}c][v_1 \cdots v_{s-1}c] &= ([u_1 \cdots u_{r-1}c][v_1 \cdots v_{s-1}])c \\
&= [u_1 \cdots u_{r-1}u_r v_1 \cdots v_{s-1}]c \\
&= [u_1 \cdots u_{r-1}u_r v_1 \cdots v_{s-1}v_s].
\end{aligned}
$$

Hence (20.6.23,24,25) establish (20.6.22) in every case. This proves the associativity of R_1, and so proves our theorem.

20.7. Doubly Transitive Groups and Near-Fields.

A certain class of groups is intimately related to projective planes. This is the class of doubly transitive groups in which only the identity fixes two letters. We shall need an additional hypothesis which may not be necessary but is required for our proof. This is condition (3) or (3′) of the following theorem:

THEOREM 20.7.1. *Suppose that G is a permutation group on letters $c_0, c_1, \cdots, c_{n-1}$ such that*

1) *G is doubly transitive.*
2) *Only the identity fixes two letters.*
3) *At most one element taking c_i into c_j displaces all letters, or (3′) n is finite.*

Then the identity and the elements of G displacing all letters form a transitive normal Abelian subgroup A. G is isomorphic to the group of linear substitutions $x \to xm + b$ in a near field K. Conversely, the linear substitutions $x \to xm + b$, $m \neq 0$ of a near field K yield a group satisfying (1)(2), the linear substitutions being regarded as a permutation group on the elements of K. If for $m \neq 0, 1$, $xm + b = x$ always has a solution, (3) is satisfied, and we may take $x = c$ and $y = xm + b$ as the finite lines of a plane coordinatized by the near field K. A permutation of G $\begin{pmatrix} c_0, & c_1, & \cdots, & c_{n-1} \\ d_0, & d_1, & \cdots, & d_{n-1} \end{pmatrix}$, regarded as a permutation of the elements of K, corresponds to a line of π whose points are (c_i, d_i) $i = 0, \cdots, n-1$.

Proof: We first prove the purely group theoretical part of the theorem. We wish to show that the identity and the elements of G displacing all letters form a transitive normal Abelian subgroup. We prove a number of lemmas.

LEMMA 20.7.1. *There exists one and only one element of order* 2 *in* G *which interchanges a specified pair of letters* (i, j).

Since G is doubly transitive, one such element g must exist. Here g^2, fixing two letters, is the identity. A second element h with this property would be such that gh^{-1} fixes two letters, whence $gh^{-1} = 1$, $g = h$.

LEMMA 20.7.2. *The elements of order* 2 *are in a single conjugate class.*

An element of order 2 fixes at most one letter. When $n = 2$, there is only one element. For $n \geq 3$, then g and h of order 2 must both displace some letter i, $g = (i, j) \cdots$, $h = (i, k) \cdots$. An x in G with $x = \begin{pmatrix} i, j, \cdots \\ i, k, \cdots \end{pmatrix}$ will be such that $x^{-1}gx = (ik) \cdots = h$. If any element of the class of elements of order 2 fixes a letter, then all do. We may subdivide into two cases:

CASE 1. The elements of order 2 displace all letters.

CASE 2. Every element of order 2 fixes a letter.

LEMMA 20.7.3. *In Case* 2 *there is one and only one element of order* 2 *fixing a given letter.*

As before, $g = (i, j) \cdots$ is transformed into $h = (i, k) \cdots$ by $x = \begin{pmatrix} i, j, \cdots \\ i, k, \cdots \end{pmatrix}$. But if g and h both fix the same letter s, then x must also fix s and $x \neq 1$ fixes both i and s, contrary to our hypothesis (2). If g fixes a letter s, then we may find an element of order 2 fixing a specified letter t by transforming g by any element carrying s into t.

Note that this lemma implies that if $g = (ij)(s) \cdots$, then g is in the center of the subgroup H_s fixing s.

LEMMA 20.7.4. *The product of two different elements of order* 2 *is an element displacing all letters.*

If $g^2 = 1$, $h^2 = 1$, $g \neq h$, suppose to the contrary that gh fixes a letter i. By Lemma 20.7.3, g and h cannot both fix i ,whence neither can fix i. But then we shall have $g = (i, j) \cdots$, $h = (j, i) \cdots$ and $gh = (i)(j) \cdots = 1$, and so $g = h$, contrary to assumption. It is thus impossible for gh to fix any letter.

LEMMA 20.7.5. *In G there is one and only one element, displacing all letters, which takes a given i into a give $j \neq i$.*

In Case 1 an element $g = (i, j) \cdots$ gives such an element. In Case 2, take $g = (i) \cdots$ of order 2 and $h = (i, j) \cdots$, and by Lemma 20.7.4, gh takes i into j and displaces all letters. Hence at least one such element exists. Hypothesis (3) says that there is at most one such element. We observe that hypothesis (3′) implies Lemma 20.7.5. Since G is doubly transitive on n letters, the subgroup fixing a letter c_0 is of index n, and the subgroup H_0 fixing c_0 is transitive on the remaining $n - 1$ letters and, since only the identity fixes two letters, H_0 is of order $n - 1$. Thus G is of order $n(n - 1)$. Hence the elements taking i into j form a left coset of H_i, the subgroup fixing i, and so there are exactly $n - 1$ such elements. For given three letters i, j, k, there is in G by double transitivity one element $g = \begin{pmatrix} i, & k, & \cdots \\ j, & k, & \cdots \end{pmatrix}$ in G and only one, since only the identity fixes two letters. For given i, j there are exactly $n - 2$ choices for k, and this leaves exactly one of the $n - 1$ elements taking i into j, which displaces all letters.

LEMMA 20.7.6. *In Case* 1 *every element displacing all letters is of order* 2 *and these together with the identity form a normal Abelian subgroup.*

Clearly, $g = (i, j) \cdots$ is one, and hence is the only element displacing all letters taking i into j. If $g^2 = 1$, $h^2 = 1$, then if $g = h$, $gh = 1$, while if $g \neq h$, then gh displaces all letters of order 2, $(gh)^2 = 1$, whence $hg = gh$ and the elements of order 2 together with the identity form an Abelian group A. By Lemma 20.7.2, A is a normal subgroup. This proves the lemma.

LEMMA 20.7.7. *In Case* 2 *a given element g displacing all letters may be written as the product $g = ab$ of two elements of order* 2, *where either a or b may be taken arbitrarily.*

Suppose $a^2 = 1$ and a fixes the letter i and that g takes i into j. Choose $b = (i, j) \cdots$; then $b^2 = 1$ and $g = ab$, since ab displaces all letters and takes i into j. Similarly, suppose we are given g and b with $b^2 = 1$, where b fixes k. If g takes i into k, then with $a = (i, k) \cdots$, we have $g = ab$.

LEMMA 20.7.8. *In Case* 2 *the product abc of three elements of order* 2 *is again of order* 2 *and $abc = cba$.*

If $a = b$, the lemma is trivial. If $b \neq a$, then $ab = g = dc$, where by Lemma 20.7.7, $d^2 = 1$. Hence $abc = dc^2 = d$. As $d = d^{-1}$, $abc = c^{-1}b^{-1}a^{-1} = cba$.

LEMMA 20.7.9. *In Case 2 the elements displacing all letters, together with the identity, form a normal Abelian subgroup.*

Let g and h displace all letters. By Lemmas 20.7.4 and 20.7.7, $g = ab$, $h = cd$ with a, b, c, d of order 2. Writing $h = eb$ with $e^2 = 1$, we have $gh^{-1} = ae$, whence if $e = a$, $gh^{-1} = 1$, or if $a \neq e$, gh^{-1} displaces all letters. Hence the elements displacing all letters, together with the identity, form a subgroup A. Using Lemma 20.7.8, $gh = (abc)d = (cba)d = c(bad) = c(dab) = hg$, and hence the group A is Abelian. Since a conjugate of an element displacing all letters also displaces all letters, A is a normal subgroup.

We shall now construct an algebraic system S whose elements shall be the letters $c_0, c_1, \cdots, c_{n-1}$ permuted by G. We arbitrarily designate one of them as 0 and another as 1, say, $c_0 = 0$, $c_1 = 1$. In S we define an addition

$$y = x + b, \tag{20.7.1}$$

if, and only if, in the subgroup A we have the permutation:

$$A_b = \begin{pmatrix} 0, & \cdots, & x, & \cdots \\ b, & \cdots, & y, & \cdots \end{pmatrix}. \tag{20.7.2}$$

In S we define a multiplication

$$y = xm, \tag{20.7.3}$$

if, and only if, in the subgroup H_0 fixing 0 we have the permutation

$$M_m = \begin{pmatrix} 0, & 1, & \cdots, & x, & \cdots \\ 0, & m, & \cdots, & y, & \cdots \end{pmatrix}. \tag{20.7.4}$$

Addition is well defined, since in A there is a unique element taking 0 into b. Multiplication by an $m \neq 0$ is well defined, since in H_0 there is a unique element taking 1 into m. If we have in A

$$A_a = \begin{pmatrix} 0 & \cdots & x \\ a & \cdots & y \end{pmatrix} \tag{20.7.5}$$

$$A_b = \begin{pmatrix} 0 & \cdots & a & \cdots & y \\ b & \cdots & c & \cdots & z \end{pmatrix},$$

then

(20.7.6) $$A_a A_b = A_c = \begin{pmatrix} 0 & \cdots & x & \cdots \\ c & \cdots & z & \cdots \end{pmatrix}.$$

From our definition of addition we have

(20.7.7) $$c = a + b, \quad y = x + a, \quad z = y + b, \quad z = x + c.$$

This gives

(20.7.8) $$(x + a) + b = x + (a + b),$$

the associative law of addition. Clearly, the identity is

$A_0 = \begin{pmatrix} 0 & \cdots & 1 & \cdots & x & \cdots \\ 0 & \cdots & 1 & \cdots & x & \cdots \end{pmatrix}$, and we have the laws

(20.7.9) $$x + 0 = 0 + x = x.$$

Moreover, if u is such that $A_u = \begin{pmatrix} 0 & \cdots & a & \cdots \\ u & \cdots & 0 & \cdots \end{pmatrix}$, then $a + u = 0$ and $u = -a$. Moreover, since A is an Abelian group, if $A_a A_b = A_b A_a = A_c$, then

(20.7.10) $$c = a + b = b + a,$$

and addition is commutative. Hence our addition is an Abelian group, and indeed if we put $a \rightleftarrows A_a$, addition in S is isomorphic to A.

In the same way, reasoning on the permutations of H_0, we may show that the nonzero elements of S form a group under multiplication. For zero we already have the rule $0 \cdot m = 0$, and we arbitrarily set $m0 = 0$ and $00 = 0$. Now let g be an arbitrary permutation of G. If $(0)g = b$, write $g_1 = gA_b^{-1}$. Then g_1 fixes 0 and is an element of H_0, say, $g_1 = M_m$. Hence

(20.7.11) $$g = M_m A_b.$$

Here if $(x)g = y$, then

(20.7.12) $$(x)g = y = xm + b, \quad m \neq 0.$$

The representation (20.7.11) is unique, since the identity is the only element common to A and H_0, and so the permutations of G must, by (20.7.12), be the linear substitutions in S.

(20.7.13) $$x \to xm + b, \quad m \neq 0.$$

We note the following relations:

$$M_m M_t = M_{mt},$$
$$M_m^{-1} A_1 M_m = A_m. \tag{20.7.14}$$

The first is immediate, the second follows because $M_m^{-1}A_1M_m$ displaces all letters and takes 0 into m, whence it must be A_m. It then follows that

$$M_t^{-1} A_m M_t = A_{mt}, \tag{20.7.15}$$

or in terms of the operations of S,

$$(xt^{-1} + m)t = x + mt. \tag{20.7.16}$$

Putting $x = yt$ in this relation, we get

$$(y + m)t = yt + mt, \tag{20.7.17}$$

the distributive law. Hence, in S, addition is a group, multiplication is a group for elements different from zero, and the right distributive law (20.7.17) holds. Hence S is a near-field, and the permutations of G are the linear substitutions in S.

$$x \to xm + b, \quad m \neq 0, \tag{20.7.18}$$

where, because of the laws of the near-field, if $\alpha: x \to xm_1 + b_1$ and $\beta: x \to xm_2 + b_2$, then

$$\alpha\beta: x \to x(m_1m_2) + b_1m_2 + b_2. \tag{20.7.19}$$

Conversely, suppose we are given a near-field S. The linear substitutions (20.7.18) of S will form a group G with composition as in (20.7.19). Let $r \neq s$ be two different elements of S. Then

$$g: x \to x(r - s) + s \tag{20.7.20}$$

is an element of G such that $(0)g = s$, $(1)g = r$. Hence G is doubly transitive. An element of G fixing two letters is conjugate to an element fixing 0 and 1. But if $x \to xm + b$ fixes 0 and 1, we find, in turn, $b = 0$, $m = 1$, and so this is the identity. Hence the identity is the only element of G fixing two letters. The substitution $x \to x + (c - b)$ displaces all letters and takes a given b into a given c. If $xm + b = x$ for $m \neq 0, 1$ always has a solution, then the only permutations displacing all letters are the additions $x \to x + t$, and of these only $x \to x + (c - b)$ takes a given b into a given c. Hence (3) is satisfied. But then $(xm + b)r = xr$ or $xs + t = xr$ always has a solution (which is clearly unique), and this is just the condition that S, regarded as a Veblen-Wedderburn system, is the coordinate system of a plane whose finite lines are $x = c$ and $y = xm + b$. This completes the proof of all parts of the theorem.

We have noted that condition (3) is automatically satisfied if G is a finite permutation group. Thus the determination of all finite doubly transitive groups in which only the identity fixes two letters is equivalent to the determination of all finite near-fields. This determination has been made by Zassenhaus [2]. We shall now discuss this.

Let K be a finite near-field. If the multiplication in K is an Abelian group, then K satisfies both distributive laws and is a finite field $GF(p^r)$. This possibility need not be further discussed, and we shall assume from here on that the multiplication in K is non-Abelian. In this case it is not possible that the left distributive law should hold, since by the Wedderburn theorem, K would then be a finite associative division ring and so a finite field $GF(p^r)$. We shall follow the notation of the preceding theorem. K is a finite near-field with n elements, and G is a doubly transitive group of degree n in which only the identity fixes two letters. G is the group of linear substitutions:

(20.7.21) $$g\colon x \to xm + b, \quad m \neq 0.$$

A is the Abelian normal subgroup of G consisting of the identity and the elements displacing all letters of G. $H_0 = M$ is the subgroup fixing the letter 0. A is the additive group of K, M the multiplicative group of K^*, then $n - 1$ elements of K different from 0.

Lemma 20.7.K1. *A is an elementary Abelian group and n is a prime power $n = p^r$. The elements $\neq 1$ of A are conjugate under M.*

A is an Abelian group, and by (20.7.14), $A_m = M_m^{-1}A_1M_m$. Thus all elements $\neq 1$ of A are conjugate under M, and so, since A must contain some element of prime order, say, p, then all elements $\neq 1$ of A are of order p, and so A is elementary Abelian. Since A is regular and of order n, $n = p^r$.

Lemma 20.7.K2. *The elements of M are automorphisms of A and each automorphism $\neq 1$ fixes only the element 0 of K.*

The elements of M are the permutations $a \to am$ of the elements of K, and each of these $\neq 1$ is an automorphism of the additive group A fixing only the element 0.

Lemma 20.7.K3. *If q and s are primes, a subgroup of M of order q^2 or qs is necessarily cyclic.*

We note first that if $(x_1, \cdots, x_k)$ is a cycle in the permutation form of M_m, then $x_1m = x_2$, $x_2m = x_3$, $\cdots$, $x_km = x_1$, whence $(x_1 + \cdots + x_k)m = x_1 + \cdots + x_k$, and so if $m \neq 1$, we have $x_1 + \cdots + x_k = 0$. Now suppose that M has a subgroup W of order q^2 which is not cyclic, and so is the direct product of two groups of order q. Let W be generated by elements x and y, and consider a transitive constituent of W which necessarily has q^2 letters; the elements will be of the form

$$x = (a_1a_2 \cdots a_q)(a_{q+1} \cdots a_{2q}) \cdots (a_{q^2-q+1} \cdots a_{q^2}),$$
$$y = (a_1, a_{q+1}, a_{2q+1}, \cdots, a_{q^2-q+1}) \cdots (a_i, a_{q+i}, \cdots a_{q^2-q+i}) \cdots.$$

Here xy^j will have as its transitive constituent, involving a_1, the following:

$$(a_1, a_{2+jq}, a_{3+2jq}, \cdots, a_{q+j(q-1)q}).$$

Hence, by our observation on cycles,

$$a_1 + a_{2+jq} + \cdots + a_{q+j(q-1)q} = 0. \tag{20.7.22}$$

$$a_1 + a_{q+1} + a_{2q+1} \cdots + a_{q^2-q} = 0. \tag{20.7.23}$$

Summing (20.7.22) over $j = 0, 1, \cdots, q - 1$, and adding (20.7.23), we have every $a_i \neq a_1$ exactly once and a_1, $q + 1$ times. Hence

$$qa_1 + (a_1 + a_2 \cdots + a_{q^2}) = 0. \tag{20.7.24}$$

But summing over all the cycles for x, we have

$$a_1 + a_2 \cdots + a_{q^2} = 0, \tag{20.7.25}$$

whence

$$qa_1 = 0. \tag{20.7.26}$$

But M is of order $n - 1 = p^r - 1$, and so q is prime to p, and (20.7.26) would yield the conflict $a_1 = 0$. Hence M cannot contain a noncyclic subgroup of order q^2.

Similarly, if M contains a noncyclic group W of order qs, where $q < s$, then W is generated by elements x, y with

$$\begin{aligned} x^s &= 1 \quad y^q = 1, \quad q \mid s - 1 \\ y^{-1}xy &= x^t, \quad t^q \equiv 1 \pmod{s}. \end{aligned} \tag{20.7.27}$$

Since M is a regular group, W will have a transitive constituent of

qs letters. W contains one subgroup of order s and s subgroups of order q. Consider the cycles containing a given letter a_1, one from each of these groups:

$$\begin{array}{l} (a_1a_2 \cdots a_s) \\ (a_1b_{11} \cdots b_{1,q-1}) \\ (a_1b_{21} \cdots b_{3q-1}) \\ \cdots \quad \cdots \quad \cdots \\ \cdots \quad \cdots \quad \cdots \\ (a_1b_{s1} \cdots b_{s,q-1}) \end{array}$$

Each of the qs letters except a_1 occurs exactly once in these cycles and a_1 occurs $s + 1$ times. But the sum of all qs letters, since these are all the cycles of, say, x taken together, is zero. Hence we conclude that $sa_1 = 0$, and since $s \mid p^r - 1$, we would have $a_1 = 0$, a conflict. Hence M cannot contain a noncyclic subgroup of order qs.

Lemma 20.7.K4. *A Sylow subgroup of M of odd order is cyclic. A Sylow 2-subgroup of M is cyclic or a generalized quaternion group.*

We have shown in Theorem 12.5.2 that a p-group P is cyclic if p is odd and P contains no noncyclic subgroup of order p^2. For $p = 2$, P is either cyclic or a generalized quaternion group.

Let us assume that M has a cyclic subgroup C such that M/C is also cyclic. From Lemma 20.7.K4 and Theorem 9.4.3, M will certainly have this property unless it has a Sylow 2-subgroup which is a generalized quaternion subgroup, and will even have this property if M is the direct product of a generalized quaternion subgroup and a group of odd order. Theorem 20.7.2 gives all finite near-fields K for which M has this property. Zassenhaus has shown that there are precisely seven other finite near-fields; we shall list these. For proof the reader is referred to the original paper by Zassenhaus.

Theorem 20.7.2. *Let $q = p^h$ be a power of a prime p and let v be an integer all of whose prime factors divide $q - 1$, where we also require $v \not\equiv 0$ (mod 4) if $q \equiv 3$ (mod 4). Then with $hv = r$ we may construct a near-field K with $n = p^r$ elements from the finite field $GF(p^r)$ in the following way:*

1) *The elements of K shall be the same as the elements of $GF(p^r)$*
2) *Addition in K shall be the same as addition in $GF(p^r)$.*
3) *A product $w \circ u$ in K can be defined in terms of a product $x \cdot y$ in $GF(p^r)$, in the following way:*

Let z be a fixed primitive root of $GF(p^r)$; then if $u = z^{kv+j}$, an integer i is uniquely determined modulo v by $q^i \equiv 1 + j(q - 1)$ [mod $v(q - 1)$]. We define the product $w \circ u$ by the rule

$$w \circ u = u \cdot w^{q^i}.$$

4) *The center of K is $GF(q)$. Every near-field K with $n = p^r$ elements can be constructed in the above way from $GF(p^r)$, if K has the property that its multiplicative group M has a normal cyclic subgroup C such that M/C is cyclic.*

Apart from the finite near-fields of the preceding theorem, Zassenhaus shows that there are exactly seven others. In these cases the near-fields K are of order p^2, and it is sufficient to give generators of the multiplicative group M as matrix transformations of two generators of the additive group. We give the same numbering as Zassenhaus.

I. $n = 5^2$, $A = \begin{pmatrix} 0, & -1, \\ 1, & 0 \end{pmatrix}$ $B = \begin{pmatrix} 1, & -2 \\ -1, & -2 \end{pmatrix}$.

$$M \cong M(2, 3)$$

II. $n = 11^2$, $A = \begin{pmatrix} 0, & -1 \\ 1, & 0 \end{pmatrix}$, $B = \begin{pmatrix} 1, & 5 \\ -5, & -2 \end{pmatrix}$, $C = \begin{pmatrix} 4, & 0 \\ 0, & 4 \end{pmatrix}$.

$$M \cong M(2, 3) \times (C)$$

III. $n = 7^2$, $A = \begin{pmatrix} 0, & -1 \\ 1, & 0 \end{pmatrix}$, $B = \begin{pmatrix} 1, & 3 \\ -1, & -2 \end{pmatrix}$.

$$M \cong G_3.$$

IV. $n = 23^2$, $A = \begin{pmatrix} 0, & -1 \\ 1, & 0 \end{pmatrix}$, $B = \begin{pmatrix} 1, & -6 \\ 12, & -2 \end{pmatrix}$, $C = \begin{pmatrix} 2, & 0 \\ 0, & 2 \end{pmatrix}$.

$$M \cong G_3 \times (C)$$

V. $n = 11^2$, $A = \begin{pmatrix} 0, & -1 \\ 1, & 0 \end{pmatrix}$, $B = \begin{pmatrix} 2, & 4 \\ 1, & -3 \end{pmatrix}$.

$$M \cong M(2, 5)$$

VI. $n = 29^2$, $A = \begin{pmatrix} 0, & -1 \\ 1, & 0 \end{pmatrix}$, $B = \begin{pmatrix} 1, & -7 \\ -12, & -2 \end{pmatrix}$, $C = \begin{pmatrix} 16, & 0 \\ 0, & 16 \end{pmatrix}$.

$$M \cong M(2, 5) \times (C)$$

VII. $n = 59^2$, $A = \begin{pmatrix} 0, & -1 \\ 1, & 0 \end{pmatrix}$, $B = \begin{pmatrix} 9, & 15 \\ -10, & -10 \end{pmatrix}$, $C = \begin{pmatrix} 4, & 0 \\ 0, & 4 \end{pmatrix}$.

$$M \cong M(2, 5) \times (C).$$

Here $M(2, 3)$ is of order 24, as given by I; $M(2, 5)$ is of order 120, as given by V; and G_3 is of order 48, as given by III. $M(2, 5)$ has a center Z of order 2, and the factor group $M(2, 5)/Z$ is the simple group of order 60.

20.8. Finite Planes. The Bruck-Ryser Theorem.

In a finite plane of order n we have shown that the following properties hold:

1) There are $n^2 + n + 1$ lines.
2) There are $n^2 + n + 1$ points.
3) Each line contains $n + 1$ points.
4) Each point is on $n + 1$ lines.
5) There is one and only one line through two distinct points.
6) Two distinct lines intersect in one and only one point.

In verifying that a system is indeed a finite plane, it is convenient to know that a system of "points" and "lines" satisfying part of the above properties is indeed a finite plane of order n and satisfies the rest.

THEOREM 20.8.1. *A system of points and lines satisfying* (1), (3), (5), *or dually* (2), *and* (4), (6) *is a finite plane of order n and satisfies the remaining properties.*

Proof: Suppose a system satisfies (1), (3), (5). Let a point P_i be on m_i lines. Then P_i is joined to n other points on each of the m_i lines. But these must be all the remaining points, each counted exactly once. Hence there are in all $1 + nm_i$ points. Hence $m = m_i$ is the same for every point. Counting incidences of points on lines, we have

$$(n + 1)(n^2 + n + 1) = m(1 + mn),$$

since there are $n^2 + n + 1$ lines, each containing $n + 1$ points, and there are $1 + mn$ points each on m lines. This gives $m = n + 1$ and (2) and (4) follow. From (5) we could not have two distinct lines intersecting in more than one point. To show (6) we need only show that there is one point in which two distinct lines intersect. A point P of a given line L is on n other lines, and this is true of each of the $n + 1$ points of L. Thus L has an intersection with $n(n + 1) =$

$n^2 + n$ other lines, but these are all the remaining lines, and so (6) holds. This shows that (1), (3), (5) imply the remainder. A dual argument shows that (2), (4), (6) imply the remainder.

COROLLARY 20.8.1. *In a finite Veblen-Wedderburn system the condition* (4) *of Theorem* 20.4.6, *i.e., that if* $r \neq s$, *then* $xr = xs + t$ *has a unique solution, is a consequence of the remainder.*

For without using condition (4), we see that properties (1), (3), (5) are satisfied.

There do not exist finite planes of every integral order n. If a conjecture of Euler's is true, then there are no planes of order n where $n \equiv 2 \pmod 4$, $n \neq 2$. It was shown by Tarry [1] in 1900 by a method of trial and error that there is no plane of order 6. For every prime power $n = p^r$ there is a field $GF(p^r)$, and so, by Theorem 20.5.5, a Desarguesian plane of order p^r. There are Hall systems of orders p^{2r}, and except for order 4, these yield non-Desarguesian planes. Also the near-fields yield non-Desarguesian planes. Albert [1] has given a construction for nonassociative division rings of orders p^r, for p an odd prime, and for $r > 2$. This, of course, yields non-Desarguesian planes of these orders, by Theorem 20.4.6. We give a simple construction due to Albert for powers p^r, p odd, r odd and $r > 1$.

THEOREM 20.8.2 (ALBERT). *Let* p *be an odd prime and* r *odd,* $r > 1$. *Then from* $GF(p^r)$ *we may construct a nonassociative division algebra* N *with* p^r *elements.*

Proof: We first construct a new product (x, y) for elements of $GF(p^r)$, where p is an odd prime, $r > 1$, r odd by the rule:

$$(x, y) = \tfrac{1}{2}(xy^p + x^py). \tag{20.8.1}$$

Since $x \to x^p$ is an automorphism of $GF(p^r)$, we easily verify that the product (x, y) satisfies the distributive laws. We wish to show that if $x \neq 0$, $y \neq 0$, then $(x, y) \neq 0$. Suppose to the contrary that $x \neq 0$, $y \neq 0$ but $(x, y) = 0$. Then we find

$$xy^p = -x^py, \tag{20.8.2}$$

whence

$$y^{p-1} = -x^{p-1}. \tag{20.8.3}$$

Since r and p are both odd, $m = (p^r - 1)/(p - 1)$ is odd. Raising (20.8.3) to the mth power, we find

$$1 = y^{p^r-1} = -x^{p^r-1} = -1, \tag{20.8.4}$$

a conflict, since $p \neq 2$. Hence if $x \neq 0$, $y \neq 0$, then $(x, y) \neq 0$. Since our system is finite and has no zero divisors, it must be a quasi-group and, given $x \neq 0$, there is a unique $u \neq 0$ such that

$$x = (u, 1) = \tfrac{1}{2}(u + u^p). \tag{20.8.5}$$

Hence let us define a one-to-one mapping α by

$$u = x\alpha, \tag{20.8.6}$$

if u and x satisfy (20.8.5).

We now define a system D whose elements are those of $GF(p^r)$. Addition in D is the same as that in $GF(p^r)$, but in D we have a product $x \circ y$ given by

$$x \circ y = (x\alpha, y\alpha), \tag{20.8.7}$$

using the product of (20.8.1) and the mapping α of (20.8.6). The unit 1 of $GF(p^r)$ is the unit of D, since we verify that

$$x \circ 1 = 1 \circ x = x. \tag{20.8.8}$$

The product in D is commutative but not associative. Albert has shown that the powers of an element not in F_p will not associate.

The methods given here yield non-Desarguesian planes of orders p^r for all $r \geq 2$ and p odd, and of orders 2^r when r is even and $r \geq 4$. It has been shown that only the Desarguesian plane exists for orders 2, 3, 4, 5, 7, 8. Other finite planes are known, in particular the Hughes planes which will be discussed later, but no finite plane has been constructed except for a prime or prime power order.

Apart from Tarry's isolated result that no plane of order 6 exists, no restrictions on possible orders of planes were known until 1949, when the following major theorem was proved by Bruck and Ryser [1].

THEOREM 20.8.3. *If $n \equiv 1, 2 \pmod 4$, there cannot be a plane of order n unless n can be expressed as a sum of two integral squares, $n = a^2 + b^2$.*

Proof: The proof given here will be a modification of the original proof by the methods used by Chowla and Ryser [1]. Let $N = n^2 +$

$n + 1$. Let variables x_i, $i = 1, \cdots, N$ be associated with the points P_i $i = 1, \cdots, N$ of a plane π of order n. Let the lines of π be L_j, $j = 1, \cdots, N$. We may define incidence numbers a_{ij}, where

$$(20.8.9) \qquad \begin{aligned} a_{ij} &= 1 \quad \text{if } P_i \in L_j, \\ a_{ij} &= 0 \quad \text{if } P_i \notin L_j,\ i, j = 1, \cdots, N. \end{aligned}$$

Then the incidence matrix A of π is defined as

$$(20.8.10) \qquad A = (a_{ij})\ i, j = 1, \cdots N.$$

This incidence matrix A satisfies the basic relations:

$$(20.8.11) \qquad AA^T = A^TA = nI + S,$$

where I is the identity matrix and S is the matrix with every entry 1.

Let $AA^T = C$. Then $C = (c_{rs})$, where

$$(20.8.12) \qquad c_{rs} = \sum_{j=1}^{N} a_{rj}a_{sj}.$$

Here $c_{rr} = n + 1$, since P_r is on exactly $n + 1$ lines, since of a_{rj}, $j = 1, \cdots, N$ exactly $n + 1$ are 1 and the rest are 0. Also, if $r \neq s$, we have $c_{rs} = 1$, since $a_{rj}a_{sj} = 0$ unless both $a_{rj} = 1$, $a_{sj} = 1$. But $a_{rj} = a_{sj} = 1$ means that the line L_j contains both P_r and P_s. But given P_r and P_s there is exactly one line L_j containing both points. Hence $c_{rr} = n + 1$, $c_{rs} = 1$, $r \neq s$, and so $AA^T = nI + S$. A dual argument shows that $A^TA = nI + S$.

The relation $AA^T = nI + S$ can also be expressed in terms of quadratic forms. With the line L_j we may associate a linear form, which may also be designated by L_j without confusion. We write

$$(20.8.13) \qquad L_j = \sum_{i=1}^{N} a_{ij}x_i,\ j = 1, \cdots, N.$$

Then

(20.8.14)

$$L_1^2 + \cdots + L^2_N = n(x_1^2 + \cdots + x_N^2) + (x_1 + \cdots + x_N)^2.$$

For this we observe that in the L_j, $j = 1, \cdots N$, each x_r occurs with a coefficient 1 exactly $n + 1$ times, since each point is on $n + 1$ lines. We also observe that a cross-product $2x_rx_s$ occurs in $L_1^2 + \cdots + L_N^2$ exactly once, since there is exactly one line L_j containing both P_r and P_s. This proves the identity (20.8.14). Now suppose that $n \equiv 1, 2 \pmod 4$. Then $N = n^2 + n + 1 \equiv 3 \pmod 4$. We also observe that (20.8.14) can be written in the form:

(20.8.15)

$$L_1^2 + \cdots + L_N^2 = n\left(x_2 + \frac{x_1}{n}\right)^2 + \cdots + n\left(x_N + \frac{x_1}{n}\right)^2 + (x_2 + \cdots + x_N)^2.$$

This is easily checked, observing that the coefficient of x_1^2 on the right of (20.8.15) is $(N - 1)/n = n + 1$. Let us change variables in (20.8.15), writing

(20.8.16) $$y_1 = x_2 + \cdots + x_N, \quad y_2 = x_2 + \frac{x_1}{n}, \cdots, y_N = x_N + \frac{x_1}{n}.$$

The x's may be expressed rationally in terms of the y's. We rewrite (20.8.15) as

(20.8.17) $$L_1^2 + \cdots + L_N^2 = y_1^2 + n\, y_2^2 + n\, y_3^2 + \cdots + n\, y_N^2.$$

We now appeal to the Theorem of Lagrange that every positive integer can be written as a sum of four squares. For this see Hardy and Wright's "Theory of Numbers' " [1], p. 300; we have

(20.8.18) $$n = a_1^2 + a_2^2 + a_3^2 + a_4^2,$$

and also the celebrated identity of Lagrange,

(20.8.19)
$$\begin{aligned}
&(a_1^2 + a_2^2 + a_3^2 + a_4^2)(y_i^2 + y_{i+1}^2 + y_{i+2}^2 + y_{i+3}^2) \\
&= (a_1y_i + a_2y_{i+1} + a_3y_{i+2} + a_4y_{i+3})^2 \\
&\quad + (a_1y_{i+1} - a_2y_i + a_3y_{i+3} - a_4y_{i+2})^2 \\
&\quad + (a_1y_{i+2} - a_3y_i + a_4y_{i+1} - a_2y_{i+3})^2 \\
&\quad + (a_1y_{i+3} - a_4y_i + a_2y_{i+2} - a_3y_{i+1})^2 \\
&= z_i^2 + z_{i+1}^2 + z_{i+2}^2 + z_{i+3}^2.
\end{aligned}$$

In (20.8.19) we find that $z_i, z_{i+1}, z_{i+2}, z_{i+3}$ can be expressed rationally in terms of $y_i, y_{i+1}, y_{i+2}, y_{i+3}$. Remembering that $N \equiv 3 \pmod 4$, we may apply (20.8.18) and (20.8.19) to (20.8.17) to obtain

(20.8.20)

$$L_1^2 + \cdots + L_N^2 = z_1^2 + z_2^2 + \cdots + z_{N-2}^2 + n(z_{N-1}^2 + z_N^2).$$

We note that each of L_j, $j = 1, \cdots N$ was originally a rational (in fact an integral) linear form in the x's, and so in turn, a rational linear form in the y's and finally in the z's, where $z_1, \cdots, z_N$ are independent indeterminates. Since (20.8.20) is an identity in the z's, it will remain a valid identity if some of the z's are specialized as linear combinations of the rest. Suppose in (20.8.20) that

$$L_1 = b_1 z_1 + \cdots + b_N z_N. \tag{20.8.21}$$

Let us put $L_1 = z_1$ if $b_1 \neq 1$ and $L_1 = -z_1$ if $b_1 = 1$. One of these may be used to specialize z_1 as a rational linear combination of $z_2, \cdots, z_N$ and also to give $L_1^2 = z_1^2$, whence with this specialization.

(20.8.22)
$$L_2^2 + \cdots + L_N^2 = z_2^2 + \cdots + z_{N-2}^2 + n(z_{N-1}^2 + z_N^2).$$

Continuing, we put $L_2 = \pm z_2, \cdots L_{N-2} = \pm z_{N-2}$ to specialize $z_2, \cdots, z_{N-2}$, and we have

$$L_{N-1}^2 + L_N^2 = n(z_{N-1}^2 + z_N^2), \tag{20.8.23}$$

where L_{N-1} and L_N are rational linear forms in z_{N-1} and z_N which are independent variables. We may take z_{N-1} and z_N as positive integers which are multiples of the denominators in L_{N-1} and L_N, whence (20.8.23) becomes a relation in which all quantities are integers. But now n, an integer which is the quotient of two integers each of which is a sum of two squares, is itself a sum of two squares, a well-known result in number theory which follows from Theorem 366 in Hardy and Wright [1]. We now have

$$n = a^2 + b^2, \tag{20.8.24}$$

and our theorem is proved. There is a partial converse to our theorem:

THEOREM 20.8.4. *If $n \equiv 0, 3 \pmod 4$ or if $n \equiv 1, 2 \pmod 4$ and $n = a^2 + b^2$, then there exist rational linear forms in $x_1, \cdots, x_N$, L_j, $j = 1, \cdots N$ satisfying:*
$$L_1^2 + \cdots + L_N^2 = n(x_1^2 + \cdots + x_N^2) + (x_1 + \cdots + x_N)^2.$$
There also exists a rational $N \times N$ matrix A satisfying $A A^T = A^T A = n I + S$.

Proof: If $n \equiv 0, 3 \pmod 4$, we may use (20.8.18) and (20.8.19) to put (20.8.17) into the form:

$$L_1^2 + \cdots + L_N^2 = z_1^2 + \cdots + z_N^2, \tag{20.8.25}$$

and, of course, $L_i = z_i$ satisfies the theorem. If $n \equiv 1, 2 \pmod 4$ and $n = a^2 + b^2$, we may use the identity:

(20.8.26)
$$(a^2 + b^2)(y_i^2 + y_{i+1}^2) = (ay_i + by_{i+1})^2 + (by_i - ay_{i+1})^2$$
$$= z_i^2 + z_{i+1}^2,$$

instead of (20.8.19) to put (20.8.17) into the form (20.8.25). If

(20.8.27) $$L_j = \sum_i b_{ij}x_i, \quad j = 1, \cdots N$$

are the linear forms of our theorem, then putting $A = (b_{ij}), i, j = 1, \cdots N$, we have

(20.8.28) $$AA^T = nI + S,$$

but not in general $A^TA = nI + S$. It is more difficult to show that under the hypotheses of the theorem a rational matrix A exists which satisfies both relations $AA^T = nI + S = A^TA$. But this and even more has been shown by Hall and Ryser [1].

20.9. Collineations in Finite Planes.

If a collineation α of a plane π fixes two points, then it also fixes the line joining them, and similarly, if α fixes two lines, then it also fixes their intersection. Hence if α fixes a quadrilateral, then α fixes a proper subplane. The following theorem gives information on the possible orders of subplanes.

Theorem 20.9.1 (Bruck). *If a plane π of order n has a subplane π^* of order m, then $n = m^2$ or $n \geq m^2 + m$.*

Proof: Let L be a line of the subplane π^* and P a point of L not belonging to π^*. There are $m + 1$ points of π^* on L and m^2 points of π^* not on L. If we join P to each of the m^2 points of π^* not on L, we obtain m^2 lines through P which must all be different, since if two were the same such a line K would contain two distinct points of π^* and so be a line of π^*, whence P as the intersection of K and L would be a point of π^*, contrary to assumption. Hence through P there are at least $m^2 + 1$ lines, namely, L and the m^2 others joining P to points of π^*. Hence, since there are $n + 1$ lines through P, we must have $n \geq m^2$. If $n \neq m^2$ there will be a further line L_1 through P

not passing through any point of π^*. Now consider the intersections of L_1 with the $m^2 + m + 1$ lines of π^*. If any two of these intersections were the same point, such a point would be a point of π^*, contrary to hypothesis. Hence L_1 contains at least $m^2 + m + 1$ points, and so $n \geq m^2 + m$.

With a little care we may list the subsets S of a plane which with any two points contains the line joining them and which with any two lines contains their intersection.

First, if S contains four points, no three on a line, then S is a subplane. The remaining possible sets we call *degenerate subplanes*. These are:

1) The void set.
2) A single point P and possibly one or more lines through P.
3) A single line L and possibly one or more points on L.
4) A single point P and a single line L not passing through P.
5) The vertices and sides of a triangle.
6) A line L and a point P on L and also one or more points on L and one or more lines through P.
7) A line L containing three or more points, a single-point P not on L, and the lines joining P to the points of L.

The collineation α is a permutation of the points and also a permutation of the lines of π. Let P be the permutation of the points and Q the permutation of the lines, where we write both P and Q as $N \times N$ matrices. Here as usual $N = n^2 + n + 1$.

(20.9.1)
$$P = (p_{ij}),$$
$$Q = (q_{ij}),$$

where
$$p_{ij} = 1 \quad \text{if } P_i\alpha = P_j,$$
$$q_{ij} = 1 \quad \text{if } Q_i\alpha = Q_j,$$

and otherwise, $p_{ij} = 0$, $q_{ij} = 0$.

Then we have

(20.9.2)
$$P^{-1}AQ = A,$$

where $A = (a_{ij})$ is the incidence matrix for π. Conversely, if permutation matrices P and Q exist, satisfying (20.9.2), they determine a collineation of π.

THEOREM 20.9.2 (PARKER [1]). *The permutations P of points and Q of lines in a collineation are similar as permutations.*

Corollary 20.9.1 (Baer). *A collineation fixes the same number of points and lines.*

Proof: We note that since

$$AA^T = A^TA = nI + S, \tag{20.9.3}$$

then

$$(\text{det. } A)^2 = \text{det. } (nI + S) = (n+1)^2 n^{N-1}, \tag{20.9.4}$$

whence in particular A is nonsingular. Thus (20.9.2) becomes

$$Q = A^{-1}PA, \tag{20.9.5}$$

and so P and Q are similar as matrices. Here P and Q have the same irreducible constituents, regarded as representations of a cyclic group. But, reducing a cycle $(x_1, \cdots, x_r)$ of length r in any permutation, we find that these constituents have the characters $1, \zeta, \zeta^2, \cdots, \zeta^{r-1}$, where ζ is a primitive rth root of unity. But this says that an mth root of unity is a character of a permutation P with a multiplicity a_m, where a_m is the number of cycles of P whose length is a multiple of m. Since these multiplicities a_m are the same for both P and Q, it follows that P and Q have the same number of cycles of each length m. Hence P and Q are similar as permutations. In particular we have the corollary, which asserts that P and Q have the same number of cycles of length one, i.e., fixed elements.

Theorem 20.9.3 (Parker). *A group of collineations G of π has the same number of transitive constituents as a permutation group on the points as it has a permutation group on lines.*

Proof: Let G be of order g. Then from (20.9.5) we have G represented as a permutation group G_1 on points and G_2 on lines, and these representations are equivalent. Let χ_1, χ_2 be the respective characters:

$$\sum_{x \in G} \chi_1(x) = \sum_{x \in G} \chi_2(x). \tag{20.9.6}$$

But by Theorem 16.6.13,

$$\sum_{x \in G} \chi_1(x) = k_1 g, \quad \sum_{x \in G} \chi_2(x) = k_2 g, \tag{20.9.7}$$

where k_1 is the number of transitive constituents of G_1, and k_2 is the number of transitive constituents of G_2. Hence $k_1 = k_2$, the assertion of the theorem. Although from the preceding theorem each individual permutation of G_1 is similar as a permutation to the corresponding element of G_2, it is not in general true that G_1 and G_2 are similar as permutation groups. For example, in a Desarguesian plane the group of all collineations fixing a point P_0 contains no line fixed by all its collineations.

THEOREM 20.9.4. *A Desarguesian plane π of order $n = p^r$ has a collineation group of order $r(n^2 + n + 1)(n^2 + n)n^2(n - 1)^2$.*

Proof: In π the number of ordered quadrilaterals P_1, P_2, P_3, P_4 is $(n^2 + n + 1)(n^2 + n)n^2(n - 1)^2$, since we may choose P_1 as any of the $n^2 + n + 1$ points, P_2 as any other point, P_3 as any of the n^2 points not on P_1P_2, and P_4 as any of the $(n - 1)^2$ points not on any one of P_1P_2, P_1P_3, or P_2P_3. By Theorem 20.5.5 the collineation group G of π is transitive on quadrilaterals. The subgroup of G fixing the quadrilateral X, Y, O, I is the group of automorphisms of the coordinatizing field $GF(p^r)$, and this is of order r, as was noted in §20.6.

THEOREM 20.9.5 (SINGER [1]). *A Desarguesian plane π of order n has a collineation α of order $N = n^2 + n + 1$ which is cyclic on the points and also on the lines.*

Proof: Let $n = p^r$. Then π is coordinatized by $GF(p^r) = F$. It is convenient to represent π in terms of homogeneous coordinates. A point P will be given as

$$P = (\lambda x_1, \lambda x_2, \lambda x_3), \tag{20.9.8}$$

where x_1, x_2, x_3 are fixed elements of F not all zero, and λ ranges over all elements of F except 0. Similarly, a line L will be given as

$$L = [u_1\mu, u_2\mu, u_3\mu], \tag{20.9.9}$$

where u_1, u_2, u_3 are fixed elements not all zero, and μ ranges over all elements of F except 0. $P \in L$ if, and only if,

$$x_1u_1 + x_2u_2 + x_3u_3 = 0. \tag{20.9.10}$$

Because F is a field, we see that the incidence relation (20.9.10) is the same for any choice of λ in (20.9.8) and any choice of μ in (20.9.9). The homogeneous coordinates may be identified with the nonhomogeneous coordinates in the following way:

$$
\begin{aligned}
(\infty) &= (0, \lambda, 0),\\
(m) &= (\lambda, \lambda m, 0),\\
(a, b) &= (\lambda a, \lambda b, \lambda),\\
L_\infty &= [0, 0, \mu],\\
(x = c) &= [\mu, 0, -c\mu],\\
(y = xm + b) &= [m\mu, -\mu, b\mu].
\end{aligned}
\tag{20.9.11}
$$

We easily check that the homogeneous representation of π agrees with the nonhomogeneous representation. The field $GF(p^{3r}) = F_1$ may be considered as a cubic extension of $F = GF(p^r)$, and if w is a primitive root of F_1, every element x of F_1 has a unique expression

$$x = x_1 + x_2 w + x_3 w^2, \quad x_i \in F. \tag{20.9.12}$$

Hence if $x \neq 0$, $\lambda \in F$, $\lambda \neq 0$ elements λx of F_1 correspond to the point $(\lambda x_1, \lambda x_2, \lambda x_3)$ of π. But in F_1 the order of w is $p^{3r} - 1 = n^3 - 1$. The elements of F are the solutions in F_1 of the equation

$$x^{p^r} = x, \tag{20.9.13}$$

whence for $x \in F$, $x \neq 0$, we have, since $n = p^r$,

$$x^{n-1} = 1. \tag{20.9.14}$$

Thus F^* (the elements of F excluding 0) are the elements of the unique subgroup of order $n - 1$ of the cyclic group $\{w\}$ of order $n^3 - 1$. Hence the elements of F^* are the elements

$$w^{Ni}, \quad N = n^2 + n + 1. \tag{20.9.15}$$

Hence w^u and w^v represent the same point of π if, and only if,

$$u \equiv v \pmod{N}. \tag{20.9.16}$$

Hence the mapping α of elements $x \in F_1$,

$$x \to xw, \tag{20.9.17}$$

is a permutation of the points of π in a cycle of length N. If $P_1 = (x_1, x_2, x_3)$ and $P_2 = (y_1, y_2, y_3)$ are two distinct points, then we readily verify that the points of the line P_1P_2 are given by

$$
\begin{aligned}
&\lambda_1(x_1, x_2, x_3) + \lambda_2(y_1, y_2, y_3)\\
&\qquad = (\lambda_1 x_1 + \lambda_2 y_1, \lambda_1 x_2 + \lambda_2 y_2, \lambda_1 x_3 + \lambda_2 y_3),
\end{aligned}
\tag{20.9.18}
$$

where λ_1, λ_2 are any elements of F not both zero. Hence if $w^i = x_1 +$

$x_2w + x_3w^2$, $w^j = y_1 + y_2w + y_3w^2$, then the points of P_1P_2 are given by

$$\lambda_1 w^i + \lambda_2 w^j. \tag{20.9.19}$$

Hence the mapping $x \to xw$ takes the elements of (20.9.19) into

$$\lambda_1 w^{i+1} + \lambda_2 w^{j+1}, \tag{20.9.20}$$

and these are the points of the line joining $P_1\alpha$ and $P_2\alpha$. Hence α is a collineation of π and is a cyclic of length N on the points. It is easy to see (for example, by Theorem 20.9.2) that α is also a cycle of length N on the lines of π.

We can give a crude upper bound on the order of the group G of collineations of a plane π of order n. An ordered quadrilateral P_1, P_2, P_3, P_4 has at most $M = (n^2 + n + 1)\,(n^2 + n)n^2(n - 1)^2$ images. The subgroups H_1 of index $\leq M$ fixing P_1, P_2, P_3, P_4 fixes the subplane π_1 generated by these points. If π_1 is of order m_1 then H_1 permutes the $n - m_1$ points on a line of π_1 which are not points of π_1. The subgroup H_2 of index $\leq n - m_1$ in H_1 fixing one of these points fixes a larger subplane π_2 of order m_2, where from Theorem 20.9.2, $m_2 \geq m_1^2$. We thus have a descending sequence of subgroups $H_1 \supset H_2 \supset \cdots \supset H_s = 1$, in which H_i fixes a subplane of order m_i, where $m_{i+1} \geq m_i^2$, and $[H_i{:}H_{i+1}] < n$. Thus $s \leq \log_2 n$, and the order of G is at most $n^s M$. The collineation groups of the known non-Desarguesian planes are not as large as the groups for the Desarguesian planes of the same order, and it seems likely that this is always the case.

The following two theorems assert that if in certain specified ways the collineation group of a finite plane is large enough, then the plane is Desarguesian.

THEOREM 20.9.6 (GLEASON [1]). *If for every pair P, L with P a point on a line L of a finite plane π, the elation group $G(P, L)$ is nontrivial, then π is Desarguesian.*

Proof: By Theorem 20.4.3 if two elation groups $G(P_1, L)$ and $G(P_2, L)$ with P_1 and P_2 different points of L are nontrivial, i.e., different from the identity, then all elations with axis L form an Abelian group in which every element $\neq 1$ is of the same prime order p. By the dual of this theorem, if $G(P, L_1)$ and $G(P, L_2)$ are nontrivial, with L_1 and L_2 different lines through P, then all elations with center P

form an Abelian groups whose elements $\neq 1$ are of the same prime order p. Hence, under the hypothesis of our present theorem, every elation group $G(P, L)$ is elementary Abelian of order p or a power of p.

LEMMA 20.9.1. *Suppose that H is a group of permutations of a finite set S, and suppose that for some prime p and each $x \in S$ there exists an element of H of order p which fixes x but no other element of S. Then H is transitive.*

Proof: Consider S_1 a transitive set of S under H. For $x \in S_1$ there exists an element of order p fixing x and displacing all remaining letters in cycles of length p. Hence the number of elements in S_1 is congruent to 1 (mod p), and the number of elements in another transitive set S_2 (if there is another) is a multiple of p. But then, taking a $y \in S_2$ by the same argument, the number of elements in S_1 is a multiple of p. This is a conflict, and so there is only one transitive set and H is transitive on S.

LEMMA 20.9.2. *Suppose for a line L of a finite plane the elation groups $G(P_i, L)$ for all $P_i \in L$ have the same order $h > 1$. Then π is a translation plane with respect to L.*

Proof: Let π be of order n. Any two of the $n + 1$ groups $G(P_i, L)$, each of order h, have only the identity in common, and together their elements form the translation group $T(L)$. Hence the order of $T(L)$ is $t = (n + 1)(h - 1) + 1$. Since only the identity of $T(L)$ can fix a point not on L, $T(L)$ permutes the n^2 points in sets of t points, whence t divides n^2. Write

$$n^2 = tm = [(n + 1)(h - 1) + 1]m. \tag{20.9.21}$$

Since $h > 1$, we have $m < n$. On the other hand, taking (20.9.21) modulo $n + 1$, we have

$$n^2 \equiv 1 \equiv m \pmod{n + 1}. \tag{20.9.22}$$

But $m \equiv 1 \pmod{n + 1}$ and $m < n$ together yield $m = 1$, $t = n^2$, whence $T(L)$ is transitive on the n^2 points of π not on L, and so π is a translation plane with respect to L.

We can now prove our theorem. Take a fixed line L of π. For each point $P \in L$ the elation group $G(P, M)$, where $M \neq L$ is another line through P, contains an element of order p fixing P and mapping L onto itself but displacing all other points of L. Hence, by Lemma

20.9.1, the group $G(L)$ of all collineations fixing L is transitive on the points of L. It then follows that for the $n + 1$ points P_i of L, all elation groups $G(P_i, L)$, being conjugate under $G(L)$, have the same order h. By Lemma 20.9.2 it follows that π is a translation plane with respect to L. But L can be taken as any line of π. Thus π is a translation plane for every line L, and by Theorem 20.5.3, π can be coordinatized by an alternative division ring. By Theorem 20.6.2 a finite alternative division ring is a field, and so π is Desarguesian.

Gleason [1] uses this theorem in the study of finite Fano planes. The Fano configuration is the configuration of the seven points and seven lines making the finite plane of order 2. A plane is a Fano plane if the diagonal points of every quadrilateral are on a line, or, what is the same thing, if every quadrilateral generates a Fano configuration. Gleason shows that every finite Fano plane is Desarguesian and that these are the finite planes over the fields $GF(2^r)$. There is not space here to prove this very interesting result.

We shall call a collineation of order 2 an *involution*.

THEOREM 20.9.7 (BAER). *Let α be an involution in a projective plane of order n. Then either* (1) $n = m^2$ *and the fixed points and lines of α form a subplane of order m, or* (2) *α is a central collineation. In case* (2) *if n is odd, α is a homology, and if n is even, α is an elation.*

Proof: We show first that every point is on a fixed line. If P is not a fixed point then $P\alpha \neq P$ and α fixes the line $PP\alpha$, which is therefore a fixed line through P. If P is a fixed point, join P to Q, another point. It may be that $L = PQ$ is a fixed line. If not, $Q\alpha \neq Q$ and $Q\alpha \notin PQ$. Here $L\alpha = PQ\alpha$. Then if R is a third point on L, $R\alpha \in L\alpha$. Then α interchanges the lines $Q\alpha R$ and $QR\alpha$ whose intersection S is another fixed point different from P. In this case PS is a fixed line through P. By a dual argument every line passes through a fixed point.

The line joining a pair of fixed points is a fixed line and the intersection of two fixed lines is a fixed point. Hence if there exists four fixed points, no three on a line, the fixed elements of α form a proper subplane π_1 of π. Let us suppose this to be the case and suppose that π_1 is of order m. Then by Theorem 20.9.1, $n \geq m^2$, and following the proof of this theorem, we see that if $n > m^2$, there is a line of π which does not pass through any point of π_1. But we have shown that every line of π contains a fixed point. Hence we cannot have $n > m^2$, and so $n = m^2$. This proves alternative (1) of the theorem.

Now suppose that there are not four fixed points, no three on a line. What is the configuration of the fixed points? We show first that there is a line containing three fixed points. A line L_1 contains a fixed point P_1. Choose a line L_2 not through P_1. Then L_2 contains a fixed point $P_2 \neq P_1$. We now have two fixed points P_1, P_2, and the line L joining them is a fixed line. Choose a third point Q on L. If Q is fixed, L is the line we seek. If Q is not a fixed point, a line L_3 through Q contains a fixed point P_3 not on L. We now have a triangle P_1, P_2, P_3 of fixed points. Consider a line L_4 not through any one of P_1, P_2, P_3. L_4 contains a fixed point P_4. If P_4 is not on one of the lines P_1P_2, P_1P_3, or P_2P_3, then P_1, P_2, P_3, P_4 are four fixed points, no three on a line, and this is the situation covered in the first alternative. Hence P_4 is on one of these lines, and we have a line containing three fixed points.

We now have a line L containing three fixed points $P_1P_2P_3$. If there were as many as two fixed points not on L, we would have a quadrilateral of fixed points, the situation of the first alternative. Hence there is either one fixed point P not on L or none. Consider now any point $P_i \in L$. There is a line K through P_i different from L, and, if there is a fixed point P not on L, different from PP_i. K contains a fixed point but, by our choice, no fixed point not on L. Hence the fixed point on K is P_i, whence it follows that every point P_i of L is a fixed point. Since α fixes every point of L, α is a central collineation with axis L, the assertion of our second alternative. There are n^2 points of π not on L, and α is of order 2. Hence if n is odd, α fixes a point not on L and is a homology. If n is even, α fixes an even number of points not on L, and so at least fixes two if it fixes any. Hence in this case α cannot fix any point not on L and is an elation. This completes the proof of all parts of the theorem.

The following is a slight improvement of a Theorem of Ostrom [1], who made the further assumption that n is odd.

THEOREM 20.9.8 (OSTROM). *If the collineation group of a finite projective plane π of order n, where n is not a square, is doubly transitive on the points of π, then π is Desarguesian.*

Proof: Let G be the collineation group of π. By hypothesis G is doubly transitive on the $N = n^2 + n + 1$ points of π. Since $N(N-1)$ divides the order of G, G must contain an element of order 2, an involution α. Since n is not a square, by Theorem 20.9.7 it follows that α is an elation if n is even and that α is a homology if n is odd.

LEMMA 20.9.3. *There is an elation in* G.

Proof: If n is even, an involution α is an elation. Hence we need consider only the case in which n is odd. Consider an involution α which is a homology and let its center be the point P and its axis be the line L. Let A be a point of L and $A_1 \neq P$ be a point not on L. Then in G there is an element σ which takes P into P and A into A_1. Then $\beta = \sigma^{-1}\alpha\sigma$ is an involution whose center is P and whose axis K passes through A_1, and so is different from L. Then $\alpha\beta$ is a central collineation, since it fixes all lines through P. If $\rho = \alpha\beta$ fixes any line T not through P, suppose $T_1 = T\alpha$. Then β must also interchange T and T_1, and if $T \neq T_1$, ρ must fix both T and T_1, whence by Theorem 20.4.1, $\rho = 1$ and $\alpha = \beta$, a conflict, since α and β are involutions with different axes. But if $T = T_1$, then T is the axis of α and also the axis of β, again a conflict, since α and β had different axes. Hence ρ fixes no lines not through P, and so ρ is an elation. This proves our Lemma.

We may now consider an elation ρ with center P on an axis L. Let P_i be any other point of L. Then in G there is an element σ interchanging P and P_i. Then σ fixes L. Hence the group $G(L)$ of collineations fixing L is transitive on the points of L, and so for all points P_i of L the elation groups $G(P_i, L)$ are of the same order h and $h > 1$, since we had an elation ρ with center P on L. From Lemma 20.9.2 of Theorem 20.9.6, π is a translation plane with respect to the axis L. But since G is doubly transitive on points, any two points of L can be mapped onto two points of any other line K by an appropriate element of G. Hence π is also a translation plane with respect to K, and so is a Moufang plane. But as was shown in proving Theorem 20.9.6, this means that π is Desarguesian. A paper by A. Wagner, as yet unpublished, shows that Theorem 20.9.8 is valid without any restriction on n.

There is a generalization of the incidence matrix of a plane due to D. R. Hughes [3]. If we are given a plane π and a group G of collineations of π, this is a matrix whose entries are elements from the group ring G^* of G, G^* being taken over the integers or over any field whose characteristic does not divide the order of G. Analogues of the incidence equations (20.9.3) can be obtained. From Theorem 20.9.3 we recall that the number of transitive sets of lines under G is the same as the number of transitive sets of points. Let us call this number w. We list our notation:

π, a given projective plane of order n.
G, a group of collineations of π of order g.
P_i, $i = 1, \cdots, w$, a fixed representative of the ith transitive set of points.
L_j, $j = 1, \cdots, w$, a fixed representative of the jth transitive set of lines.
H_i, subgroup of G fixing P_i, of order h_i.
T_j, subgroup of G fixing L_j, of order t_j.
$D_{ij} = \{x \mid x \in G, P_i x \in L_j\}$, a set of d_{ij} elements of G.

(20.9.23)
$$\delta_{ij} = \sum x, \ x \in D_{ij},$$
$$\delta^*_{ij} = \sum x^{-1}, \ x \in D_{ij},$$
$$D = (\delta_{ij}) \ i, j = 1, \cdots, w, \text{ a matrix over } G^*.$$
$$D' = (\delta^*_{ij})^T i, j = 1, \cdots, w, \text{ a matrix over } G^*.$$
$$\rho_i = \sum x, \ x \in H_i, \ i = 1, \cdots, w.$$
$$\tau_j = \sum x, \ x \in T_j, \ j = 1, \cdots, w.$$
$$\gamma = \sum x, \ x \in G.$$
$$S = w \times w \text{ matrix with every entry } \gamma.$$

We also use several diagonal matrices:

(20.9.24)
$$\begin{aligned} C_1 &= \text{diag. } (h_1^{-1}, h_2^{-1}, \cdots, h_w^{-1}). \\ C_2 &= \text{diag. } (t_1^{-1}, t_2^{-1}, \cdots, t_w^{-1}). \\ P &= \text{diag. } (\rho_1, \rho_2, \cdots, \rho_w). \\ L &= \text{diag. } (\tau_1, \tau_2, \cdots, \tau_w). \end{aligned}$$

We observe that a knowledge of the sets D_{ij} of elements x of G such that $P_i x \in L_j$ completely determines the incidences in π, for every point of π can be written $P_i u$ for some $i = 1, \cdots, w$ and some $u \in G$, and similarly, every line is of the form $L_j v$. Moreover, $P_i u \in L_j v$ if, and only if, $P_i uv^{-1} \in L_j$ or if $uv^{-1} \in D_{ij}$. Hence a knowledge of D completely determines π. If G is merely the identity, we see that D is the incidence matrix A for π.

THEOREM 20.9.9. *Given a plane π of order n and a group G of collineations of π of order g, the collineation matrix D satisfies the following relations:*

$$(20.9.25)\qquad \begin{aligned} DC_2D' &= nP + S,\\ D'C_1D &= nL + S,\\ D\,C_2S &= (n+1)S,\\ S\,C_1D &= (n+1)S. \end{aligned}$$

Proof: We prove the first equation by evaluating the elements of $U = DC_2D'$, first those on the main diagonal, and later those off the main diagonal. If $U = (u_{r,s})\ r, s = 1, \cdots, w$, then we have first

$$(20.9.26)\qquad u_{rr} = \sum_{j=1}^{w} \frac{\delta_{rj}\delta^*_{rj}}{t_j}.$$

In (20.9.26) the terms for a single j are

$$(20.9.27)\qquad \sum \frac{xy^{-1}}{t_j},\quad x \in D_{rj},\quad y \in D_{rj}.$$

We note that for $x \in D_{rj}$ the entire coset H_rxT_j is contained in D_{rj}. We consider the left cosets of H_r in G:

$$(20.9.28)\qquad G = H_r + H_rx_2 + \cdots + H_rx_{v_r},\quad v_rh_r = g.$$

For an $h \in H_r$ the equation $xy^{-1} = h$ or $x = hy$ with $x, y \in D_{rj}$ holds for every $y \in D_{rj}$ with an appropriate $x \in D_{rj}$, since $H_ry \subseteq D_{rj}$. Hence for a given $h \in H_r$ there are d_{rj} choices $x, y \in D_{rj}$ such that $xy^{-1} = h$. Hence in (20.9.26) the coefficient of h is $\sum_j d_{rj}/t_j$ But d_{rj} is the number of x's such that $P_rx \in L_j$ or $P_r \in L_jx^{-1}$. For $x \in D_{rj}$ the number of distinct lines in the set L_jx^{-1} is d_{rj}/t_j. But P_r is on a total of $n + 1$ lines. Hence

$$(20.9.29)\qquad \sum_j \frac{d_{rj}}{t_j} = n + 1.$$

Hence in (20.9.26) the coefficient of $h \in H_r$ is $n + 1$.

Now consider an equation $xy^{-1} = z, z \notin H_r$. Here P_r, P_rz are distinct points and so lie on a unique line L_mv, where m and the coset T_mv are uniquely determined. If for some j both $x \in D_{rj}$, $y \in D_{rj}$, then P_ry and $P_rzy = P_rx \in L_mvy$. But $P_ry \in L_j$, $P_rx \in L_j$, and $P_rx \neq P_ry$. Hence

$L_m vy = L_j$, whence we must have $j = m$, $vy \in T_m$. Hence in (20.9.26) the element z arises only for the summand with $j = m$, and here with x, $y \in D_{rm}$, we have $xy^{-1} = z$ for every $y \in D_{rm}$ so that $L_m y^{-1} = L_m v = P_r P_r z$ and an $x \in D_{rm}$ determined by $x = zy$. But these y's are such that y^{-1} is in the coset $T_m v$, and there are exactly t_m of these. Hence in (20.9.26) the coefficient of z is $t_m/t_m = 1$. Thus we have in (20.9.26) the coefficient of an $h \notin H_r$ as $n + 1$ and of a $z \in H_r$ as 1. Hence we have established the correctness of the first equation in (20.9.25) so far as the main diagonal is concerned. For the off-diagonal terms in $U = DC_2D'$ we have

(20.9.30) $$u_{rs} = \sum_{j=1}^{w} \frac{\delta_{rj}\delta^*_{sj}}{t_j},$$

and the terms for a single j are

(20.9.31) $$\sum \frac{xy^{-1}}{t_j}, \quad x \in D_{rj}, \quad y \in D_{sj}.$$

Here for any $z \in G$, the points $P_r z$ and P_s are distinct and lie on a unique line $L_m v$, where m and the coset $T_m v$ are uniquely determined. Here if $xy^{-1} = z$, where for some j, $x \in D_{rj}$, $y \in D_{sj}$, then $P_r x = P_r zy$ and $P_s y$ lie on $L_m vy$. But $P_r x \neq P_s y$ both lie on L_j. Hence $L_m vy = L_j$, whence $j = m$ and $L_j y^{-1} = L_m v = P_r z P_s$. But these y's are such that y^{-1} is an element of $T_m v$, and there are t_m of these. Also for each $y^{-1} \in T_m v$, $L_m vy = L_m$ and $P_r zy \in L_m$, whence $x = zy \in D_{rm}$. Hence the coefficient of any z in u_{rs} reduces to t_m/t_m, and so $u_{rs} = \sum z$, $z \in G$, $u_{rs} = \gamma$, and this completes the proof of the first relation in (20.9.25).

The second relation in (20.9.25) is the dual of the first, and its proof can be carried out in the same fashion.

In calculating $DC_2S = V = (v_{rs})$, we find

(20.9.32) $$v_{rs} = \sum_j \frac{\delta_{rj}}{t_j}\gamma = \sum_j \frac{d_{rj}}{t_j}\gamma,$$

but by (20.9.29) this is $(n + 1)\gamma$. This proves the third relation, and the fourth is dual and may be proved in the same way.

Proceeding from the relations (20.9.25), Hughes has obtained restrictions on the possible collineations in planes similar to the restrictions of the Bruck-Ryser theorem. The proof depends (as did the original proof of the Bruck-Ryser theorem) on the deep results of Hasse-Minkowski on the rational equivalence of quadratic forms. In particular he finds the following:

THEOREM 20.9.10 (HUGHES). *Let π be a plane of order n for which the Bruck-Ryser conditions on n are satisfied. Let G be a group of collineations of π of odd prime order p. Let the number u of fixed points be even. Then a necessary condition that such a collineation exists is that the equation:*

$$x^2 = ny^2 + (-1)^{(p-1)/2}pz^2$$

have a solution in integers x, y, z not all zero.

The same result holds for a collineation group G of odd order g (instead of p) if every element $\neq 1$ of G displaces the same points.

Hughes' theorem, like the Bruck-Ryser theorem, denies the existence of certain collineations but does not, of course, guarantee the existence of collineations which do satisfy the conditions.

The main content of the following theorem is that if a plane π has a certain group G of collineations, then π must have still further specific collineations. We assume that G is of a simple type. Explicitly we shall assume that G is transitive and regular on the $N = n^2 + n + 1$ points of π, and also that G is Abelian. This result was first proved by Hall [3] with G a group of order N cyclic on the N points of π. Bruck [1] extended this to the study of cases in which G is transitive and regular but had to assume in addition that G is Abelian to obtain the same result. Hoffman [1] obtained a similar result, assuming that G is cyclic on the $n^2 - 1$ points of π not on L_∞ and different from the origin.

Suppose that we have a group G of collineations of a plane π of order n, where G is Abelian and transitive and regular on the N points of π. In this case if P is a fixed point of π, every point has a unique representation Px, $x \in G$. If an integer t is prime to N, then $x \rightarrow x^t$, all $x \in G$ is trivially an automorphism of G. If, further, for every $x \in G$, $Px \rightarrow Px^t$ is a collineation of π, we say that t is a *multiplier* of π. Trivially, the multipliers form a multiplicative group modulo N.

THEOREM 20.9.11. *If a plane π of order n has a collineation group G which is Abelian and transitive and regular on the N points of π, then any prime p which divides n is a multiplier of π.*

Proof: Under the hypothesis there is only a single transitive constituent for points and so also for lines. There is a single representative point $P = P_1$ and a single representative line $L = L_1$, and if $D_{11} = \{x_1, x_2, \cdots, x_{n+1}\}$, $x_i \in G$, then Px_i, $i = 1, \cdots, n+1$ are the points of L_1. Then $x_1u, \cdots, x_{n+1}u$, $u \in G$ are the points of G. Here we have $D = \delta_{11}$, $D' = \delta^*_{11}$.

(20.9.33)
$$D = x_1 + \cdots + x_{n+1},$$
$$D' = x_1^{-1} + \cdots + x_{n+1}^{-1}.$$

C_2 and C_1 reduce to the identity. The first two relations (20.9.25) take the form

$$DD' = D'D = n\cdot 1 + \gamma.$$

The last two relations in (20.9.25) say only that there are $n + 1$ elements in D and D'. To show that $Px \to Px^p$ is a collineation of π, we must show that $Px_1^p, Px_2^p, \cdots, Px_{n+1}^p$ are on a line. For this we need to show that for some $u \in G$,

(20.9.34)
$$D^{(p)} = x_1^p + \cdots + x_{n+1}^p = (x_1 + \cdots + x_{n+1})u,$$

since the points of an arbitrary line Lu are $Px_1u, Px_2u, \cdots, Px_{n+1}u$. Conversely, if (20.9.34) holds, then $Px_1^p, \cdots, Px_{n+1}^p$ are the points of Lu, whence generally $P(x_1v)^p, \cdots, P(x_{n+1}v)^p$ are the points of Luv^p, and so $Px \to Px^p$ is a collineation and p is a multiplier. For this theorem we take the group ring G^* to be the group ring of G over the integers. G^* (mod p) is the ring G^* with coefficients reduced modulo p. We have

(20.9.35)
$$D^{(p)} = x_1^p + \cdots + x_{n+1}^p \equiv (x_1 + \cdots + x_{n+1})^p = D^p \pmod{p},$$

since the multinomial coefficients are multiples of the prime p and since G is Abelian. The assumption that G is Abelian enters at this point and also in saying as above that $(x_iv)^p = x_i^pv^p$ and that $x \to x^p$ is an automorphism of G. We note that since $p|n$ and $N = n^2 + n + 1$, we have $(p, N) = 1$. Since $p|n$, we have, from (20.9.34)

(20.9.36)
$$DD' \equiv \gamma \pmod{p}.$$

Multiplying by D^{p-1}, we have

(20.9.37)
$$D^pD' \equiv D^{p-1}\gamma \equiv (n+1)^{p-1}\gamma \equiv \gamma \pmod{p}.$$

Hence, from (20.9.35),

(20.9.38) $$D^{(p)}D' \equiv \gamma \pmod{p}.$$

From this we may write

(20.9.39) $$D^{(p)}D' = \gamma + pR,$$

where (and this is vital to our proof) the coefficients of the group elements in R are non-negative integers, for in $D^{(p)}D'$ all coefficients are non-negative, and by (20.9.38), every term $a_i u_i$, $u_i \in G$ has $a_i \equiv 1 \pmod{p}$ $a_i \geq 0$. Thus $a_i \geq 1$ and $(a_i - 1)/p$ is a non-negative integer, this being the coefficient of u_i in R. Now $x \to x^{-1}$, $x \in G$ is an automorphism of G, and hence determines an automorphism $h \to h'$ for $h \in G^*$ and $D \to D'$ under this automorphism. Applied to (20.9.39), this yields

(20.9.40) $$DD'^{(p)} = \gamma + pR'.$$

Moreover, $x \to x^p$ is an automorphism of G and determines an automorphism $h \to h^{(p)}$ of G^*. Applied to (20.9.34), this yields

(20.9.41) $$D^{(p)}D'^{(p)} = n \cdot 1 + \gamma.$$

The product of the left-hand sides of (20.9.34) and (20.9.41) is the same as that for (20.9.39) and (20.9.49). Hence, putting equal the products of the right-hand sides, we have

(20.9.42) $$(n \cdot 1 + \gamma)^2 = (\gamma + pR)(\gamma + pR').$$

The homomorphism of G^* into the integers determined by $x \to 1$, $x \in G$, applied to (20.9.39), gives

(20.9.43) $$(n + 1)^2 = n^2 + n + 1 + pR(1),$$

where $R \to R(1)$ in the homomorphism. Thus $pR(1) = n$, and also $pR'(1) = n$. But in G^*, $pR\gamma = pR(1)\gamma = n\gamma$. Using this in (20.9.42), we find

(20.9.44) $$n^2 \cdot 1 = (pR)(pR').$$

But since pR and pR' have non-negative coefficients, this will be impossible if there is more than one nonzero term in pR. Hence $pR = bu$ for some integer b and $u \in G$. But $b = pR(1) = n$, whence $pR = nu$. Substituting in (20.9.39), we have

(20.9.45) $$D^{(p)}D' = \gamma + nu.$$

Multiplying by D, and using (20.9.34), we have

(20.9.46) $$\begin{aligned} D^{(p)}D'D &= \gamma D + nDu, \\ D^{(p)}(n + \gamma) &= (n + 1)\gamma + nDu, \\ nD^{(p)} + (n + 1)\gamma &= (n + 1)\gamma + nDu. \end{aligned}$$

This now gives:

(20.9.47) $$D^{(p)} = Du.$$

But this is precisely the relation (20.9.34) which we needed, and our theorem is proved.

As an illustration of the power of this theorem, consider a plane of order 8 with a (necessarily cyclic) collineation group of order 73. Points may be represented as residues modulo 73. The multiplier is 2, and if $a_1 \cdots a_9$ are the points of a line, then $2a_1 \cdots 2a_9$ are $a_1 + s, \cdots, a_9 + s$ in some order for an appropriate s. Then the points $a_1 - s, \cdots, a_9 - s$ are on a line fixed by the multiplier 2. If one of these residues is 1, then the multiplier 2 gives us the complete set of points on a line 1, 2, 4, 8, 16, 32, 37, 55, 64 (mod 73). Any other set fixed by 2 differs from this by a constant factor and gives the same plane. The plane is the Desarguesian plane.

Hughes has proved a further result which is at once more special and more refined than Theorem 20.9.10.

THEOREM 20.9.12. *A plane π of order n, where $n \equiv 2$ (mod 4), $n > 2$, cannot have an involution.*

Proof: Suppose that π is a plane of order n, where $n \equiv 2$ (mod 4), $n > 2$, which possesses an involution b. Then by Theorem 20.9.7, since n is even and not a square, b is an elation. Let M be the axis and $C \in M$ the center of the elation. Let Q_i, $i = 1 \cdots n$ be the remaining points on M, and K_i, $i = 1 \cdots n$ the remaining lines through C. Write $n = 2m$, where m is odd. The n^2 lines of π not through C can be broken up into $n^2/2 = 2m^2$ classes of two lines, where a class with the line L also contains Lb. In each class choose one line L_i, $i = 1, \cdots, 2m^2$. Similarly, the n points other than C on a line K_i can be broken up into $n/2 = m$ classes with respect to b. In each class choose a point and name these points P_{ij}, $j = 1, \cdots, n/2 = m$. We now define incidence numbers a_{ij}^k by the rule:

(20.9.48)
$$\begin{aligned} a_{ij}{}^k &= +1 && \text{if } P_{ij} \in L_k, \\ a_{ij}{}^k &= -1 && \text{if } P_{ij}b \in L_k, \\ a_{ij}{}^k &= 0 && \text{(otherwise).} \end{aligned}$$

LEMMA 20.9.3. $$\sum_k (a_{ij}{}^k)^2 = n.$$

LEMMA 20.9.4. $$\sum_k a_{ij}{}^k a_{st}{}^k = 0 \qquad \text{if } (i, j) \neq (s, t).$$

Proof of Lemma 20.9.3. The point P_{ij} is on n lines either L_k or L_kb, so Lemma 20.9.3 is immediate.

Proof of Lemma 20.9.4. If $i = s$, $j \neq t$, the points P_{ij} and $P_{ij}b$ all lie on K_i and no two on any other line, whence the sum is zero. If $i \neq s$, let $P_{ij}P_{st}$ be L_qx, $P_{ij}P_{st}b$ be L_ry, where x and y are 1 or b. Now $r \neq q$, for if $r = q$, $x = y$, then $L_qx = L_ry$ contains P_{st} and $P_{st}b$, which are distinct points on K_s, a conflict. But if $r = q$, $x = yb$, then $L_qx = L_ryb$ contains the distinct points P_{ij} and $P_{ij}b$, which lie on K_i, a conflict. Hence $r \neq q$. But then

$$a_{ij}{}^q = a_{st}{}^q, \quad a_{ij}{}^q a_{st}{}^q = +1,$$
and
$$a_{ij}{}^r = -a_{st}{}^r, \quad a_{ij}{}^r a_{st}{}^r = -1.$$

Thus the nonzero terms of Lemma 20.9.4 can be paired so that the sum of a pair is zero. Hence the sum of Lemma 20.9.4 is zero.

From our lemma the incidence numbers $a_{ij}{}^k$ can be formed into a $2m^2 \times 2m^2$ matrix:

(20.9.49) $$A = (a_{ij}{}^k) \qquad \text{with } ij \text{ giving row, } k \text{ column,}$$

where, by our Lemma, A satisfies

(20.9.50) $$AA^T = nI.$$

Let us define numbers b_{ik} by the rule

(20.9.51) $$b_{ik} = \sum_{j=1}^{m} a_{ij}{}^k. \quad i = 1, \cdots, n, \; k = 1 \cdots 2m^2.$$

Then every b_{ik} is $+1$ or -1, since every line L_k intersects K_i in exactly one point, either P_{ij} or $P_{ij}b$, and so exactly one of $a_{ij}{}^k$ is different from 0. The $n \times 2m^2$ matrix B,

(20.9.52) $$B = (b_{ik}) \; i = 1 \cdots n, \quad k = 1 \cdots 2m^2,$$

is such that its first row is the sum of the first m rows of A, its second row is the sum of the second m rows of A, and so on. Since from (20.9.50) different rows of A had an inner product of zero, the same holds for the rows of B. We may multiply the columns of B by $+1$ or -1 without changing inner products, and this we shall do so as to make the first row of B consist exclusively of $+1$'s. Since $n > 2$, there are at least three rows in B, and rearranging the columns of B, the first three rows of B take the form:

$$(20.9.53) \qquad \begin{array}{|cc|cc|}
\multicolumn{2}{|l}{+1, \quad \cdots \quad , +1} & \multicolumn{2}{|l|}{+1, \quad \cdots \quad , +1} \\
\hline
\multicolumn{1}{|c|}{+1, \cdots, +1} & +1, \cdots, +1 & \multicolumn{1}{|c|}{-1, \cdots, -1} & -1, \cdots, -1 \\
\hline
\multicolumn{1}{|c|}{+1, \cdots, +1} & -1, \cdots, -1 & \multicolumn{1}{|c|}{+1, \cdots, +1} & -1, \cdots, -1 \\
\multicolumn{1}{c}{r} & \multicolumn{1}{c}{s} & \multicolumn{1}{c}{t} & \multicolumn{1}{c}{u}
\end{array}$$

Since the inner product of the second and third lines with the first is zero, we have $r + s = t + u$, $r + t = s + u$. Here $r + s + t + u = 2m^2$, and so:

$$(20.9.54) \qquad r + s = t + u = m^2, \quad r + t = s + u = m^2,$$

whence

$$(20.9.55) \qquad u = r, \quad s = t = m^2 - r.$$

Since the inner product of the second and third rows is also zero, we have $r + u = s + t = m^2$. But this gives

$$(20.9.56) \qquad 2r = m^2,$$

which is a conflict because $n \equiv 2 \pmod 4$, $n = 2m$, and m is odd. Hence π cannot have an involution, and our theorem is proved. This result can be obtained from the incidence relations of (20.9.25) by appropriate renumbering and mapping G^* onto the integers by the homomorphism $1 \to 1$, $b \to -1$.

An example of a non-Desarguesian plane of order 9 was given by Veblen and Wedderburn [1]. This example has been shown by Hughes [2] to be a particular case of an infinite class. Let $q = p^r$ be a power of an odd prime p. Then we have shown that there exists a near-field K of order q^2 whose center Z is the field $GF(q) = GF(P^r)$. The Hughes planes are of order q^2.

Definition of Hughes Planes: A point P is the set of triples $P = (xk, yk, zk)$, x, y, z fixed elements of K not all zero and $k \neq 0$, an arbitrary element of K. The Theorem of Singer (20.9.5) gives us a mapping:

(20.9.57)
$$\begin{aligned} x &\rightarrow a_{11}x + a_{12}y + a_{13}z \ , \\ y &\rightarrow a_{21}x + a_{22}y + a_{23}z \ , \\ z &\rightarrow a_{31}x + a_{32}y + a_{33}z \ , \end{aligned}$$

where $a_{ij} \in Z$, such that

(20.9.58)
$$(x, y, z) = P \rightarrow PA = (a_{11}x \cdots, \cdots, a_{33}z)$$

is a collineation α of order $m = q^2 + q + 1$ in the Desarguesian plane of order q with coordinates from Z. The Hughes plane is given by extending the collineation α to points with coordinates from K.

We have base lines L_t given by equation

(20.9.59)
$$x + ty + z = 0.$$

Here we take either $t = 1$ or $t \notin Z$, but otherwise t is an arbitrary element of K. This gives $1 + (q^2 - q) = q^2 - q + 1$ base lines. We define an incidence $P = (xk, yk, zk) \in L_t$ if, and only if, x, y, z satisfy (20.9.59). By the associativity of multiplication in K and the right distributive law, then from (20.9.59) we also have

(20.9.60)
$$0 = (x + ty + z)k = xk + t(yk) + zk,$$

and so our incidence rule $P \in L_t$ does not depend on which representative of P is chosen to satisfy (20.9.59). Further lines $L_t\alpha^i$, $i = 0, \cdots, m - 1$ are defined symbolically, and we say

(20.9.61)
$$PA^i \in L_t\alpha^i \ i = 0, \cdots, m - 1,$$

if, and only if, $P \in L_t$.

It is not true that the points of $L_t\alpha^i$ satisfy a linear equation. To find the points on L_t, we may in (20.9.59) take x and y arbitrarily, except for taking both to be zero, and determine z from the equation. This give $q^4 - 1$ triples of which sets of $q^2 - 1$ represent the same point. Hence L_t contains $q^2 + 1 = n + 1$ distinct points. Hence, also, $L_t\alpha^i$ contains $n + 1$ points. We have $(q^2 - q + 1)(q^2 + q + 1) = q^4 + q^2 + 1 = n^2 + n + 1$ lines in all, each containing $n + 1$ points. There are $n^2 + n + 1$ points in all. Thus to show that we have a projective plane, it is sufficient to show that any two distinct lines have a unique point in common. The mapping $P \rightarrow PA$ is one to one and has period $m = q^2 + q + 1$. If $\{P\}_s$ is the set of points on the base line L_s, $L_s\alpha^i$ contains the set of points $\{P\}_s A^i$, and $L_t\alpha^j$ contains the set of points $\{P\}_t A^j$. Hence to show that $L_s\alpha^i$ and $L_t\alpha^j$ have a unique point in common, it is sufficient to show that L_s and $L_t\alpha^{j-i} = L_t\alpha^h$ (where exponents of α are taken modulo m) have a unique point in common.

Let $P = (x, y, z)$ be a point of $L_t\alpha^h$. Then PA^{-h} is a point of L_t, and conversely. Then if

(20.9.62) $(x, y, z)A^{-h}$

$$= (b_{11}x + b_{12}y + b_{13}z,\ b_{21}x + b_{22}y + b_{23}z,\ b_{31}x + b_{32}y + b_{33}z),$$

then the condition that $P = (x, y, z)$ should lie on $L_t\alpha^h$ is

$$(b_{11}x + b_{12}y + b_{13}z) + t(b_{21}x + b_{22}y + b_{23}z) + (b_{31}x + b_{32}y + b_{33}z) = 0. \tag{20.9.63}$$

If (x, y, z) is on L_s, then we have

$$x + sy + z = 0. \tag{20.9.64}$$

We must show that, apart from a factor k on the right, (20.9.63) and (20.9.64) have a unique solution (x, y, z). We solve (20.9.64) for x and substitute in (20.9.63). This gives

$$uy + az + t(vy + bz) = 0, \tag{20.9.65}$$

where

$$\begin{aligned} u &= b_{12} + b_{32} - (b_{11} + b_{31})s, \\ v &= b_{22} - b_{21}s, \\ a &= b_{13} + b_{33} - (b_{11} + b_{31}), \\ b &= b_{23} - b_{21}. \end{aligned} \tag{20.9.66}$$

Note that $a, b \in Z$, but in general u, v are not in Z. There are three cases to consider in finding solutions of (20.9.65).

CASE 1: $b \neq 0$. Here (20.9.65) can be written as

$$(b^{-1}a + t)(vy + bz) + (u - b^{-1}av)y = 0, \tag{20.9.67}$$

using the fact that a and b^{-1} are in the center. If both coefficients $b^{-1}a + t$ and $u - b^{-1}av$ should be zero, then $t \in Z$, $t = 1$, and $a + b = 0$, $u + v = 0$. But then from $u + v = 0$ we have

$$b_{12} + b_{22} + b_{32} = (b_{11} + b_{21} + b_{31})s, \tag{20.9.68}$$

whence $s \in Z$, and so $s = 1$. Now $a + b = 0$ gives

$$b_{13} + b_{23} + b_{33} = b_{11} + b_{21} + b_{31}. \tag{20.9.69}$$

But with both $s = 1$ and $t = 1$ this says that (20.9.63) and (20.9.64) represent the same line in the Desarguesian plane π_1 over $GF(q)$. But this is not possible unless the $L_s = L_1$, $L_t\alpha^i = L_1$ because the matrix A was of order $m = q^2 + q + 1$ as a collineation of π_1. Hence not both coefficients are zero in (20.9.67). Thus if $b^{-1}a + t \neq 0$, an arbitrary value for y determines $vy + bz$ uniquely, and since $b \neq 0$, it determines z uniquely. If $b^{-1}a + t = 0$, then $u - b^{-1}av \neq 0$, and so $y = 0$, whence z is arbitrary. Thus y and z are determined uniquely apart from a right factor, and so in turn from (20.9.64), x is determined uniquely by y and z. Thus (20.9.63) and (20.9.64) are satisfied by a unique point (xk, yk, zk). This gives the desired unique solution in Case 1.

CASE 2: $b = 0$, $a \neq 0$. Here (20.9.65) becomes

$$(u + tv)y + az = 0. \tag{20.9.70}$$

Since $a \neq 0$, (20.9.70) and (20.9.64) are satisfied by a unique point (xk, yk, zk).

CASE 3: $b = 0$, $a = 0$. Here we have

$$\begin{aligned} b_{13} + b_{33} &= b_{11} + b_{31}, \\ b_{23} &= b_{21}, \end{aligned} \tag{20.9.71}$$

and we see that the point $P = (k, 0, -k)$ satisfies both (20.9.65) and (20.9.64). Also, from (20.9.71), we see by (20.9.62) that

$$PA^{-h} = (k, 0, -k)A^{-h} = (b_{11} - b_{13})(k, 0, -k) = P, \tag{20.9.72}$$

where $b_{11} - b_{13} \neq 0$, since A^{-h} is not singular. But since A^h fixes the point P of π_1, we see that $h \equiv 0 \pmod{m}$, and so $L_t\alpha^h$ is L_t. Hence our lines are now L_s and L_t where surely $s \neq t$. For these, $x + sy + z = 0$ and $x + ty + z = 0$, and so, clearly, $P = (k, 0, -k)$ lies on both these lines but no other point does. Thus in every case any two distinct lines have a unique intersection, and we have proved that they form a projective plane. We state this as a theorem.

THEOREM 20.9.13 (HUGHES). *Given a near-field K of order q^2 whose center is $GF(q) = Z$, $q = p^r$, p an odd prime, and the mapping A of (20.9.57) of order $q^2 + q + 1$ as a collineation of the Desarguesian* *plane of order* q. *Then lines* $L_t\alpha^i$ *containing points* PA^i *by the rules* (20.9.59) *and* (20.9.61) *form a projective plane* π *of order* q^2.

Hughes has shown that if the near-field K is not the field $GF(q^2)$, then the plane π is not only non-Desarguesian but also is not a Veblen-Wedderburn plane over any coordinate system.

Bibliography

Albert, A. A.

[1] *On nonassociative division algebras*, Trans. Amer. Math. Soc. *72* (1952), 296–309.

Baer, R.

[1] *Erweiterung von Gruppen und ihrer Isomorphismen*, Math. Zeit. *38* (1934), 375–416.

[2] *The decomposition of enumerable primary Abelian groups into direct summands*, Quart. J. of Math. *6* (1935), 217–221.

[3] *The decomposition of Abelian groups into direct summands*, Quart. J. of Math. *6* (1935), 222–232.

[4] *Types of elements and the characteristic subgroups of Abelian groups*, Proc. London Math. Soc. *39* (1935), 481–514.

[5] *The subgroup of elements of finite order of an Abelian group*, Ann. of Math. *37* (1936), 766–781.

[6] *Dualism in Abelian groups*, Bull. Amer. Math. Soc. *43* (1937), 121–124.

[7] *Duality and commutativity of groups*, Duke Math. J. *5* (1939), 824–838.

[8] *The significance of the system of subgroups for the structure of a group*, Amer. J. of Math. *61* (1939), 1–44.

[9] *Abelian groups that are direct summands of every containing Abelian group*, Bull. Amer. Math. Soc. *46* (1940), 800–806.

[10] *Homogeneity of projective planes*, Amer. J. of Math. *64* (1942), 137–152.

[11] *Klassification der Gruppenerweiterungen*, J. reine angew. Math. *187* (1949), 75–94.

[12] *Supersoluble groups*, Proc. Amer. Math. Soc. *6* (1955), 16–32.

Bethe, H. A.

[1] *Termaufspaltung in Kristallen*, Ann. der Physik (5) *3* (1929), 133–208.

Birkhoff, Garrett

[1] *Lattice Theory*, Colloquium publications, Amer. Math. Soc., vol. XXV, rev. ed., 1948.

Birkhoff, G., and MacLane, S.

[1] *A Survey of Modern Algebra*, The Macmillan Co., revised edition, 1953.

Bruck, R. H.

[1] *Difference sets in a finite group*. Trans. Amer. Math. Soc. *78* (1955), 464–481.

Bruck, R. H., and Kleinfeld, E.

[1] *The structure of alternative division rings*, Proc. Amer. Math. Soc. *2* (1951), 878–890.

Bruck, R. H., and Ryser, H. J.

[1] *The non-existence of certain finite projective planes*. Can. J. Math. *1* (1949), 88–93.

Burnside, W.

[1] *On an unsettled question in the theory of discontinuous groups*, Quart. J. Pure and Appl. Math. *33* (1902), 230–238.

[2] *Theory of Groups of Finite Order*, Cambridge Univ. Press, 2nd ed., 1911.

Chowla, S., and Ryser, H. J.

[1] *Combinatorial problems*, Can. J. Math. *2* (1950), 93–99.

Eckmann, B.

[1] *Cohomology of groups and transfer*, Ann. of Math. *58* (1953), 481–493.

Eilenberg, S., and MacLane, S.

[1] *Cohomology theory in abstract groups I*, Ann. of Math. *48* (1947), 51–78.

[2] *Cohomology theory in abstract groups II*, Ann. of Math. *48* (1947), 326–41.

Federer, H., and Jonsson, B.

[1] *Some properties of free groups*, Trans. Amer. Math. Soc. *68* (1950), 1–27.

Frobenius, G.

[1] *Über auflösbare Gruppen IV* Berl. Sitz. (1901), 1223–1225.

[2] *Über einen Fundamentalsatz der Gruppentheorie*, Berl. Sitz. (1903), 987–991.

Gaschütz, W.

[1] *Zur Erweiterungstheorie der Endlichen Gruppen*, J. reine angew. Math. *190* (1952), 93–107.

Gleason, A. M.

[1] *Finite Fano planes*, Amer. J. Math. *78* (1956), 797–807.

Grün, O.

[1] *Beiträge zur Gruppentheorie I*, J. reine angew. Math. *174* (1935), 1–14.

Hall, Marshall, Jr.

[1] *Group rings and extensions*, Ann. of Math. *39* (1938), 220–234.

[2] *Projective planes*, Trans. Amer. Math. Soc. *54* (1943), 229–277. Correction, Trans. Amer. Math. Soc. *65* (1949), 473–474.

[3] *Cyclic projective planes*, Duke Math. J. *14* (1947), 1079–1090.

[4] *Coset representation in free groups*, Trans. Amer. Math. Soc. *67* (1949), 421–432.

[5] *Subgroups of finite index in free groups*, Can. J. Math. *1* (1949), 187–190.

[6] *A basis for free Lie rings and higher commutators in free groups*, Proc. Amer. Math. Soc. *1* (1950), 575–581.

[7] *Subgroups of free products*, Pacific J. Math. *3* (1953), 115–120.

[8] *On a theorem of Jordan*, Pacific J. Math. *4* (1954), 219–226.

[9] *Solution of the Burnside problem for exponent 6*, Proc. Nat. Acad. Sci. *43* (1957), 751–753.

Hall, M. Jr., and Rado, T.

[1] *On Schreier systems in free groups*, Trans. Amer. Math. Soc. *64* (1948), 386–408.

Hall, M. Jr., and Ryser, H. J.

[1] *Normal completions of incidence matrices*, Amer. J. Math. *76* (1954), 581–589.

Hall, Philip

[1] *A note on soluble groups*, J. London Math. Soc. *3* (1928), 98–105.

[2] *A contribution to the theory of groups of prime-power order*. Proc. London Math. Soc. *36* (1933), 29–95.
[3] *On a Theorem of Frobenius*, Proc. London Math. Soc. *40* (1936), 468–501.

Hall, P., and Higman, G.
[1] *The p-length of a p-soluble group, and reduction theorems for Burnside's problem*, Proc. London Math. Soc. (3) *7* (1956), 1–42.

Hardy, G. H., and Wright, E. M.
[1] *An Introduction to the Theory of Numbers*, The Clarendon Press, Oxford, 1938.

Higman, Graham
[1] *On finite groups of exponent five*, Proc. Camb. Phil. Soc. *52* (1956), 381–390.

Hirsch, K. A.
[1] *On infinite soluble groups*, Proc. London Math. Soc. (2) *44* (1938), 53–60.
[2] *On infinite soluble groups II*, *Ibid.* *44* (1938), 336–344.
[3] *On infinite soluble groups III*, *Ibid.* *49* (1946), 184–194.

Hoffman, A. J.
[1] *Cyclic affine planes*, Can. J. Math. *4* (1952), 295–301.

Hölder, O.
[1] *Zuruckführung einer beliebigen algebraischen Gleichung auf eine Kette von Gleichungen*, Math. Ann. *34* (1889), 26–56.

Holyoke, T. C.
[1] *On the structure of multiply transitive permutation groups*, Amer. J. Math. *74* (1952), 787–796.

Hughes, D. R.
[1] *Regular collineation groups*, Proc. Amer. Math. Soc. *8* (1957), 159–164.
[2] *A class of non-Desarguesian projective planes*, Can. J. Math. *9* (1957), 378–388.
[3] *Generalized incidence matrices over group algebras*, Ill. J. Math. *1* (1957), 545–551.
[4] *Collineations and generalized incidence matrices*, Trans. Amer. Math. Soc. *86* (1957), 284–296.

Huppert, B.
[1] *Normalteiler und maximale Untergruppen endlicher Gruppen*, Math. Zeit. *60* (1954), 409–434.

Iwasawa, K.
[1] *Über die endlichen Gruppen und die Verbände ihrer Untergruppen*, J. Univ. Tokyo *43* (1941), 171–199.

Jordan, C.
[1] *Commentaire sur Galois*, Math. Ann. *1* (1869), 141–160.
[2] *Recherches sur les substitutions*, J. Math. Pures Appl. (2) *17* (1872), 351–363.

Kaplansky, I.
[1] *Infinite Abelian Groups*, University of Michigan Press, 1954.

Kleinfeld, E.
[1] *Alternative division rings of characteristic 2*, Proc. Nat. Acad. Sci. USA *37* (1951), 818–820.

[2] *Right alternative rings*, Proc. Amer. Math. Soc. *4* (1953), 939–944.

Kostrikin, A. I.

[1] *Solution of the restricted Burnside problem for exponent 5*, Izv. Akad. Nauk SSSR, Ser. mat. *19* (1955), 233–244. (In Russian.)

Krull, W.

[1] *Über verallgemeinerte endliche Abelsche Gruppen*, Math. Zeit. *23* (1925), 161–196.

[2] *Theorie und Anwendung der verallgemeinerten Abelschen Gruppen*, Sitz Heidelberg. Akad. Wiss. (1926), 1–32.

Kurosch, A.

[1] *Die Untergruppen der freien Produkte von beliebigen Gruppen*, Math. Ann. *109* (1934), 647–660.

[2] *The Theory of Groups*, 2nd. ed., translated from the Russian by K. A. Hirsch, two volumes, Chelsea Publishing Co., New York, 1955.

Levi, F. W.

[1] *Über die Untergruppen der freien Gruppen*, Math. Zeit. *32* (1930), 315–318.

Levi, F. W., and van der Waerden, B. L.

[1] *Über eine besondere Klasse von Gruppen*, Abh. Math. Sem. Hamburg *9* (1933), 154–158.

MacLane, S.

[1] *A conjecture of Ore on chains in partially ordered sets*, Bull. Amer. Math. Soc. *49* (1943), 567–568.

[2] *Cohomology theory in abstract groups, III*, Ann. of Math. *50* (1949), 736–761.

Magnus, W.

[1] *Über Bezhieungen zwischen höheren Kommutatoren*, J. reine angew. Math. *177* (1937), 105–115.

[2] *On a theorem of Marshall Hall*, Ann. of Math. *40* (1939), 764–768.

[3] *A connection between the Baker-Hausdorff formula and a problem of Burnside*, Ann. of Math. *52* (1950), 111–126.

Mann, H. B.

[1] *On certain systems which are almost groups*, Bull. Amer. Math. Soc. *50* (1944), 879–881.

Meier-Wunderli, H.

[1] *Note on a basis for higher commutators*, Commentarii Math. Helvetici, *16* (1951), 1–5.

Miller, G. A.

[1] *Limits of the degree of transitivity of substitution groups*, Bull. Amer. Math. Soc., *22* (1915), 68–71.

Moufang, R.

[1] *Alternativkörper und der Satz vom vollständigen Vierseit*, Abh. Math. Sem. Hamburg, *9* (1933), 207–222.

Neumann, B. H.

[1] *Die Automorphismengruppe der freien Gruppen*, Math. Ann. *107* (1932), 367–386.

[2] *On the number of generators of a free product*, J. London Math. Soc. *18* (1943), 12–20.

Neumann, H.
[1] *Generalized free products with amalgamated subgroups I*, Amer. J. Math. *70* (1948), 590–625.
[2] *Generalized free products with amalgamated subgroups II*, *Ibid.* *71* (1949), 491–540.

Nielsen, J.
[1] *Om Regnig med ikke-kommutative Faktorer og dens Anvendelse i Gruppeteorien*, Mat. Tidsskrift B (1921), 77–94.

Noether, E.
[1] *Hyperkomplexe Zahlen und Darstellungstheorie*, Math. Zeit. *30* (1929), 641–692.

Ore, O.
[1] *Direct Decompositions*, Duke Math. J. *2* (1936), 581–596.
[2] *On the theorem of Jordan-Hölder*, Trans. Amer. Math. Soc. *41* (1937), 266–275.
[3] *Chains in partially ordered sets*, Bull. Amer. Math. Soc. *49* (1943), 558–566.

Ostrom, T. G.
[1] *Double transitivity in finite projective planes*, Can. J. Math. *8* (1956), 563–567.

Parker, E. T.
[1] *On collineations of symmetric designs*, Proc. Amer. Math. Soc. *8* (1957), 350–351.

Pauli, W.
[1] *Zur Quantenmechanik des Magnetischen Elektrons*, Zeit. für Physik, *43* (1927), 601–623.

Pickert, G.
[1] *Projektive Ebenen*, Springer (1955).

Pontrjagin, L.
[1] *Topological Groups*, translated from the Russian by Emma Lehmer, Princeton University Press, 1939.

Prüfer, H.
[1] *Theorie der abelschen Gruppen I*, Math. Zeit. *20* (1924), 165–187.
[2] *Theorie der abelschen Gruppen II*, Math. Zeit. *22* (1925), 222–249.

Remak, R.
[1] *Über die Zerlegung der endlichen Gruppen in direkte unzerlegbare Faktoren*, J. reine angew. Math. *139* (1911), 293–308.
[2] *Über die Zerlegung der endlichen Gruppen in direkte unzerlegbare Faktoren*, J. reine angew. Math. *153* (1923), 131–140.

Sanov, I. N.
[1] *Solution of Burnside's problem for exponent 4*, Leningrad State Univ. Ann. *10* (1940), 166–170. (In Russian.)

San Soucie, R. L.
[1] *Right alternative division rings of characteristic two*, Proc. Amer. Math. Soc. *6* (1955), 291–296.

Schmidt, O.
[1] *Über die Zerlegung endlicher Gruppen in direkte unzerlegbare Faktoren*, Izvestiya Kiev Univ. (1912), 1–6.

Schreier, O.
[1] *Über die Erweiterung von Gruppen, I*, Monats. für Math. u. Phys. *34* (1926), 165–180.
[2] *Über die Erweiterung von Gruppen II*, Abh. Math. Sem. Hamburg, *4* (1926), 321–346.
[3] *Die Untergruppen der freien Gruppen*, Abh. Math. Sem. Hamburg, *5* (1927), 161–183.
[4] *Über den Jordan-Hölderschen Satz*, Abh. Math. Sem. Hamburg, *6* (1928), 300–302.

Singer, J.
[1] *A theorem in finite projective geometry and some applications to number theory*, Trans. Amer. Math. Soc. *43* (1938), 377–385.

Skornyakov, L. A.
[1] *Alternative fields*, Ukrain. Mat. Zur. *2* (1950), 70–85. (In Russian.)
[2] *Right alternative fields*, Izv. Akad. Nauk SSSR, Ser. Mat. *15* (1951), 177–184. (In Russian.)

Suzuki, M.
[1] *Structure of a group and the structure of its lattice of subgroups*, Ergebnisse der Mathematik und ihrer Grenzgebiete, *10* (1956), Springer, Berlin.

Tarry, G.
[1] *Le problème des 36 officiers*, C. R. Assoc. Fr. Av. Sci. (1900), 122–123; and (1901), 170–203.

Ulm, H.
[1] *Zur Theorie der abzählbar unendlichen abelschen Gruppen*, Math. Ann. *107* (1933), 774–803.

van der Waerden, B. L.
[1] *Moderne Algebra*, 2nd ed., Berlin, 1940.

Veblen, O., and Wedderburn, J. H. M.
[1] *Non-Desarguesian and non-Pascalian geometries*. Trans. Amer. Math. Soc. *8* (1907), 379–388.

Veblen, O., and Young, J. W.
[1] *Projective Geometry*, vol. 1, Ginn and Co., 1910.

Wedderburn, J. H. M.
[1] *On the direct product in the theory of finite groups*, Ann. of Math. *10* (1909), 173–176.

Wielandt, H.
[1] *Abschatzungen für den Grad einer Permutationsgruppe von vorgeschriebenem Transitivitätsgrad*, Schriften des Math. Sem. und des Inst. für angew. Math. der Univ. Berlin, *2* (1934), 151–174.
[2] *Eine Verallgemeinerung der invarianten Untergruppen*, Math. Zeit. *45* (1939), 209–244.
[3] *p-Sylowgruppen und p-Faktorgruppen*, J. reine angew. Math. *182* (1940), 180–193.

Witt, E.
[1] *Über die Kommutativität endlicher Schiefkörper*, Abh. Math. Sem. Hamburg, *8* (1931), 413.
[2] *Treue Darstellung Liescher Ringe*, J. reine angew. Math. *177* (1937), 152–160.

Zassenhaus, H.
[1] *Zum Satz von Jordan-Hölder-Schreier,* Abh. Math. Sem. Hamburg *10* (1934), 106–108.
[2] *Über endliche Fastkörper,* Abh. Math. Sem. Hamburg, *11* (1936), 187–220.
Zorn, M.
[1] *Theorie der alternativen Ringe,* Abh. Math. Sem. Hamburg, *8* (1931), 123–147.

INDEX

INDEX OF SPECIAL SYMBOLS

A number of the symbols used in this book are standard. Among these are: the symbols for set inclusion $A \supseteq B$, A includes B, $A \supset B$, A includes B properly, $A \subseteq B$, A is contained in B, $A \subset B$, A is properly contained in B; $a \in A$, a belongs to the set A; $a \mid b$, a divides b, $a \equiv b \pmod{m}$, a is congruent to b modulo m. Following standard usage, a line through a symbol indicates a denial of the relation, thus: $p \nmid s$, p does not divide s, $y \notin G$, y does not belong to G.

5

5

G